Metabolism
of
Polycyclic Aromatic Hydrocarbons
in the
Aquatic Environment

Editor

Usha Varanasi, Ph.D.

Director
Environmental Conservation Division
Northwest and Alaska Fisheries Center
National Marine Fisheries Service
National Oceanic and Atmospheric Administration
U.S. Department of Commerce
Seattle, Washington

CRC Press, Inc.
Boca Raton, Florida

Library of Congress Cataloging-in-Publication Data
Metabolism of polycyclic aromatic hydrocarbons in the aquatic environment/editor. Usha Varanasi.

   p. cm.
  Bibliography: p.
  Includes index.
  ISBN 0-8493-6844-8
  1. Polycyclic aromatic hydrocarbons — Metabolism. 2. Polycyclic
aromatic hydrocarbons — Environmental aspects. 3. Water — Pollution —
Environmental aspects.   I. Varanasi, Usha.
QH545.H92M48 1989   88-10461
591.2 4 — dc19   CIP

International Standard Book Number 0-8493-6844-8
Library of Congress Card Number 88-10461
Printed in the United States

PREFACE

During the past decade, our knowledge of polycyclic aromatic hydrocarbons (PAH) in the aquatic environment has advanced substantially to encompass studies of bioavailability, metabolism, subsequent toxic effects, and their ecological consequences. The impetus for these studies has come from diverse developments, including increased transport of petroleum across major waterways of the world and consequent concern about accidental spills, reports of epizootics of diseases in fish populations from PAH-contaminated areas, and a general awareness of the problems of marine pollution, including implications towards human health.

In this book, recent advances in the areas of PAH biogeochemistry and bioaccumulation, microbial degradation, enzymes of activation and detoxication, metabolism of PAH, and laboratory and field studies on carcinogenic/toxic effects, are presented. Additionally, important similarities and differences in metabolism of PAH by aquatic and terrestrial organisms are discussed. These chapters also illustrate that although considerable progress has been made in certain areas of PAH metabolism in the aquatic environment, the field is relatively unexplored and many exciting possibilities exist for future investigations. For example, information is needed on PAH metabolism by birds and mammals, the two very important compartments of the aquatic environment. Moreover, because many aquatic organisms reside in confined waters, where they may be exposed to xenobiotics over long periods of time, they may serve as good models for studies to establish cause and effect relationships between exposure to xenobiotics and subsequent biological effects. Further, the ability to integrate information from controlled laboratory studies with epizootological data on the same species gives added strength to their use in such studies. Hence, a chapter is devoted to an evaluation of epizootological evidence assessing the role of PAH as etiologic agents in piscine hepatocarcinogenesis. Another chapter evaluates the use of fish models in experimental carcinogenesis, even though much of this information is not obtained using PAH. Except for this specific emphasis, the underlying theme of the book is to show how biological transport, bioaccumulation, disposition, and toxicity of PAH in the aquatic environment are influenced by the ability or inability of organisms to metabolize these environmental pollutants.

Usha Varanasi

THE EDITOR

Usha Varanasi, Ph.D., is Director of the Environmental Conservation Division, Northwest and Alaska Fisheries Center, National Marine Fisheries Service, National Oceanic and Atmospheric Administration, (NOAA), U.S. Department of Commerce, Seattle, Washington.

Dr. Varanasi received her B.Sc. degree from Bombay University in 1961, obtained her M.S. in Organic Chemistry in 1963 from the California Institute of Technology, and her Ph.D. in Physical Organic Chemistry in 1967 from the University of Washington. She is a Research Professor of Chemistry, Seattle University and an Affiliate Professor of Chemistry, University of Washington, Seattle, Washington.

Dr. Varanasi serves on the Board of Directors of the Society of Environmental Toxicology and Chemistry, and is a member of the American Association for Cancer Research, American Chemical Society, American Society for Biochemistry and Molecular Biology, American Society of Pharmacology and Experimental Therapeutics, and the American Women in Science. She also serves as an advisor to the NOAA National Research Council Research Associateship Program, a reviewer for the physiology and molecular biology program of the National Science Foundation, and a member of the Ad hoc Committee for EPA Test Rules Development. She served as the Scientific Program Chair on the International Symposium "Toxic Chemicals and Aquatic Life: Science and Management" in 1986.

Dr. Varanasi has authored over 100 technical papers including book chapters. Her current research interests are to evaluate the biological transport and biochemical and biological effects of chemical contaminants in aquatic organisms.

REVIEWERS

R. H. Adamson, Ph.D. (Ch. 6)
Director
Division of Cancer Etiology
National Cancer Institute
Bethesda, Maryland

J. W. Anderson, Ph.D. (Ch. 1)
Director
Southern California Coastal Water
  Research Project
Long Beach, California

T. Andersson, Ph.D. (Ch. 5)
Postdoctoral Fellow
Department of Zoophysiology
Zoological Institute
University of Göteborg
Göteborg, Sweden

F. A. Beland, Ph.D. (Ch. 2)
Deputy Director
Division of Biochemical Toxicology
National Center for Toxicological
  Research
Food and Drug Administration
Jefferson, Arkansas

J. R. Bend, Ph.D. (Ch. 5)
Chairman
Department of Toxicology and
  Pharmacology
University of Western Ontario
London, Ontario
Canada

J. J. Black, Ph.D. (Ch. 8)
Senior Cancer Research Scientist
Department of Experimental Biology
Roswell Park Memorial Institute
Buffalo, New York

J. M. Capuzzo, Ph.D. (Ch. 9)
Associate Scientist
Biology Department
Woods Hole Oceanographic Institution
Woods Hole, Massachusetts

L. Forlin, Ph.D. (Ch. 5)
Associate Professor
Department of Zoophysiology
Zoological Institute
University of Göteborg
Göteborg, Sweden

R. H. Adamson, Ph.D. (Ch. 6)
Director
Division of Cancer Etiology
National Cancer Institute
Bethesda, Maryland

P. P. Fu, Ph.D. (Ch. 2)
Senior Research Scientist
National Center for Toxicological
  Research
Food and Drug Administration
Jefferson, Arkansas

P. L. Grover, Ph.D., D.Sc. (Ch. 6)
Chief, Carcinogenesis Section
Chester Beatty Laboratories
The Institute of Cancer Research
Royal Cancer Hospital
London, England

A. M. Guarino, Ph.D. (Ch. 4)
Chief, Fishery Research Branch
Food and Drug Administration
Dauphin Island, Alabama

J. C. Harshbarger, Ph.D. (Ch. 8)
Director
Registry of Tumors in Lower Animals
National Museum of Natural History
Smithsonian Institution
Washington, D.C.

D. E. Hinton, Ph.D. (Ch. 7)
Professor
School of Veterinary Medicine
Department of Medicine
University of California, Davis
Davis, California

R. J. Huggett, Ph.D. (Ch. 1)
Chairman
Department of Chemical Oceanography
Virginia Institute of Marine Science
School of Marine Science
College of William and Mary
Gloucester Point, Virginia

J. J. Lech, Ph.D. (Ch. 8)
Professor
Department of Pharmacology and
  Toxicology
Medical College, University of Wisconsin
Milwaukee, Wisconsin

R. F. Lee, Ph.D. (Ch. 3)
Professor
Skidaway Institute of Oceanography
Savannah, Georgia

D. V. Parke, Ph.D. (Ch. 9)
Professor and Head
Department of Biochemistry
University of Surrey
Guildford, England

J. B. Pritchard, Ph.D. (Ch. 4)
Research Physiologist
Laboratory of Pharmacology
National Institute of Environmental
  Health Sciences
Research Triangle Park, North Carolina

J. J. Stegeman, Ph.D. (Ch. 3)
Senior Scientist
Biology Department
Woods Hole Oceanographic Institution
Woods Hole, Massachusetts

D. R. Thakker, Ph.D. (Ch. 6)
Head of Bioorganic Mechanism Section
Department of Drug Metabolism
Glaxo Inc.
Research Triangle Park, North Carolina

CONTRIBUTORS

George S. Bailey, Ph.D.
Professor
Department of Food Science and
   Technology
Oregon State University
Corvallis, Oregon

William M. Baird, Ph.D.
Professor of Medicinal Chemistry
School of Pharmacy and Pharmacal
   Sciences
Purdue University
West Lafayette, Indiana

Paul C. Baumann, Ph.D.
Field Leader
Research Station
National Fisheries Contaminant Research
   Laboratory
U.S. Fish and Wildlife Service
Columbus, Ohio

Donald R. Buhler, Ph.D.
Professor and Chairman
Toxicology Program and Marine
   Freshwater Biomedical Sciences Center
Oregon State University
Corvallis, Oregon

Carl E. Cerniglia, Ph.D.
Director of Microbiology
National Center For Toxicological
   Research
Food and Drug Administration
Jefferson, Arkansas

John W. Farrington, Ph.D.
Professor
Environmental Sciences Program
University of Massachusetts, Boston
Boston, Massachusetts

Gary L. Foureman, Ph.D.
Research Associate
Department of Biochemistry and
   Biophysics
Oregon State University
Corvallis, Oregon

Douglas E. Goeger, Ph.D.
Postdoctoral Fellow
Department of Food Science and
   Technology
Oregon State University
Corvallis, Oregon

Michael A. Heitkamp, Ph.D.
Research Microbiologist
National Center for Toxicological
   Research
Food and Drug Administration
Jefferson, Arkansas

Jerry D. Hendricks, Ph.D.
Professor
Department of Food Science and
   Technology
Oregon State University
Corvallis, Oregon

Margaret O. James, Ph.D.
Associate Professor
J. Hillis Miller Health Center
College of Pharmacy
University of Florida
Gainesville, Florida

David R. Livingstone, Ph.D.
Principal Scientific Officer
Plymouth Marine Laboratories
Prospect Place, West Hoe
Plymouth, England

Anne E. McElroy, Ph.D.
Assistant Professor
Environmental Sciences Program
University of Massachusetts, Boston
Boston, Massachusetts

Michael N. Moore, Ph.D.
Project Leader
Plymouth Marine Laboratories
Prospect Place, West Hoe
Plymouth, England

Marc Nishimoto, Ph.D.
Research Chemist
Environmental Conservation Division
Northwest and Alaska Fisheries Center
National Marine Fisheries Service
National Oceanic and Atmospheric
  Administration
Seattle, Washington

Teresa A. Smolarek, Ph.D.
Research Scientist
School of Pharmacy and Pharmacal
  Sciences
Purdue University
West Lafayette, Indiana

John E. Stein, Ph.D.
Supervisory Research Chemist
Environmental Conservation Division
Northwest and Alaska Fisheries Center
National Marine Fisheries Service
National Oceanic and Atmospheric
  Administration
Seattle, Washington

John M. Teal, Ph.D.
Senior Scientist
Woods Hole Oceanographic Institution
Woods Hole, Massachusetts

Usha Varanasi, Ph.D.
Director
Environmental Conservation Division
Northwest and Alaska Fisheries Center
National Marine Fisheries Service
National Oceanic and Atmospheric
  Administration
Seattle, Washington

John Widdows, Ph.D.
Principal Scientific Officer
Plymouth Marine Laboratories
Prospect Place, West Hoe
Plymouth, England

David E. Williams, Ph.D.
Assistant Professor
Department of Food Science and
  Technology
Oregon State University
Corvallis, Oregon

ACKNOWLEDGMENT

My special thanks go to the many colleagues and peers who enthusiastically accepted my invitations to write or review various chapters and who were graciously responsive to deadlines and editorial comments. While it is impossible to mention everyone who has helped me through various phases of the book, I do want to thank Professor Robert Huggett of the Virginia Institute of Marine Science for recommending my name to CRC Press to author/edit this book, and my close associates, Drs. John Stein, Marc Nishimoto, Tracy Collier, and William Reichert for helpful discussions, useful advise, and cheerful support. Also, I thank many of my colleagues at the National Marine Fisheries Service and the National Ocean Service of the National Oceanic and Atmospheric Administration (NOAA) for supporting pollution research by the Environmental Conservation (EC) Division. I am also grateful to Ms. Bich-Thuy Le Eberhart, Mr. Tom Hom, Ms. Shirley Perry, Mr. Herbert Sanborn, and Ms. Lisa Wirick of the EC Division, and Ms. Gail Siani of the NOAA General Counsel Office for help with manuscript preparation and contractual agreements.

TABLE OF CONTENTS

Chapter 1

BIOAVAILABILITY OF POLYCYCLIC AROMATIC HYDROCARBONS IN THE AQUATIC ENVIRONMENT*

Anne E. McElroy, John W. Farrington, and John M. Teal

TABLE OF CONTENTS

---

* This Chapter was originally prepared in 1985.

I. FOREWORD

This chapter serves two purposes. The primary goal is to review what is known about the bioavailability of polynuclear aromatic hydrocarbons or polycyclic aromatic hydrocarbons (PAH) and their metabolites in the aquatic environment. Emphasis will be placed on the physical/chemical, biological, and environmental factors which drive these processes. Second, this discussion will provide a rationale for why it is important to study metabolism when considering the fate of PAH in aquatic systems. The chapter is divided into three main sections. The first introduces PAH, their sources, transformations, and sinks, and the physical/chemical factors which influence their cycling in the environment. The second section reviews information gathered from field studies on the bioavailability of PAH to aquatic organisms. Information derived from laboratory investigations on processes influencing bioavailability is discussed in the third section. Due to the complexity of the many parameters which influence the fate of compounds such as PAH in the environment, studies of these processes in the field are required to obtain a realistic assessment of the environmental fate of PAH. For the same reasons it is very difficult and often impossible to do controlled experiments in the field. In organizing this chapter we chose to focus on field-collected data on PAH bioavailability, and then to supplement this discussion with information derived from laboratory studies. This organization may seem awkward at times, but we think it best serves the purposes of the chapter within the context of this book.

II. INTRODUCTION

A. Physical/Chemical Properties of PAH

PAH generally refer to hydrocarbons containing two or more fused benzene rings.[1] Figure 1 gives structures and names of some common PAH. Classically, the term hydrocarbon referred exclusively to compounds containing only hydrogen and carbon atoms. There are many polycyclic aromatic compounds which also contain heteroatoms such as nitrogen, sulfur, and oxygen which are frequently referred to as PAH.[2] These heteroatomic compounds are chemically and functionally closely related to "true" PAH, and in this chapter are included with them.

When dealing with the behavior of organic compounds in the aquatic environment, it is of paramount importance to consider their solubility. Only recently[1,2] have experimentally determined solubilities for many PAH become available.[3-5] Table 1, taken from May,[4] summarizes some of the aqueous solubility data available for aromatic hydrocarbons. Although PAH as a group are considered to be hydrophobic, they possess a wide range of solubilities.

May[4] discusses reasons for some of the disagreements between the sets of data, e.g., impure standards and varying sensitivities of measurement techniques. Temperature and ionic strength can influence solubilities in aqueous systems. May reported effects of temperature on solubilities of PAH as ranging from approximately a two- to a fivefold increase in solubility between 5 and 30°C, depending on the specific PAH. Setschenow[6] deduced an empirical relationship for the effect of salinity (or ionic strength in the case of non-seawater systems) on solubility of a given chemical:

$$\log S_o/S_s = K_s C_s \tag{1}$$

S_o = concentration of solute in fresh water, S_s = concentration of solute in salt water, C_s = molar salt concentration, and K_s = Setschenow constant for each solute.
May[4] experimented with NaCl solutions of 0 to 0.75 mol/l and obtained K_s values for PAH as given in Table 2.

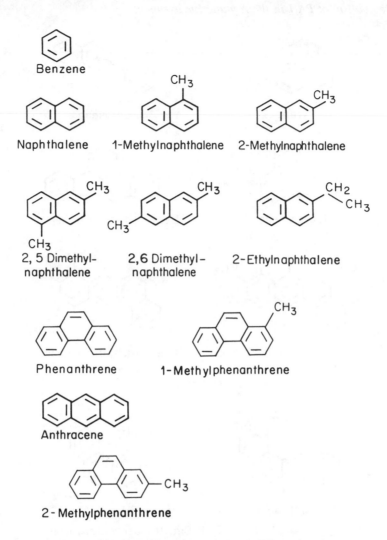

FIGURE 1.   Structures and common names of aromatic compounds discussed in the text. Figure includes benzene and "true" alkyl and unsubstituted PAH, as well as heteroatom aromatic compounds and substituted PAH.

Very few actual measurements of PAH solubility in natural seawater are available. Whitehouse[7,8] used filtered seawater and fresh water that had been photooxidized to remove dissolved organic matter to study the combined influence of temperature and salinity on PAH solubility. He confirmed May's finding that temperature had only a small (two- to fivefold) influence on solubility. He also demonstrated an inverse relationship between salinity and solubility, as might be expected from the salting-out theory, but found that the effect is even less than that produced by temperature, being at most a factor of 2 over a salinity range of 0 to 36‰ (Figure 2). The minimal influence of salinity on solubility of PAH as determined by laboratory experiments is consistent with the field data of Readman et al.,[9] which showed no correlation of PAH concentrations with increasing salinity from 0 to about 33‰ on a transect of water samples from the Tamar Estuary in the U.K. Higher molecular weight PAH were significantly correlated with suspended solids along the same transect. This phenomenon is consistent with the fact that compounds with very low water solubilities have a tendency to leave the truly dissolved state and bind to available "solid" material.

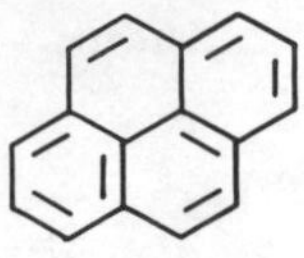

Pyrene

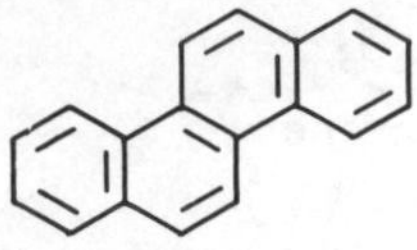

Chrysene

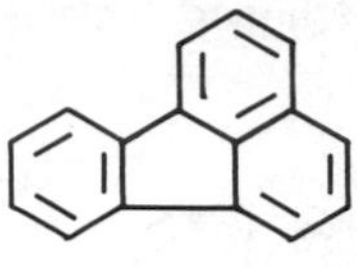

Fluoranthene

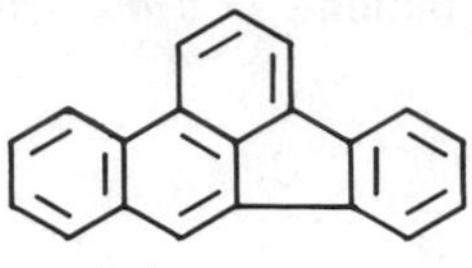

Benzo [b] fluoranthene

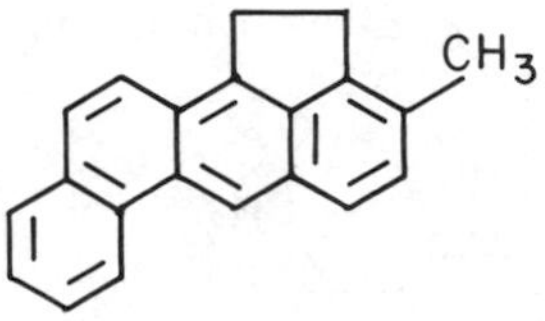

3-Methylcholanthrene

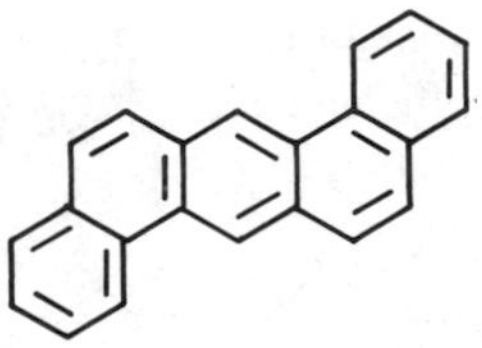

Dibenz [a, h] anthracene

FIGURE 1B.

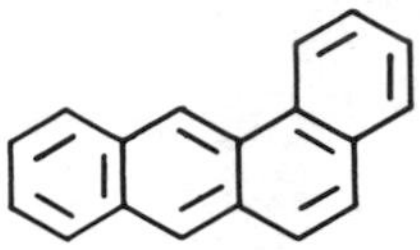

Benzanthracene

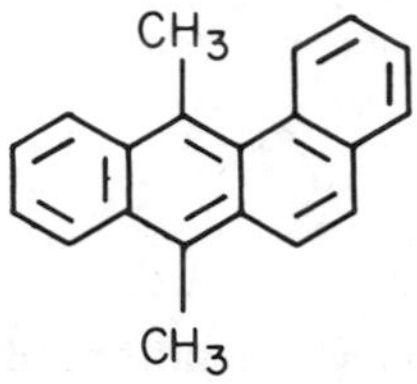

7, 12 Dimethylbenzanthracene

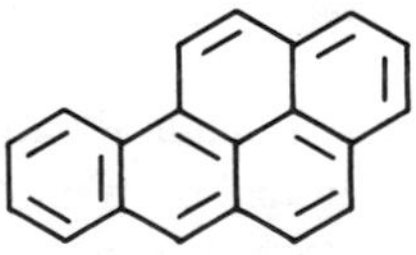

Benzo [a] pyrene

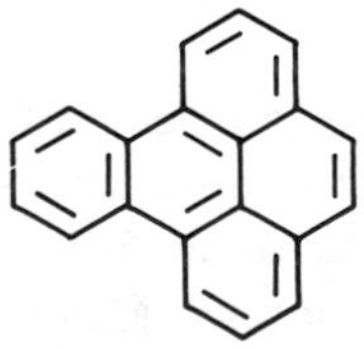

Benzo [e] pyrene

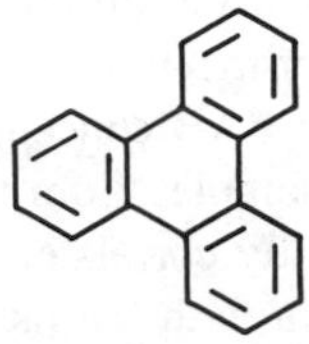

Triphenylene

FIGURE 1C.

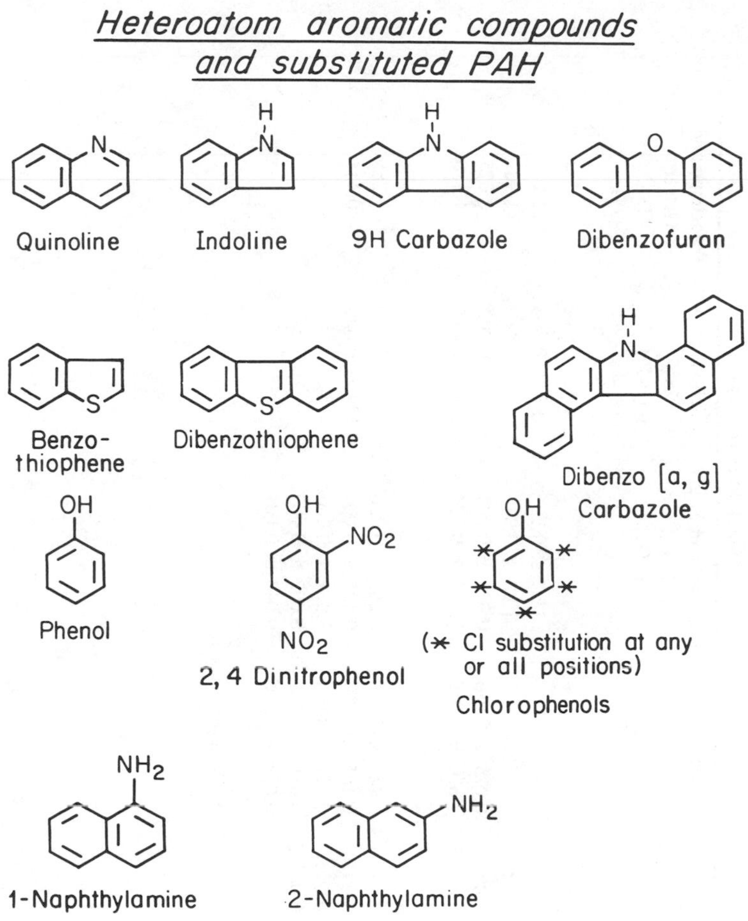

Heteroatom aromatic compounds and substituted PAH

Quinoline   Indoline   9H Carbazole   Dibenzofuran

Benzo-thiophene   Dibenzothiophene   Dibenzo [a, g] Carbazole

Phenol   2, 4 Dinitrophenol   Chlorophenols

(✳ Cl substitution at any or all positions)

1-Naphthylamine   2-Naphthylamine

FIGURE 1D.

The rudimentary nature of knowledge of factors influencing solubility of the wide range of PAH types, especially the higher molecular weight PAH with greater than four rings, is pointed out in the anomalous behavior of benz(a)anthracene in seawater[7] (Figure 2). There are no solubility data available for the more extensively alkylated PAH which are common in many crude oils.[10] Furthermore, multiple PAH solute investigations are lacking and, thus, we do not know how the presence of numerous PAH and other hydrocarbons in oil-water mixtures influences solubilities of individual PAH. There is a growing body of literature on theoretical calculations of solubility from molecular parameters such as molar volume and molecular length which is beyond the scope of this chapter, and we refer the reader to May[4,5] and Banerjee,[11] among others.

For many neutral compounds including PAH, sorption behavior can be represented as an equilibrium between the aqueous phase and the nonaqueous phase as described by a partition coefficient:[12-14]

$$K_p = [A_s]/[A_w] \qquad (2)$$

where K_p is the equilibrium partition coefficient for a particular species, A_s is the activity of the compound in the nonaqueous phase, and A_w is the activity of the compound in the truly dissolved phase. K_p is related to fundamental properties of the compound (sorbate)

Table 1

THE AQUEOUS SOLUBILITIES OF SOME AROMATIC HYDROCARBONS AS DETERMINED BY SEVERAL INVESTIGATORS

Compound	Mol wt	Solubilities (mg/kg) This work 25°C	Solubilities (mg/kg) This work 29°C	Davis et al. 29°C (1942)	MacKay and Shui 25°C (1977)	Schwarz 25°C (1977)	Wauchope and Getzen 25°C (1972)
Benzene	78.1	1791 ± 10			31.7 ± 0.2	30.3 ± 0.3	31.2
Naphthalene	128.2	31.69 ± 0.23			1.98 ± 0.04		1.90
Fluorene	166.2	1.685 ± 0.005					
Anthracene	178.2	0.0446 ± 0.0002	0.0570 ± 0.003	0.075 ± 0.005	0.073 ± 0.005	0.041 ± 0.0003	0.075
Phenanthrene	178.2	1.002 ± 0.011	1.220 ± 0.013	1.600 ± 0.050	1.290 ± 0.070	1.151 ± 0.015	1.180
2-Methylanthracene	192.3	0.0213 ± 0.003					
1-Methylphenanthrene	192.3	0.269 ± 0.003					
Fluoranthene	202.3	0.206 ± 0.002	0.264 ± 0.002	0.240 ± 0.020	0.260 ± 0.020		0.265
Pyrene	202.3	0.132 ± 0.001	0.162 ± 0.001	0.165 ± 0.007	0.135 ± 0.005	0.129 ± 0.002	0.148
Benzanthracene	228.3	0.0094 ± 0.0001	0.0122 ± 0.0001	0.011 ± 0.001	0.014 ± 0.0002		
Chrysene	228.3	0.0018 ± 0.00002	0.0022 ± 0.00003	0.0015 ± 0.0004	0.002 ± 0.0002		
Triphenylene	228.3	0.0066 ± 0.0001		0.038 ± 0.005	0.043 ± 0.001		

From May, W. E., Wasik, S. P., and Freeman, D. H., *Anal. Chem.*, 50, 997, 1978. With permission.

Table 2
SETSCHENOW CONSTANTS
FOR SOME AROMATIC
HYDROCARBONS AT 25°C

Compound	K_s (l/mol)
Benzene	0.175 ± 0.006
Naphthalene	0.213 ± 0.001
Fluorene	0.267 ± 0.005
Anthracene	0.238 ± 0.004
Phenanthrene	0.275 ± 0.010
2-Methylanthracene	0.336 ± 0.006
1-Methylphenanthrene	0.211 ± 0.018
Pyrene	0.286 ± 0.003
Fluoranthene	0.339 ± 0.010
Chrysene	0.336 ± 0.010
Triphenylene	0.216 ± 0.010
Benzanthracene	0.354 ± 0.002

From May, W. E., Wasik, S. P., and Freeman, D. H., *Anal. Chem.*, 50, 997, 1978. With permission.

and the sorbent material. Partitioning of a compound in a model test system containing octanol and water generates a partition coefficient, K_{ow}, which can be related to the solubility, S, of the compound:[15]

$$\log K_{ow} = a - b (\log S) \tag{3}$$

Likewise, the degree to which a particular nonpolar organic compound will sorb to a particular sorbant is influenced strongly by the organic carbon content of the sorbent. Given these relationships, it has been possible to predict the partitioning behavior of many nonpolar compounds including some PAH on a range of sorbants, mostly soils, if either the compound's water solubility or K_{ow}, or the sorbant's organic carbon content, f_{oc}, is known:[13,14,16,17]

$$K_{oc} = f_{oc}k_{ow} \tag{4}$$

where K_{oc} is the organic carbon normalized partition coefficient.

Several researchers have shown that there is also a relationship between K_{oc} and K_{ow} which can be expressed as:

$$\log K_{oc} = a (\log K_{ow}) + b \tag{5}$$

However, values of a and b in this equation vary and seem to be related to type of compound (e.g., PAH, chlorinated nonpolar compounds) and type of organic coating on the sorbant[14,17,18] (Table 3). Further investigation is needed to reveal what specific properties of sorbants and sorbates affect this relationship.

Dissolved organic matter (DOM) has been postulated to increase the solubility of PAH in seawater.[19] Colloids are part of the organic matter present in what is usually termed DOM. According to Hiemenez,[20] the term colloid refers to any particle with some linear dimension between 10^{-7} cm (10 Å) and 10^{-4} cm (1 μm). The high capacity of colloidal organic material (0.4 to 1.2 nm) harvested from estuarine waters to sorb hydrophobic pollutants in

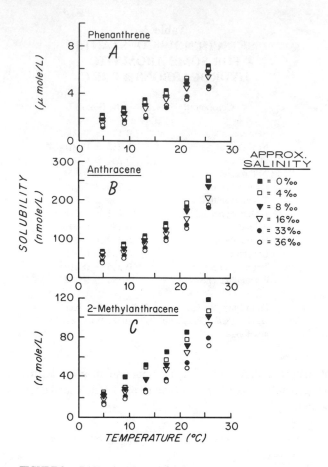

FIGURE 2.   PAH solubility as a function of temperature at various salinities. (Redrawn from Whitehouse, B. G., The Partitioning of Polynuclear Aromatic Hydrocarbons into the Dissolved Phase of the Aquatic Environment, Ph.D. thesis, Dalhousie University, Halifax, N.S., 1983. With permission.)

general,[21] and PAH in particular,[22] has been demonstrated. Means and Wijayaratne have investigated the effect of pH and salinity on partitioning of the pesticides atrazine and linuron and the PAH anthracene to estuarine colloids.[21,22] K_{oc} (the partition coefficient normalized to organic carbon content of the sorbate) was maximal at the pH and salinity of the water from which the colloids were harvested. For atrazine either increasing (to 9) or decreasing (to 5) pH from an ambient value of 8 reduced K_{oc} by fourfold. Decreasing pH from ambient values had an even smaller effect on the K_{oc} of anthracene. The effect of salinity on the sorptive capacity of these colloids was less pronounced. A decrease in salinity from 20 to 8‰ increased the K_{oc} of atrazine by less than twofold. Although the effects of salinity on solubility have been studied for a few compounds, the results thus far indicate that they are limited (less than an order of magnitude).

The influence of colloidal organic matter in interstitial waters of natural sediment has only recently begun to be investigated. Results of investigations on the influence of interstitial colloids on partitioning of polychlorinated biphenyls (PCB), similar in many properties to PAH, have shown that observed distributions of most PCB in pore waters were consistent with a three-phase, solution-colloid-solid equilibrium partitioning model.[23,24] Eadie et al.[25] have reported initial data on PAH concentrations in interstitial waters from natural sediments.

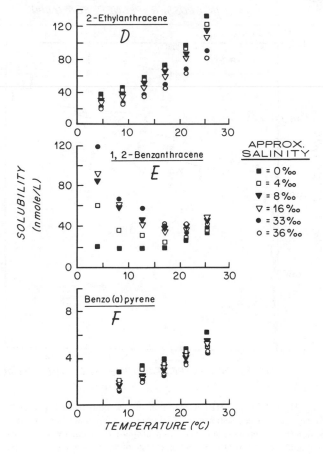

FIGURE 2B.

Table 3
REGRESSION PARAMETERS OF PUBLISHED
LOG K_{oc} = a LOG K_{ow} + b

Compounds	a	b	r	Ref.
PAH	1.00	−0.32	0.990	14
PCB and chlorinated benzenes	0.904	−0.78	0.994	18
Chlorinated alkenes and benzenes	0.72	0.49	0.975	17

They did not find strong correlations between either sediment PAH concentrations or DOC and PAH concentration in pore water. The relative abundance of the seven PAH they analyzed in pore waters was not proportional to their water solubility. Incomplete phase separation of fine-grained (<1 μm) particulate material was offered as a possible explanation for this observation. The contribution of ''nonsettling or nonfilterable'' particulates to an analytical artifact in the determination of partitioning behavior of hydrophobic compounds, has been recently discussed by Gschwend and Wu.[26]

The discussion above referred primarily to nonpolar compounds, including PAH. Solubility and sorptive behavior of PAH containing certain polar functional groups (e.g., NH_2, OH, COOH) should be more affected by factors such as ionic strength, pH, and ionic properties of available sorbants. Westall et al.[27] have found pH and ionic strength to have a marked

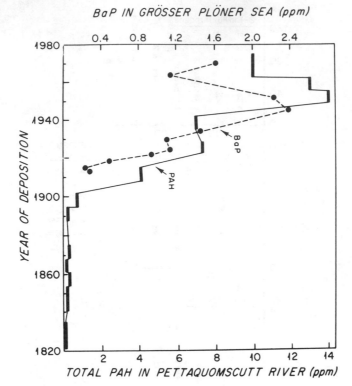

FIGURE 3.. Total PAH concentration (dry weight) in Pettaquomscutt River, Rhode Island sediment core sections as a function of date of deposition (horizontal bars, left scale). Benzo(a)pyrene abundance in the Gösser Plöner Sea sediment as a function of date of deposition (cross in circle, right scale). (From Hites, R. A., LaFlamme, R. E., Windsor, J. G., Jr., Farrington, J. W., and Euser, W. G., *Geochim. Cosmochim. Acta,* 44, 873, 1980. With permission.)

influence on sorption behavior of chlorophenols and dinitrophenols. Though these results should be applicable to higher molecular weight compounds, the large hydrophobic region of many PAH may mitigate the influence of a polar moiety. Means et al.[28] looked at sorption behavior of carboxy- and amino-substituted PAH on 14 different soils. Partition coefficients were primarily correlated with organic carbon content of the soils, with the exception of two amino-PAH that showed enhanced sorption on soils with low carbon- and high clay contents.

B. Sources and Sinks of PAH in the Aquatic Environment

PAH are synthesized by some bacteria, plants, and fungi,[1,2] and are released by natural processes such as marine seeps[29] and forest and grass fires.[30] Still, it is generally accepted that anthropogenic activity is responsible for the majority of PAH released into the environment.[2,31] Near urban areas, because of the ubiquitous occurrence of PAH in sediments, their composition, and typical depth profiles in sediments corresponding to anthropogenic fossil fuel utilization (Figure 3),[31] the primary source of PAH for these aquatic ecosystems is thought to be atmospheric deposition of PAH produced from pyrolysis of fossil fuels.[32-35] It would be extremely difficult to differentiate anthropogenically derived PAH from PAH derived from natural sources pre-dating man's use of fossil fuels by analysis of

Table 4
INPUT OF PETROLEUM HYDROCARBONS INTO THE MARINE ENVIRONMENT (MTA)

Source	Probable range	Best estimate
Natural sources		
Marine seeps	0.02—2.0	0.2
Sediment erosion	0.005—0.5	0.05
(Total natural sources)	(0.025—2.5)	(0.25)
Offshore production	0.04—0.06	0.05
Transportation		
Tanker operations	0.4—1.5	0.7
Dry-docking	0.02—0.05	0.03
Marine terminals	0.01—0.03	0.02
Bilge and fuel oils	0.2—0.6	0.3
Tanker accidents	0.3—0.4	0.4
Nontanker accidents	0.02—0.04	0.02
(Total transportation)	(0.95—2.62)	(1.47)
Atmosphere	0.05—0.5	0.3
Municipal and industrial wastes and runoff		
Municipal wastes	0.4—1.5	0.7
Refineries	0.06—0.6	0.1
Nonrefining industrial wastes	0.1—0.3	0.2
Urban runoff	0.01—0.2	0.12
River runoff	0.01—0.5	0.04
Ocean dumping	0.005—0.02	0.02
(Total wastes and runoff)	(0.585—3.12)	(1.18)
Total	1.7—8.8	3.2

Note: MTA = million tons annually. The total best estimate, 3.2 MTA, is a sum of the individual best estimates. A value of 0.3 was used for the atmospheric inputs to obtain the total, although this best estimate is only a center point between the range limits and cannot be supported rigorously by the data and calculations used for estimation of this input.

From National Research Council, *Oil in the Sea, Inputs, Fates, and Effects*, National Academy Press, Washington, D.C., 1985. With permission.

sediments in the deep sea, due to the extremely slow rate of sediment deposition in the open ocean (approximately 1 mm/1000 yr). Outside of urban areas, there is a paucity of direct measurements of PAH in the atmosphere and related calculation of flux to the aquatic environment.[2,10] Gagosian et al.[36] were unable to detect (< 5 pg/m^3) individual PAH in uncontaminated air at Einewetok Atoll, Marshall Islands in the Pacific Ocean.

A general picture of the importance of various sources to the marine environment, including atmospheric contributions, can be estimated from the recent compilation of petroleum hydrocarbon inputs by the NRC[10] (Table 4), although PAH constitute only a small fraction of the complex mixture known as petroleum hydrocarbons. In coastal marine environments,

Table 5
**ANNUAL INPUTS OF PAH VIA URBAN RUNOFF OF THE UPPER
NARRAGANSETT BAY WATERSHED (kg/year)**

Compound	Residential	Commercial	Industrial	Highway	Total
Naphthalene	0.18	1.7	4.8	2.6	9.3
2-Methylnaphthalene	7.5	1.0	12.1	4.9	25.5
1-Methylnaphthalene	0.35	0.79	14.3	1.9	17.3
Biphenyl	0.28	0.55	3.9	0.76	5.5
2-Ethylnaphthalene	0.02	0.66	12.6	2.2	15.4
Fluorene	0.04	0.44	9.7	7.9	18.1
Dibenzothiophene	0.18	0.30	20.4	3.9	24.8
Phenanthrene	1.7	2.1	32.4	32.2	68.4
Fluoranthene	20.8	9.4	40.7	101.2	172.1
Pyrene	12.0	6.7	32.4	35.1	86.2
Benz(a)anthracene	2.9	1.7	16.3	58.7	79.2
Chrysene	5.2	7.7	8.8	86.0	107.7
Benzo(e)pyrene	5.7	3.3	4.8	10.1	23.9
Benzo(a)pyrene	4.8	1.1	14.4	12.4	32.7
Sum PAH	55.1	37.4	22.8	360.0	681.0
f_2	2,200	3,069	10,500	13,200	38,000
Total hydrocarbons	36,250	31,880	444,700	152,700	665,300
Sum PAH total hydrocarbons (%)	0.15	0.12	0.05	0.23	0.10
f_2 hydrocarbons/total hydrocarbons (%)	6.1	9.6	4.3	8.6	5.7

From Hoffman, E. J., Mills, G. L., Latimer, J. S., and Quinn, J. G., *Environ. Sci. Technol.*, 18, 580, 1984. With permission.

municipal and industrial wastes and runoff and riverine inputs are important sources. Specific measurements of PAH flux from these sources have been made in several urban environments[37-39] (Table 5). A good estimate of the total PAH input from these sources is extremely difficult due to the diffuse nature of the source and the importance of aperiodic events such as storms.[38]

Because of their hydrophobic nature, PAH in the aquatic environment rapidly become associated with particles[40,41] and are deposited in sediments. The importance of sediments as reservoirs for PAH has been well documented in natural[2,10,33,35,42] as well as experimental[43-47] systems. Small-scale physical processes such as local turbulence, and large-scale physical forces such as currents which redistribute particulate material will also affect the distribution of PAH. The influence of physical processes on the distribution of pollutants has been discussed in a general manner in numerous papers, books, and reviews dealing with discharge or dumping of wastes in the ocean.[10,48-52] Work on the interactions between dissolved and particulate phases for certain radionuclides concerning water transport also has a bearing on this problem.[53-55]

Sediment-mixing by burrowing organisms (bioturbation) can have a significant effect on PAH distribution in the sediment, especially near the sediment-water interface.[56-59] For example, Bender and Davis[60] measured weight-specific expulsion rates from the deposit-feeding bivalve, *Yoldia limatula*, in the laboratory as a function of temperature, and calculated

Table 6
CALCULATED AND MEASURED RATE CONSTANTS FOR PHOTOLYSIS OF AROMATICS IN SUNLIGHT AT 40° N LATITUDE[a]

PAH	Measured	Winter	Spring	Summer	Fall	S/W[b]
Benz(a)anthracene	6.0×10^{-5} (spring)	1.4×10^{-4}	2.2×10^{-4}	3.8×10^{-4}	2.2×10^{-4}	2.7
Benzo(a)pyrene	1.8×10^{-4} (winter)	1.8×10^{-4}	2.8×10^{-4}	3.9×10^{-4}	2.3×10^{-4}	2.2
Quinoline	4.0×10^{-7} (summer)	5.0×10^{-8}	2.3×10^{-7}	3.6×10^{-7}	1.3×10^{-7}	7.2
Benzo(f)quinoline	3.7×10^{-4} (summer)	1.5×10^{-4}	2.4×10^{-4}	4.8×10^{-4}	2.0×10^{-4}	3.2
9H-carbazole	6.6×10^{-5} (winter)	6.5×10^{-5}	1.0×10^{-4}	2.0×10^{-4}	9.0×10^{-5}	3.1
7H-dibenzo(o,g) carbazole	5.2×10^{-4} (winter)	2.3×10^{-4}	3.9×10^{-4}	5.0×10^{-4}	3.2×10^{-4}	2.2
Benzo(b)thiophene	6.9×10^{-7} (summer)	2.3×10^{-8}	2.7×10^{-7}	5.7×10^{-7}	1.2×10^{-7}	25
Dibenzothiophene	1.0×10^{-6} (spring)	2.9×10^{-7}	1.1×10^{-6}	1.5×10^{-6}	9.1×10^{-7}	5.2

[a] K_{pE} in s^{-1}.
[b] S/W = summer K_{pE}/winter K_{pE}.

From Payne, J. R. and Phillips, C. R., *Environ. Sci. Technol.*, 19, 569, 1985. With permission.

that a 14-mm-long animal in Narragansett Bay, Rhode Island would resuspend 440 g dry weight of fine sediment annually. Gordon et al.[61] estimated that observed densities of the polychaete, *Arenicola marina* could remove all sediment-bound petroleum hydrocarbons from the Chedabucto Bay oil spill in a period of 2 to 4 years. Lee et al.[62] and Gardner et al.[63] found that the polychaete, *Capitella* increased the removal of PAH from sediment in microcosm experiments. The effect was most pronounced with PAH of greater than three rings. Karickhoff and Morris[64] found bioturbation by oligochaetes increased flux of sediment-sorbed chlorinated hydrocarbons to the water column by a factor of 4 to 6. The large polychaete, *Nereis virens* had a similar effect on flux of sediment-sorbed benz(a)anthracene.[45] The effects of bioturbation on the distribution of PAH would be expected to be significant in areas where physical disturbances are either small or infrequent.

C. Transformations of PAH Leading to Altered Reactivity

In addition to the metabolism of PAH by eukaryotic organisms, which will be discussed in the remainder of this book, we must mention other important processes which transform PAH in the environment. PAH are photoreactive in the atmosphere[65] and in the upper parts of the water column.[66] The photochemistry of PAH in water has been the subject of two recent reviews.[68,69] A compilation of rate constants for photolysis of several PAH are shown in Table 6. Mesocosm studies on the fate of PAH in the CEPEX enclosures showed photooxidation was a significant factor in PAH removal from the water column.[69] Hinga,[47] using the MERL mesocosms to investigate the fate of dimethylbenzanthracene and benz(a)anthracene in coastal marine environments, reported that photooxidation of PAH is a major process in this system. Lee and Ryan[70] and Hinga[47] suggested that photooxidation of PAH enhanced their susceptibility to microbial mineralization. Investigations on the fate and effects of anthracene in a freshwater stream have shown photooxidation to be important to both the removal from the system and toxicity to fish.[71]

Once transformed into polar metabolites, PAH should be more mobile in an aquatic system. However, investigations on the residence time of radiolabeled PAH metabolites produced in microcosm experiments have indicated tht oxygenated PAH metabolites can be retained in marine sediments for long periods. Hinga et al.[46] reported the presence of benz(a)anthracene metabolites in sediment samples up to 220 days after addition of the parent compound to the water column of a MERL mesocosm. In a smaller microcosm experiment, McElroy[45] also found metabolites of benz(a)anthracene to persist in sediment samples for at least 1 month. Observations on the presence of PCB and DDT metabolites in sediments from the California Bight[72,73] suggest that metabolites of these hydrophobic aromatic compounds have significant residence times in marine benthic systems. The persistence of metabolites or reaction products for weeks to months may reflect some partitioning (sorption) or removal of these compounds to particulate or colloidal phases where they are less available for further chemical or biochemical transformation. Little, if any, data are available to support further comment or speculation on this point.

Microbial metabolism of PAH will be covered in Chapter 2. We point out that microbial, photo, and eukaryotic transformation processes may be interactive. Eukaryotic organisms may be exposed to PAH metabolites produced by either microbial- or photooxidation. Almost nothing is known about the biogeochemical mobility of PAH metabolites. Eukaryotic organisms and enteric or associated bacteria may metabolize PAH in a "cooperative" manner. Burrowing organisms should enhance microbial metabolism of PAH by introducing oxygen and nutrients into anaerobic sediments. The significance of these interactive processes in the environment has yet to be investigated.

We have presented a general introduction to factors influencing the fate of PAH in the environment. These processes are summarized in Figure 4. Many different processes occur simultaneously, making detailed analysis of their individual impact difficult. With this in mind, we will go on to review the field and laboratory data on the bioavailability of PAH to aquatic organisms.

III. BIOAVAILABILITY OF PAH TO AQUATIC ORGANISMS: FIELD STUDIES

A. Levels of PAH Observed in Aquatic Organisms

Levels of individual PAH have been measured in aquatic organisms worldwide. Reported values range from undetectable quantities (approximately 0.01 µg/kg dry weight) to values in excess of 5000 µg/kg dry weight for individual PAH in tissues of aquatic organisms. Although individual PAH are not routinely measured in many monitoring studies, a good picture of levels observed in aquatic organisms from a variety of environments around the world can be found in the following references.[74-83] Concentrations of a representative 5-ring PAH, benzo(a)pyrene, in a large number of different species from all over the world compiled by Neff[1] are reproduced in part in Table 7.

In general, elevated concentrations of PAH can be correlated with the proximity of the organism to areas receiving chronic hydrocarbon discharge.[75-78,83] The majority of PAH measurements have been made on bivalve mollusks. Bivalves have the advantage of being sessile and attached, they can rapidly accumulate PAH, and have little capacity for PAH metabolism (see discussion in later chapters). PAH concentrations in fish have been less well surveyed, but are usually low relative to invertebrates inhabiting the same environment.[1,44,75,76,78-80] The low body burdens of PAH in fish are believed to be due to their ability to rapidly metabolize PAH (as discussed in other chapters).

B. Correlations Between Concentrations Observed in Organisms and Levels Found in Their Habitat

Sediments appear to be a major source of PAH and other hydrophobic pollutant compounds to aquatic organisms. Gossett et al.[85] measured concentrations of 27 organic compounds

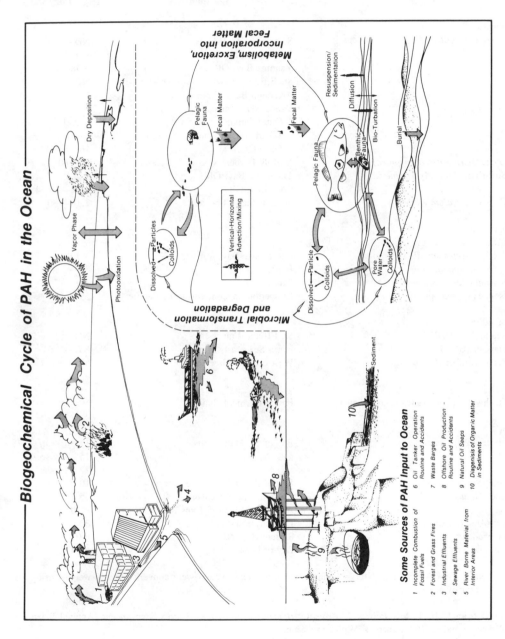

FIGURE 4. Biogeochemical processes affecting PAH in marine ecosystems.

Table 7
CONCENTRATIONS OF 3,4-BENZOPYRENE IN TISSUES OF AQUATIC ORGANISMS

Organism	Location	[BP] (μg/kg dry wt)
Mollusks		
Mussels (*Mytilis edulis*)	Seine Estuary, France	ND—380
	Tillamook Bay, OR	<0.4—67.4
	Yaquina Bay, OR	0.48—120.8
	St. Effiam, France	ND
	Arcachon Basin, France	5.0
	Vancouver, Canada	
	Outer harbor	8 ± 1
	Wharf, marina, and dock areas	72 ± 20
	False Creek	168 ± 24
Oyster (*Crassostrea virginica*)	Norfolk Harbor, VA	20—60
Soft-shell clam (*Mya arenaria*)	Tillamook Bay, OR	1.2
	Coos Bay, OR	1.32—26.64
Cockle (*Cardium edule*)	Seine Estuary, France	220—780
Crustaceans		
Crabs		
(*Callinectes sapidus*)	Chesapeake Bay, VA	<2.0
(*Maia squinada*)	Brest, France	3.5
Shrimp (*Penaeus aztecus*)	Palacios, TX	<4
Isopods (*Lygia* sp.)	Clipperton Lagoon, equatorial Pacific	536
Echinoderms		
Sea urchin (UNID)	St. Effiam, France	ND
Starfish (UNID)	North Sea coast, France	ND—126
Sea cucumber (UNID)	West coast of Greenland	ND
Annelids		
Freshwater oligochaetes (*Tublifex* sp.)	Freshwater pool, Italy	50
Fish		
Mullet (*Mugil chelo*)	Bay of St. Malo, France	17.3 (flesh)
		155.0 (viscera)
		22.0 (scales)
Sole (*Solea solea*)	Bay of St. Malo, France	10.0
Cod (*Gadus* sp.)	Holsteinborg, Greenland	15
	Atlantic, 40 km off Toms River, NJ	<10
Eel (*Anguilla* sp.)	Dunkerque, France	30
Menhaden	Raritan Bay, NJ	6.0
Sardine	Bay of Naples, Italy	65.4

Note: ND = not detectable; UNID = unidentified; wet weight values were converted to dry weights using a factor of 4.

Adapted from Neff, J. M., *Polycyclic Aromatic Hydrocarbons in the Aquatic Environment,* Applied Science Publishers, London, 1979, chap. 4. With permission.

(including DDT, DDD, DDE, many organochlorines, PCB, naphthalene, benzene, and phenol) in effluent waters, sediments, fish livers, and invertebrates near a waste-water treatment outfall in southern California. Sediment and tissue concentrations were positively correlated with each other and with K_{ow}, but were negatively correlated with effluent concentrations. This may be due to an overlap or similarity in the partitioning constants for

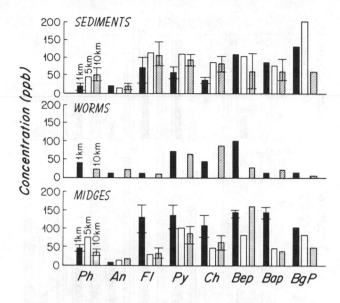

FIGURE 5. Concentrations of individual PAH in sediments (top panel, normalized to dry weight), oligochaete worms (middle panel, normalized to wet weight), and chironomid midges (bottom panel, normalized to wet weight) from sites 1-, 5-, and 10 km from a large coal-fired power plant on Lake Erie in Michigan. In the top panel, bars represent one standard deviation of separate sediment sample extracts. In the bottom panel, bars represent one standard deviation of multiple analyses of single extracts. Ph, phenanthrene; An, anthracene; Fl, fluoranthene; Py, pyrene; Ch, chrysene; BeP, benzo(e)pyrene; BaP, benzo(a)pyrene; BgP, benzo (ghi) perylene. (Redrawn from Eadie, B. J., Faust, W., Gardner, W. S., and Nalepa, T., *Chemosphere*, 11, 185, 1982. With permission.)

uptake in organisms and sorption to sediments. Excessive bioconcentration of PAH is rarely observed when comparing concentrations of PAH in organisms to those in nearby sediments.

Eadie et al.[25,86] measured concentrations of PAH in a number of benthic invertebrates, sediments, and, in some cases, pore waters in several sites in Lakes Michigan and Erie (Figures 5 and 6). Concentrations of PAH in sediments and organisms at three sites on a river transect 1, 5, and 10 km fron a coal-fired power plant were highly variable, both between stations and between individual PAH at each station (Figure 5).[25] A trend for relative enrichment of the largest PAH (benzo[ghi]perylene) vs. the two smallest PAH (phenanthrene and anthracene) was observed in sediments at the two sites closest to the plant. Relative to the sediment, oligochaete worms were depleted of almost all PAH, especially fluoranthene, benzo(a)pyrene, and benzo(ghi)perylene. PAH profiles in midges were very different, closely resembling that observed in the sediments, except for benzo(ghi)perylene, where a depletion was observed. The enhanced ability of worms to metabolize certain PAH was offered as an hypothesis to explain these differences.

Figure 6 compares PAH concentrations in sediment, pore waters, and in the amphipod, *Pontoporeia* at three sites with low, medium, and high organic carbon sediment.[86] A clear trend for increased concentrations of all PAH with increased organic carbon content (or decreased particle size) was observed. This trend was not evident in pore water samples. With the exception of fluoranthene and pyrene, PAH concentrations in *Pontoporeia* seemed to be correlated with organic carbon content of the sediment at that site. In both studies, tissue/sediment PAH ratios ranged from about 1 to 10, while *Pontoporeia*/pore water PAH ratios ranged from approximately 100 to 1000. Comparing the absolute values of these data is problematic without statistical analysis and because of probable incomplete phase separation during preparation of pore water for analysis. Nevertheless, these data represent an excellent first step towards deciphering the contribution of sediment organic carbon, pore

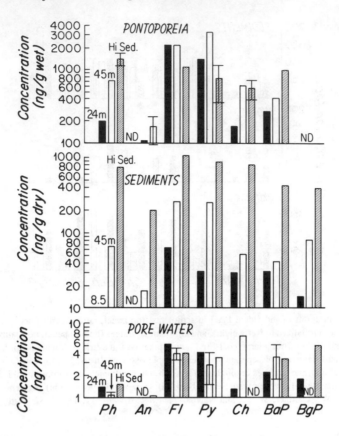

FIGURE 6.   Concentrations of individual PAH on a log scale in the benthic amphipod *Pontoporeia hoyi* (top panel), sediments (middle panel), and pore water (bottom panel) at three stations in southeastern Lake Michigan having different sediment characteristics. At the 24 m depth station (solid bars) sediment was coarse sand and 0.4% organic carbon. The 45 m depth station (open bars) sediment was a silty sand containing 1.7% organic carbon. The third station noted as "Hi Sed." (hatched bars) was in a high depositional area and had sediments predominantly (90%) comprised of particles of < 64 μm in grain size and contained 3.3% organic carbon. PAH abbreviations are the same as described in Figure 5. (From Eadie, B. J., Landrum, P. F., and Faust, W., *Chemosphere*, 11, 847, 1982. With permission.)

water, and species differences to the processes determining bioavailability of PAH to benthic organisms. Concentrations of phenanthrene, benz(a)anthracene, and benzo(a)pyrene measured by Black et al.[80] in sediment, fish, and invertebrates collected from the Hershey River downstream from a former creosote plant show similar trends (Figure 7). With the exception of lamprey eels, PAH concentrations in organisms on a wet weight basis were always less than those in the sediment. PAH concentrations in brown trout and white suckers showed the greatest depletion relative to sediment concentrations, consistent with their ability to metabolize PAH.

Feeding strategies of the organism have an influence on bioaccumulation of hydrocarbons and PAH. Boehm[87] found dramatic differences in the patterns of uptake and depuration of petroleum hydrocarbons, including PAH, between a deposit-feeding (*Macoma balthica*) and suspension-feeding (*Mytilus edulis*) bivalve in an area impacted by the *Tsesis* oil spill. Behavior of the petroleum hydrocarbons in *Mytilus* reflected the behavior of oil in the water column. Conversely, in *Macoma*, uptake and depuration were apparently driven by concentrations of hydrocarbons in the sediment. Similar results have been obtained by Shaw and Wiggs,[88] who measured hydrocarbons in *Mytilus* and *Macoma* in a bay in Alaska polluted

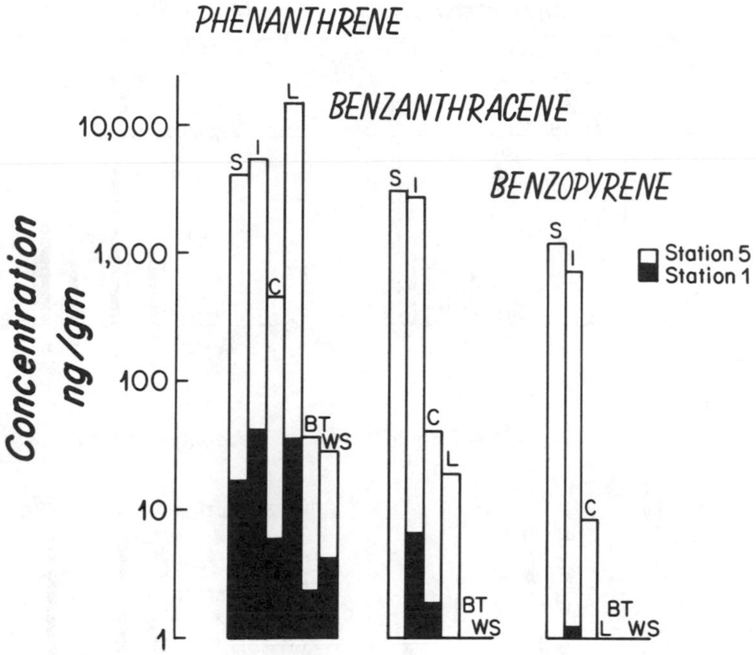

FIGURE 7. Concentration (wet weights) of three individual PAH on a log scale for sediment and biota from the Hershey River, Michigan Station 1 was located upstream of a former creosote plant and station 5 was located downstream of the plant site. Fish samples consisted of muscle tissue pooled from >3 animals. Crayfish samples consisted of muscle tissue pooled from 10 to 20 animals. Insects were analyzed whole. S, sediments; I, insects; C, crayfish; L, lampreys; BT, brown trout; WS, white suckers. (Redrawn from Black, J. J., Har, T. F., Jr., and Evans, E., *Polynuclear Aromatic Hydrocarbons*, Cooke, M. and Dennis, A. J., Eds., Battelle Press, Columbus, Ohio, 1980, 343.)

by both petroleum and coal. The distribution of hydrocarbons in *Mytilus* resembled those from petroleum; in *Macoma*, those from detrital coal. These data indicate that feeding type can significantly alter which hydrocarbons are accumulated by organisms living in the same sediment.

The mixture of hydrocarbons present in most coastal sediments indicates that pyrolysis of fossil fuels is the major source.[31,33,35] However, comparisons of the distribution of hydrocarbons accumulated within benthic organisms and the sediment they live in, indicate that not all hydrocarbons in sediment are equally available for bioaccumulation. Farrington et al.[44] compared PAH profiles in mussels and polychaetes collected in the northeastern U.S. with PAH profiles in sediments from the same locations (Figure 8). PAH in the organisms were indicative of both pyrogenic and petroleum sources, while PAH in the sediments contained PAH with a predominantly pyrogenic origin. These results indicate that PAH in oil are more bioavailable than PAH released through high temperature pyrolysis. PAH of pyrogenic origin (with the exception of creosote leaking from pilings) may be more tightly bound to particulate matter, due to their high temperature of formation or subsequent reactions, creating tight bonds between PAH and particulates in the atmosphere.[44]

Data on PAH accumulation in mussels and oysters collected as part of the U.S. Environmental Protection Agency Mussel Watch Program are best explained by invoking this differing bioavailability hypothesis.[77] Mussels collected in relatively pristine areas contained PAH indicative of a pyrogenic source, either from local sources such as creosote pilings and small boat motors or from long-range transport of combustion source material. The

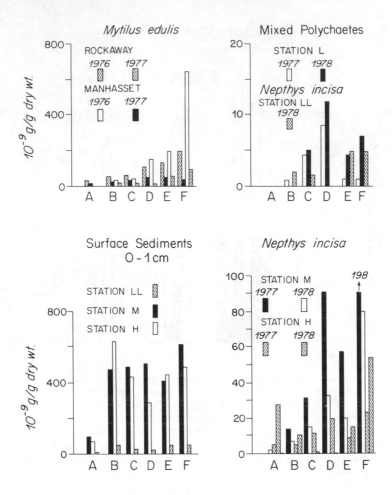

FIGURE 8.   Concentration of individual PAH in samples of the mussel *Mytilus edulis,* mixed polychaetes, the polychaete *Nephtys incisa* and surface sediments collected from several stations in the New York Bight. Organisms were collected in 1977 and 1978. Surface sediments were collected in 1978 only. Station H was in a sewage sludge and dredge spoil dump site. Stations M and LL were in the Hudson trough and contained fine-grained sediments. Station L was on the north side of the trough and also contained fine-grained sediment. A, C - 2 naphthalene; B, phenanthrene; C, C - 1 phenanthrenes; D, C - 2 phenanthrenes; E, fluoranthene; F, pyrene. (From Farrington, J. W., Teal, J. M., Tripp, B. W., Livramento, J. B., and McElroy, A., Biogeochemistry of Petroleum Components at Sediment-Water Interface. Final report prepared for DOE, BLM, March, 1983. With permission.)

relative contribution of petroleum-derived PAH in the mussels was greater at collection sites near urban areas, indicating their greater bioavailability.

C. Acute Vs. Chronic Exposures

Differences in accumulation and depuration of PAH as a function of length of exposure have been observed in field studies. The release of benzo(a)pyrene and other petroleum hydrocarbons from mussels removed from an area contaminated with creosote was slow, with a half-life for benzo(a)pyrene release being 16 days at a temperature of 7 to 9°C.[89] Boehm and Quinn[90] reported a loss of only 30% of the hydrocarbons (including PAH) in 120 days from the hard shell clam, *Mercenaria mercenaria* transported from a highly polluted Providence River site to a relatively clean site in Narragansett Bay, Rhode Island. Depuration from organisms exposed to acute PAH exposures during oil spills is rapid during the first few days, and then proceeds at a much slower rate similar to that observed after chronic exposure.[89,91-94] It is important to note that spill situations may produce extremely high but transient levels of aromatic hydrocarbons and oil droplets. Therefore, initial depuration may seem extremely rapid due to the onset of depuration or release of oil droplets which were never really absorbed. Nevertheless, these observations support earlier laboratory investigations (which will be discussed below) suggesting segregation of accumulated hydrocarbons into compartments within an organism with different rates of elimination.[95]

D. Effect of Physiological Condition on Accumulation

The prevailing view that organisms accumulate hydrocarbons in proportion to their lipid content, raises the question of what happens to accumulated hydrocarbons in organisms which drastically alter their lipid composition and content during spawning. Mix et al.[96] followed benzo(a)pyrene concentrations in somatic and gonadal tissue in mussels chronically exposed to hydrocarbons in the field, and found that gametogenesis and spawning did not significantly alter whole-body PAH concentrations. Similar results have been more recently reported for different populations of scallops (*Pecten maximus*) in France,[81] and for mussels of the Laguna Veneta region of Italy.[97] Seasonal variation in PAH concentrations in both somatic and gonadal tissue was observed, but they did not correlate solely with the gametogenic cycle. In a 4-year study of aromatic and saturated hydrocarbon content in a rocky intertidal marine community, Clark[98] found concentrations of aromatic hydrocarbons in five different organisms to be significantly different depending on season of collection, although no distinct seasonal trends were observed. These results indicate that ''total'' lipid content is not the only factor governing PAH accumulation, storage, and retention in organisms. It is possible and even likely that interspecific differences in metabolism and reproduction may make further generalizations problematic.

E. Evidence for the Presence of Metabolites Even When the Parent Compound is No Longer Detectable

The discussion above has been limited to data on the bioaccumulation of parent PAH alone. The majority of work on PAH metabolites has been done in the laboratory using radioisotopes. Until very recently, analytical constraints have made analysis of PAH metabolic products in field samples very difficult. Routinely used extraction and ''cleanup'' techniques are designed to get rid of polar impurities, a fraction that includes PAH metabolites. When extraction conditions are made broad enough to include some metabolites, they become difficult to detect in the myriad of other compounds in the extract. Consequently, very little data exist on the presence of PAH metabolites in field samples.

Krahn et al.[99] have developed high performance liquid chromatography (HPLC) flourescence techniques which can be used to indicate the presence of metabolites of several PAH in biological samples. A modification of this technique was used to estimate relative concentrations of naphthalene, phenanthrene, dimethylnaphthalene, benzo(a)pyrene, and biphenyl metabolites in the bile of many samples of English sole (*Parophrys vetulus*) from both reference and polluted sites in Puget Sound, Washington.[100] The sum of PAH metabolites detected in fish bile from polluted sites averaged >10 times that observed in fish bile from the reference sites. Gas/liquid chromatography-mass spectrometry analysis of extracts of hydrolyzed bile of English sole from one of the polluted sites indicated that concentrations of individual metabolites of fluorene, phenanthrene, anthracene, biphenyl, and dimethyl-naphthalene ranged from 90 to 19,000 ng/g wet weight. Although these concentrations are substantial, their significance is difficult to assess, as the metabolite and water content of the bile is dependent on a number of factors such as nutritional status and physiological state.

Electron paramagnetic resonance (EPR) spectroscopy is another technique that has been used to indicate the presence of PAH metabolites in English sole (*P. vetulus*) with hepatic neoplasms collected from Puget Sound.[101,102] As indicated by the intensity of the EPR signal, liver microsomes prepared from feral diseased fish had statistically higher concentrations of organic free radicals than microsomes from fish with no evidence of hepatic disease.[102] Incubation of different fractions of sediment extracts (alkanes, aromatic PCB, polar) from this area with microsomes prepared from healthy English sole yielded detectable levels of similar organic free radicals only when microsomes were incubated with the aromatic fraction of the sediment. The type of EPR spectra generated from the microsomes of diseased fish was consistent with that which would be generated by a PAH bound to protein or DNA. In a subsequent study,[102] based on comparison of EPR spectra from diseased fish with synthetic standards, Roubal and Malins identified the free radicals in the fish to be *N*-oxyl metabolites of nitrogen heterocycles. The hypothesis that fish can metabolize these compounds was supported by data from an experiment where similar free radicals were observed in the bile of healthy fish injected with carbazole.

Gossett et al.[72] and Brown et al.[73] have reported direct chemical evidence of oxygenated metabolites of DDT and PCB in sediments, scorpion fish livers, and shrimp muscle collected in the California Bight. Specialized chemical-extraction techniques coupled with gas chromatography with electron capture detection indicated that oxygenated PCB comprised 89 to 99% of total PCB in these samples. Four different tri- to penta-chloro-biphenylols were identified by mass spectrometry of these samples. These data suggest that the presence of metabolic products of PAH and related compounds in marine organisms may be a widespread phenomenon.

Although there are very few direct measurements of PAH metabolites in field samples as mentioned above, their presence can be inferred from other data. The mutagenic potential of PAH is believed to arise only after metabolic activation to reactive metabolites,[103,104] and several groups have proposed using elevated mixed function oxygenase (MFO) activity as an indicator of PAH exposure in fish.[105-107] The Ames test for mutagenicity[108,109] has also been used as a screening tool to detect the presence of PAH metabolites in biological samples. Although it was not used directly for this purpose, Kurelec et al.[106] used the Ames test to detect the presence of metabolically activated pollutant compounds in water samples using fish liver homogenates as the test system. Exposure to mutagenic PAH such as chrysene, benzo(a)pyrene, dimethylbenzanthracene, and 3-methylcholanthrene was positively correlated with microsomal aryl hydrocarbon hydroxylase activity and mutagenic potencies of liver microsomes in carp.[110] Significant correlations observed by Malins et al.[84,101,111] between neoplasia in fish and the presence of PAH in their environment are also suggestive of PAH exposure.

Bioaccumulation is influenced by many processes, including availability and metabolism. Comparisons between bioaccumulation of refractory compounds such as PCB and that of PAH may give a good indication whether or not PAH have been accumulated and then metabolized, even when it is not possible to measure metabolites directly. This approach assumes that availability of PAH and PCB is similar, and that metabolism is the only factor affecting their relative bioaccumulation. The potential utility of this approach is supported by the information in Table 8 compiled by Connor.[112] Connor observed that tissue/sediment concentration ratios of PAH and chlorinated aromatic hydrocarbons (CAH) obtained in surveys of Puget Sound and the New York Bight were correlated with published reports of the MFO activity of these organisms.

F. Trophic Transfer of PAH and/or Metabolites

As it is usually difficult to determine exactly on what and how much field-collected organisms have been feeding, there are very little field data to either support or refute the potential for and importance of trophic transfer of PAH in the aquatic environment. Certainly, biomagnification via trophic pathways, as has been documented with pesticides in mammals,

Table 8
RATIO OF ANIMAL/SEDIMENT RATIOS FOR CAH/PAH AND MFO ACTIVITY FOR VARIOUS PHYLA IN PUGET SOUND (PS) AND THE NEW YORK BIGHT (NYB)

Animal (no. of samples)	CAH/PAH ratio		MFO activity (pmol min^{-1} mg^{-1})
	Median	Range	
Fish			
Winter flounder			213 ± 15
NYB winter flounder liver (1)	3400		
NYB winter flounder flesh (4)	800	120—2000	
PS sole liver (6)	2200	790—3300	
Polychaetes			
Sand worm			89 ± 24
NYB polychaetes (1)	210		
PS "worms" (4)	25	6—45	
Crustaceans			
Blue crab hepatopancreas			42 ± 15
PS crab hepatopancreas (5)	1300	450—2900	
NYB lobster digestive gland (1)	590		
NYB lobster flesh (3)	100	95—120	
NYB grass shrimp (1)	28		
PS shrimp (6)	19		
Mollusks			
Oyster			8 ± 2
Mussel digestive gland			3 ± 1
NYB sea scallops (3)	48	10—76	
NYB mussels (2)	50	32—65	
PS clams (7)	12	1—49	

Note: The geometric mean of the animal/sediment ratios for each PAH and CAH compound was used to calculate an overall CAH/PAH ratio.

From Connor, M. S., *Environ. Sci. Technol.,* 18, 31, 1984. With permission.

birds, and nonaquatic organisms, has not been observed with PAH in aquatic systems. Malins et al.[84] found that PAH exposure in fish, as indicated by the presence of PAH metabolites in the bile, was more closely correlated with PAH concentrations in their gut contents than PAH concentrations in the sediments where the fish were collected. Maccubbin et al.[113] found concentrations of a number of PAH in the stomach contents of white suckers (*Catastomus commersoni*) which were correlated with incidence of neoplasia in this fish. However, in this case, PAH concentrations in the gut contents of the fish also correlated with PAH concentrations in the sediment, making the uptake route ambiguous. The contribution of PAH on sediment in the guts of either predator or prey organisms is another potential confounding factor in determining true PAH accumulation through dietary pathways. Because of these uncertainties in field-collected organisms, investigation of trophic transfer of PAH is probably more tractable in a controlled laboratory situation.

Table 9
**TISSUE-TO-SEDIMENT RATIOS (TSR) FOR AROMATIC
HYDROCARBONS IN BENTHIC ORGANISMS EXPOSED TO
DUWAMISH RIVER DELTA SEDIMENT**

Species	Sum of aromatic hydrocarbons		
	3-ring	4-ring	5-ring
Eohaustorius washingtonius	0.050 ± 0.003	0.510 ± 0.027	0.180 ± 0.023
Rheopoxynius abronius	0.042 ± 0.003	0.200 ± 0.035	0.036 ± 0.029
Macoma nasuta	0.022 ± 0.005	0.078 ± 0.006	0.053 ± 0.002

Note: TSR = (ng/g wet weight of tissue per ng/g wet weight of sediment). Values are expressed
as means ± standard deviation. Each value represents the mean of three pooled samples
each containing 37 individuals for *M. nasuta*, two pooled samples each containing 100 for
E. washintonianus, and three pooled samples each containing 60 for *R. abronius.*

Adapted from Varanasi, U., Reichert, W. L., Stein, J. E., Brown, D. W., and Sanborn, H. R.,
Environ. Sci. Technol., 19, 836, 1985. With permission.

IV. LABORATORY INVESTIGATIONS ON THE FACTORS INFLUENCING PAH ACCUMULATION BY AQUATIC ORGANISMS

A. Accumulation and Depuration are Influenced by Physical/Chemical Characteristics of PAH

Experiments measuring uptake and depuration of individual PAH from water into organisms have demonstrated that bioaccumulation is positively correlated with physical/chemical properties of the PAH such as molecular weight and octanol/water partition coefficients.[114-119] This relationship has been observed for both nonpolar PAH (as cited above) and aromatic amines.[120] At equilibrium, the degree to which an organism will bioaccumulate PAH and other lipophilic organics can be roughly predicted knowing its physical/chemical properties.[15,117,121-123] A compilation of a number of the equations which have been empirically derived correlating water solubility (WS), octanol/water partition coefficient (K_{ow}), soil adsorption coefficient (K_{oc}, the partition coefficient normalized to organic carbon content of the soil), and bioconcentration factor (BCF) in fish, can be found in a review on the subject by Kenaga.[123] It should be noted that these equations were developed from bioconcentration data obtained in relatively short-term laboratory studies. In addition, although these relationships are designed to be general predictors of bioconcentration potential of organic compounds, most of the data have been generated using chlorinated hydrocarbons and fish. Metabolism and active excretion processes can drastically reduce the bioaccumulation of compounds such as PAH. In some cases, the relative availability of PAH does not seem to be a simple function of physical/chemical properties. Varanasi et al.[140] found 4-ring PAH to be more available than either 3- or 5-ring PAH for accumulation from contaminated sediments by two species of amphipod (*Rheopoxynius abronius* and *Pandalus platyceros*) and one species of clam (*Macoma nasuta*) in Puget Sound (Table 9). It is possible that the larger PAH are less available due to their size, even though partitioning would favor their bioaccumulation.

Pruell et al.[119] measured bioconcentration of PCB and PAH in mussels (*Mytilus edulis*) exposed to sediments from polluted (exposed) and relatively clean (control) areas of Narragansett Bay, Rhode Island in a laboratory-dosing system which maintains a constant load of suspended particulates. BCF (concentration in the organism/concentration in the dissolved phase) displayed a log/log correlation with the octanol/water partition coefficient for five

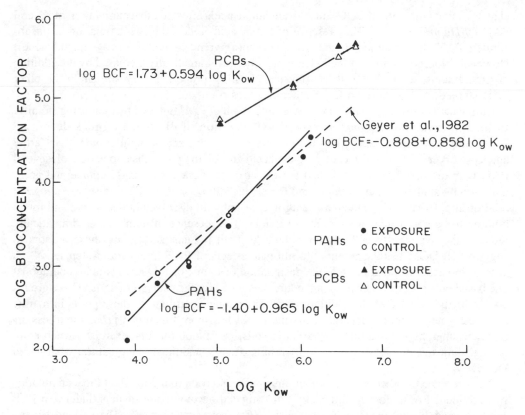

FIGURE 9. Bioconcentration factors (concentration in mussels normalized to wet weight/concentration in dissolved phase) of individual PAH and PCB congeners in *Mytilus edulis* plotted as a function of K_{ow}. Solid symbols represent values from mussels exposed for 20 d and open symbols represent values from mussels held in control tanks for 20 days. Solid lines and corresponding regressions are derived from this study. Broken line and regression represents the relationship derived by Geyer et al.[117] (From Pruell, R. J., Lake, J. L., Davis, W. R., and Quinn, J. G., *Mar. Biol.*, 91, 497, 1986. With permission.)

PAH and four PCB congeners (Figure 9). For comparison, the relationship observed between BCF and K_{ow} in mussels described by Geyer et al.[117] has also been included in this figure. The PAH data come very close to the relationship described by Geyer, but the PCB data are better described with an equation with the same slope, but with a different intercept. Since Geyer's equation was generated from bioaccumulation data on a wide range of compounds including many pesticides, naphthalene, and PCB in mussels, the explanation for the discrepancy is not apparent. Some aspect of the dosing system in Pruell's experiment may have imposed differential availability between PAH and PCB, or there may be some difference between the way mussels accumulate or release these two classes of compounds that is independent of their K_{ow}.

B. Influence of Exogenous Factors

In general, decreasing temperature has been found to increase bioaccumulation of PAH.[124-128] This increase is due to a number of factors including decreased rates of uptake, depuration, and metabolism. When both parent compound and metabolites are taken into account, the effect of temperature is much less pronounced.[124,128]

It is well known that DOM affects bioaccumulation of PAH by aquatic organisms. Sanborn and Malins[129] found that bovine serum albumin (BSA) added to the medium drastically reduced accumulation of naphthalene by larval spot shrimp (*Pandalus platyceros*). Leversee et al.[130] found both natural DOC and Aldrich humic acids affected bioaccumulation of several PAH by *Daphnia magna*. The presence of humic acids reduced bioaccumulation of unsubstituted PAH in proportion to their K_{ow} (benzo(a)pyrene > anthracene > naphthalene). However, bioaccumulation of methylcholanthrene was dramatically increased by the addition of humic acids, even though it has physical/chemical properties very similar to the other PAH studied. No explanation for this paradox was offered.

Humic acids were also found to increase salting-out of PAH initiated by increasing salinity. In bluegill sunfish (*Lepomis macrochirus*), addition of dissolved humics decreased benzo(a)pyrene accumulation, but had no effect on anthracene accumulation.[131] McCarthy and co-workers[132,133] have recently investigated the binding and disassociation of several PAH with dissolved humic acids and the effect of humic acids on PAH uptake and accumulation by sunfish *L. macrochirus* and the cladoceran *D. magna*. They observed a log/log relationship between K_{ow} and an association constant with dissolved humic acids, and found binding to be completely and rapidly reversible. The presence of humic acids dramatically reduced the availability of PAH for uptake by both organisms, leading these authors to suggest that DOM has the potential to mitigate effects of PAH in aquatic systems.

It is important to point out that the designation of humic or fulvic acids is an operational one based on a specific extraction procedure developed for the analysis of soils. These groups do not include all DOM. In addition, different sources contribute to these pools in marine and freshwater systems. Generally, marine DOM has more functional groups and less aromaticity than humic and fulvic material in soils.[134] Caution must be exercised when comparing studies using DOM such as Aldrich humic acids, and those using naturally harvested DOM.

The presence of other organic contaminants has also been found to affect bioaccumulation of individual PAH. Fortner and Sick[135] investigated accumulation of naphthalene, a PCB mixture, and benzo(a)pyrene by the oyster (*Crassostrea virginica*). They found several instances where multiple components had antagonistic effects on accumulation. Gruger et al.[136] found that exposure to naphthalene influenced the type of 2,6-dimethylnaphthalene metabolites produced *in vivo* by the starry flounder *(Platicthys stellatus)*. Stein et al.[137] exposed English sole (*Parophrys vetulus*) to sediments labeled with ³H-benzo(a)pyrene and ¹⁴C-Aroclor 1254® either singly or together, and found that accumulation of benzo(a)pyrene-derived radioactivity was enhanced when the fish were exposed to both compounds simultaneously relative to exposure to the PAH alone. Considering the fact that marine organisms are almost never exposed to just one pollutant compound at a time, it is important to understand the interactive nature of exposure to multiple compounds.

C. Influence of Endogenous Factors

Metabolism of PAH by aquatic organisms is covered in detail in the following chapters. However, when discussing bioaccumulation, we must consider the metabolic potential of the organism. Quantifying only the concentration of the parent compound can lead to gross underestimation of true uptake in organisms capable of metabolizing various PAH. Investigations monitoring accumulation of both parent compound and metabolic products in a number of organisms have clearly demonstrated this effect.[128,138-141]

Even when metabolism is taken into account, in general, PAH body burdens are greater in organisms which cannot metabolize them. Varanasi et al.[140] compared uptake of PAH from contaminated sediments which had also been spiked with ³H-benzo(a)pyrene by several species of benthic organisms, including two deposit-feeding amphipods (*Rheopoxynius abronius* and *Eohaustorius washingtonianus*), a deposit-feeding clam (*Macoma natusa*), a shrimp (*Pandalus platyceros*), and the English sole (*Parophrys vetulus*). With the exception that the amphipods achieved higher tissue PAH concentrations than the clam, body burdens of PAH in these organisms were inversely correlated with the degree of ³H-benzo(a)pyrene

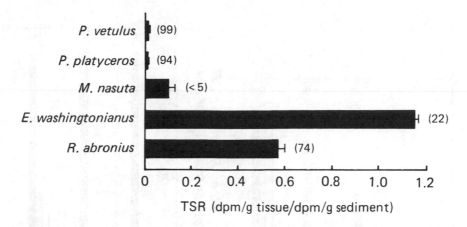

FIGURE 10. Tissue-to-sediment ratios (TSR); concentration of benzo(a)pyrene-derived radioactivity (BaP and its metabolites) in whole-body/BaP-derived radioactivity primarily as BaP in sediment extracts normalized to wet weights) for fish (*P. vetulus,*) shrimp (*P. platyceros*), clams (*M. nasuta*), and two species of amphipods (*E. washingtonianus* and *R. abronius*). Values presented as means ± 1 S.D. The numbers in parentheses at the top of each histogram represent the percentage of total radioactivity recovered present as BaP metabolites in each organism and give a comparative measure of each organism's ability to metabolize BaP. (Adapted from Varanasi, U., Reichert, W. L., Stein, J. E., Brown, D. W., and Sanborn, H. R., *Environ. Sci. Technol.*, 19, 836, 1985. With permission.)

metabolism (Figure 10). Differences in the feeding strategies between the amphipods and the clams were offered as a possible explanation for this apparent paradox, although avoidance behavior of the clams (e.g., valve closure) could also produce the observed results.

In addition to metabolism, there are other endogenous factors that influence PAH accumulation. Stegeman and Teal[95] found that accumulation of petroleum hydrocarbons by two populations of the oyster, *Crassostrea virginica*, was positively correlated with their fat contents. This observation, coupled with earlier work by Hamelink,[142] among others, has led to the general theory that aquatic organisms accumulate lipophilic substances in proportion to lipid pools within the body. However, bioconcentration can be influenced by the type of lipid available. Schneider[143] found that difference between PCB concentrations in different organs of cod could be eliminated if residues were normalized to the neutral lipid or "fat" (mostly triglycerides) content of the organ (Figure 11). Large differences between PCB content of cod gonads and fillets vs. liver were still evident when residues were normalized to extractable lipid content alone. We are not aware of similar data comparing PAH residues as a function of lipid class, but would expect them to behave as PCB in this case.

Reproductive state can also influence accumulation and release of PAH. Rossi and Anderson[144] found that while both male and female polychaetes (*Neanthes arenaceodentata*) accumulated diaromatic hydrocarbons to a similar extent, male worms released them more rapidly than the female worms, who retained the PAH until spawning when they were very rapidly depurated. In this case, the PAH seemed to be closely associated with the yolk lipids of the worms.

PAH uptake and depuration have been shown to be affected by whether or not an organism is feeding. Leversee et al.[139] reported more rapid elimination of benzo(a)pyrene from the midge, *Chironomus riparius* when the animals were fed. How an animal feeds affects bioaccumulation just as it does uptake. Roesijadi et al.[145] found that two deposit-feeding species, *Phascolosoma agassizi* and *Macoma inquinata*, accumulated more diaromatic PAH from oil-contaminated sediment than a suspension-feeding clam, *Protothaca staminea*.

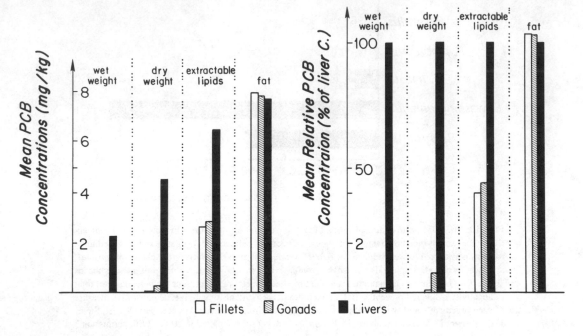

FIGURE 11.   Left panel: mean PCB concentration in cod fillets (open bars), gonads (hatched bars) and livers (solid bars) normalized to wet weight, dry weight, extractable lipids (material soluble in 10% water in methanol), and fat (material soluble in *n*-hexane). Right panel: the same data normalized to liver concentration.) Redrawn from Schneider, R., *Meeresforsch.*, 29, 69, 1962. With permission.)

D. Influence of Source on Bioavailability

Numerous laboratory studies have shown that aquatic organisms can accumulate PAH from the water column, from sediments, and from their diet (see Neff[1] for review). However, it is also clear from these studies that the bioavailability of PAH from different sources is not equivalent. Accumulation of dissolved PAH has been well documented with BCF (concentration in organism on a wet weight basis/concentration in water) ranging up to 10,000.[116] The relative availability of PAH from other sources is less clear. It is generally believed that PAH sorbed to sediments are only available to organisms after undergoing desorption into the dissolved phase. However, as discussed in the introduction, there will always be a dynamic exchange between PAH in the solid and dissolved phase. Altering the concentration in either phase will initiate redistribution of the PAH between the phases. The presence of organic matter in the water often presents a third phase, colloids, whose influence is not easily measured. Adequately assessing the importance of uptake from these various compartments is an extremely difficult task requiring comparisons between concentration in all phases, total amounts transferred, and, particularly, rates of transfer between compartments. Because sediments serve as the primary reservoir for PAH (see discussion in Introduction), the availability of sediment-sorbed PAH is of particular interest.

Work on bioavailability of sediment-sorbed PAH has generally shown that accumulation from sediment sources is possible,[43-45,119,125,140,145-150] although Rossi[151] reported that the polychaete *Neanthes arenaceodentata* was unable to accumulate naphthalenes from labeled sediments. Several investigators have compared accumulation of PAH added to the water column vs. that of PAH added to sediments; or accumulation of PAH in sediments with organisms placed in the sediment vs. organisms held above the sediment. Unfortunately, it is very difficult in most of these studies to discriminate and compare the contributions of either source.

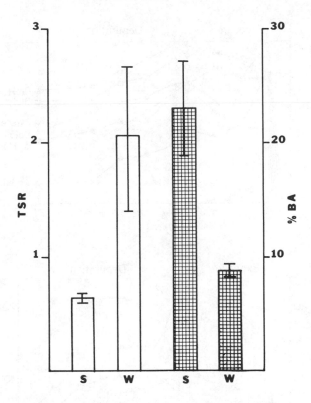

FIGURE 12.   Bioaccumulation and metabolism of benzanthracene (BaA) in *N. virens* exposed to ^{14}C-labeled BaA in benthic microcosms for 1 week where BaA was bound to the sediment reservoir prior to placement in the microcosms (S) or where BA was added to the water column and secondarily associated with surface sediment (W). Tissue-to-sediment concentration ratios (TSR); BaA-derived radioactivity (parent and metabolites) in *N. virens*/ BaA-derived radioactivity primarily parent in surface sediments normalized to dry weight. % BA shows the percentage of BA-derived activity in *N. virens* still present as the parent compound. Values are expressed as means ± 1 S.D. of 6(S) or 3(W) individual samples of four *N. Virens* each. (Adapted from McElroy.[45])

Despite these limitations, several generalizations about the bioavailability of PAH in the dissolved vs. particulate state can be made. Uptake from the dissolved state seems to occur more rapidly. Roesijadi et al.[145] found that accumulation of PAH by a suspension-feeding clam reached equilibrium within 40 days, while PAH concentrations in two deposit feeders exposed to the same oiled sediment did not reach equilibrium within 60 days. McElroy[45] investigated bioaccumulation of ^{14}C-benz(a)anthracene by the polychaete, *Nereis virens* in benthic microcosms where the PAH was introduced directly to the water column or already sorbed to fine sediments. After 1 week, tissue/surface sediment ratios for *N. Virens* and the degree to which accumulated residues were metabolized were significantly higher when the benz(a)anthracene was introduced via the water column (Figure 12). Obana et al.[118] compared accumulation and depuration of seven PAH by the short-necked clam, *Tapes japonica*, exposed to PAH added to the water column in a recirculating aquarium, or in sediments collected from an urban harbor heavily contaminated with PAH. The clams reached apparent equilibrium with water concentrations within 1 day in the first case, while concentrations of most individual PAH were still increasing in clams exposed to contaminated sediment for 7 days (Figure 13). Considering the reversible nature of PAH/particle association (except for bound pyrogenic PAH), the differential availability of dissolved vs. sorbed PAH should primarily be a kinetic consideration. Particularly for the larger PAH, the particulate reservoir should be the primary source, even though actual uptake may occur via a dissolved pathway.

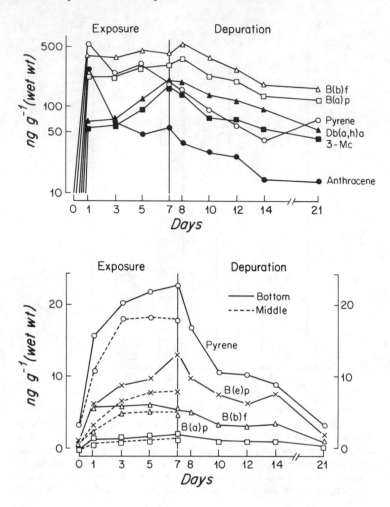

FIGURE 13.   Top panel: uptake and release of PAH by short-necked clams exposed to PAH dissolved in artificial sea water. Bottom panel: uptake and release of PAH by short-necked clams exposed to sediments contaminated with PAH. Solid line shows data from clams placed on the sediment. Dashed line shows data for clams held in a cage 10 cm off the sediment surface. B(b)F, benzo(b)fluoranthene; B(a)P, benzo(a)pyrene; B(e)P, benzo(e)pyrene; Db(a,h)A, dibenz(a,h)anthracene; 3-Mc, 3-methylcholanthrene. (Redrawn from Obana, H., Hori, S., Nakamura, A., and Kashimoto, T., *Water Res., 17*, 1183, 1983. With permission.)

Landrum and Scavia[152] investigated the influence of sediment on uptake, depuration, and biotransformation of anthracene by the amphipod, *Hyalella acteca*. The presence of natural sediment had no significant effect on initial uptake rates, but body burdens were still increasing after 8 h of anthracene exposure in the presence of sediment, while concentrations were beginning to plateau in the absence of sediment, or with a sandy substrate. Depuration rates were faster in the presence of sediment. Biotransformation was also influenced by the presence of natural sediment. A lower percentage of bound anthracene-derived residues was found in animals in sediment. Landrum and Scavia[152] estimated that anthracene associated with sediment and pore water contributed 77% of the amphipods' steady-state body burden. Because of the combined source of organic contaminants to benthic organisms, they cautioned against the use of organism/water BCF to describe bioaccumulation in the benthos.

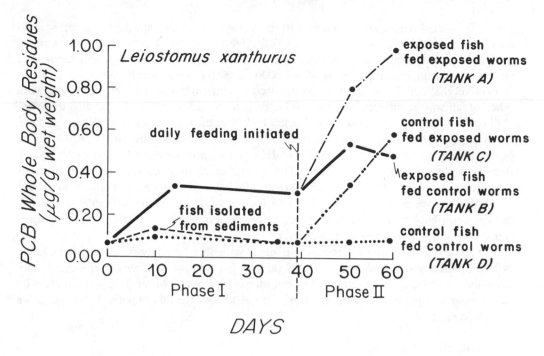

FIGURE 14. Whole-body PCB residues in spot (*Leiostomus xanthurus*) exposed to PCB in the laboratory. Phase I: fish exposed directly to contaminated sediment (solid line), fish kept isolated from contaminated sediment by screens (dashed line), control fish. During this time, worms (*N. virens*) were also exposed to contaminated sediments. During phase II, fish were moved to new tanks as described in the Figure. (Redrawn from Rubinstein, N. I., Gilliam, W. T., and Gregory, N. R., *Aquat. Toxicol.*, 5, 33, 1984. With permission.)

E. Trophic Transfer

Aquatic organisms have been shown to be capable of accumulating PAH via their diet.[45,153-157] In cases where uptake from food vs. sediments has been compared, dietary route appears to be more efficient.[45,155] McElroy[45] compared bioaccumulation and metabolism of ^{14}C-labeled benz(a)anthracene by the omnivorous deposit-feeding polychaete in microcosms where the PAH was introduced either already sorbed to sediments or in a prepared protein-based diet. The isotope in the prepared diet was more rapidly metabolized than isotope that bound to sediments. Rubinstein et al.[158] conducted an elegant experiment looking at bioaccumulation of PCB by a demersal fish (*Leiostomus xanthurus*) from a contaminated sediment, and from worms (*N. virens*) fed contaminated sediment. As shown in Figure 14, fish were able to accumulate PCB either directly from the sediment or from ingesting contaminated worms. The greater efficiency of the dietary route is evident during phase II of the study, where fish exposed to both contaminated sediment and fed contaminated worms accumulated more PCB than fish exposed to contaminated sediment and fed control worms. Even though this study was done on PCB, the experimental design made it possible to clearly differentiate between the sediment and dietary contribution of this group of aromatic hydrocarbons to bioaccumulation in a predatory fish, and points out the potential importance of this pathway.

Studies comparing direct uptake from solution to that from dietary routes present a less consistent picture. Dobroski and Epifanio[153] found the efficiency of ^{14}C-labeled benzo(a)pyrene uptake directly from the water column to be greater than from contaminated algae in larval bivalves. However, they concluded that under natural conditions where contaminant concentrations in the food would be likely to be higher than that in the water column, dietary sources would be more important. Lu et al.[141] compared uptake of radioactivity derived from ^{14}C-labeled benzo(a)pyrene added to the water column by fish, mosquito larvae, and snails exposed either as single species or collectively in a model ecosystem. In the single species exposures, no bioaccumulation was observed in fish, while mosquito larvae and snails attained radioactivity levels 40 to 2000 times that observed in the water column, respectively. When the organisms were exposed together in the model ecosystem along with other organisms at different trophic levels, BCF increased dramatically for all groups including fish (from 0 to almost 10^3). Clark[98] looked at bioaccumulation of ^{14}C-naphthalene in an experimental food chain consisting of diatoms, mussels, and snails. Although he observed trophic transfer, comparisons of BCF between dietary uptake and direct uptake from the water column led him to conclude that partitioning from seawater across membranes is a much more important route for PAH accumulation. However, as pointed out in the discussion above, even though efficiency of transfer of dissolved organics may be greater, the net contribution of sediment and dietary reservoirs may make them a more important source from an environmental standpoint.

The metabolic and physical processes involved with digestion may facilitate accumulation of hydrophobic compounds such as PAH directly from particles. If this were true, dietary accumulation would be expected to be both more efficient and faster than accumulation by passive partitioning across body surfaces. To our knowledge this hypothesis has yet to be investigated.

F. A Question of Spiking

Much of the data discussed above was generated in experiments where natural sediments were spiked with PAH with or without radiotracers. In a recent experiment where several benthic organisms were exposed to sediments from an urban estuary which had been spiked with ^{3}H-benzo(a)pyrene, Varanasi et al.[140] found that tissue/sediment ratios were larger for the radioisotope than the naturally occurring benzo(a)pyrene residues. These laboratory data support the theories advanced previously,[43,44,159] suggesting that not all PAH measured by vigorous chemical-extraction techniques are equally bioavailable to organisms. Aromatic hydrocarbons from petroleum released into the aquatic environment may be less tightly bound to particles than aromatic hydrocarbons released by pyrolysis which enter the marine environment already associated with particulates. Additional work is needed to determine what factors influence the degree of association between hydrophobic pollutant compounds such as PAH and different types of particulate material.

V. DIRECTIONS FOR FUTURE RESEARCH

The preceding discussion highlights many areas needing additional research. In conclusion, we would like to point out just a few of the more important questions from an environmental standpoint. The physical properties of PAH/sorbate interactions are very poorly understood at present, although results of recent investigations are encouraging. A better understanding of these interactions and what factors influence them would help us to better predict biogeochemical cycles of PAH in the aquatic environment. Considering the influence of DOM on partitioning behavior and bioavailability, it will also be important to determine the impact on PAH/sorbate interactions of the large quantities of man-made and natural dispersants entering the aquatic environment from various sources such as sewer effluents and plankton blooms.

There is also a paucity of information available on heteroatomic PAH in the environment, despite the known toxicity of several of these compounds. Their chemistry and analysis are more complex than that of the neutral PAH, requiring advanced analytical techniques and more complex interpretation schemes. Increased use of some of the new energy-producing technologies such as coal liquification and shale oil could increase the input of heteroatomic PAH to the environment.[10] The fate and transport of other polar compounds such as PAH metabolites and photooxidation products in the environment are also not well understood at present. Little is known about their residence times, although preliminary data suggest that they are not ephemeral compounds.

Elucidation of the myriad of factors which influence the cycling of PAH in the environment and their bioavailability is an extremely difficult and tedious process. Therefore, it is necessary to use what is known of the physical/chemical properties and the chemical and biological processes that affect their cycling to generate predictive relationships. These relationships must be sensitive to site-specific factors of the environment and the dynamics of the system. Metabolism is a prime example of just one of the complex and continually changing factors that affect PAH in the environment.

ACKNOWLEDGMENTS

The authors wish to thank Drs. Bruce Brownawell, Judith McDowell Capuzzo, Jay Means, John Stegeman, and Jean Whelan for valuable discussion on the topic. Financial support from the National Oceanic and Atmospheric Administration (Contract 83-ABD-00012) to AEM and JMT; the National Academy of Sciences/National Academy of Engineering, Environmental Protection Agency Visiting Investigators Program to AEM; and Andrew W. Mellon Foundation support through the Coastal Research Center at the Woods Hole Oceanographic Institution to JWF are greatfully acknowledged. This is contribution number 6394 of the Woods Hole Oceanographic Institution.

REFERENCES

1. **Neff, J. M.**, *Polycyclic Aromatic Hydrocarbons in the Aquatic Environment*, Applied Science Publishers, London, 1979, chap. 4.
2. National Research Council, *Polycyclic Aromatic Hydrocarbons: Evaluation of Sources and Effects*, National Academy Press, 1983.
3. **Mackay, D. and Shiu, W. Y.**, Aqueous solubility of polynuclear aromatic hydrocarbons, *J. Chem. Eng. Data*, 22, 399, 1977.
4. **May, W. E., Wasik, S. P., and Freeman, D. H.**, Determination of the solubility behavior of some polycyclic aromatic hydrocarbons in water, *Anal. Chem.*, 50, 997, 1978.
5. **May, W. E.**, The solubility behavior of polycyclic aromatic hydrocarbons in aqueous systems, in *Petroleum in the Marine Environment*, Petrakis, L. and Weiss, F. T., Eds., American Chemical Society, Washington, D. C., 1980, chap. 7.
6. **Setschenow, J. Z.**, *J. Phys. Chem.*, 4, 117, 1898.
7. **Whitehouse, B. G.**, The Partitioning of Polynuclear Aromatic Hydrocarbons into the Dissolved Phase of the Aquatic Environment, Ph.D. thesis, Dalhousie University, Halifax, N.S., Canada, 1983.
8. **Whitehouse, B. G.**, The effects of temperature and salinity on the aqueous solubility of polynuclear aromatic hydrocarbons, *Mar. Chem.*, 14, 319, 1984.
9. **Readman, J. W., Mantoura, F. C., Rhead, M. M., and Brown, L.**, Aquatic distribution and heterotrophic degradation of polycyclic aromatic hydrocarbons (PAH) in the Tamar estuary, *Estuarine Coastal Shelf Sci.*, 14, 369, 1982.
10. National Research Council, *Oil in the Sea Inputs, Fates, and Effects*, National Academy Press, Washington, D. C., 1985.
11. **Banerjee, S.**, Calculation of water solubility of organic compounds with UNIFAC-derived parameters, *Environ. Sci. Technol.*, 19, 369, 1985.
12. **Leo, A., Hansch, C., and Elkins, D.**, Partition coefficients and their uses, *Chem. Rev.*, 71, 525, 1971.
13. **Karickhoff, S. W., Brown, D. S., and Scott, T. A.**, Sorption of hydrophobic pollutants on natural sediments, *Water Res.*, 13, 241, 1979.

14. **Means, J. C., Hassett, J. J., Woods, S. G., and Banwart, W. L.,** Sorption properties of energy-related pollutants and sediments, in *Polynuclear Aromatic Hydrocarbons,* Jones, P. W. and Leber, P., Eds., Ann Arbor Science Publishers, Ann Arbor, MI, 1979, 327.

15. **Chiou, C. T., Freed, V. H., Schmedding, D. W., and Kohnert, R. L.,** Partition coefficient and bioaccumulation of selected organic chemicals, *Environ. Sci. Technol.,* 11, 475, 1977.

16. **Mackay, D., Mascarenhas, R., and Shiu, W. Y.,** Aqueous solubility of polychlorinated biphenyls, *Chemosphere,* 9, 257, 1980.

17. **Schwarzenbach, R. P. and Westall, J.,** Transport of nonpolar organic compounds from surface water to groundwater. Laboratory sorption studies, *Environ. Sci. Technol.,* 15, 1360, 1981.

18. **Chiou, C. T., Porter, P. E., and Schmedding, D. W.,** Partition equilibria of nonionic organic compounds between soil organic matter and water, *Environ. Sci. Technol.,* 17, 227, 1983.

19. **Boehm, P. D. and Quinn, J. G.,** Solubilization of hydrocarbons by dissolved organic matter in seawater, *Geochim. Cosmochim. Acta,* 37, 2459, 1973.

20. **Hiemenz, P. C.,** *Principles of Colloid and Surface Chemistry,* Marcel Dekker, New York, 1977, 1.

21. **Means, J. C. and Wijayaratne, R.,** Role of natural colloids in the transport of hydrophobic pollutants, *Science,* 215, 968, 1982.

22. **Witjayaratne, R. D. and Means, J. C.,** Sorption of polycyclic aromatic hydrocarbons by natural estuarine colloids, *Mar. Environ. Res.,* 11, 77, 1980.

23. **Brownawell, B. J. and Farrington, J. W.,** Partitioning of PCBs in marine sediments, in *Marine and Estuarine Geochemistry,* Sigleo, A. C. and Hattori, A., Eds., Lewis Publishing , Chelsea, MI, 1985, 97.

24. **Brownawell, B. J. and Farrington, J. W.,** Biogeochemistry of PCBs in interstitial waters of a coastal marine sediment, *Geochim. Cosmochim. Acta,* 50, 157, 1986.

25. **Eadie, B. J., Landrum, P. F., and Faust, W.,** Polycyclic aromatic hydrocarbons in sediments, pore water and the amphipod *Pontoporeia hoyi* from Lake Michigan, *Chemosphere,* 11, 847, 1982.

26. **Gschwend, P. M. and Wu, S.,** On the constancy of sediment-water partition coefficients of hydrophobic organic pollutants, *Environ., Sci. Technol.,* 19, 90, 1985.

27. **Westall, J. C., Leuenberger, C., and Schwarzenbach, R. P.,** Influence of pH and ionic strength on the aqueous-nonaqueous distribution of chlorinated phenols, *Environ. Sci. Technol.,* 19, 193, 1985.

28. **Means, J. C., Woods, S. G., Hassett, J. J., and Banwart, W. L.,** Sorption of amino- and carboxy-substituted polynuclear aromatic hydrocarbons by sediments and soils, *Environ. Sci. Technol.,* 16, 93, 1982.

29. **Wilson, R. D., Monaghan, P. H., Osanik, A., Priu, L. C., and Rogers, M. A.,** Natural marine oil seepage, *Science,* 184, 857, 1977.

30. **Youngblood, W. W. and Blumer, M.,** Polycyclic aromatic hydrocarbons in the environment: homologous series in soils and recent marine sediments, *Geochim. Cosmochim. Acta,* 39, 1303, 1975.

31. **Hites, R. A., LaFlamme, R. E., Windsor, , J. G., Jr., Farrington, J. W., and Euser, W. G.,** Polycyclic aromatic hydrocarbons in an anoxic sediment core from the Pettaquamscutt River (Rhode Island, U.S.A.), *Geochim. Cosmochim. Acta,* 44, 873, 1980.

32. **LaFlamme, R. E. and Hites, R. A.,** The global distribution of polycyclic aromatic hydrocarbons in recent sediments, *Geochem. Cosmochim. Acta,* 42, 289, 1978.

33. **Wakeham, S. G. and Farrington, J. W.,** Hydrocarbons in contemporary aquatic sediments, in *Contaminants and Sediments Vol. I,* Baker, R. A., Ed., Ann Arbor Science Publishers, Ann Arbor, MI, 1980, 3.

34. **Tan, Y. L. and Heit, M.,** Biogenic and abiogenic polynuclear aromatic hydrocarbons in sediments from two remote Adirondack lakes, *Geochim. Cosmochim. Acta,* 45, 2267, 1981.

35. **Gschwend, P. M. and Hites, R. A.,** Fluxes of polycyclic aromatic hydrocarbons to marine and lacustrine sediments in the northeastern United States, *Geochim. Cosmochim. Acta,* 45, 2359, 1981.

36. **Gagosian, R. B., Peltzer, E. T., and Zafiriou, O. C.,** Atmospheric transport of continentally derived lipids to the tropical North Pacific, *Nature,* 291, 312, 1981.

37. **Eganhouse, R. P. and Kaplan, I. R.,** Extractable organic matter in urban stormwater runoff. I. Transport dynamics and mass emission rates, *Environ. Sci. Technol.,* 15, 310, 1981.

38. **Hoffman, E. J., Mills, G. L., Latimer, J. S., and Quinn, J. G.,** Urban runoff as a source of polycyclic aromatic hydrocarbons to coastal waters, *Environ. Sci. Technol.,* 18, 580, 1984.

39. **Prahl, F. G., Creclius, E., and Carpenter, R.,** Polycyclic aromatic hydrocarbons in Washington coastal sediments: an evaluation of atmospheric and riverine routes of introduction, *Environ. Sci. Technol.,* 18, 687, 1984.

40. **Gearing, J. N., Gearing, P. J., Wade, T., Quinn, J. G., McCarthy, M. B., Farrington, J. W., and Lee, R. F.,** The rates of transport and fates of petroleum hydrocarbons in a controlled marine ecosystem and a note on analytical variability, in *Proc. 1979 Oil Spill Conf. (Prevention, Behavior, Control and Cleanup),* American Petroleum Institute, Washington, D.C., 1979, 555.

41. **Gearing, P. J., Gearing, J. N., Pruell, R. J., Wade, T. L., and Quinn, J. G.,** Partitioning of no. 2 fuel oil in controlled estuarine ecosystems, sediments and suspended particulate matter, *Environ. Sci. Technol.,* 14, 1129, 1980.

42. National Research Council, *Petroleum in the Marine Environment*, National Academy Press, Washington, D.C., 1975, 107.

43. **Farrington, J. W., Tripp, B. W., Teal, J. M., Mille, G., Tjessum, K., Davis, A. C., Livramento, J., Hayward, N., and Frew, N. M.**, Biogeochemistry of aromatic hydrocarbons in the benthos of microcosms, *Toxicol. Environ. Chem.*, 5, 331, 1982.

44. **Farrington, J. W., Teal, J. M., Tripp, B. W., Livramento, J. B., and McElroy, A.**, Biogeochemistry of Petroleum Components at the Sediment-Water Interface, final report prepared for DOE, BLM, March 1983.

45. **McElroy, A. E.**, Benz(a)anthracene in Benthic Marine Environments: Bioavailability, Metabolism, and Physiological Effects on the Polychaete *Nereis virens*, Ph.D. thesis, Massachusetts Institute of Technology and the Woods Hole Oceanographic Institution, Woods Hole, MA, 1985.

46. **Hinga, K. R., Lee, R. F., Farrington, J. W., Pilson, M. E. Q., Tjessum, K., and Davis, A. C.**, Biogeochemistry of benzanthracene in an enclosed marine ecosystem, *Environ. Sci. Technol.*, 14, 1136, 1980.

47. **Hinga, K. R.**, The Fate of Polycyclic Aromatic Hydrocarbons in *Enclosed Marine Ecosystems*, Ph.D. thesis, University of Rhode Island Graduate School of Oceanography, Narragansett, RI, 1984.

48. **Park, P. K., Kester, D. R., Duedall, I. W., and Ketchum, B. H.**, *Wastes in the Ocean Volume 3*, John Wiley & Sons, New York, 1983, 522.

49. **Duedall, I. W., Ketchum, B. H., Park, P. K., and Kester, D. R.**, *Wastes in the Ocean Volume 1*, John Wiley & Sons, New York, 1983.

50. **Kester, D. R., Ketchum, B. H., Duedall, I., W., and Park, P. K.**, *Wastes in the Ocean Volume 2*, John Wiley & Sons, New York, 1983.

51. **Boehm, P. D.**, Coupling of organic pollutants between the estuary and continental shelf and the sediments and water column in the New York Bight region, *Can. J. Fish. Aquat. Sci.*, 40, 262, 1983.

52. **O'Connor, T. P., Okubo, A., Champ, M. A., and Park, P. K.**, Projected consequences of dumping sewage sludge at a deep ocean site near New York Bight, *Can. J. Fish. Aquat. Sci.*, 40, 228, 1983.

53. **Bacon, M. P., Brewer, P. G., Pencer, D. W., Marray, J. W., and Goddard, J.**, The behavior of lead-210, polonium-210, manganese and iron in the Cariaco Trench, *Deep Sea Res.*, 27A, 119, 1980.

54. **Sholkovitz, E. R.**, The geochemistry of plutonium in fresh and marine water environments, *Earth Sci. Rev.*, 19, 95, 1983.

55. **Broecker, W. S. and Peng, T. S.**, *Tracers in the SEA*, Lamont-Doherty Geological Observatory, Columbia University, Palisades, 1982.

56. **Wood, L. W.**, Role of oligochaetes in the circulation of water and solutes across the mud-water interface, *Verh. Int. Verin.*, 19, 1530, 1975.

57. **Aller, R. C.**, Experimental studies of changes produced by deposit feeders on pore water, sediment, and overlying water chemistry, *Am. J. Sci.*, 278, 1185, 1978.

58. **Aller, R. C. and Yingst, J. Y.**, Biogeochemistry of tube-dwellings: a study of the sedentary polychaete *Amphitrite ornata* (Leidy), *J. Mar. Res.*, 36, 201, 1978.

59. **Fisher, J. B., Lick, W. J., and McCall, P. L.**, Vertical mixing of lake sediments by tubificid oiligochaetes, *J. Geophys. Res.*, 85, 3997, 1980.

60. **Bender, K. and Davis, W. R.**, The effect of feeding by *Yoldia limatula* on bioturbation, *Ophelia*, 23, 91, 1984.

61. **Gordon, D. C., Dale, J., Jr., and Keizer, P. D.**, Importance of sediment working by the deposit-feeding polychaete *Arenicola marina* on the weathering rate of sediment-bound oil, *Can. J. Fish. Aquat. Sci.*, 35, 591, 1978.

62. **Lee, R. F., Gardner, W. S., Anderson, J. W., Blaylock, J. W., and Barwell-Clarke, J.**, Fate of polycyclic aromatic hydrocarbons in controlled ecosystem enclosures, *Environ. Sci. Technol.*, 12, 832, 1978.

63. **Gardner, W. S., Lee, R. F., Tenore, K. R., and Smith, L. W.**, Degradation of selected polycyclic aromatic hydrocarbons in coastal sediments: importance of microbes and polychaete worms, *Water Air Soil Pollut.*, 11, 339, 1979.

64. **Karickhoff, S. W. and Morris, K. R.**, Impact of tubificid oligochaetes on pollutant transport in bottom sediments, *Environ. Sci. Technol.*, 19, 51, 1985.

65. **Zafiriou, O. C.**, Marine organic photochemistry previewed, *Mar. Chem.*, 5, 497, 1977.

66. **Zepp, R. G. and Schlotzhauer, P. F.**, Photoreactivity of selected aromatic hydrocarbons in water, in *Polynuclear Aromatic Hydrocarbons*, Jones, P. W. and Leber, P., Eds., Ann Arbor Science Publishers, Ann Arbor, MI, 1979, 141.

67. **Zafiriou, O. C., Joussot-Dubien, J., Zepp, R. G., and Zika, R. G.**, Photochemistry of natural waters, *Environ. Sci. Technol.*, 18, 358A, 1984.

68. **Payne, J. R. and Phillips, C. R.**, Photochemistry of petroleum in water, *Environ. Sci. Technol.*, 19, 569, 1985.

69. **Lee, R. F. and Takahashi, M.,** The fate and effects of petroleum in controlled ecosystem enclosures, *Rapp. P.-v. Reun. Cons. Int. Explor. Mer.,* 171, 150, 1977.

70. **Lee, R. F. and Ryan, C.,** Microbial and photochemical degradation of polycyclic aromatic hydrocarbons in estuarine waters and sediments, *Can. J. Fish. Aquat. Sci.,* 40, 86, 1983.

71. **Landrum, P. R., Bartell, S. M., Giesy, J. P., Leversee, G. J., Bowling, J. W., Haddock J., LaGory, K., Gerould, S., and Bruno, M.,** Fate of anthracene in an artificial stream: a case study, *Ecotoxicol. Environ. Saf.,* 8, 183, 1984.

72. **Gossett, R. W., Brown, D. A., McHugh, S. R., and Westcott, A. M.,** Measuring the oxygenated metabolites of chlorinated hydrocarbons, in *Southern California Coastal Water Research Biennial Report 1983 — 1984,* Bascom, W., Ed., Southern California Coastal Water Research Project, Long Beach, 1984, 155.

73. **Brown, D., Gossett, R. W., Hershelman, P., Ward, C. F., Westcott, A. M., and Cross, J. N.,** Municipal wastewater contaminants in the southern California Bight. I. Metal and organic contaminants in sediments and organisms, *Mar. Environ. Res.,* 18, 291, 1986.

74. **Mix, M. C.,** Chemical carcinogens in bivalve mollusks from Oregon estuaries, Environmental Protection Agency Rep. EPA-600/3-79-034, 1979.

75. **Pancirov, R. J. and Brown, R. A.,** Polynuclear aromatic hydrocarbons in marine tissues, *Environ. Sci. Technol.,* 11, 989, 1977.

76. **Malins, D. C., McCain, B. B., Brown, D. W., Sparks, A. K., and Hodgins, H. O.,** Chemical contaminants and biological abnormalities in central and southern Puget Sound, NOAA technical memorandum OMPA-2, Boulder, CO, 1980.

77. **Farrington, J. W., Risebrough, R. W., Parker, P. L., Davis, A. C., de Lappe, B., Winters, J. K., Boatwright, D., and Frew, N. M.,** Hydrocarbons, polychlorinated biphenyls, and DDE in mussels and oysters from the U.S. coast, 1976 — 1978 the mussel watch, Woods Hole Oceanographic Inst. Tech. Rep. WHOI-82-42, 1982.

78. **Knutzen, J. and Sortland, B.,** Polycyclic aromatic hydrocarbons (PAH) in some algae and invertebrates from moderately polluted parts of the coast of Norway, *Water Res.,* 16, 421, 1982.

79. **Bogovski, P., Veldre, I., Itra, A., and Paalme, L.,** Polynuclear aromatic hydrocarbons in Estonian water, sediments, and aquatic organisms, in *Carcinogenic Polynuclear Aromatic Hydrocarbons in the Marine Environment,* Richards, N. L. and Jackson, B. L., Eds., EPA-600/9-82-013, 1982, 260.

80. **Black J. J., Har, T. F., Jr., and Evans, E.,** HPLC studies of PAH pollution in a Michigan trout stream, in *Polynuclear Aromatic Hydrocarbons,* Cooke, M. and Dennis, A. J., Eds., Battelle Press, Columbus, OH, 1980, 343.

81. **Friocourt, M. P., Bodennec, G., and Berthou, F.,** Determination of polyaromatic hydrocarbons in scallops (Pecten maximum) by UV fluorescence and HPLC combined with UV and fluorescence detectors, *Bull. Environ. Contam. Toxicol.,* 34, 228, 1985.

82. **Hungspreugs, M., Silpipat, S., Tonapong, C., Lee, R. F., Windom, H. L., and Tenore, K. R.,** Heavy metals and polycyclic hydrocarbon compounds in benthic organisms of the Upper Gulf of Thailand, *Mar. Pollut. Bull.,* 15, 213, 1984.

83. **Bender, M. E. and Huggett, R. J.,** Polynuclear aromatic hydrocarbons residues in shell fish: species variation and apparent intraspecific differences, in *Progressive States of Malignant Neoplastic Growth,* Kaiser, H. E., Ed., Martinus Nijhoff, in press.

84. **Malins, D. C., McCain, B. B., Brown, D. W., Chan, S.-L., Myers, M. A., Landahl, J. T., Prohaska, P. C., Friedman, A. J., Rhodes, L. D., Burrows, D. G., Gronhard, W. D. , and Hodgins, H. O.,** Chemical pollutants in sediments and diseases in bottom-dwelling fish, *Environ. Sci. Technol.,* 18, 705, 1984.

85. **Gossett, R. W., Brown, D. A., and Young, D. R.,** Predicting the bioaccumulation of organic compounds in marine organisms using octanol/water partition coefficients, *Mar. Pollut. Bull.,* 14, 387, 1983.

86. **Eadie, B. J., Faust, W., Gardner, W. S., and Nalepa, T.,** Polycyclic aromatic hydrocarbons in sediments and associated benthos in Lake Erie, *Chemosphere,* 11, 185, 1982.

87. **Boehm, P. D., Barak, J., Fiest, D., and Elskus, A.,** The analytical chemistry of *Mytilus edulis, Macoma balthica,* sediment trap and surface sediment samples, in *The Tsesis Oil Spill,* Kineman, J. J., Elmgren, R., and Harrison, S., Eds., DOC/NOAA, 1980, 219.

88. **Shaw, D. G. and Wiggs, J. N.,** Hydrocarbons in the intertidal environment of Kachemak Bay, Alaska, *Mar. Pollut. Bull.,* 11, 297, 1980.

89. **Dunn, B. P. and Stich, H. F.,** Release of the carcinogen benzo(a)pyrene from environmentally contaminated mussels, *Bull. Environ. Contam. Toxicol.,* 15, 398, 1976.

90. **Boehm, P. D. and Quinn, J. G.,** The persistence of chronically accumulated hydrocarbons in the hard shell clam *Mercenaria, Mar. Biol.,* 44, 227, 1977.

91. **Disalvo, L. H., Guard, H. E., and Hunter, L.,** Tissue hydrocarbon burden of mussels as potential monitors of environmental hydrocarbon insult, *Environ. Sci. Technol.,* 9, 247, 1975.

92. **Farrington, J. W., Davis, A. C., Frew, N. M., and Rabin, K. S.,** No. 2 fuel oil compounds in *Mytilus edulis* retention and release after an oil spill, *Mar. Biol.,* 66, 15, 1982.

93. **Grahl-Nielsen, O., Staveland, J. T., and Wilhelmsen, S.,** Aromatic hydrocarbons in benthic organisms from coastal areas polluted by Iranian crude oil, *J. Fish. Res. Board Can.,* 35, 615, 1978.

94. **Marchand, M. and Cabane, F.,** Hydrocarbures dans les moules et les huitres, *Rev. Int. Oceanogr. Med.,* 59, 1980.

95. **Stegeman, J. J. and Teal, J. M.,** Accumulation, release and retention of petroleum hydrocarbons by the oyster *Crassostrea virginica, Mar. Biol.,* 22, 37, 1973.

96. **Mix, M. C., Hemingway, S. J., and Schaffer, R. L.,** Benzo(a)pyrene concentrations in somatic and gonad tissues of bay mussels, *Mytilus edulis, Bull. Environ. Contam. Toxicol.,* 28, 46, 1982.

97. **Nasci, C. and Fossato, V. U.,** Studies on physiology of mussels and their ability in accumulating hydrocarbons and chlorinated hydrocarbons, *Environ. Technol. Lett.,* 3, 273, 1982.

98. **Clark, R. C., Jr.,** The Biogeochemistry of Aromatic and Saturated Hydrocarbons in a Rocky Intertidal Marine Community in the Strait of Juan de Fuca, Ph.D. thesis, University of Washington, Seattle, 1983.

99. **Krahn, M. M., Brown, D. W., Collier, T. K., Friedman, A. J., Jenkins, R. G., and Malins, D. C.,** Determination of naphthalene and its metabolites in biological systems by liquid chromatography with fluorescence detection, *Chromatogr. News,* 8, 29, 1980.

100. **Krahn, M. M., Myers, M. S., Burrows, D. G., and Malins, D. C.,** Determination of metabolites of xenobiotics in the bile of fish from polluted waterways, *Xenobiotica,* 14, 633, 1984.

101. **Malins, D. C., Myers, M. S., and Roubal, W. T.,** Organic free radicals associated with idiopathic liver lesions of English sole *(Parophrys vetulus)* from polluted marine environments, *Environ. Sci. Technol.,* 17, 679, 1983.

102. **Roubal, W. T. and Malins, D. C.,** Free radical derivatives of nitrogen heterocycles in liver of English sole *(Parophrys vetulus)* with hepatic neoplasms and other liver lesions, *Aquat. Toxicol.,* 6, 87, 1985.

103. **Sims, P. and Grover, P. L.,** Epoxides in polycyclic aromatic hydrocarbon metabolism and carcinogenesis, *Adv. Can. Res.,* 20, 165, 1974.

104. **Jerina, D. M. and Daley, J. W.,** Arene oxides: a new aspect of drug metabolism, *Science,* 185, 573, 1974.

105. **Payne, J. F.,** Field evaluation of benzopyrene hydroxylase induction as a monitor for marine petroleum pollution, *Science,* 191, 945, 1976.

106. **Kurelec, B., Matijasevic, Z., Rijaved, M., Alacevic, M., Britvic, S., Muller, W. E. G., and Zahn, R. K.,** Induction of benzo(a)pyrene monooxygenase in fish and the salmonella test as a tool for detecting mutagenic/carcinogenic xenobiotics in the aquatic environment, *Bull. Environ. Contam. Toxicol.,* 21, 799, 1979.

107. **Spies, R. B., Felton, J. S., and Dillard, L.,** Hepatic mixed-function oxidases in California flatfishes are increased in contaminated environments and by oil and PCB ingestion, *Mar. Biol.,* 70, 117, 1982.

108. **Ames, B. M., Durston, W. E., Yamaski, E., and Leu, F. D.,** Carcinogens as mutagens: a simple test system containing liver homogenates for activation and bacteria for detection, *Proc. Natl. Acad. Sci. U.S.A.,* 70, 2281, 1973.

109. **Ames, B. M., McCann, J., and Yamasaki, E.,** Methods for detecting carcinogens and mutagens with the salmonella/mammalian-microsome mutagenicity test, *Mutat. Res.,* 31, 346, 1975.

110. **Protic-Sablijic, M. and Kurelec, B.,** High mutagenic potency of several polycyclic aromatic hydrocarbons induced by liver postmitochondrial fractions from control and xenobiotic-treated immature carp, *Mutat. Res.,* 118, 177, 1983.

111. **Malins, D. C., Krahn, M. M., Myers, M. S., Rhodes, L. D., Brown, D. W., Krone, C. A., McCain, B. B., and Chan, S.-L.,** Toxic chemicals in sediments and biota from a creosote-polluted harbor: relationships with hepatic neoplasms and other hepatic lesions in English sole *(Parophyrs vetulus), Carcinogenesis,* 6, 1463, 1985.

112. **Connor, M. S.,** Fish/sediment concentration ratios for organic compounds, *Environ. Sci. Technol.,* 18, 31, 1984.

113. **Maccubin, A. E., Black, P., Trzeciak, L., and Black, J. J.,** Evidence for polynuclear aromatic hydrocarbons in the diet of bottom-feeding fish, *Bull. Environ. Contam. Toxiccl.,* 34, 876, 1985.

114. **Neff, J. M., Cox, B. A., Dixit, D., and Anderson, J. W.,** Accumulation and release of petroleum-derived aromatic hydrocarbons by four species of marine animals, *Mar. Biol.,* 38, 279, 1976.

115. **Roubal, W. T., Collier, T. K., and Malins, D. C.,** Accumulation and metabolism of carbon-14 labelled benzene, naphthalene, and anthracene by young coho salmon *(Oncorhynchus kisutch), Arch. Environ. Contam. Toxicol.,* 5, 513, 1977.

116. **Southworth, G. R., Beauchamp, J. J., and Schmieder, P. K.,** Bioaccumulation potential of polycyclic aromatic hydrocarbons in *Daphnia pulex, Water Res.,* 12, 973, 1978.

117. **Geyer, H., Sheehan, P., Ktozias, D., Freitag, D., and Korte, F.,** Prediction of ecotoxicological behavior of chemicals: relationship between physico-chemical properties and bioaccumulation of organic chemicals in the mussel *Mytilus edulis, Chemosphere,* 11, 1121, 1982.

118. **Obana, H., Hori, S., Nakamura, A., and Kashimoto, T.,** Uptake and release of polynuclear aromatic hydrocarbons by short-necked clams *(Tapes japonica), Water Res.,* 17, 1183, 1983.

119. **Pruell, R. J., Lake, J. L., Davis, W. R., and Quinn, J. G.,** Uptake and depuration of organic contaminants by blue mussels, *Mytilus edulis,* exposed to environmentally contaminated sediment, *Mar. Biol.,* 91, 497, 1986.

120. **Southworth, G. R., Beauchamp, J. J., and Schmieder, P. K.,** Bioaccumulation potential and acute toxicity of synthetic fuels effluent in freshwater biota: azaarenes, *Environ. Sci. Technol.,* 12, 1062, 1978.

121. **Neely, W. B., Branson, D. R., and Blau, G. E.,** Partition coefficient to measure bioconcentration potential of organic chemicals in fish, *Environ. Sci. Technol.,* 13, 1113, 1974.

122. **Vieth, G. D., DeFoe, D. L., and Bergstedt, B. V.,** Measuring and estimating the bioconcentration factor of chemicals in fish, *J. Fish. Res. Board Can.,* 36, 1040, 1979.

123. **Kenaga, E. E. and Goring, C. A. I.,** Relationship between water solubility, soil sorption, octanol-water partitioning, and concentration of chemicals in biota, in *Aquatic Toxicology , ASTM STP 707,* Eaton, J. G., Parrish, P. R., and Hendricks, A. C., Eds., American Society for Testing and Materials, 1980, 78.

124. **Varanasi, U., Gmur, D. J., and Reichert, W. L.,** Effect of environmental temperature on naphthalene metabolism by juvenile starry flounder *(Platichthys stellatus), Arch. Environ. Contam. Toxicol.,* 10, 203, 1981.

125. **Fucik, K. W., Armstrong, H. W., and Neff, J. M.,** Uptake of naphthalenes by the clam, *Rangia cuneata,* in the vicinity of an oil separator platform in Trinity Bay, Texas, in *1977 Oil Spill Conf. (Prevention, Behavior, Control, and Cleanup),* American Petroleum Institute, Washington, D.C., 1977, 637.

126. **Harris, R. P., Berdugo, V., O'Hara, S. C. M., and Corner, E. D. S.,** Accumulation of 14C-1-naphthalene by an oceanic and an estuarine copepod during long-term exposure to low-level concentrations, *Mar. Biol.,* 42, 187, 1977.

127. **Collier, T. K., Thomas, L. C., and Malins, D. C.,** Influence of environmental temperature on disposition of dietary naphthalene in coho salmon *(Oncorhynchus kisutch):* isolation and identification of individual metabolites, *Comp. Biochem. Physiol.,* 61C, 23, 1978.

128. **Gerould, S., Landrum, P., and Giesy, J. P.,** Anthracene biocencentration and biotransformation in chironomids: effects of temperature and concentration, *Environ. Pollut.,* in press.

129. **Sanborn, H. R. and Malins, D. C.,** Toxicity and metabolism of naphthalene: a study with marine larval invertebrates, *Proc. Soc. Exp. Biol. Med.,* 154, 151, 1977.

130. **Leversee, G. J., Landrum, P. F., Giesy, J. P., and Fannin, T.,** Humic acids reduce bioaccumulation of some polycyclic aromatic hydrocarbons, *Can. J. Fish. Aquat. Sci.,* 40, 63, 1983.

131. **Spacie, A., Landrum, P. F., and Leversee, G. J.,** Uptake, depuration, and biotransformation of anthracene and benzo(a)pyrene in bluegill sunfish, *Ecotoxicol. Environ. Saf.,* 7, 330, 1983.

132. **McCarthy, J. F. and Jimenez, B. D.,** Interactions between polycyclic aromatic hydrocarbons and dissolved humic material: binding and dissociation, *Environ. Sci. Technol.,* 19, 1072, 1985.

133. **McCarthy, J. F., Jimenez, B. D., and Barbee, T.,** Effect of dissolved humic material on accumulation of polycyclic aromatic hydrocarbons: structure-activity relationships, *Aquat. Toxicol.,* 7, 15, 1985.

134. **Steurmer, D. and Harvey, G. R.,** Humic substances from seawater, *Nature,* 250, 480, 1974.

135. **Fortner, A. R. and Sick, L. V.,** Simultaneous accumulations of naphthalene, a PCB mixture, and benzo(a)pyrene, by the oyster, *Crassostrea virginica, Bull. Environ. Contam. Toxicol.,* 34, 256, 1985.

136. **Gruger, E. H., Jr., Schnell, J. V., Fraser, P. S., Brown, D. W., and Malins, D. C.,** Metabolism of 2,6-dimethylnaphthalene in starry flounder *(Platichthys stellatus)* exposed to naphthalene and p-cresol, *Aquat. Toxicol.,* 1, 37, 1981.

137. **Stein, J. E., Hom, T., and Varanasi, U.,** Simultaneous exposure of English sole *(Parophrys vetulus)* to sediment-associated xenobiotics. I. Uptake and disposition of ^{14}C-polychlorinated biphenyls and ^{3}H-benzo(a)pyrene, *Mar. Environ. Res.,* 13, 97, 1984.

138. **McElroy, A. E.,** *In vivo* metabolism of benz(a)anthracene by the polychaete *Nereis virens, Mar. Environ. Res.,* 17, 133, 1985.

139. **Leversee, G. J., Giesy, J. P., Landrum, P. F., Gerould, S., Bowling, J. W., Fanwin, T. E., Haddock, J. D., and Bartell, S. M.,** Kinetics and biotransformation of benzo(a)pyrene in *Chironomus riparius, Arch. Environ. Contam. Toxicol.,* 11, 25, 1982.

140. **Varanasi, U., Reichert, W. L., Stein, J. E., Brown, D. W., and Sanborn, H. R.,** Bioavailability and biotransformation of aromatic hydrocarbons in benthic organisms exposed to sediment from an urban estuary, *Environ. Sci. Technol.,* 19, 836, 1985.

141. **Lu, P.-Y., Metcalf, R. L., Plummer, N., and Mandel, D.,** The environmental fate of three carcinogens: benzo(a)pyrene, benzidine, and vinyl chloride evaluated in laboratory model ecosystems, *Arch. Environ. Contam. Toxicol.,* 6, 129, 1977.

142. **Hamelink, J. L., Waybrant, R. C., and Ball, R. C.,** A proposal: exchange equilibria control the degree chlorinated hydrocarbons are biologically magnified in lentic environments, *Trans. Am. Fish. Soc.,* 100, 207, 1971.

143. **Schneider, R.,** Polychlorinated biphenyls (PCBs) in cod tissues from the Western Baltic: significance of equilibrium partitioning and lipid composition in the bioaccumulation of lipophilic pollutants in gill-breathing animals, *Meeresforsch.,* 29, 69, 1982.

144. **Rossi, S. S. and Anderson, J. W.,** Accumulation and release of fuel-oil-derived diaromatic hydrocarbons by the polychaete *Neanthes arenaceodentata, Mar. Biol.,* 39, 51, 1977.

145. **Roesijadi, G., Anderson, J. W., and Blaylock, J. W.,** Uptake of hydrocarbons from marine sediments contaminated with Prudhoe Bay crude oil: influence of feeding type of test species and availability of polycyclic aromatic hydrocarbons, *J. Fish. Res. Board Can.,* 35, 608, 1978.

146. **Anderson, J. W., Moore, L. J., Blaylock, J. W., Woodruff, D. L., and Kiesser, S. L.,** Bioavailability of sediment-sorbed naphthalenes to the sipunculid worm, *Phascolosoma agassizii,* in *Fates and Effects of Petroleum Hydrocarbons in Marine Organisms and Ecosystems,* Wolfe, D. A., Ed., Pergamon Press, New York, 1977, 276.

147. **Lyes, M. C.,** Bioavailability of a hydrocarbon from water and sediment to the marine worm *Arenicola marina, Mar. Biol.,* 55, 121, 1979.

148. **McCain, B. B., Hodgins, H. O., Gronlund, W. D., Hawkes, J. W., Brown, D. W., and Myers, M. S.,** Bioavailability of crude oil from experimentally oiled sediments to English sole *(Parophrys vetulus),* and pathological consequences, *J. Fish. Res. Board Can.,* 35, 657, 1978.

149. **Varanasi, U. and Gmur, D. J.,** Hydrocarbons and metabolites in English sole *(Parophrys vetulus)* exposed simultaneously to [^{3}H] benzo(a)pyrene and [^{14}C] naphthalene in oil-contaminated sediment, *Aquat. Toxicol.,* 1, 49, 1981.

150. **Augenfeld, J. M., Anderson, J. W., Riley, R. G., and Thomas, B. J.,** The fate of polyaromatic hydrocarbons in an intertidal sediment exposure system: bioavailability to *Macoma inquinata* (Mollusca: Pelecypoda) and *Abarenicola pacifica* (Annelida: polychaeta), *Mar. Environ. Res.,* 7, 31, 1982.

151. **Rossi, S. S.,** Bioavailability of petroleum hydrocarbons from water, sediment, and detritus, to the marine annelid *Neanthes arenaceodentata,* in *Proc. 1977 Oil Spill Conf. (Prevention, Behavior, Control, Cleanup),* American Petroleum Institute, Washington, D.C., 1977, 621.

152. **Landrum, P. F. and Scavia, D.,** Influence of sediment on anthracene uptake, depuration, and biotransformation by the amphipod *Hyalella azteca, Can. J. Fish. Aquat. Sci.,* 40, 298, 1983.

153. **Dobroski, D. J., Jr. and Epifanio, C. E.,** Accumulation of benzo(a)pyrene in a larval bivalve via trophic transfer, *Can. J. Fish. Aquat. Sci.,* 12, 2318, 1980.

154. **Lee, R. F., Ryan, C., and Neuhauser, M. L.,** Fate of petroleum hydrocarbons taken up from food and water by the blue crab *Calinectes sapidus, Mar. Biol.,* 37, 363, 1976.

155. **Corner, E. D. S., Harris, R. P., Kilvington, C. C., and O'Hara, S. C. M.,** Petroleum compounds in the marine food web: short-term experiments on the fate of naphthalene in *Calanus, J. Mar. Biol. Assoc. U.K.,* 56, 121, 1976.

156. **Dillon, T. M.,** Dietary accumulation of dimethylnaphthalene by the grass shrimp *Paleomonetes pugio* under stable and fluctuation temperature, *Bull. Environ. Contam. Toxicol.,* 28, 149, 1982.

157. **Malins, D. C. and Roubal, W. T.,** Aryl sulfate formation in sea urchins *(Strongelocentrotus drobachiensis)* ingesting marine algae *(Fucus distichilus)* containing 2,6-dimethyl naphthalene, *Environ. Res.,* 27, 290, 1982.

158. **Rubinstein, N. I., Gilliam, W. T., and Gregory, N. R.,** Dietary accumulation of PCBs from a contaminated sediment source by a demersal fish *(Leiostomus xanthurus), Aquat. Toxicol.,* 5, 33, 1984.

159. **Farrington, J. W., Wakeham, S. G., Livramento, J. B., Tripp, B. W., and Teal, J. M.,** Aromatic hydrocarbons in New York Bight polychaetes: UV-fluorescence analysis and GC/GCMS analysis, *Environ. Sci. Technol.,* 20, 69, 1986.

Chapter 2

MICROBIAL DEGRADATION OF POLYCYCLIC AROMATIC HYDROCARBONS (PAH) IN THE AQUATIC ENVIRONMENT*

Carl E. Cerniglia and Michael A. Heitkamp

TABLE OF CONTENTS

I. INTRODUCTION

Microorganisms play an important role in recycling of elements such as carbon, nitrogen, phosphorus, oxygen, and sulfur and in the degradation of organic compounds to carbon dioxide and water in nature. The cyclic conversion of organic carbon to carbon dioxide with the concomitant reduction of molecular oxygen involves the combined metabolic activities of many different microorganisms. The remarkable catalytic ability of microorganisms to degrade a wide array of organic compounds has created an explosion of interest in the use of microorganisms in pollution abatement and in mitigating environmental damage. Microorganisms have adapted to the wide array of natural and synthetic organic compounds emitted into the environment and possess enzymes that utilize some of these compounds as nutrients.

Generally, the biodegradation of naturally occurring compounds is relatively faster than chemicals released from anthropogenic sources. The recalcitrance of man-made compounds may be due to substituents and structural moieties on the molecule that differ from those in natural products; thus, indigenous microorganisms have not evolved the enzyme systems to deal with these novel chemical structures. Another plausible explanation is that microorganisms may have the enzymatic capability to degrade the pollutant, but the compound may not be present in sufficient levels to induce the necessary degradative enzymes in the micoorganisms. Therefore, the microbial degradation of organic compounds in the environment depends, in part, on the structural similarities between synthetic and naturally occurring compounds.[1]

In recent years, the concern about the presence of polycyclic aromatic hydrocarbons (PAH) in air, soil, and water systems has increased, since this important class of chemicals is carcinogenic in experimental animals and a potential health risk to man.[2-4] A summary of the structure, toxicity, and genotoxicity of PAH commonly found in soils and aquatic ecosystems is given in Table 1.

PAH are mainly formed as products from the combustion of fossil fuels and also occur as natural constituents of unaltered fossil fuels. Due to their hydrophobic properties and limited water solubility, PAH tend to adsorb to particulates and eventually migrate to the sediments in river, lake, estuarine, and marine waters. Industrial effluents from coal gasification and liquification processes, waste incineration, coke, carbon black, and other petroleum-derived products also add to the high input of PAH in urban terrestrial and marine sediments.[7-15] PAH have been identified as genotoxic pollutants in freshwater and coastal sediments.[16] The concentration of PAH in sediments depends upon the site's distance from industrialized regions, anthropogenic activities, and the various PAH transport mechanisms. Johnson and Larsen[17] made a worldwide comparison of the total concentration of PAH in marine and freshwater sediments. The PAH levels ranged from 5 ppb (nanograms of PAH per gram of sediment) for an undeveloped area in Alaska to 1.79×10^6 ppb for an oil refinery outfall in Southampton, England. Sediments from other industrialized areas ranged from 198 to 232,000 ppb. PAH enter the biosphere through various routes such as accidental discharges of fossil fuels, direct aerial fallout, chronic leakage, industrial and sewage discharges, and surface water runoff. A variety of processes including volatilization, adsorption, chemical oxidation, photo-decomposition, and biodegradation are important mechanisms for environmental loss of PAH. A schematic representation of the sources, occurrence, accumulation, and disposition of PAH in the environment is shown in Figure 1. This review will deal with the microbial degradation of PAH.

II. PRESENT KNOWLEDGE OF THE MICROBIAL DEGRADATION OF PAH

Studies on the microbial metabolism of aromatic hydrocarbons were initiated almost 80 years ago when Sohngen[18] and Stormer[19] isolated bacteria capable of utilizing aromatic

Table 1
TOXICOLOGICAL CHARACTERISTICS[a] OF SELECTED PAH

Chemical	Toxicity	Carcinogenicity	Genotoxicity
2-Rings			
Naphthalene	Algae, 24 h, 50% ↓ , 33 ppm Fish, 96 h, TLm = 1—2 ppm Rat, LD_{50} = 306—600 ppm Rabbit, LD_{50} = 800 ppm	–	–
1-,2-Methyl-naphthalene	Fish, 48 h, LD_{50} = 8—9 ppm	–	–
3-Rings			
Acenaphthene	ND	–, ?	+ Ames
Anthracene	Mouse, LD_{50} = 430 ppm	–	–
Fluorene	ND	–	–
Phenanthrene	Mouse, LD_{50} = 700 ppm	–/?	–
4-Rings			
Benz(a)anthracene	ND	+	+ Ames + SCE + UDS
Chrysene	Mouse, LD_{50} = 320 ppm	+	+ Ames + SCE + CA
7,12-Dimethyl-benz(a)anthracene	ND	+	+ Ames

Table 1 (continued)
TOXICOLOGICAL CHARACTERISTICS[a] OF SELECTED PAH

Chemical	Toxicity	Carcinogenicity	Genotoxicity
Fluoranthene	Mouse, LD_{50} = 500 ppm Rat, LD_{50} = 2000 ppm Rabbit, LD_{50} = 3180 ppm	+/?	+ Ames
Pyrene	Mouse, LD_{50} = 514—678 ppm	−, ?	+/? Ames + UDS + SCE
5-Rings Benzo(a)pyrene	Mouse, LD_{50} = 250 mg/kg Embryotoxic Teratogenic	+	+ Ames + UDS + SCE + CA + DA
Dibenz(a,h)anthracene	ND	+	+ Ames + *E. coli* + DNA damage + CA
3-Methyl-cholanthrene	ND	+	+ Ames
Perylene	ND	−, ?	+ Ames + CA
6-Rings Indeno-(1,2,3-c,d)pyrene	ND	+	+ Ames
7-Rings Coronene	ND	+/−, ?	+ Ames

Table 1 (continued)
TOXICOLOGICAL CHARACTERISTICS[a] OF SELECTED PAH

Note: Ames = *Salmonella typhimurium* reversion assay, CA = chromosomal abberations, DA = DNA adducts, SCE = sister chromatid exchange, TLm = median threshold limit, UDS = unscheduled DNA synthesis, ND = no data, ? = inadequate/inconclusive, and LD$_{50}$ = lethal dose for 50%.

[a] Data compiled from References 5 and 6.

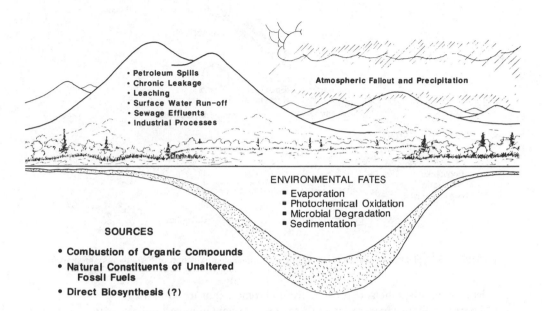

FIGURE 1.   Schematic representation of the sources, occurrence, accumulation, and disposition of PAH in the environment.

hydrocarbons as the sole source of carbon. Since that time there have been numerous studies on the metabolism, biochemistry, genetics, and regulation of PAH degradation. In reviewing the literature, some general statements can be made about our present knowledge of the microbial degradation of PAH:

1.   A wide variety of bacteria, fungi, and algae have the ability to metabolize PAH.
2.   Hydroxylation of unsubstituted PAH always involves the incorporation of molecular oxygen.
3.   Prokaryotic microorganisms metabolize PAH by an initial dioxygenase attack to *cis*-dihydrodiols that are further oxidized to dihydroxy products.
4.   Eukaryotic microorganisms use monooxygenases to initially attack PAH to form arene oxides followed by the enzymatic addition of water to yield *trans*-dihydrodiols.
5.   PAH with more than three condensed benzene rings do not serve as substrates for microbial growth, though they may be subject to cometabolic transformations.
6.   Fungi hydroxylate PAH as a prelude to detoxification, whereas bacteria oxidize PAH as a prelude to ring fission and assimilation.
7.   Many of the genes coding for PAH degradation are plasmid associated.
8.   Lower weight PAH such as naphthalene degrade rapidly, whereas higher weight PAH such as benz(a)anthracene or benzo(a)pyrene are quite resistant to microbial attack.
9.   Most rapid biodegradation of PAH occurs at the water/sediment interface and degradation rates can be influenced by environmental factors.

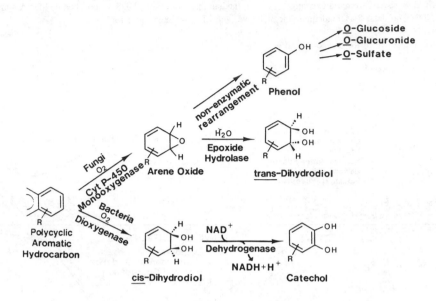

FIGURE 2.   Typical monooxygenase vs. dioxygenase mediated oxidation of PAH by microorganisms.

10.   Microbial adaptations can occur from chronic exposure to PAH.
11.   There are higher biodegradation rates in PAH-contaminated sediments than in pristine
       sediments.

We will focus on these topics in this review on the microbial aspects of PAH degradation.
It is important to determine the capacity of microorganisms to metabolize PAH and potentially
detoxify these persistent pollutants. The reader is also referred to other reviews for background information and additional details on the microbial degradation of PAH.[20-32]

III. GENERAL ASPECTS OF PAH–DEGRADATIVE PATHWAYS

Molecular oxygen is essential to catalyze initial hydroxylation of PAH by aerobic microorganisms. Mono- and dioxygenases are two groups of enzymes which are important to the
microbial catabolism of PAH.[33] Dioxygenases incorporate both atoms of the oxygen molecule
into the PAH. This reaction is the major mechanism for the initial oxidative attack on PAH
by bacteria which leads to the formation of dihydrodiols that are in the *cis*-configuration[30,31]
(Figure 2). The dioxygenase that catalyzes these initial reactions is a multicomponent enzyme
system that consists of a flavoprotein, a ferredoxin, and an iron sulfur protein.[34-36] In many
organisms the ability to degrade PAH is mediated by plasmids that carry both structural and
regulatory genes.[37] The second step in the bacterial oxidation of PAH is the rearomatization
of the *cis*-dihydrodiol via a dehydrogenase to form a dihydroxylated intermediate. This
reaction is highly stereoselective.[38] Dihydroxylation of the benzene nucleus is a prerequisite
for cleavage of the aromatic ring.[22] Enzymatic fission of the aromatic ring is also catalyzed
by dioxygenases. For cleavage to occur, both hydroxyl groups must be placed either *ortho*
or *para* to each other. If the hydroxyl substitutents are placed *ortho* to each other, then

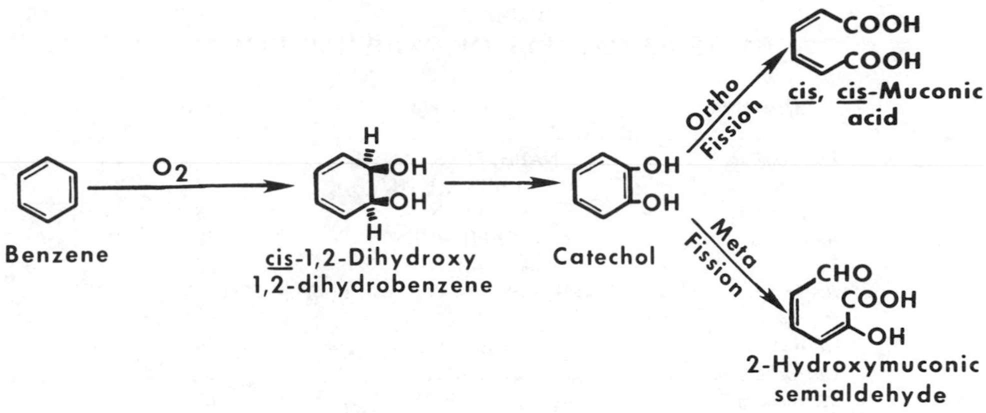

COOH
COOH
cis, cis-Muconic acid

Ortho Fission

H
OH
OH
H

Benzene
O_2
cis-1,2-Dihydroxy 1,2-dihydrobenzene
Catechol

OH
OH

Meta Fission

CHO
COOH
OH
2-Hydroxymuconic semialdehyde

FIGURE 3. The *ortho* and *meta* pathways for aromatic ring cleavage of catechol.

oxygenolytic ring cleavage can occur either between the two hydroxyl groups by *ortho* or intradiol cleaving dioxygenases or adjacent to the two hydroxyl groups by *meta* or extradiol cleavage dioxygenases.[32] These different aromatic ring fission pathways are shown in Figure 3.

In contrast to bacteria, eukaryotic microorganisms oxidize PAH via a cytochrome P-450 monooxygenase by incorporating one atom of the oxygen molecule into the PAH, the other being reduced to water.[25,26] An arene oxide is formed which then becomes a substrate for further metabolism by enzymatic hydration to form a dihydrodiol with a *trans*-configuration (Figure 2). The arene oxide can also undergo isomerization to form phenols which can be conjugated with sulfate, glucuronic acid, glucose, and glutathione. These reactions are similar to those occurring in fish and terrestrial mammals. Figure 2 illustrates the different initial oxidative reactions in the microbial metabolism of PAH. It seems that fungi hydroxylate PAH as a prelude to detoxification, whereas bacteria oxidize PAH to dihydroxylated compounds as a prelude to ring fission and assimilation.[22] Recent studies with blue-green[39] and green algae[40-42] have suggested that the dichotomy of oxidation mechanisms between prokaryotic and eukaryotic microorganisms may be different in photoautotrophs, since both *cis*- and *trans*-dihydrodiols have been detected as transformation products from algal incubations with various PAH.

Since sediments are a major sink for the PAH in aquatic environments, there has been increasing interest in determining the persistence of PAH in anoxic environments. The anaerobic biodegradation of monoaromatic compounds containing oxygen as ring substitutent via anaerobic respiration, fermentation, and photometabolism, has been reported.[43-47] Recently, Vogel and Grabic-Galic[47] described the anaerobic degradation of benzene and toluene to CO_2 and nonvolatile intermediates by acclimated methanogenic cultures. They demonstrated that the oxygen incorporated into cresol and phenol, formed from the anaerobic transformation of toluene and benzene, respectively, was derived from water. This is the first report describing oxygen incorporation derived from water and not molecular oxygen into an unsubstituted aromatic hydrocarbon. These results are significant, since many of the monoaromatic hydrocarbons shown to be biodegradable under anaerobic conditions contain oxygen in ring substituents. Since many unsubstituted PAH are not degraded in the water column and persist in high concentration in the sediments, anaerobic catabolism of PAH could be an important pathway in the removal of PAH in anaerobic ecosystems.

Table 2
BACTERIA CAPABLE OF OXIDIZING PAH

Organism	Substrate	Ref.
Pseudomonas	NAPH, PHEN, ANTH	48—56
Flavobacteria	PHEN, ANTH	57
Alcaligenes	PHEN	58
Aeromonas	NAPH, PHEN	59
Vibrio	PHEN	55
Beijerenckia	NAPH, PHEN, ANTH, BA, B(a)P	60—62
Bacillus	NAPH	63
Nocardia	PHEN, ANTH	49
Corynebacteria	NAPH	64
Micrococcus	PHEN	65

Note: NAPH = naphthalene, PHEN = phenanthrene, ANTH = anthracene,
BA = benz(a)anthracene, and B(a)P = benzo(a)pyrene.

IV. DEGRADATIVE PATHWAYS OF SELECTED PAH IN PROKARYOTES AND EUKARYOTES

Although many PAH and alkylated PAH have been identified and quantitatively characterized from aquatic sediments, the biodegradability of only a few members of this large class of compounds has been shown to occur by metabolism in any biological system. The ability of microorganisms to completely degrade naphthalene, phenanthrene, and anthracene to carbon dioxide (mineralization) has been extensively investigated. The use of pure cultures of bacteria (Table 2) isolated after enrichment from natural samples has enabled researchers to elucidate the metabolic pathways for PAH catabolism. However, microorganisms have not been isolated which have the capacity to completely degrade PAH containing more than three aromatic rings, although microorganisms have been isolated and identified which have the ability to partially metabolize (transform) these compounds to oxidized products.

In this review, we will describe the microbial metabolism of naphthalene, phenanthrene, anthracene, and benzo(a)pyrene by pure cultures of bacteria, filamentous fungi, yeasts, cyanobacteria (blue-green algae), diatoms, and other eukaryotic algae in some detail.

A. Naphthalene

The bacterial mineralization of naphthalene is achieved by the sequence of reactions illustrated in Figure 4. Most of the studies on the bacterial degradation of naphthalene have been conducted with microorganisms from the genus *Pseudomonas*.[48-50,52,56] The initial step involves the incorporation of both atoms of molecular oxygen to form (+)-*cis*-1R,2S-dihydroxy-1,2-dihydronaphthalene.[52,66] The naphthalene dioxygenase has been characterized by Ensley and co-workers.[67] The *cis*-naphthalene dihydrodiol is dehydrogenated to 1,2-dihydroxynaphthalene by a (+)-*cis*-naphthalene dehydrogenase.[38] Davies and Evans[50] reported that 1,2-dihydroxynaphthalene was enzymatically cleaved by a dioxygenase from a *Pseudomonas* sp. to yield *cis*-2'-hydroxybenzalpyruvate. An aldolase catalyzes the cleavage of *cis*-2'-hydroxybenzalpyruvate to pyruvate and salicylaldehyde, the latter of which is subsequently oxidized to salicylate by a dehydrogenase. Further oxidation of salicylate by salicylate hydroxylase yields catechol, which can undergo either *ortho* or *meta* fission depending on the bacterial species. *Aeromonas* sp. has also been shown to metabolize naphthalene by a similar degradative sequence[55] (Figure 4).

A wide taxonomic and phylogenetic spectrum of fungi can transform naphthalene to metabolites that are similar to those identified in mammalian enzyme and laboratory animal

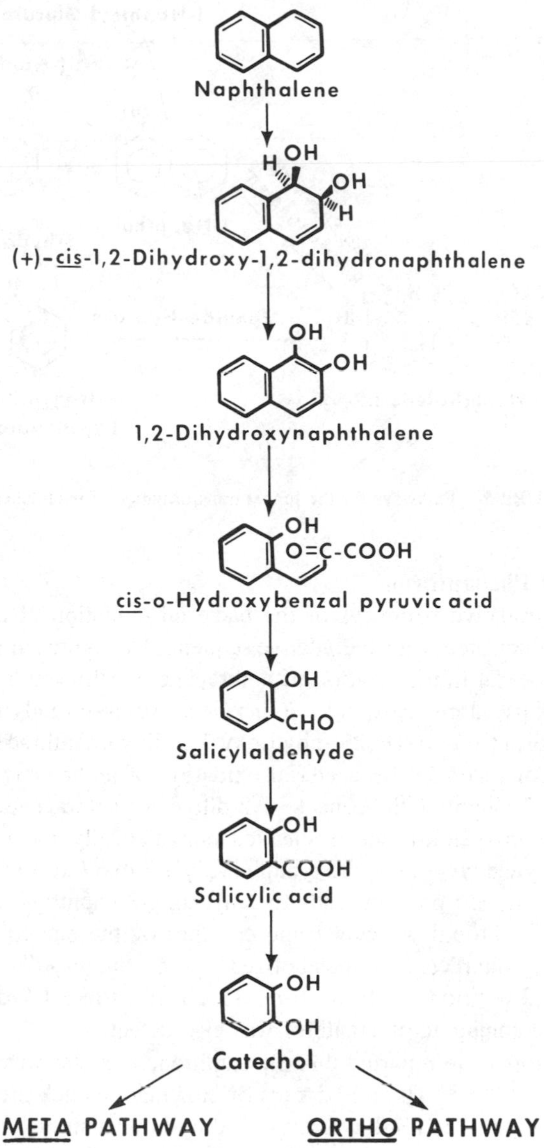

FIGURE 4. Pathway for the bacterial degradation of naphthalene.

studies.[68-70] Fungi oxidize naphthalene via a cytochrome P-450 monooxygenase to form naphthalene 1,2-oxide. This arene oxide can also undergo rearrangement to form 1-naphthol via the NIH shift mechanism. 1-Naphthol is further metabolized to 4-hydroxy-1-tetralone or conjugated with glucuronic acid or sulfate[71-74] (Figure 5).

Cyanobacteria, grown photoautotrophically in the presence of naphthalene, oxidize naphthalene predominantly to 1-naphthol.[75-78] The mechanism of 1-naphthol formation is similar to that reported for fungal and mammalian metabolism systems. Interestingly, *cis*-1,2-dihydroxy-1,2-dihydronaphthalene is also formed from cyanobacterial oxidation of naphthalene as a minor metabolite. The *cis*-naphthalene dihydrodiol is a common intermediate in the bacterial metabolism of naphthalene (Figure 4) and suggests multiple pathways for the metabolism of PAH in cyanobacteria.

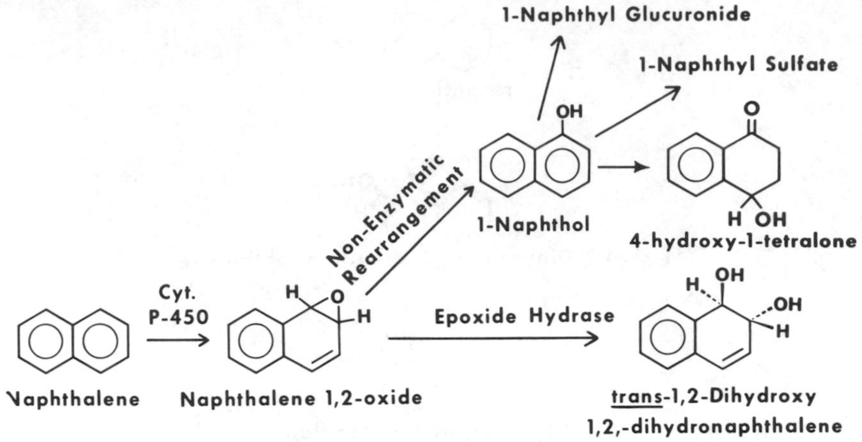

FIGURE 5.   Pathways for the fungal transformation of naphthalene.

B. Anthracene and Phenanthrene

The complete degradative pathways of the bacterial oxidation of anthracene and phenanthrene have been elucidated. The degradative sequences are quite similar to those described for naphthalene. Bacterial mineralization of anthracene is achieved by the pathway shown in Figure 6. Various pseudomonads and a *Beijerinckia* strain initially oxidize anthracene in the 1,2-position to form (+)-*cis*-1R,2S-dihydroxy-1,2-dihydroanthracene.[54,60] As illustrated in Figure 6, the second step in the bacterial oxidation of anthracene is the conversion of *cis*-1,2-dihydroxy-1,2-dihydroanthracene to 1,2-dihydroxyanthracene. Evans et al.[51] have shown that 1,2-dihydroxyanthracene is cleaved enzymatically by a dioxygenase from a *Pseudomonas* sp. to yield *cis*-4-(2-hydroxynaphth-3-yl)-2-oxo-but-3-enoic acid. Further metabolism of this ring fission product leads to 2-hydroxy-3-naphthoic acid. This compound is further metabolized through salicylate and catechol by the enzymes of the naphthalene pathway (Figure 4). Anthracene is metabolized by the fungus, *Cunninghamella elegans* primarily in the 1,2-position to form *trans*-1S,2S-dihydroxy-1,2-dihydroanthracene[79,80] (Figure 6). A sulfate conjugate of 1-anthrol was also detected.

Various investigators have reported that phenanthrene is metabolized by bacteria by two different pathways.[51,54-59,65,81] The initial sites of enzymatic attack are in the 1,2- and 3,4-positions to form (+)-*cis*-1R,2S-dihydroxy-1,2-dihydrophenanthrene and (+)-*cis*-3S,4R-dihydroxy-3,4-dihydrophenanthrene[54,61] (Figure 7). The major isomer formed is *cis*-3,4-dihydroxy-3,4-dihydrophenanthrene. *Pseudomonads* and a *Nocardia* strain oxidized *cis*-3,4-dihydroxy-3,4-dihydrophenanthrene to 3,4-dihydroxyphenanthrene which is subsequently cleaved and converted to 1-hydroxy-2-naphthoic acid. This ring cleavage product is decarboxylated oxidatively to give 1,2-dihydroxynaphthalene which could then enter the naphthalene pathway[51] (Figure 4). *Aeromonos*, *Vibrio*, *Alcaligenes*, and *Micrococcus* strains use an alternate pathway for 1-hydroxy-2-naphthoic acid catabolism.[55,59,65] These microorganisms oxidize 1-hydroxy-2-naphthoic acid via *ortho*-phthalic acid to protocatechuic acid (Figure 7).

Although there have been numerous studies on the bacterial oxidation of phenanthrene, very little is known about the fungal or algal oxidation of phenanthrene.[80] The fungus, *C. elegans*, initially oxidizes phenanthrene in the 1,2-, 3,4-, and 9,10-positions to form dihydrodiols with a *trans*-configuration (Figure 8). The major enantiomers of each of the dihydrodiols have an S,S-absolute configuration. A recent report on the cyanobacterial oxidation of phenanthrene indicates that *Agmenellum quadruplicatum* strain PR-6 and *Os-*

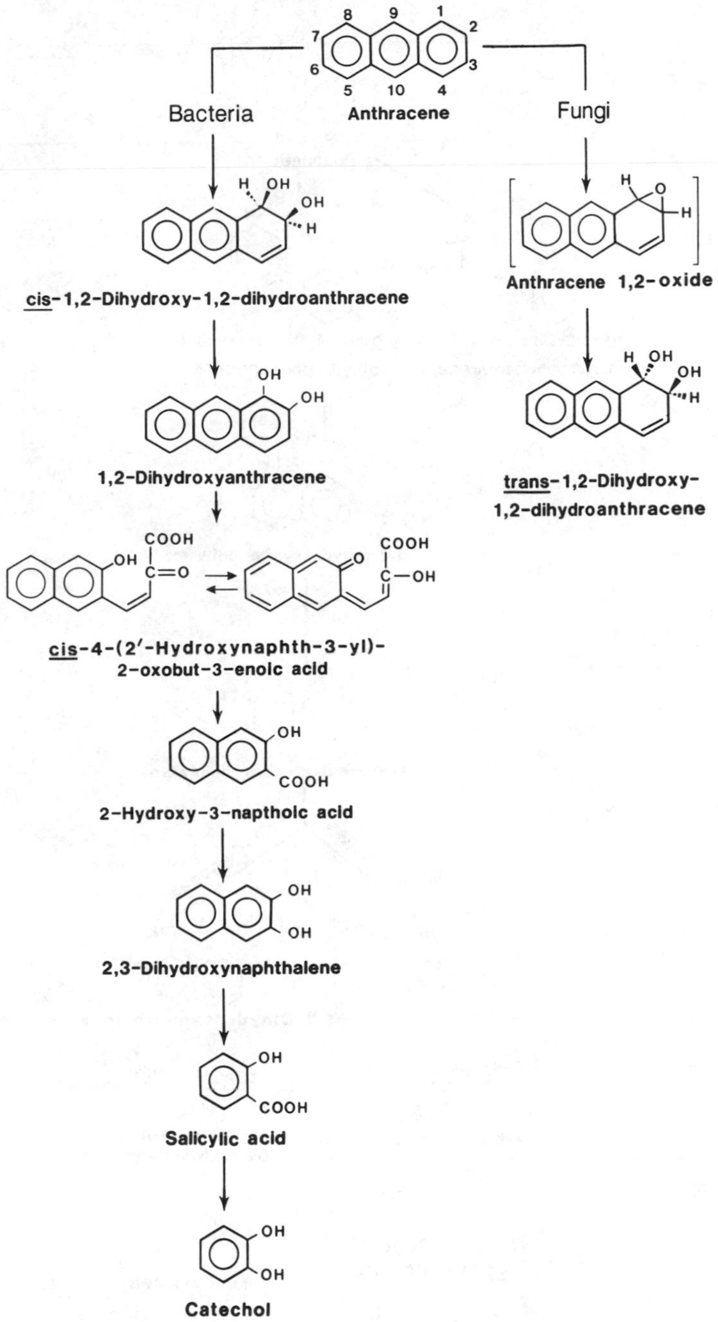

FIGURE 6. Pathway for the bacterial and fungal degradation of anthracene.

cillatoria sp. strain JCM oxidized phenanthrene predominately to *trans*-9,10-dihydroxy-9,10-dihydrophenanthrene[114] This is quite similar to mammalian cytochrome P-450 monooxygenase-mediated reactions.[82-84]

C. Benzo(a)pyrene

Little is known about the bacterial oxidation of PAH containing more than three aromatic rings. Although bacteria capable of utilizing tetra- and pentacyclic aromatic hydrocarbons

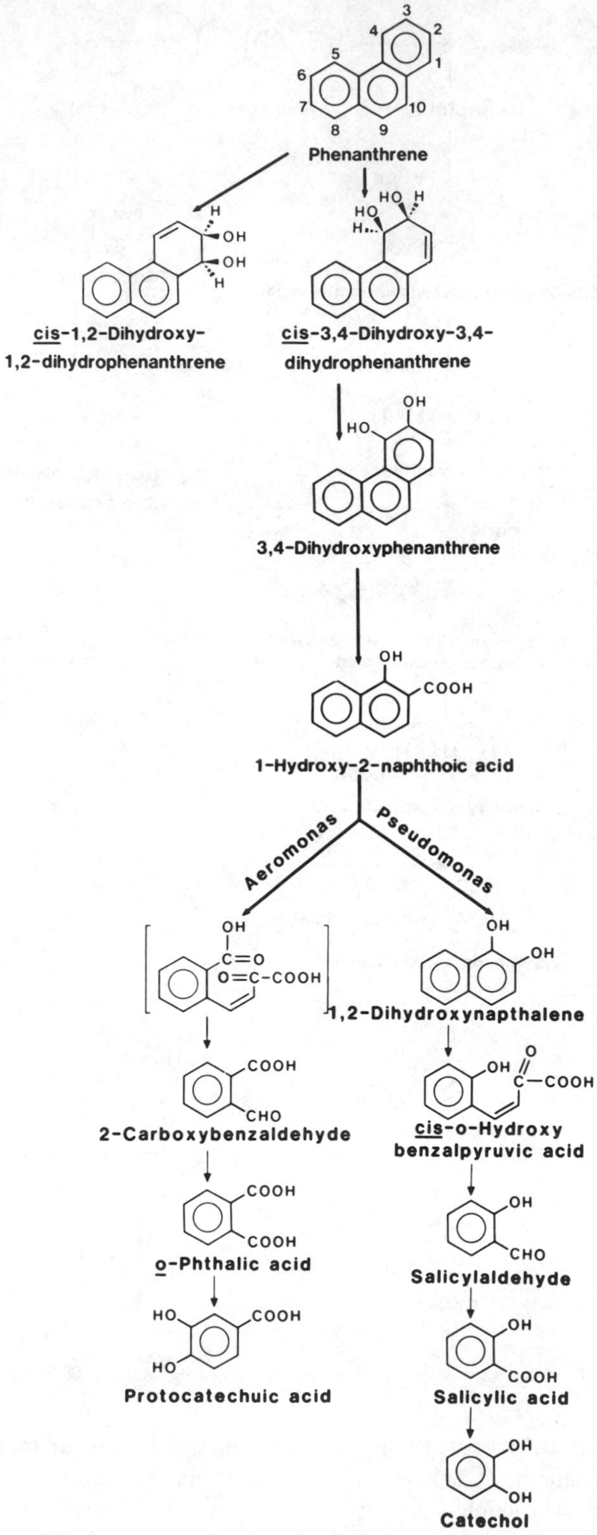

FIGURE 7.   Pathways for the bacterial degradation of phenanthrene.

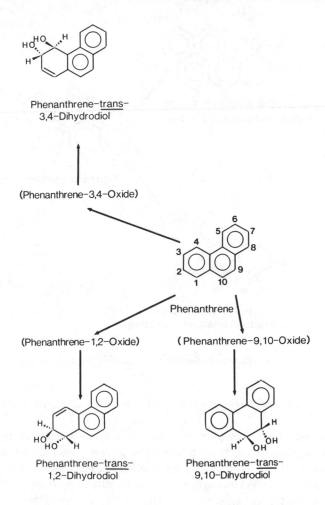

FIGURE 8. Pathway for the fungal transformation of phenanthrene.

as the sole source of carbon and energy have never been demonstrated, microorganisms can oxidize these extremely insoluble PAH when grown on an alternative carbon source. For example, Gibson and colleagues[62] showed that a mutant strain of *Beijerinckia* sp., grown on succinate in the presence of biphenyl, oxidized the potent carcinogen benzo(a)pyrene to *cis*-9,10-dihydroxy-9,10-dihydrobenzo(a)pyrene and *cis*-7,8-dihydroxy-7,8-dihydrobenzo(a)pyrene (Figure 9). Nothing is known concerning the identification of the ring cleavage products and the reaction sequence for the bacterial mineralization of benzo(a)pyrene.

Several fungi have been shown to oxidize benzo(a)pyrene by mechanisms similar to those reported in mammals.[85-91] *C. elegans* metabolized benzo(a)pyrene to various phase I and II metabolites including epoxides, phenols, *trans*-dihydrodiols, dihydrodiol epoxides, tetraols, quinones, and sulfate and glucuronide conjugates[85-87] (Figure 10). These results are of environmental and toxicological significance, since the fungus, *C. elegans*, metabolized benzo(a)pyrene to compounds which have been postulated to be proximate or ultimate carcinogenic metabolites in higher organisms.[86,87] However, the overall effect of incubation of benzo(a)pyrene with *C. elegans* was to transform this genotoxic compound into detoxified derivatives,[92] such as sulfate conjugates. *Saccharomyces cerevisiae, Neurospora crassa, C. bainieri, C. elegans, Aspergillus ochraceus*, and other yeast strains contain a cytochrome P-450 monooxygenase that catalyzes the hydroxylation of benzo(a)pyrene to *trans*-

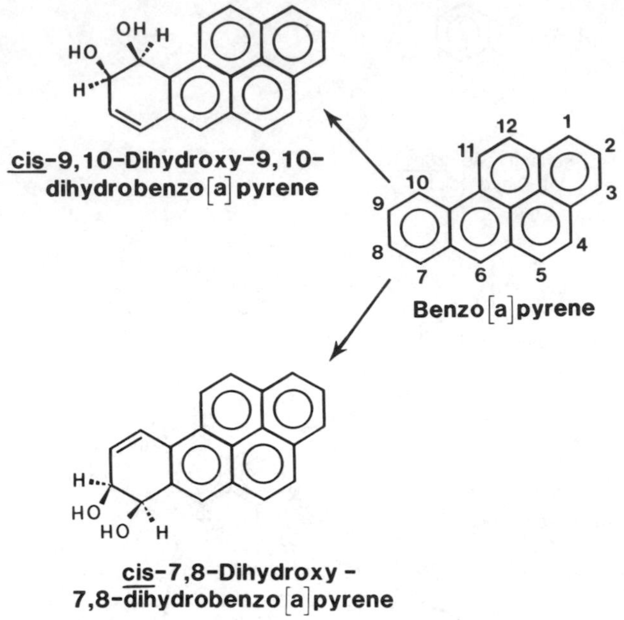

FIGURE 9.   Pathway for the bacterial oxidation of benzo[a]pyrene.

7,8-dihydroxy-7,8-dihydrobenzo(a)pyrene, 3-hydroxybenzo(a)pyrene, and 9-hydroxy-benzo(a)pyrene.[88-91]

Algae have recently been shown to have the capacity to metabolize zo(a)pyrene. The freshwater green alga *Selenastrum capricornutum,* grown photoautotrophically, oxidized benzo(a)pyrene to *cis*-4,5-dihydroxy-4,5-dihydrobenzo(a)pyrene, *cis*-7,8-dihydroxy-7,8-dihydrobenzo(a)pyrene, *cis*-9,10-dihydroxy-9,10-dihydrobenzo(a)pyrene, and *cis*-11,12-dihydroxy-11,12-dihydrobenzo(a)pyrene.[40,41] The *cis*-11,12-dihydrodiol was the predominant isomer. The formation of *cis*-dihydrodiols suggests a dioxygenase pathway similar to that found in prokaryotes rather than the monooxygenase pathway found in other eukaryotes (Figure 2). In a subsequent study Schoeny et al.[42] found that ethyl acetate extractable benzo(a)pyrene metabolites formed by *S. capricornutum* were mutagenic in the *Salmonella typhimurium* mutation assay.

V. ENVIRONMENTAL FATE OF PAH IN AQUATIC ECOSYSTEMS

Lower molecular weight PAH, such as naphthalene and its alklylated derivatives, are relatively water soluble, but PAH containing three or more fused benzene rings, due to their hydrophobic nature, are bound and transported by fine particles and dissolved organic matter in aquatic environments.[93] For this reason, sediments serve as natural repositories for most PAH in aquatic ecosystems. The aquatic distribution and sediment deposition rates have been determined in sediment: water experimental test systems for a few PAH.[94,95] Although PAH may undergo volatilization, photodecomposition, and degradation while in open waters, most sediment deposition occurs below the photolytic zone, and the primary factor affecting the persistence of deposited PAH is microbial degradation (Figure 1). The most rapid biodegradation of deposited PAH occurs at the water/sediment interface. For example, Gardner et al.[96] reported that degradation of PAH in coastal sediments was more rapid in upper surfaces than in lower layers, which indicates that buried PAH may be very persistent. Sediments contain complex communities of bacteria and fungi. The ability of these diverse

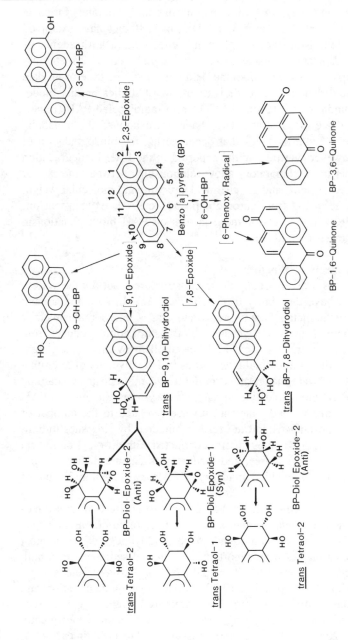

FIGURE 10. Pathway for the fungal transformation of benzo[a]pyrene.

assemblages of microorganisms to metabolize PAH is determined by their combined enzymatic capability (constitutive or inducible) to degrade fused aromatic ring systems, which is affected by several environmental factors.

The metabolism and cometabolism of PAH by pure cultures of microorganisms have been studied for almost 80 years; however, PAH degradation rates by mixed microbial cultures in natural waters and sediments have only been examined in the last 10 years. A summary of reported biodegradation rates (half-lives, turnover rates, or mineralization rates) for some PAH in water and sediments is presented in Table 3. To date, most studies have examined the biodegradation of naphthalene, anthracene, phenanthrene, benz(a)anthracene, or benzo(a)pyrene. Some limited data are available for the biodegradation of 2-methylnaphthalene, dimethylbenz(a)anthracene, pyrene, fluorene, and chrysene, but little or no data exist for other common PAH and their alkylated derivatives. Most studies predict biodegradation rates for PAH based upon the evolution of $^{14}CO_2$ from radiolabeled PAH (mineralization) or upon the disappearance of substrate. They have not examined the metabolic pathways or biological activities of intermediates that occur during the biodegradation of PAH in natural ecosystems. However, various studies reported in the PAH biodegradation literature and recent experiments in our laboratory show relationships between the rate of PAH biodegradation in aquatic ecosystems and the chemical structure and molecular weight of PAH, microbial adaptations occurring from chronic exposure to PAH, oxygen tension, redox potential, and temperature. The effects of these factors on PAH biodegradation in aquatic ecosystems will be reviewed individually.

A. Chemical Structure and Molecular Weight

Fused polyaromatic ring systems are very stable chemical structures, but a wide variety of bacteria, fungi, and algae do have the ability to metabolize PAH. In natural sediment and water, the concerted, synergistic metabolism by mixed cultures of microorganisms may result in enhanced PAH metabolism, since intermediary biotransformation products from one microorganism may serve as substrates for catabolic metabolism by other microorganisms. Recognition of the complexity of microbial interactions and the diversity of microbial populations in natural sediment and water has resulted in the use of microcosms containing natural components to predict PAH biodegradation in natural ecosystems. In general, PAH biodegradation rates in natural sediment and water are inversely related to the number of fused benzene rings in the aromatic nucleus and are further hindered by ring substitution, such as alkylation. Although comparisons among studies conducted with different biological components, experimental designs, temperatures, and incubation times are difficult, some relationships between chemical structure and biodegradation of PAH are evident in the studies reviewed in Table 3.

Since several microorganisms have been isolated that are able to utilize two- and three-ringed PAH, such as naphthalene and phenanthrene, as sole sources of carbon and energy, it is not surprising that these PAH have been found to be readily degradable in most ecosystems. Naphthalene degradation has been reported in water and sediment from both pristine and oil-contaminated ecosystems.[94,97,98,104,106] Naphthalene is relatively water soluble (31.3 mg/l) and has such a high vapor pressure (0.23 mmHg) that biodegradation and volatilization in open waters may be important processes affecting its fate in aquatic ecosystems. The addition of a third fused-benzene ring (phenanthrene or anthracene) greatly decreases water solubility (30 to 700 times lower), vapor pressure (330 to 1180 times lower), and microbial degradation rates. For example, in the four studies listed in Table 3 which measured both naphthalene and anthracene degradation,[94,97,102,105] anthracene degraded 2 to 50 times slower than naphthalene.

None of the studies reviewed in Table 3 directly compared the degradation of naphthalene, 2-methylnaphthalene, and phenanthrene in natural sediments. However, recent experiments in our laboratory with microcosms containing sediment and water from different aquatic ecosystems demonstrate the effects of alkylation on the mineralization of low molecular weight PAH.[107] For example, Figure 11 shows that a methyl-substituted naphthalene (2-methylnaphthalene) is mineralized significantly slower than phenanthrene, a three-ringed PAH, in both DeGray Reservoir (Figure 11A, a pristine freshwater ecosystem) and Redfish Bay (Figure 11B, an oil-exposed estuarine ecosystem). A similar relationship for the mineralization of these three PAH has been reported in mesocosms containing water from Narragansett Bay, Rhode Island.[104]

The presence of polar residues of PAH has been reported by a few researchers, but the identification of these metabolites and the chemical pathways for the biodegradation of PAH have not been determined. However, we recently reported *cis*-1,2-dihydroxy-1,2-dihydronaphthalene, 1-naphthol, salicylic acid, and catechol as metabolites of naphthalene in sediment/water microcosms.[108] These results represent the first example illustrating that naphthalene metabolism in sediments from freshwater and estuarine ecosystems is identical to the studies reported previously for bacterial pure cultures (Figure 4). However, similar studies have not been conducted with PAH with an aromatic nucleus containing three or more fused benzene rings.

To date, no microorganisms have been isolated that are capable of utilizing PAH with an aromatic nucleus containing four or more fused benzene rings as sole sources of carbon and energy. Thus, the degradation of larger carcinogenic PAH probably results from cometabolic reactions that occur very slowly in natural ecosystems. Herbes and Schwall[97] reported an insignificant accumulation of intermediates during the degradation of high molecular weight PAH in sediments. We observed a similar lack of accumulation of degradative intermediates in sediments for PAH containing four or more fused benzene rings in recent experiments at our laboratory. Our results support the hypotheses of Herbes and Schwall,[97] that the rate-limiting steps in the biodegradation of high molecular weight PAH are the initial ring-oxidation reactions and that intermediate metabolites are further oxidized almost as rapidly as they are produced.

Benzo(a)pyrene, a highly carcinogenic five-ringed PAH, is degraded very slowly in natural sediments[96,97,106] and water.[103] Lee et al.[94] reported no degradation of benzo(a)pyrene in water from control or oil-treated ecosystem enclosures. In recent studies in our laboratory, we detected only 0.2 to 3.1% mineralization of benzo(a)pyrene in sediment/water microcosms from three different ecosystems. Recently, field sites were monitored for the disappearance of benzo(a)pyrene residues from slash burn areas[109] and from soil containing oily sludge.[110] In both studies, benzo(a)pyrene persisted much longer than lower molecular weight PAH, but residues varied widely among time points, and only minimal degradation of benzo(a)pyrene was detected. To date, the degradation of six- or seven-ringed PAH, such as indeno-(1,2,3-c,d)pyrene or coronene, has not been reported.

We recently utilized sediment/water microcosms in our laboratory for a comparative study to determine the relationships between chemical structure and PAH degradation by measuring mineralization rates for naphthalene, 2-methylnaphthalene, phenanthrene, pyrene, benzo(a)pyrene, and 3-methyl-cholanthrene.[107] We collected sediment and water from a pristine ecosystem (DeGray Reservoir, Arkansas), an ecosystem chronically exposed to agricultural chemicals (Lake Chicot, Arkansas) (data not shown), and an estuarine ecosystem which is chronically exposed to petrogenic chemicals (Redfish Bay, Texas). It is noteworthy that in all three ecosystems we observed a similar ranking of mineralization: naphthalene > phenanthrene > 2-methylnaphthalene > pyrene > 3-methylcholanthrene > benzo(a)pyrene (Figure 11). Although PAH mineralization rates varied among the different ecosystems and were related to past histories of PAH exposure, the consistent relationships between chemical structure and mineralization of PAH observed within each ecosystem suggest that biodegradation testing with a representative PAH from each ring-class can enable the estimation of mineralization rates of other PAH within similar ecosystems.[107]

Table 3
BIODEGRADATION OF PAH IN WATER AND SEDIMENT

Biological component	Experimental design	PAH	Temp.(°C)	Biodegradation rate	Ref.
Petroleum-contaminated sediments and water	0.5 g sediment, 0.5 ml water in a 20-ml vial	NAPH	12	7.1 h turnover time	97
		ANTH	12	400 h turnover time	
		BA	12	60 wk turnover time	
		B(a)P	12	178 wk turnover time	
Pristine sediments and water	0.5 g sediment, 0.5 ml water in a 20-ml vial	NAPH	12	10—400 times longer for all four PAH	97
		ANTH	12		
		BA	12		
		B(a)P			
Sediments and water amended with Prudhoe crude oil	100 ml water in a 250-ml flask	NAPH	12	0.4—0.5 µg/l/d	94
		2MNAPH	12	0.4—0.5 µg/l/d	
		ANTH	12	0.02 µg/l/d	
		BA	12	ND	
		B(a)P	12	ND	
Sediments and seawater	2 l sediment in a tray with flowing water	ANTH	20	40 ng/g sediment per 72 h	96[a]
		BA	20	500 ng/g sediment per 72 h	
		B(a)P	20	2 ng/g sediment per 72 h	
Sediments and water near an oil field	2 l of an 8:1 stirred suspension of water/sediment	NAPH	30	15—20-d half-life at 250 mV redox	98[a]
Sediments and water	Microcosms with 13 m³ water and 30 cm sediment	BA	16	24 h for 50% loss, 100% loss in 2 years	95, 99[a],
		DMBA	16	12 h for 50—95% loss, 62—100% lost in 62—365 d	100
Water	100 ml water in a 250-ml flask	PHEN	25	75% loss after 4 wk	101
		PYR	25	16.7% loss after 4 wk	
Sediments and water near coal-cok-ing discharge	0.5 g sediment, 0.5 ml water in a 20-ml vial	NAPH	21	13-h turnover time	102
		ANTH	21	62-h turnover time	
		BA	21	300-h turnover time	
Water	100 ml water in a 100-ml flask	NAPH	10	1—30-d turnover time	103
		B(a)P	10	5.5—24.6-yr turnover time	
Water	100 ml water in a 250-ml flask	NAPH	10	125—320-d half-life	104[a]
		2MNAPH	7	390—530-d half-life	
		PHEN	8	180-d half-life	

Sample	Conditions	Compound	Temp	Rate or half-life	Ref.
Water from heavily oiled river	100 ml water in a 250-ml flask	NAPH	22	14-d half-life	104[a]
		2MNAPH	22	16-d half-life	
		PHEN	27	36-d half-life	
Sediment	1 g sediment, 50 ml seawater in a 125-ml flask	ANTH	18	95—141-d half-life	104[a]
		BA	15	1100-d half-life	
		FLUO	10	37-d half-life	
		CHRY	10	510-d half-life	
Sediment from heavily oiled river	1 g sediment, 50 ml seawater in a 125-ml flask	ANTH	27	57-d half-life	104[a]
		BA	27	16-d half-life	
		CHRY	27	79-d half-life	
Water and sediment from an intertidal mudflat	2.5—5.0 ml of a sediment/water (1:2) slurry	NAPH	10	8% in 14 d	105[a]
		NAPH	20	31% in 14 d	
		NAPH	30	35% in 14 d	
		ANTH	10	8% in 28 d	
		ANTH	20	11% in 28 d	
		ANTH	30	16% in 28 d	
Water and sediment in the northern North Sea	4 × 4 × 1 cm of sediment, 500 ml of seawater	NAPH	Ambient	0.1—3.1 mg/m²/cm/d	106
			Ambient	0.1—26.8 µg/m²/cm/d	
Water and sediment from a pristine ecosystem	20 g sediment, 180 ml water in a 500-ml glass microcosm	B(a)P	22	4.4-wk half-life	107
		NAPH	22	20-wk half-life	
		2MNAPH	22	18-wk half-life	
		PHEN	22	ND	
		PYR	22	>200-wk half-life	
		3-MC	22	ND	
		B(a)P	22	ND	
Water and sediment from an ecosystem exposed to petrogenic PAH	20 g sediment, 180 ml water in a 500-ml glass microcosm	NAPH	22	2.4-wk half-life	107
		2MNAPH	22	14-wk half-life	
		PHEN	22	4-wk half-life	
		PYR	22	34-wk half-life	
		3-MC	22	87-wk half-life	
		B(a)P	22	>200-wk half-life	

Note: ND = not detected, NAPH = naphthalene, 2MNAPH = 2-methylnaphthalene, ANTH = anthracene, FLUO = fluorene, BA = benz(a)anthracene, B(a)P = benzo(a)pyrene, amb = ambient, DMBA = dimethylbenz(a)anthracene, PYR = pyrene, and CHRY = chrysene.

a   Additional rates are reported for different concentrations, temperatures, or exposure conditions.

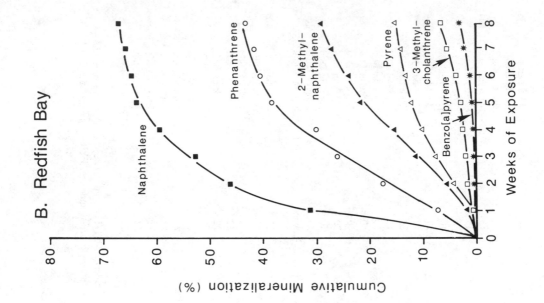

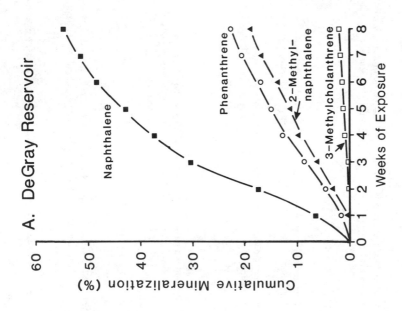

FIGURE 11. Mineralization of PAH in sediment-water microcosms from Degray Reservoir, Arkansas, (A) and Port Aransas, Texas (B).

B. Microbial Adaptations

Prolonged exposure to chemical toxicants can cause adaptations in microbial populations that result in greater resistance to toxicity or enhanced ability to utilize toxicants as substrates for metabolism or co-metabolism. These phenomena occur commonly in microbial populations in natural waters and sediments that are exposed to oil. Increases in PAH degradation are probably related to the presence and expression of catabolic enzymes encoded on plasmids in the cellular cytoplasm. Theoretically, chronic exposure to PAH may result in elevated total populations of microorganisms, selective increases in populations of micoorganisms containing constitutive PAH-degrading enzymes, or, if above threshold concentrations, induce PAH-degrading enzymes in some indigenous microorganisms. There is some evidence that extreme adaptive increases in PAH degradation can occur in sediments chronically exposed to high concentrations of degradable hydrocarbons. For example, a turnover time of only 7.1 h was reported for naphthalene in petroleum-contaminated sediments from a small stream flowing through an oil-holding facility in Tennessee.[97]

Several other studies have shown microbial adaptations and increased PAH degradation in response to chronic PAH exposure, but the actual mechanisms of these adaptations are not known.[101,102,105,106] It may be that microbial adaptations occur as a two-step process in which acutely toxic, low molecular weight PAH initially eliminate PAH-sensitive microorganisms from mixed populations, and then continued exposure to PAH causes the selection and increased growth or activity of resistant microorganisms that are able to metabolize PAH. Naphthalene, an acutely toxic PAH, has been reported to significantly inhibit microbial glucose metabolism and thymidine incorporation in marine sediments and the growth of marine bacteria in batch cultures.[111,112]

Heterotrophic microorganisms are responsible for the catabolic degradation of organic chemicals in nature. Several experimental methods have been developed to measure the numbers, activity, and degradation potential of heterotrophic microorganisms in water and sediments. However, microbiological analyses of water and sediment used in PAH degradation studies have shown that many heterotrophic bacteria are unable to degrade PAH, and that total heterotrophic populations or activities are not always a reliable indicator of the potential for PAH degradation by microorganisms. For example, Sherrill and Sayler[101] measured heterotrophic and phenanthrene-degrading microbial populations in freshwater sediments and reported that a large proportion of the total variation in phenanthrene degradation was correlated with densities of both total viable heterotrophic bacteria and microbial populations of phenanthrene degraders. A similar increase in anthracene-degrading microorganisms has been reported in intertidal marine sediments after an initial acute addition of anthracene, but numbers of total heterotrophic bacteria were increased only slightly after 1 month.[105] Herbes[102] measured heterotrophic and phenanthrene-degrading microbial populations in sediments in the vicinity of a coal-coking wastewater discharge and found that rate constants for anthracene transformation were not related to total heterotroph numbers and proposed that continuous inputs of PAH result in an increased ability within a microbial community to utilize certain PAH. Furthermore, elevated PAH transformation rates were observed for more than 1 year after removal of the source of PAH, which indicates that microbial communities may shift only slowly in response to changes in PAH concentrations. A similar slow reversion of adapted microbial populations was reported by Pierce et al.[113] on an estuarine beach after a fuel-oil spill. On this beach, more than 1 year was required for the proportion of naphthalene degraders to decline to precontamination levels, even though the concentration of aromatics declined by 75% within 8 days. In another study, Massie et al.[106] found that high degradation rates for naphthalene in water and sediments in the northern North Sea were not associated with high total heterotrophic activity and concluded that existing microbial populations had not increased, but had adapted to degrade oil.

C. Oxygen Tension and Redox Potential

As reviewed earlier, molecular oxygen is essential for the biodegradation of PAH by microorganisms. Although aromatic ring reduction and hydrolytic cleavage under anaerobic conditions have been reported for monoaromatic benzoates,[43-46] similar reductive catabolism has not been reported for PAH. In unstratified aquatic ecosystems, there is usually sufficient molecular oxygen dissolved in open waters to facilitate the oxidative metabolism of PAH. However, dissolved oxygen diffuses slowly into sediments and is rapidly utilized by active microorganisms in the sediment/water interface, which causes a depletion of oxygen in sediments that are only buried 1 to 2 cm deep. This explains, in part, the importance of the sediment/water interface as a site for PAH biodegradation and the persistence of PAH buried in sediments. For example, Gardner et al.[96] reported that microbial degradation of PAH was more rapid in upper surfaces than in subsurface layers of sediments.

The lack of oxygen in buried sediments causes obligate and facultative anaerobic microorganisms to utilize buried, oxygen-containing compounds as electron acceptors. This results in decreased redox potentials in buried sediments and further slows the biodegradation of PAH. Bauer and Capone[105] observed a lack of mineralization of naphthalene or anthracene in anoxic intertidal marine sediments. They also conducted an anoxic/oxic conversion experiment in which anoxic sediments were incubated for 3 months in a nitrogen atmosphere and then gassed with compressed air, and found that microorganisms in anoxic sediments readily mineralized naphthalene and anthracene upon the addition of oxygen. From these results they concluded that facultative PAH-degrading aerobes exist under anoxic conditions and may help mitigate PAH toxicity in anaerobic sediments that are occasionally aerated by physical mixing or benthic photosynthesis. Although PAH-degrading aerobes may survive in anoxic sediments, Bauer and Capone[111] recently reported that anaerobic sediments are more sensitive than aerobic sediments to the toxic effects of naphthalene and anthracene.

The redox potential of submerged sediments may range from −300 mV for anoxic sediments up to +700 mV for highly aerobic sediments. Hambrick et al.[98] investigated the effect of oxidation-reduction potential on the microbial degradation of naphthalene in aquatic sediments from a salt marsh stream located near an oil field in Louisiana. ^{14}C-naphthalene was added to 2-l volumes of continuously mixed sediment/water suspensions, and an automated apparatus metered the oxygen supply to each suspension to maintain preset redox values ranging from −250 to +510 mV. They found that mineralization of naphthalene at pH values of 5.0, 6.5, and 8.0 increased with increasing oxidation-reduction potential (increased aerobiosis) and concluded that oxygen availability is an important factor for rapid hydrocarbon mineralization, which is limiting in reduced or anaerobic sediments.

D. Temperature

Temperature directly affects the rate at which PAH are degraded by microorganisms in natural ecosystems. Several investigators have reported seasonal fluxes in heterotrophic activity and PAH degradation rates with the highest activities in summer and the lowest activities in winter.[99,101,104,113] Lee and Ryan[104] suggested that high summer activity in estuarine waters and sediments exposed to PAH was due to both a higher density of bacteria and to higher metabolic activity of individual bacteria at the higher temperature. The magnitude of temperature effects on PAH degradation was demonstrated by Bauer and Capone[105] in intertidal marine sediments exposed to anthracene and naphthalene. They reported a doubling and tripling in anthracene mineralization rates at 20 and 30°C, respectively, over 10°C incubations. Similarly, they found that naphthalene mineralization rates increased 2.7 and 4.6 times over 10°C incubations at 20 and 30°C, respectively.

It is unclear whether temperature-related differences in hydrocarbon degradation result from seasonal selection of psychrophilic or mesophilic hydrocarbon-degrading microorganisms or from low-temperature suppression of the degradative capacity of stable PAH-degrading microbial populations that persist throughout the year. For example, Pierce et al.[113] monitored the biological oxygen demand of residual fuel oil in sediments from an estuarine beach exposed to an oil spill in order to demonstrate seasonal selection of populations of mesophilic and psychrophilic hydrocarbon-degrading bacteria. They reported that during the winter months, populations of psychrophilic bacteria were contained in the sediment that were capable of degrading petroleum at 5°C. However, as the temperature increased to about 12°C, predominantly psychrophilic bacterial populations were replaced by mesophilic populations, which predominated during the summer months. These seasonal population shifts between psychrophilic and mesophilic bacterial populations resulted in temporary reductions in the percentage of hydrocarbon-degrading bacteria during the spring and fall when the ambient temperature was about 12°C.

However, Sherrill and Sayler[101] examined the effects of temperature on phenanthrene degradation in freshwater environments. Phenanthrene degradation was not detected at extremes in incubation temperature (5 and 45°C), but was directly related in increases in incubation temperature between 15 and 37°C, with the maximum rate observed at 37°C. Although biodegradation of phenanthrene was directly related to the temperature of incubation, they found that it was not correlated to ambient site temperatures, since samples collected during the winter months exhibited phenanthrene degradation potentials as high as those of samples collected during warmer weather when both were incubated at 25°C. They suggested that microbial populations capable of degradation are present during the winter months, but their degradative capacity is suppressed at lower temperatures.

VI. CONCLUDING REMARKS

We have described how PAH undergo microbial metabolism in natural systems as well as illustrated the multiple oxidative pathways and mechanisms of oxidation in the microbial catabolism of PAH. Future research considerations about the biodegradation and detoxification of PAH should include:

1. Studies comparing the degradation of PAH in freshwater and marine environments. Are the metabolic pathways different between marine bacteria and pure cultures derived from freshwater and terrestrial environments?
2. Studies on the assessment of the environmental risk and impact of PAH in coastal ecosystems and their potential for biodegradation. Are the biodegradation products more bioavailable and/or toxic than the parent compound?
3. Studies on the biodegradation and toxicity of the photochemical oxidation products of PAH
4. Studies on the environmental fate of alkylated and nitrated PAH since these compounds have been shown to be more genotoxic than unsubstituted analogs
5. Studies on the identification of ring cleavage products and the exact reaction sequences for the bacterial oxidation of PAH that contain more than three aromatic rings
6. Studies on the transport, uptake, and site of PAH hydroxylation in microbial cells
7. Studies on the use of biotechnology to construct a microorganism that can degrade multiringed PAH such as BaP
8. Studies on the importance of anaerobic microorganisms in the degradation of PAH

REFERENCES

1. **Hutzinger, O. and Veerkamp, W.**, Xenobiotic chemicals with pollution potential, in *Microbial Degradation of Xenobiotics and Recalcitrant Compounds,* Leisinger, T., Hutter, R., Cook, A. M., and Nuesch, J., Eds., Academic Press, New York, 1981, 3.
2. **National Academy of Sciences**, *Polycyclic Aromatic Hydrocarbons: Evaluation of Sources and Effects,* National Academy Press, Washington, D.C., 1983.
3. **Dipple, A.**, Polynuclear aromatic carcinogens, in *Chemical Carcinogens,* Searle, C. E., Ed., ACS Monogr. Ser. 3, American Chemical Society, Washington, D.C., 1976, 245.
4. **Sims, P.**, The metabolic activation of chemical carcinogens, *Br. Med. Bull.,* 36, 11, 1980.
5. **Verschueren, K.**, *Handbook of Environmental Data on Organic Chemicals,* Van Nostrand Reinhold, New York, 1983, 1310.
6. International Agency for Research on Cancer, Polynuclear aromatic compounds. I. Chemical, environmental, and experimental data, in *IARC Monographs on the Evaluation of the Carcinogenic Risk of Chemicals to Humans,* World Health Organization, Lyon, France, 1983, 95.
7. **Blumer, M.**, Polycyclic aromatic compounds in nature, *Sci. Am.,* 234, 34, 1976.
8. **Grimmer, G.**, Sources and occurrence of polycyclic aromatic hydrocarbons, in *Polycyclic Aromatic Hydrocarbons,* Vol. 3, Egan, H., Ed., IARC Publ. 29, 1979, 163.
9. **LaFlamme, R. E. and Hites, R. A.**, The global distribution of polycyclic aromatic hydrocarbons in recent sediment, *Geochim. Cosmochim. Acta,* 42, 289, 1978.
10. **Hites, R. A., LaFlamme, R. E., and Farrington, J. W.**, Sedimentary polycyclic aromatic hydrocarbons: the historical record, *Science,* 198, 829, 1977.
11. **Eadie, B. J., Faust, W., Gardner, W. S., and Nalepa, T.**, Polycyclic aromatic hydrocarbons in sediments and associated benthos in Lake Erie, *Chemosphere,* 11, 185, 1982.
12. **Youngblood, W. W. and Blumer, M.**, Polycyclic aromatic hydrocarbons in the environment: homologous series in soils and recent marine sediments, *Geochim. Cosmochim. Acta,* 39, 1303, 1975.
13. **Larsen, P. F., Gadbois, D. F., Johnson, A. C., and Doggett, L. F.**, Distribution of polycyclic aromatic hydrocarbons in the surficial sediments of Casco Bay, Maine, *Bull. Environ. Contam. Toxicol.,* 30, 530, 1983.
14. **Jackim, E. and Lake, C.**, Polynuclear aromatic hydrocarbons in estuarine and nearshore environments, in *Estuarine Interactions,* Wiley, M. L., Ed., Academic Press, New York, 1980, 415.
15. **John, E. D., Cooke, M., and Nickless, G.**, Polycyclic aromatic hydrocarbons in sediments taken from the Severn Estuary drainage system, *Bull. Environ. Contam. Toxicol.,* 22, 653, 1979.
16. **West, W. R., Smith, P. A., Booth, G. M., Wise, S. A., and Lee, M. L.**, Determination of genotoxic polycyclic aromatic hydrocarbons in a sediment from the Black River, Ohio, *Arch. Environ. Contam. Toxicol.,* 15, 241, 1986.
17. **Johnson, A. C. and Larsen, D.**, The distribution of polycyclic aromatic hydrocarbons in the surficial sediments of Penobscot Bay (Maine, U.S.A.) in relation to possible sources and to other sites worldwide, *Mar. Environ. Res.,* 15, 1, 1985.
18. **Sohngen, N. L.**, Benzin, Petroleum, Paraffinol und Paraffin als Kohlenstoff und Energiequelle für Mikroben., *Zentralbl. Bakteriol. Parasitenkd. Abt. 2,* 37, 595, 1913.
19. **Stormer, K.**, Über die Wirkung des Schwefelkohlenstoffs und ähnlicher Stoff auf den Boden, *Zentralbl. Bakteriol. Parasitenkd. Abt. 2,* 20, 282, 1908.
20. **Hou, C. T.**, Microbial transformation of important industrial hydrocarbons, in *Microbial Transformations of Bioactive Compounds,* Vol. 1, Rosazza, J. P., Ed., CRC Press, Boca Raton, FL, 1982, chap. 4.
21. **Fewson, C. A.**, Biodegradation of aromatics with industrial relevance, in *Microbial Degradation of Xenobiotics and Recalcitrant Compounds,* Leisinger, T., Hutter, R., Cook, A. M., and Nuesch, J., Eds., Academic Press, New York, 1981, 141.
22. **Dagley, S.**, New perspectives in aromatic catabolism, in *Microbial Degradation of Xenobiotics and Recalcitrant Compounds,* Leisinger, T., Hutter, R., Cook, A. M., and Nuesch, J., Eds., Academic Press, New York, 1981, 181.
23. **Hopper, D. J.**, Microbial degradation of aromatic hydrocarbons, in *Developments in Biodegradation of Hydrocarbons,* Vol. 1, Watkinson, R. J., Ed., Applied Science Publishers, London, 1978, 85.
24. **Cripps, R. E. and Watkinson, R. J.**, Polycyclic aromatic hydrocarbons: metabolism and environmental aspects, in *Developments in Biodegradation of Hydrocarbons,* Vol. 1, Watkinson, R. J., Ed., Applied Science Publishers, London, 1978, 113.
25. **Cerniglia, C. E.**, Microbial metabolism of polycyclic aromatic hydrocarbons, in *Advances in Applied Microbiology,* Vol. 30, Laskin, A., Ed., Academic Press, New York, 1984, 31.
26. **Cerniglia, C. E.**, Aromatic hydrocarbons: metabolism by bacteria, fungi and algae, in *Biochemical Toxicology,* Vol. 3, Hodgson, E., Bend, J. R., and Philpot, R. M., Eds., Elsevier/North-Holland, New York, 1981, 321.

27. **Atlas, R. M.,** Microbial degradation of petroleum hydrocarbons: an environmental perspective, *Microbiol. Rev.,* 45, 180, 1981.

28. **Sims, R. C. and Overcash, M. R.,** Fate of polynuclear aromatic compounds (PNAs) in soil-plant systems, *Residue Rev.,* 88, 1, 1983.

29. **Chapman, P. J.,** Degradation mechanisms, in *Microbial Degradation of Pollutants in Marine Environments,* Bourquin, A. W. and Pritchard, P. H., Eds., EPA-600/9-79-012, Gulf Breeze, FL, 1979, 28.

30. **Gibson, D. T.,** Microbial transformation of aromatic pollutants, in *Transformation and Biological Effects,* Hutzinger, O., Van Lelyveld, I. H., and Zoeteman, B. C. J., Eds., Pergamon Press, New York, 1978, 187.

31. **Gibson, D. T.,** Microbial degradation of hydrocarbons, *Toxicol. Environ. Chem.,* 5, 237, 1982.

32. **Dagley, S.,** Catabolism of aromatic compounds by microorganisms, *Adv. Microb. Physiol.,* 6, 1, 1971.

33. **Wiseman, A. and King, D. J.,** Microbial oxygenases and their potential application, in *Topics in Enzyme and Fermentation, Biotechnology,* Vol. 6, Wiseman, A., Ed., Halstead Press, New York, 1982, 151.

34. **Yeh, W. K., Gibson, D. T., and Liu, T.,** Toluene dioxygenase: a multicomponent enzyme system, *Biochem. Biophys. Commun.,* 78, 401, 1977.

35. **Subramanian, V., Liu, T., Yeh, W. K., and Gibson, D. T.,** Toluene dioxygenase: purification of an iron-sulfur protein by affinity chromatography, *Biochem. Biophys. Res. Commun.,* 91, 1131, 1979.

36. **Crutcher, S. E. and Geary, P. J.,** Properties of the iron-sulfur proteins of the benzene dioxygenase system from *Pseudomonas putida, Biochem. J.,* 177, 393, 1979.

37. **Williams, P. A.,** Genetics of biodegradation, in *Microbial Degradation of Xenobiotics and Recalcitrant Compounds,* Leisinger, T., Hutter, R., Cook, A. M., and Nuesch, J., Eds., Academic Press, New York, 1981, 97.

38. **Patel, T. R. and Gibson, D. T.,** Purification and properties of (+)-*cis*-naphthalene dihydrodiol dehydrogenase of *Pseudomonas putida, J. Bacteriol.,* 119, 879, 1974.

39. **Narro, M. L., Cerniglia, C. E., Gibson, D. T., and Van Baalen, C.,** The oxidation of aromatic compounds by microalgae, in *Environmental Regulation of Microbial Metabolism,* Kulaev, I. S., Dawes, E. A., and Tempest, D. W., Eds., Academic Press, New York, 1985, 249.

40. **Cody, T. E., Radike, M. J., and Warshawsky, D.,** The phytotoxicity of benzo[a]pyrene in the green alga, *Selenastrum capricornutum, Environ. Res.,* 35, 122, 1984.

41. **Lindquist, B. and Warshawsky, D.,** Identification of the 11,12-dihydro-11,12-dihydroxybenzo[a]pyrene as a major metabolite produced by the green alga, *Selenastrum capricornutum, Biochem. Biophys. Res. Commun.,* 130, 7, 1985.

42. **Schoeny, R., Cody, T., Radike, M., and Warshawsky, D.,** Mutagenicity of algal metabolites of benzo[a]pyrene for *Salmonella typhimurium, Environ. Mutagenesis,* 7, 839, 1985.

43. **Evans, W. C.,** Biochemistry of the bacterial catabolism of aromatic compounds in anaerobic environments, *Nature,* 270, 17, 1977.

44. **Healy, J. B., Jr. and Young, L. Y.,** Anaerobic biodegradation of eleven aromatic compounds to methane, *Appl. Environ. Microbiol.,* 38, 84, 1979.

45. **Kaiser, J. P. and Hanselmann, K. W.,** Fermentative metabolism of substituted monoaromatic compounds by a bacterial community from anaerobic sediments, *Arch. Microbiol.,* 133, 185, 1982.

46. **Guyer, M. and Hegeman, G.,** Evidence for a reductive pathway for the anaerobic metabolism of benzoate, *J. Bacteriol.,* 99, 906, 1969.

47. **Vogel, T. M. and Grabic-Galic, D.,** Incorporation of oxygen from water into toluene and benzene during anaerobic fermentative transformation, *Appl. Environ. Microbiol.,* 52, 200, 1986.

48. **Walker, N. and Wiltshire, G. H.,** The breakdown of naphthalene by a soil bacterium, *J. Gen. Microbiol.,* 8, 273, 1953.

49. **Treccani, V., Walker, N., and Wiltshire, G. H.,** The metabolism of naphthalene by soil bacteria, *J. Gen. Microbiol.,* 11, 341, 1954.

50. **Davies, J. I. and Evans, W. C.,** Oxidative metabolism of naphthalene by soil pseudomonads, *Biochem. J.,* 91, 251, 1964.

51. **Evans, W. C., Fernley, H. N., and Griffiths, E.,** Oxidative metabolism of phenanthrene and anthracene by soil pseudomonads; the ring fission mechanism, *Biochem. J.,* 95, 819, 1965.

52. **Jeffrey, A. M., Yeh, H. J. C., Jerina, D. M., Patel, T. R., Davey, J. F., and Gibson, D. T.,** Initial reactions in the oxidation of naphthalene by *Pseudomonas putida, Biochemistry,* 14, 575, 1975.

53. **Dean-Raymond, D. and Bartha, R.,** Biodegradation of some polynuclear aromatic petroleum components by marine bacteria, *Dev. Ind. Microbiol.,* 16, 97, 1975.

54. **Jerina, D. M., Selander, H., Yagi, H., Wells, M. C., Davey, J. F., Mahadevan, V., and Gibson, D. T.,** Dihydrodiols from anthracene and phenanthrene, *J. Am. Chem. Soc.,* 98, 5988, 1976.

55. **Kiyohara, H. and Nagao, K.,** The catabolism of phenanthrene and naphthalene by bacteria, *J. Gen. Microbiol.,* 105, 69, 1978.

56. **Barnsley, E. A.,** Bacterial oxidation of naphthalene and phenanthrene, *J. Bacteriol.,* 153, 1069, 1983.

57. **Colla, C., Biaggi, C., and Treccani, V.,** Ricerche sul metabolismo ossidativo microbico dell'anthracene e fenantrene. II. Isolamento e carraterizzazione del 3,4-diidro-3,4-diossifenanthrene, *Ann. Microbiol. Ed. Enzimol.,* 9, 1, 1959.

58. **Kiyohara, H., Nagao, K., Kuono, K., and Yano, K.,** Phenanthrene degrading phenotype of *Alcaligens faecalis* AFK2, *Appl. Environ. Microbiol.,* 43, 458, 1982.

59. **Kiyohara, H., Nagao, K., and Nomi, R.,** Degradation of phenanthrene through *o*-phthalate by an *Aeromonas* sp., *Agric. Biol. Chem.,* 40, 1075, 1976.

60. **Akhtar, M. N., Boyd, D. R., Thompson, N. J., Koreeda, M., Gibson, D. T., Mahadevan, V., and Jerina, D. M.,** Absolute stereochemistry of the dihydroanthracene-*cis*- and *trans*-1,2-diols produced from anthracene by mammals and bacteria, *J. Chem. Soc.,* p. 2506, 1975.

61. **Koreeda, M., Akhtar, M. N., Boyd, D. R., Neill, J. D., Gibson, D. T., and Jerina, D. M.,** Absolute stereochemistry of *cis*-1,2-, *trans*-1,2 and *cis*-3,4-dihydrodiol metabolites of phenanthrene, *J. Org. Chem.,* 43, 1023, 1978.

62. **Gibson, D. T., Mahadevan, V., Jerina, D. M., Yagi, H., and Yeh, H. J. C.,** Oxidation of the carcinogens benzo[a]pyrene and benz[a]anthracene to dihydrodiols by a bacterium, *Science,* 189, 295, 1975.

63. **Cerniglia, C. E., Freeman, J. P., and Evans, F. E.,** Evidence for an arene oxide-NIH shift pathway in the transformation of naphthalene to 1-naphthol by *Bacillus cereus, Arch. Microbiol.,* 138, 283, 1984.

64. **Dua, R. D. and Meera, S.,** Purification and characterization of naphthalene oxygenase from *Corynebacterium renale, Eur. J. Biochem.,* 120, 461, 1981.

65. **Ghosh, D. K. and Mishra, A. K.,** Oxidation of phenanthrene by a strain of *Micrococcus:* evidence of protocatechuate pathway, *Curr. Microbiol.,* p. 219, 1983.

66. **Catterall, F. A., Murray, K., and Williams, P. A.,** The configuration of the 1,2-dihydroxy-1,2-dihydronaphthalene formed by the bacterial metabolism of naphthalene, *Biochem. Biophys. Acta,* 237, 361, 1971.

67. **Ensley, B. D., Gibson, D. T., and Laborde, A. L.,** Oxidation of naphthalene by a multicomponent enzyme system from *Pseudomonas* sp. strain NCIB 9816, *J. Bacteriol.,* 149, 948, 1982.

68. **Ferries, J. P., Fasco, M. J., Stylianopoulou, F. L., Jerina, D. M., Daly, J. W., and Jeffrey, A. M.,** Mono-oxygenase activity in *Cunninghamella bainieri:* evidence for a fungal system similar to liver microsomes, *Arch. Biochem. Biophys.,* 156, 97, 1973.

69. **Smith, R. V. and Rosazza, J. P.,** Microbial models of mammalian metabolism, *Arch. Biochem. Biophys.,* 161, 551, 1974.

70. **Cerniglia, C. E., Herbert, R. L., Szaniszlo, P. J., and Gibson, D. T.,** Fungal transformation of naphthalene, *Arch. Microbiol.,* 117, 135, 1978.

71. **Cerniglia, C. E. and Gibson, D. T.,** Metabolism of naphthalene by *Cunninghamella elegans, Appl. Environ. Microbiol.,* 34, 363, 1977.

72. **Cerniglia, C. E. and Gibson, D. T.,** Metabolism of naphthalene by cell extracts of *Cunninghamella elegans, Arch. Biochem. Biophys.,* 186, 121, 1978.

73. **Cerniglia, C. E., Freeman, J. P., and Mitchum, R. K.,** Glucuronide and sulfate conjugation in the fungal metabolism of aromatic hydrocarbons, *Appl. Environ. Microbiol.,* 43, 1070, 1982.

74. **Cerniglia, C. E., Althaus, J. R., Evans, F. E., Freeman, J. P., Mitchum, R. K., and Yang, S. K.,** Stereochemistry and evidence for an arene oxide-NIH shift pathway in the fungal metabolism of naphthalene, *Chem. Biol. Interact.,* 44, 119, 1983.

75. **Cerniglia, C. E., Gibson, D. T., and Van Baalen, C.,** Algal oxidation of aromatic hydrocarbons: formation of 1-naphthol from naphthalene by *Agmenellum quadruplicatum* strain PR-6, *Biochem. Biophys. Res. Commun.,* 88, 50, 1979.

76. **Cerniglia, C. E., Van Baalen, C., and Gibson, D. T.,** Metabolism of naphthalene by the cyanobacterium *Oscillatoria* sp. strain JCM, *J. Gen. Microbiol.,* 116, 485, 1980.

77. **Cerniglia, C. E., Gibson, D. T., and Van Baalen, C.,** Oxidation of naphthalene by cyanobacteria and microalgae, *J. Gen. Microbiol.,* 116, 495, 1980.

78. **Cerniglia, C. E., Gibson, D. T., and Van Baalen, C.,** Naphthalene metabolism by diatoms isolated from the Kachemak Bay region of Alaska, *J. Gen. Microbiol.,* 128, 987, 1982.

79. **Cerniglia, C. E.,** Initial reactions in the oxidation of anthracene by *Cunninghamella elegans, J. Gen. Microbiol.,* 128, 2055, 1982.

80. **Cerniglia, C. E. and Yang, S. K.,** Stereoselective metabolism of anthracene and phenanthrene by the fungus *Cunninghamella elegans, Appl. Environ. Microbiol.,* 47, 119, 1984.

81. **Rogoff, M. H. and Wender, I.,** Microbiology of coal. I. Bacterial oxidation of phenanthrene, *J. Bacteriol.,* 73, 264, 1957.

82. **Boyland, E. and Sims, P.,** The metabolism of phenanthrene in rabbits and rats: dihydroxy compounds and related glucosiduric acids, *Biochem. J.,* 84, 571, 1962.

83. **Chaturapit, S. and Holder, C. M.,** Studies on the hepatic microsomal metabolism of ^{14}C-phenanthrene, *Biochem. Pharmacol.,* 27, 1865, 1978.

84. **Nordquist, M., Thakker, D. R., Vyas, K. P., Yagi, H., Levin, W., Ryan, D. E., Thomas, P. E., Conney, A. H., and Jerina, D. M.,** Metabolism of chrysenes and phenanthrenes to bay-region diol epoxides by rat liver enzymes, *Mol. Pharmacol.,* 19, 168, 1981.

85. **Cerniglia, C. E. and Gibson, D. T.,** Oxidation of benzo[a]pyrene by the filamentous fungus *Cunninghamella elegans, J. Biol. Chem.,* 254, 12174, 1979.

86. **Cerniglia, C. E. and Gibson, D. T.,** Fungal oxidation of benzo[a]pyrene and (±)-*trans*-7,8-dihydroxy-7,8-dihydrobenzo[a]pyrene: evidence for the formation of benzo[a]pyrene 7,8-diol-9,10-epoxide, *J. Biol. Chem.,* 255, 5159, 1980.

87. **Cerniglia, C. E. and Gibson, D. T.,** Fungal oxidation of (±)-9,10-dihydroxy-9,10-dihydrobenzo[a]pyrene: formation of diastereomeric benzo[a]pyrene 9,10-diol-7,8-epoxides, *Proc. Natl. Acad. Sci. U.S.A.,* 77, 4554, 1980.

88. **Cerniglia, C. E. and Crow, S. A.,** Metabolism of aromatic hydrocarbons by yeasts, *Arch. Microbiol.,* 129, 9, 1981.

89. **Lin, W. S. and Kapoor, M.,** Induction of aryl hydrocarbon hydrooxylase in *Neurospora crassa* by benzo[a]pyrene, *Curr. Microbiol.,* 3, 177, 1979.

90. **Ghosh, D. K., Dutta, D., Samanta, T. B., and Mishra, A. K.,** Microsomal benzo[a]pyrene hydroxylase in *Aspergillus ochraceus* TS: assay and characterization of the enzyme system, *Biochem. Biophys. Res. Commun.,* 113, 497, 1983.

91. **Wiseman, A. and Woods, L. F. J.,** Benzo[a]pyrene metabolites formed by the action of yeast cytochrome P-450/P-448, *J. Chem. Biotechnol.,* 29, 320, 1979.

92. **Cerniglia, C. E., White, G. L., and Heflich, R. H.,** Fungal metabolism and detoxification of polycyclic aromatic hydrocarbons, *Arch. Microbiol.,* 143, 105, 1985.

93. **Morehead, N. R., Eadie, B. J., Lake, B., Landrum, P. D., and Berner, D.,** The sorption of PAH onto dissolved organic matter in Lake Michigan waters, *Chemosphere,* 15, 403, 1986.

94. **Lee, R. F., Gardner, W. D., Anderson, J. W., Blaylock, J. W., and Barwell-Clarke, J.,** Fate of polycyclic aromatic hydrocarbons in controlled ecosystem enclosures, *Environ. Sci. Technol.,* 12, 832, 1978.

95. **Lee, R. F., Hinga, K., and Almquist, G.,** Fate of radiolabeled polycyclic aromatic hydrocarbons and pentachlorophenol in enclosed marine ecosystems, in *Marine Mesocosms, Biological and Chemical Research in Experimental Ecosystems,* Grice, G. D. and Reeve, M. R., Eds., Springer-Verlag, New York, 1982, chap. 9.

96. **Gardner, W. D., Lee, R. F., Tenore, K. R., and Smith, L. W.,** Degradation of selected polycyclic aromatic hydrocarbons in coastal sediments: importance of microbes polychaete worms, *Water Air Soil Pollut.,* 11, 339, 1979.

97. **Herbes, S. E. and Schwall, L. R.,** Microbial transformation of polycyclic aromatic hydrocarbons in pristine and petroleum-contaminated sediments, *Appl. Environ. Microbiol.,* 35, 306, 1978.

98. **Hambrick, G. A., DeLaune, R. D., and Patrick, W. H.,** Effect of estuarine sediment pH and oxidation-reduction potential on microbial hydrocarbon degradation, *Appl. Environ. Microbiol.,* 40, 365, 1980.

99. **Hinga, K. R., Pilson, M. E. Q., Lee, R. F., Farrington, J. W., Tjessem, K., and Davis, A. C.,** Biogeochemistry of benzanthracene in an enclosed marine ecosystem, *Environ. Sci. Technol.,* 14, 1136, 1980.

100. **Hinga, K. R., Pilson, M. E. Q., Almquist, G., and Lee, R. F.,** The degradation of 7,12-dimethylbenz[a]anthracene in an enclosed marine ecosystem, *Mar. Environ. Res.,* 18, 79, 1986.

101. **Sherrill, T. W. and Sayler, G. S.,** Phenanthrene biodegradation in freshwater environments, *Appl. Environ. Microbiol.,* 39, 172, 1980.

102. **Herbes, S. E.,** Rates of microbial transformation of polycyclic aromatic hydrocarbons in water and sediments in the vicinity of a coal-coking wastewater discharge, *Appl. Environ. Microbiol.,* 41, 20, 1981.

103. **Readman, J. W., Mantoura, R. F. C., Rhead, M. M., and Brown, L.,** Aquatic distribution and heterotrophic degradation of polycyclic aromatic hydrocarbons (PAH) in the Tamar Estuary, *Estuarine Coastal Shelf Sci.,* 14, 369, 1982.

104. **Lee, R. F. and Ryan, C.,** Microbial and photochemical degradation of polycyclic aromatic hydrocarbons in estuarine waters and sediments, *Can. J. Fish. Aquat. Sci.,* 40, 86, 1983.

105. **Bauer, J. E. and Capone, D. G.,** Degradation and mineralization of the polycyclic aromatic hydrocarbons anthracene and naphthalene in intertidal marine sediments, *Appl. Environ. Microbiol.,* 50, 81, 1985.

106. **Massie, L. C., Ward, A. P., and Davies, J. M.,** The effects of oil exploration and production in the northern North Sea. I. Microbial biodegradation of hydrocarbons in water and sediments, 1978—1981, *Mar. Environ. Res.,* 15, 235, 1985.

107. **Heitkamp, M. A. and Cerniglia, C. E.,** The effects of chemical structure and exposure on the microbial degradation of polycyclic aromatic hydrocarbons in freshwater and estuarine ecosystems, *Environ. Toxicol. Chem.,* 6, 535, 1987.

108. **Heitkamp, M. A., Freeman, J. P., and Cerniglia, C. E.,** Napthalene biodegradation in environmental microcosms: estimates of degradation rates and characterization of metabolites, *Appl. Environ. Microbiol.,* 53, 129, 1987.
109. **Sullivan, T. J. and Mix, M. C.,** Persistence and fate of polynuclear aromatic hydrocarbons deposited on slash burn sites in the Cascade mountains and coast range of Oregon, *Arch. Environ. Contam. Toxicol.,* 14, 187, 1985.
110. **Bossert, I., Kachel, W. M., and Bartha, R.,** Fate of hydrocarbons during oily sludge disposal in soil, *Appl. Environ. Microbiol.,* 47, 763, 1984.
111. **Bauer, J. E. and Capone, D. G.,** Effects of four aromatic organic pollutants on microbial glucose metabolism and thymidine incorporation in marine sediments, *Appl. Environ. Microbiol.,* 49, 828, 1985.
112. **Calder, J. A. and Lader, J. H.,** Effect of dissolved aromatic hydrocarbons on the growth of marine bacteria in batch culture, *Appl. Environ. Microbiol.,* 32, 95, 1976.
113. **Pierce, R. H., Cundell, A. M., and Traxler, R. W.,** Persistence and biodegradation of spilled residual fuel oil on an estuarine beach, *Appl. Microbiol.,* 29, 646, 1975.
114. **Narro, M.,** personal communication.

Chapter 3

BIOTRANSFORMATION AND DISPOSITION OF PAH IN AQUATIC INVERTEBRATES

Margaret O. James

TABLE OF CONTENTS

I. INTRODUCTION

The ability of invertebrates to take up polycyclic aromatic hydrocarbons (PAH) from polluted aquatic environments has been well documented (see References 1 and 2 and Chapter 1 in this Volume). Indeed, based on the results of several studies of the uptake and depuration of PAH by mussels, oysters, and clams, consideration has been given to using lower in vertebrates such as bivalve mollusks as indicators of marine pollution.[1] It is important from the standpoint of human health to understand the disposition (absorption, biotransformation, distribution, excretion) of potentially carcinogenic PAH in invertebrates, especially those used as human food, since this class of carcinogens must be biotransformed (metabolized) to active metabolites in order to exert mutagenic, toxic, and carcinogenic effects. Some *in vivo* studies have shown that PAH metabolites are poorly excreted by certain marine invertebrates (see Section II) and, thus, certain invertebrates exposed to PAH could be dietary sources of potentially bioactive PAH metabolites, even when the parent PAH has been eliminated. Additionally, it is clear that species which rapidly and extensively metabolize PAH could not be used as indicators of pollution, since it would be inappropriately expensive to monitor for PAH metabolites (even if all metabolites were known) as well as for parent PAH. Thus, studies are needed to determine the capacity of potential indicator species to biotransform PAH. Finally, an understanding of PAH biotransformation in invertebrates may help predict the toxicity of PAH to invertebrates themselves, since many of the toxic effects of PAH are mediated by metabolites which bind to important cellular macromolecules.

Several approaches have been made to determine if invertebrate species biotransform PAH, but the quality of studies in the literature varies considerably. Probably the best approach is to conduct both *in vivo* and *in vitro* studies in each species of interest, for the following reasons. *In vivo* studies will normally provide good evidence as to whether or not metabolism occurs, and could provide information on rates of elimination of parent PAH, rates of formation and elimination of metabolites, and formation of adducts of metabolites to macromolecules. *In vitro* studies are necessary to understand metabolic pathways for formation of PAH metabolites, especially biologically active metabolites which may be unstable *in vivo; in vitro* studies are also useful in demonstrating that adducts to macromolecules can be formed, and in obtaining structural information on the adducts. There are, however, several problems associated with study of PAH biotransformation in marine invertebrates. These will be discussed in the relevant sections of this chapter, but two problems affect all studies. Since PAH can be decomposed photochemically and microbially in seawater,[1,2] *in vivo* and *in vitro* studies should correct for (*in vivo* studies) or eliminate (*in vitro* studies) these factors. As will be discussed, many invertebrates have very low inherent capability for xenobiotic biotransformation. Thus, rates of PAH biotransformation *in vivo* by invertebrates may be close to rates of breakdown by other pathways. *In vitro* studies should be easier to interpret, but a major problem is that many published methods for determining rates of PAH biotransformation were designed in studies with mammalian species, and are insufficiently sensitive for conducting studies with invertebrates.

II. *IN VIVO* METABOLISM OF PAH

All invertebrates studied to date have been shown to be capable of absorbing PAH from PAH-containing environments, and subsequently eliminating PAH when transferred to uncontaminated water. There have, however, been few rigorous studies of metabolism of PAH *in vivo*, or investigation of PAH metabolite levels in tissues of invertebrates. It has usually proved necessary to use radiolabeled PAH to get meaningful results for metabolite identification, and, thus, good studies of PAH metabolism *in vivo* have all been conducted in the laboratory. Three routes of administration have commonly been used: injection of radiolabeled PAH into the circulation, feeding, or exposure in the water or to sediment. Each route has advantages and disadvantages, but since metabolism of PAH in any species, and especially invertebrates, is slow, it is unlikely that there would be any major route-dependent qualitative differences in metabolites found.

A. Arthropods

Many edible species fall into this phylum, and because there is concern that there is a health risk for humans eating PAH-contaminated animals, there have been several studies of PAH metabolism in arthropods, especially crustaceans. *In vivo* studies have been especially necessary to demonstrate biotransformation in crustacea, because *in vitro* studies, especially in hepatopancreas fractions, do not always indicate the true capability of the animals to metabolize PAH (see Section III for explanation).

1. Benzo(a)pyrene

The first work in this area was conducted by Lee et al.[3] who showed that ^{3}H-benzo(a)pyrene was absorbed by blue crabs, *Callinectes sapidus,* from water or from food. This study showed that hepatopancreas was the major organ of uptake, regardless of exposure route. Hydroxybenzo(a)pyrene and more polar metabolites were found in hepatopancreas, gills, blood, stomach, muscle, and gonads, and the extent of metabolism increased with time. The position of hydroxylation and the nature of the polar metabolites were not investigated. It was noteworthy that even 20 days after transfer to clean water, the hemolymph and hepatopancreas of crabs exposed to benzo(a)pyrene contained unmetabolized benzo(a)pyrene as well as metabolites, indicating slow metabolism by the blue crabs.[3]

Bend et al.[4,5] studied the fate of ^{14}C-benzo(a)pyrene after a single intrapericardial dose of 1 mg/kg to the lobster *Homarus americanus.* Hepatopancreas contained 80 to 90% of the total radiolabel left in animals at all of the times studied, and even 6.5 weeks after the dose more than 80% of the ^{14}C in hepatopancreas was unchanged benzo(a)pyrene. The major metabolites in hepatopancreas were phenols, primarily a metabolite coeluting with 3-hydroxybenzo(a)pyrene, and only small amounts of conjugated (nonhexane extractable) metabolites were found. Low amounts of ^{14}C were present in tail muscle at all points studied, and this ^{14}C was partly unchanged benzo(a)pyrene and partly phenolic metabolites. The rate of excretion of ^{14}C from lobsters was very slow, with an estimated half-life of more than 2 months. This slow excretion by lobster was due, at least in part, to the slow rate of biotransformation.

The first study to examine temperature effects on the *in vivo* fate of benzo(a)pyrene was conducted in the Florida spiny lobster *Panulirus arqus.*[5,6] The ^{14}C-benzo(a)pyrene (1 mg/kg) was injected into the pericardial sinus of spiny lobsters acclimatized and maintained at summer water temperatures (26.5 to 29°C) or winter temperatures (13.5 to 16.5°C). This study showed that excretion of radiolabel was more rapid at the higher water temperature and overall elimination half-lives of benzo(a)pyrene-derived radioactivity were 1.1 and 2.3 weeks at summer and winter water temperatures, respectively (Figure 1). Since these studies were actually conducted in summer and winter, it is possible that other seasonally varying factors could influence excretion rates. As with the lobster and blue crab, the hepatopancreas contained most of the dose remaining in the spiny lobster (6). Tail muscle (the major edible tissue) contained relatively small concentrations of ^{14}C, accounting for about 5% of the radiolabel left in the animal at any time. Excretion of ^{14}C from both hepatopancreas and tail muscle (Figure 2) closely paralleled excretion from the whole animal (Figure 1). The spiny lobster metabolized benzo(a)pyrene relatively rapidly in both seasons, with less than 5% of the hepatopancreas ^{14}C as parent benzo(a)pyrene at 3 days after the dose in either summer or winter (Figure 3). Ethyl acetate-extractable hepatopancreas metabolites (polar metabolites shown in Figure 3) were analyzed by reverse-phase high-performance liquid chromatography

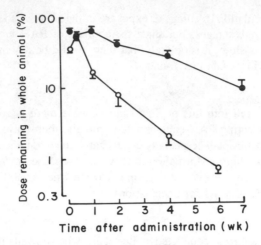

FIGURE 1. Effect of temperature on the elimination of benzo(a)pyrene (BaP) derived radioactivity by the spiny lobster, *Panulirus argus*. Doses of ^{14}C-benzo(a)pyrene (1 mg/kg) were administered intrapericardially, and animals were maintained in summer (○) or winter (●) when seawater temperatures were 26.5 to 29°C and 13.5 to 16.5°C, respectively. Results are mean ± S.E. for three or four lobsters at each time point.

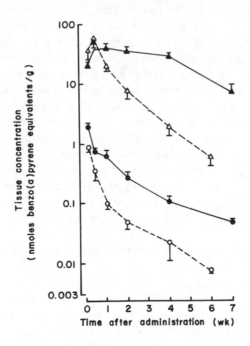

FIGURE 2. Depuration of radioactivity from the hepatopancreas (▲,△) and tail muscle (●,○) of spiny lobster at various times after intrapericardial injection of ^{14}C-benzo(a)pyrene. Lobsters were dosed and maintained during successive summer (broken line) and winter (solid line) seasons when seawater temperatures were 26.5 to 29°C and 13.5 to 16.5°C, respectively. Results are mean ± S.E. for three or four lobsters at each time point.

(HPLC) and found to coelute with the major vertebrate benzo(a)pyrene metabolites,[7] namely, 1-, 3-, 7-, and 9-hydroxybenzo(a)pyrenes, 1,6-, 3,6-, and 6,12-quinones, and 7,8-, 4,5-, and 9,10-dihydrodiols. Treatment of the nonextractable benzo(a)pyrene metabolites with sulfatase and β-glucuronidase also yielded phenols and dihydrodiols, indicating the presence of glycoside and sulfate conjugates. Some ^{14}C remained in the aqueous phase after hydrolysis

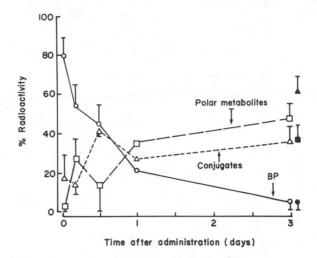

FIGURE 3. Analysis of ^{14}C-benzo(a)pyrene and metabolites in the hepatopancreas of spiny lobsters maintained at 26 to 29°C (open symbols) or 13.5 to 16.5°C (closed symbols). Results are expressed as mean ± S.E. of the percentage recovered radioactivity as benzo(a)pyrene, benzo(a)pyrene conjugates, or polar metabolites. The number of lobsters analyzed at each time point were 2 (1-h), 5 (4-h), 3 (12-h), 1 (1-d), 3 (3-d summer) and 3 (3-d winter).

with sulfatase and β-glucuronidase, and may include glutathione conjugates, but this has not yet been established. In all reverse-phase HPLC studies with hepatopancreas extracts it was very important to add unlabeled metabolite standards and to quantitate metabolites by the amount of radiolabel coeluting with the known metabolites, because endogenous components of hepatopancreas had unpredictable effects on the chromatographic separation: the order of elution of metabolites was always the same, but the presence of hepatopancreas extracts affected retention times. Similar proportions of polar metabolites and conjugates were found in hepatopancreas at each season. Thus, the slower excretion of benzo(a)pyrene metabolites by spiny lobster in winter was not due to slower biotransformation, but to less efficient elimination of metabolites.

A few studies have been conducted with zooplankton which serve as an important source of food to many larger marine animals. Lee showed that the copoped *Calanus plumchrus* contained unchanged benzo(a)pyrene and hydroxylated benzo(a)pyrene after exposure in filtered seawater.[8] In a recent elegant study, Reichert et al.[9] exposed two species of benthic deposit-feeding amphipods, *Rhepoxynius abronius* and *Eohaustorius washingtonianus*, to sediment-associated ^{3}H-benzo(a)pyrene for up to 7 d, and then analyzed the whole organism for metabolites by reverse- and normal-phase HPLC. Although similar concentrations of benzo(a)pyrene-derived radioactivity were present in each species after exposure for 3 or 7 d (see Figure 4), there were important differences in the extent of biotransformation and the metabolites found. *E. washingtonianus* contained 73 to 88% of its body burden as unchanged benzo(a)pyrene, and there was no significant decrease in unmetabolized benzo(a)pyrene with time, whereas *R. abronius* contained 30 to 51% as unchanged benzo(a)pyrene and the percentage unchanged decreased with time. Analysis of organic solvent-extractable metabolites after hydrolysis with β-glucuronidase-containing sulfatase showed that both species produced 7,8- and 9,10-dihydrodiols, quinones, and 3- and 9-hydroxybenzo(a)pyrene (Figures 5 and 6). *E. washingtonianus* (Figure 5), but not *R. abronius* (Figure 6), produced the 4,5-dihydrodiol. Covalent binding of benzo(a)pyrene intermediates to the amphipods macromolecules was also studied[9] and was significantly higher in *R. abronius* (Figure 7), consistent with the more rapid biotransformation of benzo(a)pyrene to potentially active metabolites.

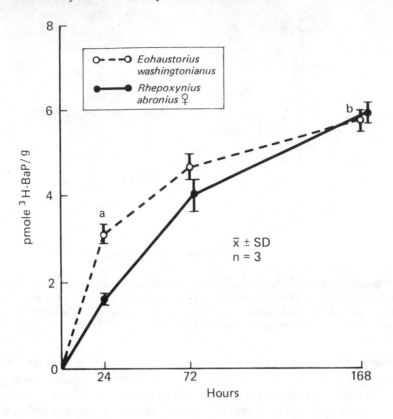

FIGURE 4.   Concentrations of ³H-benzo(a)pyrene derived radioactivity in amphipods. Points marked (a) are significantly different from *R. abronius* ($p < 0.05$), and points marked (b) are significantly different from the 24-h point ($p < 0.01$).

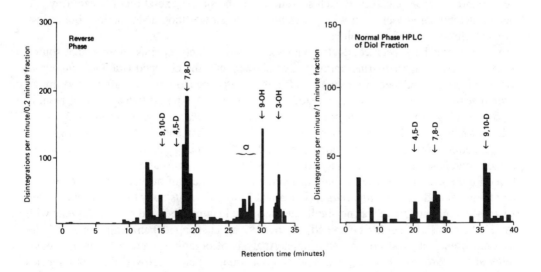

FIGURE 5.   Analysis of ³H-benzo(a)pyrene metabolites released after treatment of cell-free homogenates of *Eohaustorius washingtonianus* with β-glucuronidase and arylsulfatase. The panel on the left shows reverse phase HPLC of the organic extract after hydrolysis. The panel on the right shows normal phase HPLC analysis of the diol fraction collected from reverse-phase HPLC, and shows a clear separation of the 4,5- and 7,8-dihydrodiols. Abbreviations: 9,10-D, BaP-9,10-diol; 4,5-D, BaP-4,5-diol; 7,8-D, BaP-7,8-diol; Q, quinones (BaP-1,6- ; 3,6- ; and 6,12-quinones); 9-OH, 9-hydroxyBaP; and 3-OH, 3-hydroxyBaP.

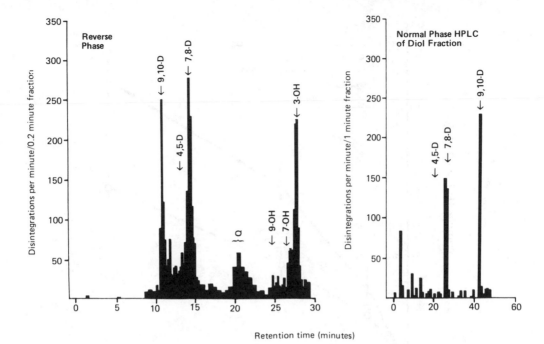

FIGURE 6. Analysis of ³H-benzo(a)pyrene metabolites released after treatment of cell-free homogenates of *Rhepoxynius abronius* with β-glucuronidase and arylsulfatase. The left panel shows reverse-phase HPLC of the organic extract after hydrolysis. The right panel shows normal phase HPLC analysis of the diol fraction collected from reverse phase HPLC. Abbreviations 9,10-D, BaP-9,10-diol; 4,5-D, BaP-4,5-diol; 7,8-D, BaP-7,8-diol; Q, quinones (BaP-1,6- ; 3,6- ; and 6,12-quinones); 9-OH, 9-hydroxyBaP; 7-OH, 7-hydroxyBaP; and 3-OH, 3-hydroxy-BaP.

Studies with the larvae of an aquatic insect, the midge, *Chironomus riparius,* showed that the larval midges could take up benzo(a)pyrene from the water and metabolize benzo(a)pyrene fairly rapidly.[10] Metabolites identified from the midge larvae and the water they were maintained in were 3- and 7-hydroxybenzo(a)pyrene, and 9,10- and 7,8-dihydro-diols of benzo(a)pyrene. There was evidence that metabolites were excreted slowly from these larval midges.

Overall, studies show that all of the arthropods studied to date can biotransform benzo(a)pyrene to potentially toxic metabolites, but that there are important species differences in the rates of formation of metabolites and the position of initial monooxygenation.

2. Other PAH

Indirect evidence that the copepod, *Calanus helgolandicus,* could metabolize naphthalene was obtained by Corner et al.,[11] but metabolites were not identified. In the crab, *Maia squinado,* Corner et al. identified several naphthalene metabolites in urine, including 1,2-dihydroxy-1,2-dihydronaphthalene and its glucoside, 1-naphthylsulfate, and 1-naphthylglucoside.[12]

Adult and larval spot shrimp, *Pandalus platyceros,* exposed in seawater to ³H or ¹⁴C-labeled naphthalene, were shown by HPLC analysis to contain metabolites in thoracic segments.[13] 1-Naphthol was a major metabolite in larvae and adults, but larvae contained more naphthyl sulfate than adults. Other metabolites found were naphthylglycoside (adults only), naphthylglucuronide, naphthoquinone, and naphthalene 1,2-dihydrodiol. There was a possibility that some of the 1,2-dihydrodiol arose from microbial metabolism of the naphthalene in seawater, since this metabolite was found when naphthalene was added to filtered seawater.[13]

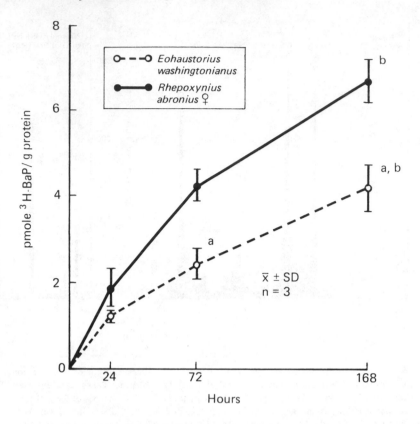

FIGURE 7.   Covalent binding of ^{3}H-benzo(a)pyrene intermediates to amphipod macromolecules. Points marked (a) are significantly different from *Rhepoxynius abronius* (p <0.01), and points marked (b) are significantly different from the 24-h point.

The fates of ^{14}C-labeled methylnaphthalene, naphthalene, and fluorene were studied in the blue crab, *Callinectes sapidus,* after feeding spiked food.[3] For each PAH, the major site of uptake of radiolabel was hepatopancreas, and in each case, polar metabolites, tentatively identified as hydroxy and dihydroxy derivatives and their conjugates, were found. Tritiated methylcholanthrene and naphthalene were taken up from seawater by copepods, amphipods, and crab zoea.[8] In copepods, naphthalene was extensively metabolized to naphthol, whereas methylcholanthrene was poorly metabolized to several metabolites including hydroxymethylcholanthrene.[8]

The metabolism of ^{14}C-phenanthrene was studied in the Norway lobster, *Nephrops norvegicus,* after force-feeding.[14,15] The hepatopancreas was not analyzed, but the green gland, gonads, and intestines were found to contain 9,10-dihydro-9,10-dihydroxyphenanthrene as the major metabolite, with lesser amounts of 1,2-dihydro-1,2-dihydroxy and 1-, 2-, and 9-hydroxyphenanthrenes.

The disposition of ^{14}C-phenanthrene has also been studied in the blue crab, *C. sapidus,* after force feeding sexually mature male or female crabs with ground mussels which had been exposed in water to ^{14}C-phenanthrene.[16] Differences in elimination rate constants for parent phenanthrene were found between males and females, such that elimination half-lives of phenanthrene were 9.9 and 4.5 h for males and females, respectively. Elimination of phenanthrene was due mainly to metabolism, and phenanthrene metabolites were excreted slowly in both sexes. The major phenanthrene metabolites found were dihydrodiols, especially 9,10-dihydrodiol, with little or no evidence for monohydroxylated metabolites.[17]

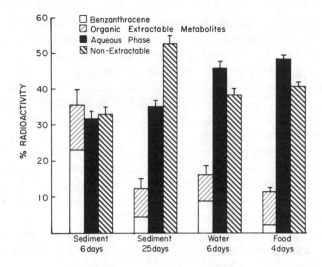

FIGURE 8. Effect of route of administration of the disposition of ^{14}C-benzanthracene in *Nereis virens*. Worms were exposed for 6 or 25 d to ^{14}C-benzanthracene mixed in the sediment, for 6 d to ^{14}C-benzanthracene in the water column, or for 4 d to a diet of ^{14}C-benzanthracene-spiked food. Worms were then homogenized and extracted first with organic solvents and then water. The organic phase was concentrated and analyzed by reverse phase HPLC, which separated unchanged benzanthracene from polar metabolites (mainly diols, tetrols, and diol-epoxides).

B. Annelids

Very few *in vivo* studies have examined the fate of PAH in annelids. Some evidence that the polychaete, *Neanthes arenaceodentata* could metabolize naphthalene was presented,[18] but there was no convincing evidence that the lugworm, *Abarenicola pacifica,* could metabolize phenanthrene, chrysene, or benzo(a)pyrene.[19] An extensive study of the ability of the polychaete *Nereis virens* to metabolize benz(a)anthracene was conducted by McElroy.[20,21] Worms were exposed to ^{14}C-benzanthracene in sediments, the water column, or food for varying periods, and then extracted with organic solvents and water. The organic extract was shown by HPLC to contain parent benz(a)anthracene and polar metabolites which comigrated with known diols, tetrols, and diol-epoxides of benz(a)anthracene. The aqueous phase was assumed to contain conjugates. About 20% of the radioactivity in the aqueous phase was sensitive to hydrolysis by a mixture of β-glucuronidase and sulfatase, but the remaining aqueous phase radioactivity was uncharacterized: it was speculated that this radioactivity may consist of glutathione or other peptide conjugates. In addition, a considerable amount of the ^{14}C was not extractable into organic solvents or water, and was presumed to be bound to macromolecules. Figure 8 shows the effect of route of administration and duration of exposure on the distribution of benzanthracene-derived radioactivity between organic solvent, water, and a nonextractable pool.[21]

C. Mollusks

Although many studies have shown that mollusks (mainly mussels, clams, and oysters) living in PAH-polluted environments attain very high body burdens of PAH (see Reference 1 for detailed discussion), there have been fewer studies designed to determine PAH metabolism by mollusks.

The results obtained from *in vivo* studies have been somewhat conflicting. The mussel,

Mytilus edulis, was shown to bioconcentrate radiolabeled benzo(a)pyrene, naphthalene, toluene, and aliphatic hydrocarbons from seawater, but no metabolites were found in mussel tissue.[22] More recently, the horse mussel, *Modiala modiolus* was shown to contain no metabolites of ^{14}C-phenanthrene after exposure in seawater.[23] In oysters, *Ostrea edulis,* exposed to ^{14}C-naphthalene, 1- and 2-naphthols were found in extracts of tissue digests.[24] It was, however, shown that these naphthol metabolites were also produced microbially, and the authors thought that there was insufficient evidence to state that the metabolites found were produced by the bivalve.[24]

It is possible that because of the large volumes of water filtered through mollusks continually, water-soluble metabolites would be rapidly excreted and, therefore, not detected in tissues. The primary metabolites of the larger multiring PAH are not, however, very water soluble, and it might be expected that if metabolites are formed, small amounts could be detected using very sensitive radiochemical techniques.

Since mollusks can accumulate high concentrations of parent hydrocarbon but contain negligible concentrations of metabolites, it appears that if mollusks do metabolize PAH, this metabolism is very slow.

D. Other Invertebrates

Very little information is available on the *in vivo* fate of PAH in other invertebrates, so the remaining studies are discussed in this section.

Malins and Roubal showed that marine algae, *Fucus distichus,* accumulated, but did not metabolize ^{3}H-2,6-dimethylnaphthalene,[25] but that an echinoderm, the sea urchin, *Strongylocentrotus droebachiensis,* fed with *Fucus* containing labeled 2,6-dimethylnaphthalene, metabolized the dimethylnaphthalene to 1- and 3-hydroxy-2,6-dimethylnaphthalenes, which were then converted to sulfates.[25] There was no evidence for hydroxylation of the methyl groups. By contrast, hydroxymethyl metabolites of 2,6-dimethylnaphthalene and other methylated aromatic hydrocarbons are major metabolites in vertebrate species.[26,27] Indirect evidence that the sea urchin, *S. purpuratus* could convert benzo(a)pyrene to genotoxic metabolites was that eggs taken from sea urchins injected with benzo(a)pyrene had lower fertilization rates than eggs from untreated sea urchins.[28] Although metabolism was not studied,[28] in all other species PAH must be biotransformed to active metabolites before causing genotoxicity.

Lindmark showed that a protozoan, the marine ciliate, *Parauronema acutum,* was unable to convert benzo(a)pyrene or benzanthracene to mutagens,[29] but radiolabeled compounds were not used, and no attempts were made to identify metabolites. Solbakken et al. studied the fate of ^{14}C-labeled naphthalene, phenanthrene, and halogenated hydrocarbons in several coral species, representatives of the phylum *Cnidaria,*[30] and found that depuration of the hydrocarbons was very slow. Metabolite production was not studied. The fate of [3]H- or ^{14}C-benzo(a)pyrene was studied in a member of the phylum *Porifera,* the marine sponge, *Tethya lyncurium.*[31] In live sponges exposed to light, the radiolabel was found in DNA, RNA, and protein fractions; little or no binding to macromolecules was found in dead sponges or live sponges kept in the dark. The results suggested that benzo(a)pyrene accumulated by the sponge was converted nonenzymatically (i.e., photochemically) to mutagens, which then bound to macromolecules.

There are many difficulties inherent in conducting *in vivo* studies of xenobiotic metabolism, particularly when working with compounds such as PAH, which are capable of photochemical degradation, in species which have inherently low metabolic capacity such as most invertebrates. Results obtained suggest that higher invertebrates, such as echinoderms, arthropods, and annelids, can metabolize PAH, whereas lower invertebrates, such as protozoa, cnidaria, and porifera, cannot. The ability of mollusks to metabolize PAH is still questionable and may vary between species in this phylum.

III. *IN VITRO* METABOLISM OF PAH

The most important enzymes for primary biotransformation of PAH are the cytochrome P-450 family of heme proteins, which catalyze the monooxygenation of many organic molecules. Monooxygenation of PAH results in formation of epoxides, or arene oxides, and phenols. Epoxide or arene oxides are further metabolized by epoxide hydrolase to dihydrodiols, or by glutathione (GSH) S-transferases to glutathione conjugates, while phenols are further metabolized by UDP-glycosyltransferases or PAP-sulfotransferases to β-D-glycosides or to sulfates. Dihydrodiols may be excreted without further biotransformation, or they may be excreted as glycoside or sulfate conjugates. Glutathione conjugates are further metabolized by peptidases and are usually excreted as N-acetylcysteine conjugates. A further discussion of enzymes involved in PAH biotransformation may be found in Reference 32 and in Chapter 4. Although reference will be made to conjugating enzymes and epoxide hydrolase, this section will focus on studies of the primary metabolism of PAH, catalyzed by the cytochrome P-450 system.

The rationale for conducting studies of PAH metabolism *in vitro* is that such studies are considerably less expensive than *in vivo* studies, and, moreover, that detailed information can be gained about the importance of different enzymes in the metabolism of PAH and the nature of the primary metabolites. Indeed, metabolite identification from *in vitro* studies is usually easier and results are less subject to ambiguous interpretation than results from *in vivo* studies (e.g., are the metabolites formed by the subject organism or by microbes in seawater?). In addition, it is usually easier to conduct studies of binding to macromolecules or activation to mutagens with *in vitro* preparations than by using the whole organism. Nevertheless, the results of *in vitro* studies must be interpreted with caution, especially since *in vitro* preparations of some invertebrate organs contain inhibitors of xenobiotic metabolizing enzymes, as discussed below.

Another general consideration is that many published studies of benzo(a)pyrene metabolism in invertebrates have not reported qualitative or quantitative data on the individual metabolites formed, or even the total metabolites, but have rather measured production of hexane- and alkali-soluble fluorescent metabolites, mainly, 3-hydroxybenzo(a)pyrene.[33] This method (measurement of fluorescent metabolites) is commonly referred to, and will be in this discussion, as aryl hydrocarbon hydroxylase (AHH) activity. For the purpose of this review studies of AHH activity will be discussed only if no other information is available for a particular species or phylum.

A. Arthropods

Of the aquatic invertebrate phyla, the most detailed studies of PAH metabolism *in vitro* have been conducted with arthropods, especially larger crustacea. The crustacean hepatopancreas, or digestive gland, is functionally similar to the vertebrate liver.[33] Since the liver microsomal fraction is quantitatively the most important primary site of biotransformation (especially monooxygenation) of PAH and other xenobiotics in vertebrates, and since *in vivo* studies have shown that hepatopancreas contains high concentrations of radiolabel after administration of PAH, it was logical to assume that the major site of PAH biotransformation (at least the first step, cytochrome P-450-dependent monooxygenation) *in vitro* would be hepatopancreas microsomes. It is now clear that hepatopancreas is the major site of cytochrome P-450-dependent monooxygenation of PAH and other xenobiotics, although initial studies with decapod crustacea (see below) suggested otherwise.

A number of studies of PAH metabolism *in vitro* in crabs and lobsters showed that preparations of hepatopancreas has lower monooxygenase activity than preparations of other organs such as stomach, gills, green glands, and gonads.[35-39] The first studies of cytochrome P-450-dependent monooxygenase activity in hepatopancreas of the lobster, *Homarus amer-*

icanus, showed the presence of cytochrome P-450, but undetectable monooxygenase activity.[40,41] It was shown that secretions from lobster hepatopancreas inhibited monooxygenase activity.[36] Similar results were obtained for hepatopancreas of the spiny lobster, *Panulirus argus.*[42] Studies by James et al.[42-44] showed that spiny lobster and blue crab hepatopancreas microsomes could support monooxygenase activity with several substrates, including benzo(a)pyrene, if an organic hydroperoxide or other oxidizing agent were used instead of NADPH, thereby bypassing the need for NADPH cytochrome P-450 reductase. In further studies it was shown that one effect of the inhibitors present in hepatopancreas microsomes and other subcellular fractions was to inhibit NADPH cytochrome P-450 reductase, possibly by proteolysis.[42,45] The inhibitors present in lobster and spiny lobster hepatopancreas microsomes could be removed by a washing procedure: when the inhibitor-free microsomes were mixed with active NADPH cytochrome P-450 reductase from any of several species the ability of the crustacean cytochrome P-450 to catalyze monooxygenation could be studied.[46-49] Similar inhibitors of monooxygenase activity have been reported in hepatopancreas microsomes from other decapod crustacea, including the blue crab, *Callinectes sapidus,*[37] and the spiny crab, *Maja crispata.*[39] There is, however, no evidence for monooxygenase inhibitors in some other crustacea which have been studied, including the barnacle and two copepod species.[50,51]

Other concerns with some of the published studies of PAH metabolism (especially AHH activity) in tissues of crustacea are that assays were carried out under nonoptimal incubation conditions: for example, incubation times of 1 h were commonly used without determining if metabolite formation was linear for this long. It is quite likely that in 1 h phenolic metabolites of benzo(a)pyrene, such as 3-hydroxybenzo(a)pyrene, may be formed and further metabolized by additional oxidation, or by conjugation.

Thus, reported rates of PAH metabolism in preparations of hepatopancreas or digestive gland from some decapod crustacea may not accurately represent the *in vivo* situation.

The benzo(a)pyrene primary metabolites formed by reductase-fortified microsomal fractions from hepatopancreas of decapod crustacea or unfortified microsomes from a cirriped crustacean are presented in Table 1. Large differences were observed in the overall rates of metabolism per gram of hepatopancreas, which reflect the *in vivo* rates of benzo(a)pyrene metabolism for the respective species, where this has been studied. Thus, spiny lobster has higher rates of benzo(a)pyrene metabolism than lobster or blue crab, as is found *in vivo.*[3,5,6] It should also be noted that the rates shown in Table 1 were measured at 30°C, which is considerably higher than the normal environmental temperature for lobster: at lower *in vitro* incubation temperatures, lower rates were found.[49] The major crustacean metabolites of benzo(a)pyrene (Table 1) are those found in vertebrate species.[7] In the spiny lobster, several minor metabolites were found which did not correspond to any of the known available benzo(a)pyrene metabolites, but these metabolites have not been identified.[52] In preparations from the lobster, blue crab, and barnacle, the major metabolites were phenolic, whereas dihydrodiols predominated in preparations from the spiny lobster (Table 1). This may reflect differences in epoxide hydrolase activity between these species.[53] It was also noteworthy (Table 1) that the selective ''P-448'' inhibitor, α-naphthoflavone,[54] inhibited benzo(a)pyrene monooxygenase activity in the blue crab and barnacle, but stimulated activity with the spiny lobster and lobster, suggesting differences in the properties of the major constitutive cytochromes P-450 between these species. In a study of benzo(a)pyrene metabolism by stomach microsomes, Singer et al.[38] found that phenols, primarily the 3-hydroxy- and 9-hydroxy-benzo(a)pyrenes, were the major metabolites, and dihydrodiols, tentatively identified as 4,5- and 7,8-dihydrodiols, were minor metabolites. Benzo(a)pyrene metabolism in stomach tissue from the blue crab was also inhibited by α-naphthoflavone.[38] Other studies of benzo(a)pyrene metabolism in crustacea have measured only formation of fluorescent metabolites, commonly

Table 1

BENZO(a)PYRENE METABOLISM IN HEPATOPANCREAS MICROSOMES FROM SEVERAL CRUSTACEAN SPECIES

| Species | Sex | Cytochrome P-450 content (nmol/mg protein) | Benzo(a)pyrene metabolism (pmol/min/mg protein)[a] | | | | | | | Effect of α-naphthoflavone | Overall metabolism (nmol/min/g HP) |
| | | | Dihydrodiols | | | Quinones | Phenols | | | | |
			9,10-	4,5-	7,8-		I	II	Total		
Spiny lobster[b]	Male	2.08	161	132	106	322	126	183	1343	Stimulates	14.16
	Female	1.95	220	119	159	283	161	61	1250	Stimulates	11.88
Lobster[c]	Male	0.55	7	26	20	86	8	89	306	Stimulates	0.44
	Female	0.46	6	17	12	69	9	130	288	Stimulates	0.37
Blue crab[d]	Female	0.32	13	5	137	49	106	138	526	Inhibits	0.35
Barnacle[e]	—	0.11	8	—	15	1	21	64	118	Inhibits	0.06

[a] Metabolites were identified by measuring the ^{14}C in fractions corresponding to the named metabolites. Phenols I includes 7- and 9-hydroxybenzo(a)pyrenes and Phenols II includes 3- and 1-hydroxybenzo(a)pyrenes. All assays were conducted at 30°C. Assays with spiny lobster, lobster, and blue crab were in the presence of added NADPH cytochrome P-450 reductase from rat or pig.

[b] Data from References 48 and 80.

[c] Data from Reference 49.

[d] James and Little.[81]

[e] Calculated from data in Reference 50.

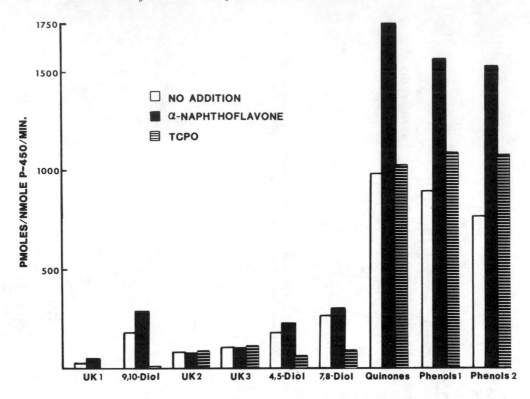

FIGURE 9.   Metabolism of benzo(a)pyrene by partially purified cytochrome P-450 from spiny lobster hepatopancreas microsomes. The spiny lobster cytochrome P-450 was mixed with NADPH-cytochrome P-450 reductase
purified from pig liver, and the mixture was incubated for 5 min with ^{14}C-benzo(a)pyrene and NADPH. Metabolites
are shown in their order of elution from reverse-phase (C_{18}) HPLC. The "UK" metabolites did not correspond to
any of the available benzo(a)pyrene metabolite standards. Phenols 1 contained 9-hydroxy- and 7-hydroxy-
benzo(a)pyrenes, and phenols 2 contained 1-hydroxy- and 3-hydroxy-benzo(a)pyrenes. Incubations were carried
out in the presence and absence of α-naphthoflavone (ANF) or trichloropropene oxide (TCPO).

termed AHH activity, without showing that fluorescent metabolites (primarily 3-hydroxy-
benzo[a]pyrene) are the major metabolites of benzo(a)pyrene in these species.[55,56] In native
spiny crab hepatopancreas microsomes, prepared in the presence of proteinase and lipase
inhibitors but without added NADPH cytochrome P-450 reductase, AHH activity was 6.6
pmol/min/mg protein.[55]

Further studies were done to purify the spiny lobster hepatopancreas microsomal cytochrome P-450 and study monooxygenase activity of the purified forms.[57] One form of
cytochrome P-450 was purified to a specific content of 12 ± 1 nmol/mg protein, but this
preparation still contained a small amount of epoxide hydrolase. This partially purified
cytochrome was reconstituted with pig liver NADPH cytochrome P-450 reductase and di-
lauroylphosphatidyl choline, and used to study monooxygenase activity with benzo(a)pyrene.
Figure 9 shows the metabolites found by HPLC analysis, shown in their relative order of
elution from a reversed-phase (C_{18}) column. Higher activity was found with the purified
cytochrome than with the microsomes (cf. Figure 9 with Table 1). Benzo(a)pyrene monooxygenase activity was also studied in the presence of α-naphthoflavone or trichloropropene
oxide (TCPO). TCPO is a potent inhibitor of microsomal epoxide hydrolase and was included
to determine if the small amounts of dihydrodiols formed by the purified P-450 were due
to contamination of the preparation with epoxide hydrolase. Dihydrodiol formation was
inhibited by TCPO, confirming that the cytochrome P-450 preparation contains small amounts

of epoxide hydrolase. The unknown metabolite UK1, which eluted just before benzo(a)pyrene 9,10-dihydrodiol, was not formed in the presence of TCPO, suggesting that this metabolite was also a dihydrodiol. As was found with microsomes (Table 1), α-naphthoflavone stimulated overall metabolism.

There are very few *in vitro* studies with PAH other than benzo(a)pyrene. Singer et al. showed that stomach microsomes from the blue crab could metabolize phenanthrene, benz(a)anthracene, dimethylbenz(a)anthracene, and chrysene as well as benzo(a)pyrene to phenolic and dihydrodiol products.[38] The highest turnover was reported with dimethylbenz(a)anthracene as substrate.[38]

The question of whether or not PAH metabolism can be induced in crustaceans by exposure to chemicals which induce cytochrome P-450 and other xenobiotic-metabolizing enzymes is still not definitively answered. To the author's knowledge, there has been only one attempt to induce crustacea with phenobarbital.[38] In this study, phenobarbital injection had no effect on AHH activity in blue crab stomach microsomes. There have been few studies with other inducing agents, and the results of those studies are not clear-cut. There was no evidence for induction by 3-methylcholanthrene of benzo(a)pyrene metabolism or ethoxyresorufin *O*-deethylase activity in reductase-fortified preparations of hepatopancreas microsomes from the spiny lobster.[48] Only one dose and time point was studied, however, so it is possible that the correct conditions were not used. In blue crabs treated with benz(a)anthracene there was no increase in stomach microsomal AHH activity. There was, however, some evidence that marsh crabs, *Sesarma cinerum* and fiddler crabs, *Uca ripax* exposed to fuel oils in the environment had higher cytochrome P-450 contents in hepatopancreas microsomes than crabs from uncontaminated areas.[58,59] In the spiny crab, *Maja crispata,* treatment with benzo(a)pyrene increased AHH activity in hepatopancreas and stomach microsomes,[39] but it was not determined if benzo(a)pyrene metabolism at positions other than the 3-hydroxy position was affected. In addition, the hepatopancreas activity was well below optimal,[39] as reported by the same workers in Reference 55. Small increases in AHH activity were reported in *Calanus helgolandicus* microsomes after exposure to naphthalene, 2-methylnaphthalene, 3-methylcholanthrene, and benzo(a)pyrene.[51] Again, only production of fluorescent metabolites was measured,[51] so it was possible that overall benzo(a)pyrene metabolism was affected differently. Further studies are needed in this area.

B. Annelids

Studies of the monooxygenase system of the polychaete worms, *Nereis virens* and *Capitella capitata* have been carried out by Lee and co-workers.[58-60] In microsomes prepared from *C. capitata,* AHH activity was not detectable, but if worms were exposed to crude oil or to benz(a)anthracene, low activity (1 to 10 pmol/min/mg microsomal protein) could be detected.[59] It was not shown that 3-hydroxybenzo(a)pyrene was the major metabolite in *C. capitata,* but in experiments with *N. virens,* positional metabolism of benzo(a)pyrene was studied by reverse-phase HPLC, and 3-hydroxybenzo(a)pyrene was shown to be the major metabolite.[60] Other *N. virens* metabolites were benzo(a)pyrene 4,5- and 7,8-dihydrodiols, 6- and 9-hydroxybenzo(a)pyrenes, and quinones.[60] AHH activity was detectable in untreated *N. virens* and was 25 ± 15 pmol/min/mg microsomal protein.[58] Benzo(a)pyrene treatment of *N. virens* did not alter AHH activity or cytochrome P-450 content, but did alter the ability of microsomes to bind SKF-525A.[58] Treatment of *N. virens* with Aroclor® 1254 did increase AHH activity as well as produce alterations in the ability of microsomes to bind SKF-525A and benzphetamine.[58]

C. Mollusks

The literature contains conflicting reports on the existence and activity of PAH-metabolizing systems in mollusks. Some investigators reported that marine bivalve mollusks (clams,

Mercenaria mercenaria; mussels, *Mytilus edulis;* and oysters, *Crassostrea virginica*) had undetectable AHH activity.[61,62] Others showed that digestive gland microsomes from the bivalve mollusk, *M. edulis,* cockle mussel, *Cardium edule,* and the gastropod mollusk periwinkle, *Littorina littorea* contained cytochromes P-450 and b_5, and low NADPH cytochrome *c* reductase and AHH activities.[63,64] Recent detailed studies of benzo(a)pyrene monooxygenase in the mussel, *M. edulis,* and two other bivalve species have been performed by Stegeman with ^{3}H-benzo(a)pyrene as substrate.[65] These studies showed that benzo(a)pyrene monooxygenase activity was present in the digestive gland (most activity), gill, foot, and gonad. Even in the digestive gland overall benzo(a)pyrene monooxygenase activity was low (9 to 11 pmol/min/mg microsomal protein) in mussels, calico clams, *Macrocallista maculata,* and Bermuda mussels, *Arca zebra.*[65] In other studies with radiolabeled benzo(a)pyrene as substrate low benzo(a)pyrene monooxygenase activity (2 to 4 pmol/min/mg homogenate protein) was found in the digestive glands of oysters, *C. virginica,* and clams, *Mercenaria mercenaria.*[66,67] Studies of the positional metabolism of benzo(a)pyrene in digestive gland microsomes from the mussel *M. edulis* showed that benzo(a)pyrene quinones (1,6-, 3,6-, and 6,12-) accounted for 65% of total metabolites, phenols for 30%, and dihydrodiols for 3% of total metabolites.[65] Despite being able to metabolize benzo(a)pyrene, enzymes from clams showed little ability to activate benzo(a)pyrene or 3-methylcholanthrene to mutagens in the Ames bacterial mutagenesis assay.[67,68] Similarly, postmitochondrial fractions from the digestive gland of the mussel, *M. edulis* could not transform benzo(a)pyrene to bacterial mutagens.[68] Possibly, the rate of formation of potentially mutagenic metabolites of PAH was too slow to induce mutation of bacteria in the time frame of the Ames test. It was interesting, although not strictly relevent to the present discussion, that mussels and clams activated aromatic amines to mutagens,[67-69] and that there was direct evidence for FAD-containing monooxygenase enzyme activity in mussel digestive gland microsomes.[62]

Evidence for seasonal variation and, possibly, for induction of the cytochrome P-450 system in mollusks has been reported.[63-65] Livingstone and Farrar showed that exposure of mussels to diesel oil in January resulted in increases in content of cytochromes P-450 and b_5 and in NADPH cytochrome *c* reductase activity, but no increase in AHH activity.[64] Periwinkles exposed to diesel oil in June had increased content of cytochrome P-450 and b_5, and increased NADPH cytochrome *c* reductase activity, but, again, AHH activity was not increased.[64] Similar rates of benzo(a)pyrene metabolism were found in digestive gland microsomes of *M. edulis* from hydrocarbon-polluted and clean areas, when the mussels were tested at the same time of year.[65] Fish from the same areas showed marked differences in hepatic microsomal benzo(a)pyrene monooxygenase activity, with those from the polluted area having higher activity.[65] There was, however, some evidence for seasonal variation in bivalve digestive gland monooxygenase activity.[64,65]

D. Other Invertebrates

There have been few direct studies of PAH metabolism in other invertebrate phyla. The ability of homogenate, postmitochondrial supernatant, or microsomes from several invertebrates to activate benzo(a)pyrene to bacterial mutagens, an indirect measure of metabolism, has been studied.[69,70] Members of certain invertebrate phyla such as protozoa, porifera, and mollusks were unable to convert benzo(a)pyrene to mutagens (although all could convert the arylamine, aminoanthracene, to mutagens).[69,70] Some, but not all, echinoderms were able to convert benzo(a)pyrene to bacterial mutagens in the Ames test, and all protochordates tested were able to convert benzo(a)pyrene to mutagens.[69]

IV. COMPARISON WITH OTHER SPECIES

A. Metabolite Profiles of Benzo(a)pyrene

Positional metabolism of benzo(a)pyrene has been studied in more species than any other

PAH. Among invertebrate species, benzo(a)pyrene primary metabolite profiles have been studied *in vitro* in several crustacean species (see Table 1), in the polychaete worm, *N. virens,*[60] and in two molluskan species, the clam and the oyster.[67] In all species studied, the major metabolites coeluted on reverse-phase and sometimes normal-phase HPLC with one or more of the known major vertebrate metabolite standards, and were then usually presumed to be identical to these known metabolites. In some species, minor, unidentified metabolites were also found (see previous section). Results showed that there was considerable diversity among the different invertebrate species in the pattern of benzo(a)pyrene primary metabolites formed. Thus, comparisons of invertebrates with other species must largely be done on a species-by-species basis. Some important metabolites for comparison are the benzo ring dihydrodiols, especially the 7,8-dihydrodiol, since this metabolite is a penultimate carcinogen and is more mutagenic than other benzo(a)pyrene metabolites.[71-77] All marine invertebrates studied formed the 7,8-dihydrodiol from benzo(a)pyrene, and in control and Aroclor®-treated clams and control oysters, this was the major benzo(a)pyrene metabolite.[67] In most of the other inverterates studied, the 7,8-dihydrodiol accounted for between 3 and 13% of the total metabolites. This is similar to the uninduced rat,[7] but differs from fish, where the 7,8-dihydrodiol usually accounts for 20 to 30% of the total metabolites.[2,78] It is not yet clear if continued exposure of invertebrates to PAH will induce synthesis of a form or forms of cytochrome P-450 which preferentially metabolize benzo(a)pyrene in the benzo ring, as is true with fish.[2] Since it is known that benzo(a)pyrene metabolites are excreted very slowly by invertebrates (see previous sections), it will be of interest to determine rates of excretion of the 7,8-dihydrodiol: if this metabolite is excreted slowly by invertebrates, shellfish may become a potential source of this proximate carcinogen, especially if eaten raw. The effect of cooking on seafood concentrations of benzo(a)pyrene and mutagenic metabolites has not, to the author's knowledge, been studied. Another metabolite of interest is the 4,5-dihydrodiol, since this metabolite is present in only small amounts, or is absent from incubations with fish liver preparations, but is an important mammalian metabolite.[2,7] Benzo(a)pyrene 4,5-dihydrodiol accounted for 6 to 10% of the total metabolites formed by spiny lobster and lobster hepatopancreas microsomes, but was not formed by the barnacle or untreated clams or oysters, and was only 1% of the total metabolites in the blue crab and mussel (see Table 1 and References 65 and 67). The *in vivo* study of Reichert et al.[9] showed that one amphipod species, *E. washingtonianus,* formed the 4,5-dihydrodiol, but another, *R. abronius,* did not.

B. Binding of PAH Metabolites to Macromolecules

There have been very few studies in this important area, and in no case have adducts of PAH metabolites to DNA, RNA, or protein been characterized. Two amphipod species, *E. washingtonianus* and *R. abronius,* which were exposed to benzo(a)pyrene contained benzo(a)pyrene-derived radioactivity bound to tissue macromolecules, and binding was higher in *R. abronius* which metabolized benzo(a)pyrene more rapidly (Figure 7).[9] It is of interest here that *R. abronius* is more sensitive to toxicants in polluted sediment than *E. washingtonianus:* this may well be related to the increased ability of *R. abronius* to metabolize PAH to reactive metabolites which can bind to tissue macromolecules.[9] Polychaete worms, *N. virens,* exposed to benz(a)anthracene, contained water- and organic solvent-insoluble radioactivity which was apparently bound to tissue.[20] Clearly, more work should be done in this area.

V. SUMMARY

A combination of *in vivo* and *in vitro* studies has shown that cytochrome P-450-dependent monooxygenation, the first step in PAH biotransformation, is usually slower in invertebrates than vertebrates, sometimes proceeding at rates so slow as to be undetectable by many standard techniques; and, furthermore, that PAH metabolites are excreted quite inefficiently by invertebrates. Although there was *in vivo* evidence for the presence of glycoside and sulfate conjugates of PAH primary metabolites, these pathways have not been studied *in vitro* in invertebrates. The presence of glutathione conjugates of primary PAH metabolites has not been directly demonstrated *in vivo*, although *in vitro* studies have shown the presence of glutathione S-transferases in invertebrates.[53] Since conjugation of primary PAH metabolites (especially glutathione conjugation) usually leads to a reduction in toxicity, more studies should be conducted to determine the roles of conjugation pathways in invertebrates.

Monooxygenation of PAH occurred most rapidly in higher invertebrates, such as arthropods, echinoderms, and annelids and very slowly or not at all in the more primitive invertebrates, such as protozoa, porifera, cnidaria, and mollusks (see Table 2). Where metabolism occurred, wide interspecies variations were observed in rates of biotransformation. It is not yet clear if the cytochrome P-450 system is not present in the lower invertebrates, or is simply not functional with the PAH as substrates. There was some evidence that in certain mollusks and annelids the cytochrome P-450 system was able to metabolize PAH only after induction by hydrocarbon mixtures or mixtures of polychlorinated biphenyls.[60,67] Other studies, however,[38,48] have failed to demonstrate that the cytochrome P-450 system of crustaceans can be induced by compounds of the 3-methylcholanthrene or tetrachlorodibenzo-*p*-dioxin (TCDD) type which induce PAH monooxygenation in vertebrate species. More information is needed on PAH biotransformation and the regulation of cytochrome P-450 in invertebrate species before clear conclusions can be drawn concerning the ability of a particular species to biotransform PAH.

A few studies have examined the influence of environmental temperature on PAH biotransformation, *in vivo* and *in vitro,* and on PAH metabolite excretion. While it is clear that PAH biotransformation proceeds *in vitro* more rapidly at higher temperatures,[42,45,49,50] there are insufficient studies to conclude that this is true *in vivo*. There is evidence that PAH metabolite excretion by the spiny lobster is more rapid at higher environmental seawater temperatures.[6] Further studies are needed to clearly delineate the effects of temperature on PAH biotransformation and excretion of PAH metabolites.

Another important area in which we have little or no information is the binding of PAH metabolites to important cellular macromolecules. Technology for studying this has improved in the past decade such that it is now possible to study adduct formation even in tissues with low capacity for biotransformation, such as extrahepatic mammalian tissues[79] and most invertebrates. Future studies of PAH disposition in invertebrates should attempt to determine the nature and extent of binding of PAH metabolites to DNA, RNA, and protein, and to correlate binding to tissue macromolecules with genotoxic or cytotoxic events.

Table 2
EXTENT OF PAH MONOOXYGENATION IN SOME AQUATIC INVERTEBRATE PHYLA

Phylum	Extent of PAH monooxygenation[a]			
	In vivo	Ref.	In vitro	Ref.
Protozoa	−	29	−	68
Porifera	−	31	−	67
Cnideria	−	8, 30	N.D.	
Annelida	++	2, 18—21	+	57—59
Mollusca	±	1, 9, 22—24	±	2, 60—67
Arthropoda	+ to +++++	2—6, 8, 10—17	+ to +++++	5, 35—52, 55—58
Echinodermata	++	25, 28	++	67
Protochordata	N.D.[b]		++	67

[a] Scale: −, no monooxygenation detected; +, very slow mcnooxygenation; +++++, rapid monooxygenation.
[b] N.D., not determined.

REFERENCES

1. **Mix, M. C.,** Polycyclic aromatic hydrocarbons in the aquatic environment: occurrence and biological monitoring, in *Reviews in Environmental Toxicology,* Vol. 50, Hodgson, E., Ed., Elsevier/North-Holland, Amsterdam, 1984, 51.
2. **Stegeman, J. J.,** Polynuclear aromatic hydrocarbons and their metabolism in the marine environment, in *Polycyclic Hydrocarbons and Cancer,* Vol. 3, Gelboin, H. V. and Ts'o, P. O. P., Eds., Elsevier, New York, 1981, 1.
3. **Lee, R. F., Ryan, C., and Neuhauser, M. L.,** Fate of petroleum hydrocarbons taken up from food and water by the blue crab, *Callinectes sapidus, Mar. Biol.,* 37, 363, 1976.
4. **Foureman, G. L., Ben-Zvi, S., Dostal, L., Fouts, J. R., and Bend, J. R.,** Distribution of ^{14}C-benzo(a)pyrene in the lobster, *Homarus americanus,* at various times after a single injection into the pericardial sinus, *Bull. Mt. Desert Isl. Biol. Lab.,* 18, 93, 1978.
5. **Bend, J. R., James, M. O., Little, P. J., and Foureman, G. L.,** *In vitro* and *in vivo* metabolism of benzo(a)pyrene by selected marine crustacean species, in *Phyletic Approaches to Cancer,* Dawe, C. J., et al., Eds., Japan Scientific Society Press, Tokyo, 1981, 179.
6. **Little, P. J., James, M. O., Pritchard, J. B., and Bend, J. R.,** Temperature-dependent disposition of ^{14}C-benzo(a)pyrene in the spiny lobster, *Panulirus argus, Toxicol. Appl. Pharmacol.,* 77, 325, 1985.
7. **Holder, G., Yagi, J., Dansette, P., Jerina, D. M., Levin, W., Lu, A. Y. H., and Conney, A. H.,** Effects of inducers and epoxide hydrase on the metabolism of benzo(a)pyrene by liver microsomes and a reconstituted system: analysis by high pressure liquid chromatography, *Proc. Natl. Acad. Sci. U.S.A.,* 71, 4356, 1974.
8. **Lee, R. F.,** Fate of petroleum hydrocarbons in marine zooplankton, in *Proc. 1975 Conf. on Prevention and Control of Oil Pollution,* American Petroleum Institute, Washington, D.C., 1975, 549.
9. **Reichert, W. L., Eberhart, B.-T. L., and Varanasi, U.,** Exposure of two species of deposit-feeding amphipods to sediment-associated ^{3}H-benzo(a)pyrene: uptake, metabolism and covalent binding to tissue macromolecules, *Aquat. Toxicol.,* 6, 45, 1985.
10. **Leversee, G. J., Giesy, J. P., Landrum, P. F., Gerould, S., Bowling, J. W., Fannin, T. E., Haddock, J. D., and Bartell, S. M.,** Kinetics and biotransformation of benzo(a)pyrene in *Chironomus riparius, Arch. Environ. Contam. Toxicol.,* 11, 25, 1982.
11. **Corner, E. D. S., Harris, R. P., Kilvington, C. C., and O'Hara, S. C. M.,** Petroleum compounds in the marine food web: short-term experiments on the fate of naphthalene in *Calanus, J. Mar. Biol. Assoc. U.K.,* 56, 121, 1976.
12. **Corner, E. D. S., Kilvington, C. C., and O'Hara, S. C. M.,** Qualitative studies on the metabolism of naphthalene in *Maia squinada* (Herbst), *J. Mar. Biol. Assoc. U.K.,* 53, 819, 1973.
13. **Sanborn, H. R. and Malins, D. C.,** The disposition of aromatic hydrocarbons in adult spot shrimp (*Pandalus platyceros*) and the formation of metabolites of naphthalene in adult and larval spot shrimp, *Xenobiotica,* 10, 193, 1980.
14. **Palmork, K. H. and Solbakken, J. E.,** Accumulation and elimination of radioactivity in the Norway lobster (*Nephrops norvegicus*) following intragastric administration of [9-^{14}C]phenanthrene, *Bull. Environ. Contam. Toxicol.,* 25, 668, 1980.
15. **Solbakken, J. E. and Palmork, K. H.,** Metabolism of phenathrene in various marine animals, *Comp. Biochem. Physiol.,* 70C, 21, 1981.
16. **Moese, M. D. and O'Connor, J. M.,** Phenanthrene kinetics in blue crab from dietary sources, *Mar. Environ. Res.,* 17, 254, 1985.
17. **Moese, M. D.,** personal communication, 1986.
18. **Rossi, S. S. and Anderson, J. W.,** Petroleum hydrocarbon resistance in the marine worm *Neanthes arenaceodentata, Mar. Biol.,* 39, 51, 1977.
19. **Augenfeld, J. M., Anderson, J. W., Kiesser, S. L., Fellingham, G. W., Riley, R. G., and Thomas, B. L.,** Exposure of *Abarenicola pacifica* to oiled sediment: effects on glycogen content and alterations in sediment-bound hydrocarbons, in *Proc. 1983 Oil Spill Conf. (Prevention, Behavior, Control, Cleanup),* American Petroleum Institute, Washington, D.C., 1983, 443.
20. **McElroy, A. E.,** *In vivo* metabolism of benz(a)anthracene by the polychaete *Nereis virens, Mar. Environ. Res.,* 17, 133, 1985.
21. **McElroy, A. E.,** Benz(a)anthracene in Benthic Marine Environments: Bioavailability, Metabolism and Physiological Effects on the Polychaete *Nereis virens,* Ph.D. thesis, Woods Hole, Oceanographic Institution and Massachusetts Institute of Technology, Woods Hole, 1985.
22. **Lee, R. F., Sauerheber, R., and Benson, A. A.,** Petroleum hydrocarbons: uptake and discharge by the marine mussel *Mytilus edulis, Science,* 177, 344, 1972.
23. **Palmork, K. H. and Solbakken, J. E.,** Distribution and elimination of [9-^{14}C]-phenanthrene in the horse mussel *(Modiola modiolus), Bull. Environ. Contam. Toxicol.,* 26, 196, 1981.

24. **Riley, R. T., Mix, M. C., Schaffer, R. L., and Bunting, D. L.,** Uptake and accumulation of naphthalene by the oyster *Ostrea edulis,* in a flow-through system, *Mar. Biol.,* 61, 267, 1981.

25. **Malins, D. C. and Roubal, W. T.,** Aryl sulfate formation in sea urchins (*Strongylocentrotus droebachiensis*) ingesting marine algae (*Fucus distichus*) containing 2,6-dimethylnaphthalene, *Environ. Res.,* 27, 290, 1982.

26. **Kaubisch, N., Daly, J. W., and Jerina, D. M.,** Arene oxides and intermediates in the oxidative metabolism of aromatic compounds. Isomerization of methyl-substituted arene oxides, *Biochemistry,* 11, 3080, 1972.

27. **Cavalieri, E., Roth, R., and Rogan, E.,** Hydroxylation and conjugation at the benzylic carbon atom: a possible mechanism of carcinogenic activation for some methyl-substituted aromatic hydrocarbons, in *Polynuclear Aromatic Hydrocarbons,* Jones, P. W. and Leber, P., Eds., Ann Arbor Science Publishers, Ann Arbor, MI, 1979, 517.

28. **Hose, J. E. and Puffer, H. W.,** Cytologic and cytogenic anomalies induced in purple sea urchin embryos (*Stongylocentrotus purpuratus S*) by parental exposure to benzo(a)pyrene, *Mar. Biol. Lett.,* 4, 87, 1983.

29. **Lindmark, D. G.,** Activation of polynuclear aromatic hydrocarbons to mutagens by the marine ciliate, *Parauronema acutum, Appl. Environ. Microbiol.,* 41, 1238, 1981.

30. **Solbakken, J. E., Knap, A. H., Sleeter, T. D., Searle, C. E., and Palmork, K. H.,** Investigation into the fate of ^{14}C-labelled xenobiotics (naphthalene, phenanthrene, 2,4,5,2',4',5'-hexachlorobiphenyl, oc-tachlorostyrene) in Bermudian corals, *Mar. Ecol. Prog. Ser.,* 16, 149, 1984.

31. **Zahn, R. K., Kurelec, B., Zahn-Daimler, G., Wuller, W. E. G., Rijavec, M., Batel, R., Given, R., Pondeljak, V., and Beyer, R.,** The effect of benzo(a)pyrene on sponges as model organisms in marine pollution, *Chem. Biol. Interact.,* 39, 205, 1982.

32. **Jakoby, W. B., Ed.,** *Enzymatic Basis of Detoxication,* Vol. 1 and 2, Academic Press, New York, 1982.

33. **Nebert, D.,** Genetic differences in microsomal electron transport: the Ah locus, in *Methods in Enzymology,* Fleischer, S. and Packer, L., Eds., Vol. 52, Academic Press, New York, 1978, 226.

34. **Vonk, H. J.,** Digestion and metabolism, in *The Physiology of Crustacea,* Vol. 1, Waterman, T. H., Ed., Academic Press, New York, 1960, 291.

35. **Khan, M. A. Q., Coello, W., Khan, A. A., and Pinto, H.,** Some characteristics of the microsomal mixed-function oxidase in the freshwater crayfish, *Cambarus, Life Sci.,* 11, 405, 1972.

36. **Pohl, R. J., Bend, J. R., Guarino, A. M., and Fouts, J. R.,** Hepatic microsomal mixed-function oxidase activity of several marine species from coastal Maine, *Drug Metab. Dispos.,* 2, 545, 1974.

37. **Singer, S. and Lee, R. F.,** Mixed function oxidase activity in the blue crab, *Callinectes sapidus:* tissue distribution and correlation with changes during molting and development, *Biol. Bull.,* 153, 377, 1977.

38. **Singer, S. C., March, P. E., Gonsoulin, F., and Lee, R. F.,** Mixed function oxygenase activity in the blue crab *Callinectes sapidus:* characterization of enzyme activity from stomach tissue, *Comp. Biochem. Physiol.,* 65C, 129, 1980.

39. **Bihari, N., Batel, R., Kurelec, B., and Zahn, R. K.,** Tissue distribution, seasonal variation and induction of benzo(a)pyrene monooxygenase activity in the crab *Maja crispata, Sci. Total Environ.,* 35, 41, 1984.

40. **Elmamlouk, T. H., Gessner, T., and Brownie, A. C.,** Occurrence of cytochrome P-450 in hepatopancreas of *Homarus americanus, Comp. Biochem. Physiol.,* 48B, 419, 1974.

41. **Elmamlouk, T. H. and Gessner, T.,** Species difference in metabolism of parathion: apparent inability of hepatopancreas fractions to produce paraoxon, *Comp. Biochem. Physiol.,* 53C, 19, 1976.

42. **James, M. O., Khan, M. A. Q., and Bend, J. R.,** Hepatic mixed-function oxidase activities in several marine species common to coastal Florida, *Comp. Biochem. Physiol.,* 62C, 155, 1979.

43. **James, M. O., Fouts, J. R., and Bend, J. R.,** Xenobiotic metabolizing enzymes in marine fish, in *Pesticides in the Marine Environment,* Khan, M. A. Q., Ed., Plenum Press, New York, 1977, 171.

44. **Bend, J. R., James, M. O., and Dansette, P. M.,** *In vitro* metabolism of xenobiotics in some marine animals, *Ann. N.Y. Acad. Sci.,* 298, 505, 1977.

45. **James, M. O.,** unpublished observations, 1981.

46. **James, M. O.,** Catalytic properties of cytochrome P-450 in hepatopancreas of the spiny lobster, *Panulirus argus, Mar. Environ. Res.,* 14, 1, 1984.

47. **James, M. O. and Shiverick, K. T.,** Cytochrome P-450-dependent oxidation of progesterone, testosterone and ecdysone in the spiny lobster, *Panulirus argus, Arch. Biochem. Biophys.,* 233, 1, 1984.

48. **James, M. O. and Little, P. J.,** 3-Methylcholanthrene does not induce *in vitro* xenobiotic metabolism in spiny lobster hepatopancreas, or affect *in vivo* disposition of benzo(a)pyrene, *Comp. Biochem. Physiol.,* 78C, 241, 1984.

49. **James, M. O., Sherman, B., Fisher, S. A., and Bend, J. R.,** Benzo(a)pyrene metabolism in reconstituted monooxygenase systems containing cytochrome P-450 from lobster (*Homarus americanus*) hepatopancreas fractions and NADPH-cytochrome P-450 reductase from pig liver, *Bull. Mt. Desert Isl. Biol. Lab.,* 22, 37, 1982.

50. **Stegeman, J. J. and Kaplan, H. B.,** Mixed-function oxygenase activity and benzo(a)pyrene metabolism in the barnacle, *Balanus eburneus* (Crustacea: cirripedia), *Comp. Biochem. Physiol.,* 68C, 55, 1981.

51. **Walters, J. M., Cain, R. B., Higgins, I. J., and Corner, E. D. S.,** Cell-free benzo(a)pyrene hydroxylase activity in marine zooplankton, *J. Mar. Biol. Assoc. U.K.,* 59, 553, 1979.

52. **Little, P. J. and James, M. O.,** unpublished observations, 1979.
53. **James, M. O., Bowen, E. R., Dansette, P. M., and Bend, J. R.,** Epoxide hydrase and glutathione S-transferase activities with selected alkene and arene oxides in several marine species, *Chem. Biol. Interact.,* 25, 321, 1979.
54. **Huang, M.-T., Johnson, E. F., Muller-Eberhard, U., Koop, D. R., Coon, M. J., and Conney, A. H.,** Specificity in the activation and inhibition by flavonoids of benzo(a)pyrene hydroxylation by cytochrome P-450 isozymes from rabbit liver microsomes, *J. Biol. Chem.,* 256, 10897, 1981.
55. **Batel, R., Bihari, N., and Zahn, R. K.,** Purification and characterization of a single form of cytochrome P-450 from the spiny crab, *Maja crispata, Comp. Biochem. Physiol.,* 83C, 165, 1986.
56. **O'Hara, S. C. M., Corner, E. D. S., Forsberg, T. E. V., and Moore, M. N.,** Studies on benzo(a)pyrene monooxygenase in the shore crab, *Carcinus maenas, J. Mar. Biol. Assoc. U.K.,* 62, 339, 1982.
57. **James, M. O. and Little, P. J.,** Characterization of cytochrome P-450 dependent mixed-function oxidation in the spiny lobster, *Panulirus argus* in *Biochemistry, Biophysics and Regulation of Cytochrome P-450,* Gustafsson, J.-A., et al., Eds., Elsevier/North-Holland Biomedical Press, Amsterdam, 1980, 113.
58. **Lee, R. F., Singer, S. C., and Page, D. S.,** Responses of cytochrome P-450 systems in marine crab and polychaetes to organic pollutants, *Aquat. Toxicol.,* 1, 155, 1981.
59. **Lee, R. F.,** Mixed function oxygenases (MFO) in marine invertebrates, *Mar. Biol. Lett.,* 2, 87, 1981.
60. **Fries, C. R. and Lee, R. F.,** Pollutant effects on the mixed function oxygenase (MFO) and reproductive systems of the marine polychaete *Nereis virens, Mar. Biol.,* 79, 187, 1984.
61. **Vandermeulen, J. H. and Penrose, W. R.,** Absence of aryl hydrocarbon hydroxylase (AHH) in three marine bivalves, *J. Fish. Res. Board Can.,* 35, 643, 1978.
62. **Kurulec, B.,** Exclusive activation of aromatic amines in the marine mussel *Mytilus edulis* by FAD-containing monooxygenase, *Biochem. Biophys. Res. Commun.,* 127, 773, 1985.
63. **Livingstone, D. R. and Farrar, S. V.,** Tissue and sub-cellular distribution of enzyme activities of mixed-function oxygenase and benzo(a)pyrene metabolism in the common mussel *Mytilus edulis* L., *Sci. Total Environ.,* 39, 209, 1984.
64. **Livingstone, D. R. and Farrar, S. V.,** Responses of the mixed function oxidase system of some bivalve and gastropod molluscs to exposure to polynuclear aromatic and other hydrocarbons, *Mar. Environ. Res.,* 17, 101, 1985.
65. **Stegeman, J. J.,** Benzo(a)pyrene oxidation and microsomal enzyme activity in the mussel (*Mytilus edulis*) and other bivalve mollusc species from the western North Atlantic, *Mar. Biol.,* 89, 21, 1985.
66. **Anderson, R. S.,** Benzo(a)pyrene metabolism in the American oyster *Crassostrea virginica,* EPA Ecol. Res. Ser. Monogr. (EPA-600/3-78-009), 1978.
67. **Anderson, R. S.,** Metabolism of a model environmental carcinogen by bivalve molluscs, *Mar. Environ. Res.,* 17, 137, 1985.
68. **Anderson, R. S. and Doos, J. E.,** Activation of mammalian carcinogens to bacterial mutagens by microsomal enzymes from a pelecypod mollusk, *Mercenaria mercenaria, Mutat. Res.,* 116, 247, 1983.
69. **Kurulec, B., Britvic, S., and Zahn, R. K.,** The activation of aromatic amines in some marine invertebrates, *Mar. Environ. Res.,* 17, 141, 1985.
70. **Lindmark, D. G.,** Activation of polynuclear aromatic hydrocarbons to mutagens by the marine ciliate *Parauronema acutum, Appl. Environ. Microbiol.,* 41, 1238, 1981.
71. **Wood, A. W., Wislocki, P. G., Chang, R. L., Levin, W., Lu, A. Y. H., Yagi, H., Hernandez, O., Jerina, D. M., and Conney, A. H.,** Mutagenicity and cytotoxicity of benzo(a)pyrene benzo-ring epoxides, *Cancer Res.,* 36, 3358, 1976.
72. **Glatt, H. R. and Oesch, F.,** Phenolic benzo(a)pyrene metabolites are mutagens, *Mutat. Res.,* 36, 379, 1976.
73. **Kapitulnik, J., Levin, W., Jerina, D. M., and Conney, A. H.,** Lack of carcinogenicity of 4-, 5-, 6-, 7-, 8-, 9-, and 10-hydroxybenzo(a)pyrene on mouse skin, *Cancer Res.,* 36, 3625, 1976.
74. **Jerina, D. M., Yagi, H., Hernandez, O., Dansette, P. M., Wood, A. W., Levin, W., Chang, R. L., Wislocki, P. G., and Conney, A. H.,** Synthesis and biologic activity of potential benzo(a)pyrene metabolites, in *Carcinogenesis, Vol. 1. Polynuclear Aromatic Hydrocarbons: Chemistry, Metabolism and Carcinogenesis,* Freudenthal, R. I. and Jones, P. W., Eds., Raven Press, New York, 1976, 91.
75. **Slaga, T. J., Bracken, W. M., Dresner, S., Levin, W., Yagi, H., Jerina, D. M., and Conney, A. H.,** Skin tumor-initiating activities of the twelve isomeric phenols of benzo(a)pyrene, *Cancer Res.,* 38, 678, 1978.
76. **Chang, R. L., Wislocki, P. G., Kapitulnik, J., Wood, A. W., Levin, W., Yagi, H., Duck Mah, H., Jerina, D. M., and Conney, A. H.,** Carcinogenicity of 2-hydroxybenzo(a)pyrene and 6-hydroxybenzo(a)pyrene in newborn mice, *Cancer Res.,* 39, 2660, 1979.
77. **Wood, A. W., Levin, W., Chang, R. L., Yagi, H., Thakker, D. R., Lehr, R. E., Jerina, D. M., and Conney, A. H.,** Bay-region activation of carcinogenic polycyclic hydrocarbons, in *Polynuclear Aromatic Hydrocarbons,* Jones, P. W. and Leber, P., Eds., Ann Arbor Science Publishers, Ann Arbor, Michigan, 1979, 531.

78. **James, M. O. and Little, P. J.,** Polyhalogenated biphenyls and phenobarbital: evaluation as inducers of drug-metabolizing activities in the sheepshead, *Archosargus probatocephalus, Chem. Biol. Interact.,* 36, 229, 1981.

79. **Everson, R. B., Randerath, E., Santella, R. M., Cefalo, R. C., Avitts, T. A., and Randerath, K.,** Detection of smoking-related covalent DNA adducts in human placenta, *Science,* 231, 54, 1986.

80. **James, M. O.,** unpublished.

81. **James, M. O. and Little, P. J.,** unpublished observations.

Chapter 4

BIOTRANSFORMATION AND DISPOSITION OF POLYCYCLIC AROMATIC HYDROCARBONS (PAH) IN FISH*

Usha Varanasi, John E. Stein, and Marc Nishimoto

TABLE OF CONTENTS

I. INTRODUCTION

During the last decade and a half, increasing attention has been focused on the ability of aquatic organisms to metabolize chemical contaminants that are present in urban waterways. Earlier studies with subcellular fractions of tissues from aquatic organisms were primarily concerned with obtaining comparative information on enzyme activities in phylogenetically diverse species.[1,2] However, during the early to mid-1970s, concern about the impact of oil spills on the aquatic environment had heightened, and, consequently, there was great interest in evaluating the uptake and disposition of petroleum constituents, especially benzenes and polycyclic aromatic hydrocarbons (PAH) in fish and shellfish. Measurements of tissue concentrations of benzenes and PAH in fish sampled from oil-impacted areas or urban estuaries showed the presence of aromatic hydrocarbons with 1-3 benzenoid rings, such as derivatives of benzene, naphthalene (NPH), biphenyl, and phenanthrene (PHN), but rarely showed detectable levels of 4- to 6-ring PAH, such as benz(a)anthracene (BaA) or benzo(a)pyrene (BaP).[3] These results, combined with the fact that 4- to 6-ring PAH are relatively water insoluble, generated the belief that once these PAH entered the aquatic environment, they would be too tightly bound to particulate matter, and hence may not be readily bioavailable to fish.[3] In the late 1970s and early 1980s, however, considerable evidence was obtained that demonstrated strong, positive associations between levels of PAH in sediments and prevalence of hepatic neoplasms in benthic fishes from polluted areas in the U.S.[4-6] These findings, along with the detection of PAH, such as BaP and BaA, in microgram-per-gram levels in urban sediments, and the results with rodents demonstrating that these PAH exert their carcinogenic and teratogenic effects only after metabolic activation, have given a strong impetus to studies of bioavailability and metabolism of carcinogenic PAH by fish.

Hence, the current interest in biotransformation and disposition of PAH in fish stems from two rather divergent perspectives: one is to evaluate the potential of food-chain transfer of toxic PAH and their metabolites, and the other is to evaluate mechanisms of toxicity of environmental pollutants in fish. In this chapter, we will first discuss endogenous and exogenous factors that affect metabolism and disposition of PAH in fish. This will be followed by a discussion of the metabolic activation and detoxication of PAH with emphasis on the disposition of phase I and phase II metabolites in various tissues and fluids, and the interactions of reactive intermediates with cellular macromolecules.

II. DISPOSITION OF PAH IN FISH

A number of chemical, biochemical, physiological, and environmental factors affect the rates of uptake, metabolism, and excretion of PAH in aquatic organisms. Further, the dose and route of exposure of PAH may significantly influence the accumulation and disposition of PAH in fish. Generalizations as to how individual factors affect PAH metabolism and disposition are difficult to formulate, because there are virtually no comprehensive studies examining the influence of a single factor, while controlling many other factors. Nevertheless, the literature does provide important insights into how dose, route of exposure, molecular size, environmental parameters (e.g., water temperature), species differences, age, sex, and exposure to other xenobiotics influence the toxicokinetics of PAH in fish.

A. Route and Length of Exposure

The variety of routes used to expose fish to PAH in laboratory studies are usually designed to mimic environmental exposure or to obtain precise information in response to a known dose. The early studies on the uptake of PAH by fish predominantly used water-borne benzenes and PAH, because of the interest in the effects of oil spills on aquatic ecosystems.[7,8] Low molecular weight PAH (e.g., derivatives of benzene and NPH) can be present in high concentrations in the water column of oil-impacted areas.[9] More recently, the uptake of PAH from sediment by benthic species has received considerable attention.[10,11] This interest stems from the observations that prevalence of liver lesions, including neoplasms, in benthic fish species are positively correlated with the levels of PAH present in sediments (see Chapter 8), and that appreciable amounts of sediment are often found both in the gastrointestinal tract of benthic fishes as well as in the food organisms (i.e., polychaetes, mollusks) that are ingested by these fish species. Thus, the issue of bioavailability and disposition of PAH is crucial in demonstrating cause-and-effect relationships. The factors affecting the bioavailability of sediment-associated PAH are discussed in Chapter 1.

Comparison of the uptake and disposition of PAH in benthic fish exposed to PAH intragastrically or to sediment-associated PAH is important, because both routes of exposure have environmental significance. Two studies conducted in our laboratories with English sole (*Parophrys vetulus*) allow comparison of the pattern of distribution of BaP and its metabolites in tissues and fluids either after a single exposure *per os (p.o.)* to [³H]BaP (2.0 mg/kg fish), or after a continuous exposure for up to 10 d to a reference sediment containing added [³H]BaP (3 μg/g dry weight of sediment).[12,13] Figures 1A and 1B show that the salient feature of both studies is the high concentrations of BaP and its metabolites in the hepatobiliary system and much lower concentrations in extrahepatic tissues, such as muscle. Moreover, the relative ranking of the concentration of BaP-derived radioactivity in tissues and bile was similar for both routes of exposure: bile>liver>blood>skin>muscle. A time-course experiment to study the uptake of [³H]BaP from water (75 ng BaP per liter) by Northern pike (*Esox lucius*) also showed that bile contained the highest concentration of radioactivity (Figure 1C).[14] Moreover, the relative ranking of the concentration of BaP-derived radioactivity in the tissues of pike was similar to that observed for English sole exposed to BaP orally or when associated with sediment.[12,13]

In the study discussed above, English sole could have accumulated BaP from sediment (1) by direct desorption from particles at epithelial membranes (i.e., skin, gastrointestinal surfaces, gills), (2) from sediment-associated water (SAW), or (3) after partial breakdown of ingested organic-carbon-rich particles and uptake of liberated BaP.[13] The significance of each possible route of uptake is not known. However, the English sole were fed uncontaminated food and did not have to browse through sediment and ingest large amounts of sediment as they may do in the process of obtaining food in their environment.[15] Thus, it is suggested that the uptake of BaP in this study was from a combination of absorption from SAW, and from incidental ingestion of sediment and direct desorption from particles at skin and gill membranes. In summary, the results from the above studies with different routes of exposure showing similarities in the disposition of BaP and its metabolites in tissues and fluids suggest that the route of exposure has little effect on the distribution of BaP and its metabolites in tissues of fish.

In addition to the routes of exposure (i.e., water-borne, sediment) that are appropriate for fish and other aquatic species, many studies also use either intragastric administration of PAH dissolved in an oil carrier (i.e., salmon oil, corn oil) or injection of PAH in various lipophilic solvents into the peritoneal cavity or muscle. Intragastric administration of PAH simulates a route by which fish may be exposed to PAH in the environment. In fact, food-chain transfer of lipophilic xenobiotics (e.g., polychlorinated biphenyls [PCB] and high molecular weight PAH) to fish may be an important route of uptake.[16,17] Results from our laboratory (Table 1) show higher levels of BaP and its metabolites in livers of starry flounder (*Platichthys stellatus*) and English sole administered these hydrocarbons via an intraperitoneal (i.p.) injection than in species exposed to the PAH intragastrically. Results of studies on the uptake of PAH in fish given intramuscular (i.m.) injections suggest that this route of exposure is better than administration by i.p. injection. For example, James and Bend showed that i.m. injections of 3-methylcholanthrene (3-MC) increased hepatic mixed-function ox-

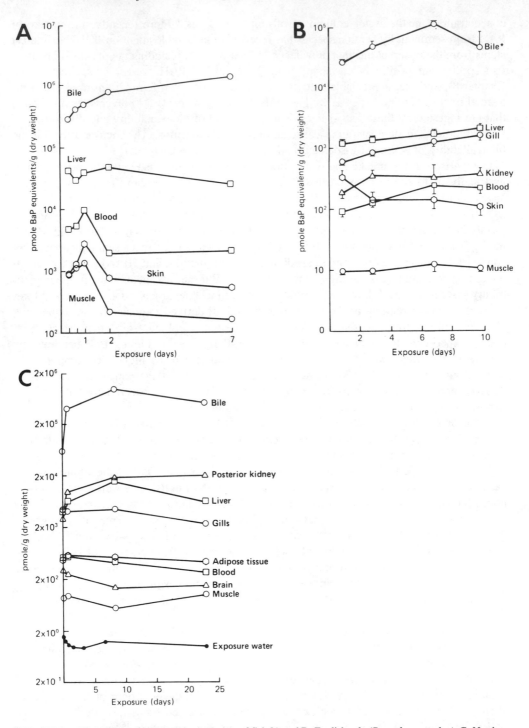

FIGURE 1.   Disposition of radioactivity in tissues of fish [A and B, English sole *(Parophrys vetulus)*; C, Northern pike *(Esox lucius)*] exposed to [³H]BaP via (A) sediment (3 μg/g sediment, dry weight), (B) diet (2 mg/kg body weight), and (C) water (75 ng BaP/l). (Adapted from Varanasi, U. and Gmur, D. J., unpublished observations, 1980. Stein, J. E., Hom, T., and Varanasi, U., *Mar. Environm. Res.,* 13, 97, 1984. Balk, L., Meijer, J., DePierre, J. W., and Appelgren, L.-E., *Toxicol. Appl. Pharmacol.,* 74, 430, 1984. With permission.)

Table 1

EFFECT OF ROUTE OF EXPOSURE ON LEVELS OF PAH AND THEIR METABOLITES IN LIVER OF FISH[a]

Aromatic Hydrocarbon[b]	Species	Dose (mg/kg)	Route of administration	% administered dose (n)[c]	Ref.
NPH	Starry flounder	0.44	i.p.	3.6 ± 0.4 (3)	18
	Starry flounder	0.44	p.o.	0.7 ± 0.5 (4)	18
BaP	Starry flounder	2.0	i.p.	3.5 ± 0.7 (6)	19
	Starry flounder	2.0	p.o.	0.6 ± 0.4 (6)	19
	English sole	2.0	i.p.	6 ± 1 (12)	19
	English sole	2.0	p.o.	0.43 ± 0.07 (9)	19
	English sole	0.1	p.o.	0.8 ± 0.2 (4)	20

[a] Fish were sacrificed at 24 h after exposure to radiolabeled PAH. Values represent the mean ± SEM.
[b] NPH and BaP.
[c] Includes PAH and metabolites.

idase (MFO) in sheepshead (*Archosargus probatocephalus*) to a greater extent than did i.p. injections of 3-MC.[21] Moreover, i.m. injections of 3-MC induced hepatic MFO in stingray, (*Dasyatis sabina*) whereas i.p. injections did not. This difference in induction of hepatic MFO in stingray was attributed to poor absorption of 3-MC in fish administered this compound by an i.p. injection. Further, Williams et al.[22] showed in winter flounder (*Pseudoplueronectes americanus*) that i.m. administration of a variety of compounds results in a faster attainment of maximum plasma levels of the test compounds when compared to administration via i.p. injection.

The route of exposure (i.p. vs. p.o.) also influences the relative proportions of NPH and its phase I and II metabolites in liver and bile of fish (Table 2). The results show consistently higher percentages of NPH and its phase I metabolites in the liver and higher percentages of phase I metabolites in bile of fish injected intraperitoneally compared to those exposed orally to NPH (Table 2). In addition, the percent administered dose in livers of fish given NPH via an i.p. injection was greater than in fish exposed orally, therefore, the actual concentrations of NPH and its phase I metabolites were higher in livers of fish given NPH intraperitoneally. In contrast, no marked differences in concentrations of NPH and its phase I metabolites were observed in bile of fish exposed intraperitoneally or orally. The higher proportions of NPH and its phase I metabolites in livers of fish exposed to NPH intraperitoneally may primarily be due to two factors: (1) in fish exposed to NPH intraperitoneally, the rate at which the liver accumulates NPH is greater than the rate at which NPH is metabolized, and (2) in fish exposed to NPH orally, a part of the dose may be metabolized in the gastrointestinal (GI) tract prior to absorption into the liver. The metabolism of PAH in the GI tract was demonstrated by a recent study showing that when BaP dissolved in lipid is fed to killifish (*Fundulus heteroclitus*), a portion of the BaP is metabolized to phase I and II metabolites in the intestine before secretion into the portal circulation and transport to the liver.[23] Therefore, the combination of the above two factors would lead to higher proportions of NPH and phase I metabolites in livers of fish exposed to NPH intraperitoneally.

As with route of exposure, the length of exposure also affects the disposition and excretion of NPH in fish. For example, the half-lives ($t_{1/2}$) for elimination of NPH-derived radioactivity from tissues and blood of rainbow trout (*Salmo gairdneri*) that were exposed to water-borne NPH for 27 d were 30 to 140 times greater than the corresponding $t_{1/2}$s for tissues and blood of trout exposed for 8 h to water-borne NPH (Table 3).[24] Further, it was shown in a study with starry flounder that the rates of decline of tissue concentrations of parent NPH were

Table 2
EFFECT OF THE ROUTE OF ADMINISTRATION ON THE DISPOSITION OF ^{14}C-NPH IN LIVER AND BILE OF STARRY FLOUNDER[a]

	Time (h)	Liver[b] i.p.	Liver[b] p.o.	Bile[c] i.p.	Bile[c] p.o.
		% Total radioactivity			
NPH	24	97	85	7	14
	168	76	19	2	1
Total NPH	24	3	15	93	86
Metabolites	168	24	81	98	99
		% Total metabolites			
Phase I metabolites	24	77	56	3	8
(nonconjugates)	168	66	38	3	9
Phase II metabolites	24	23	44	97	92
(conjugates)	168	34	62	97	91

[a]   Adapted from Reference 18. Values represent the mean.
[b]   Percent administered dose in liver at 24 and 168 h for i.p. exposure was 3.6 and 0.4, respectively, and for oral exposure was 0.7 and 0.02, respectively.
[c]   Percent administered dose in bile at 24 and 168 h for i.p. exposure was 0.49 and 2.3, respectively, and for oral exposure fish was 0.2 and 0.9, respectively.

Table 3
ELIMINATION HALF-LIVES OF ^{14}C-NPH AND ^{14}C-METHYL NPH FROM RAINBOW TROUT EXPOSED TO THESE HYDROCARBONS IN WATER[a]

Exposure	^{14}C-NPH		^{14}C-2-Methyl NPH
Concentrations (mg/l)	0.017	0.017	0.023
Duration (d)	0.33	27	26
Depuration (d)	1	35	36

Tissue	$t_{1/2}$ (h)		
Muscle	6.5	909	13[b]
			711
Liver	11.0	343	211[c]
Blood	7.5	379	23[d]

[a]   Adapted from Reference 24; measurements include both parent hydrocarbons and metabolites.
[b]   Elimination was biexponential; the upper value was calculated using data for first 2 d of depuration, whereas the lower value was calculated using data from days 4—36.
[c]   Because of variability in micrograms of ^{14}C-NPH per gram of liver, rate of elimination was calculated as a single-phase elimination, rather than biphasic elimination as for muscle.
[d]   After 4 d of depuration the amount of ^{14}C-2-methyl NPH was below limits of detection, therefore the $t_{1/2}$ for the secondary phase elimination could not be calculated.

Table 4
RATES OF DEPURATION OF NPH AND
NPH METABOLITES (NPH-M) FROM
TISSUES AND FLUIDS OF FISH[a]

Species	Tissue	Depuration rate constants (h^{-1})	
		NPH	NPH-M
Starry flounder	Liver	1.8	0.40
	Blood	1.7	0.84
	Skin	1.5	0.62
Rock sole	Liver	2.0	0.53
	Blood	0.81	0.60
	Skin	0.77	0.56

[a] Adapted from Reference 18. Fish were exposed to NPH p.o., and the rate constants were calculated over a 1-week period.

significantly greater than the rate constants for the excretion of NPH metabolites in both tissues and blood (Table 4). Thus, the slower elimination of NPH-derived radioactivity from trout after the long-term exposure may be partially explained by the above observations that unmetabolized NPH is cleared more rapidly than NPH metabolites from tissues and blood of fish (Table 4). This hypothesis is further supported by the results showing that the proportion of metabolites present in muscle after 7 to 9 d of depuration was greater than those detected during the long-term exposure of trout to NPH or 2-methylNPH.[24] For [^{14}C]2-methylNPH the differences were striking, in that only 1% of the radioactivity was present as metabolites in muscle after 26 d of exposure, but after 7 d of depuration 24% of the radioactivity was present as metabolites in muscle. The marked change in proportion of 2-methylNPH metabolites present in muscle during the early part of the depuration phase is consistent with a rapid elimination of unmetabolized 2-methylNPH and much slower elimination of metabolites. However, the differential rates of elimination of parent PAH and its metabolites do not completely explain why the $t_{1/2}$ after long-term exposure is greater than that for short-term exposure, because the percent NPH metabolites in muscle of trout exposed to NPH for 27 d was less than that after 1 d of exposure. Therefore, it is clearly evident that additional studies are required to determine the underlying biochemical/physiological mechanism for slower elimination of PAH and their metabolites in fish chronically exposed to these compounds.

Studies discussed so far should make it clear that the evaluation of factors affecting disposition of PAH is difficult, because the extent of metabolism and types of metabolites strongly affect both accumulation and excretion. If based solely on lipophilicity, it can be assumed that PAH metabolites should be more easily excreted than parent compounds via urine and bile. However, the electrophilic properties of many PAH metabolites allow them to interact covalently with cellular macromolecules and, hence, they can be preferentially retained within the cell (discussed later). Moreover, certain benzenes and NPH can be directly released from gills and skin, whereas polar metabolites, such as conjugates, may not be so easily released.[25] Therefore, the extent of accumulation and excretion of these compounds will be dependent on the balance among each of these interactions and pathways.

B. Effect of Molecular Size

Accumulation of xenobiotics by aquatic organisms should, in principle, be directly related to the lipophilicity of the xenobiotic as discussed in Chapter 1. A number of studies show that the greater the molecular weight of a PAH the higher the bioconcentration factor (BCF), thereby providing indirect evidence for a positive correlation between the BCF of PAH in fish and the lipophilicity of the PAH. For example, in coho salmon (*Oncorhynchus kisutch*) and starry flounder exposed to the water-soluble fraction of crude oil, the BCF for alkylated NPH were greater than the corresponding BCF for alkylated benzenes.[26] Roubal et al. also showed that in coho salmon exposed to radiolabeled benzene, NPH, and anthracene (ANH) administered p.o. or by i.p. injection, the total concentration of the individual PAH and its metabolites in liver and brain increased by nearly a factor of 10 with increasing molecular size.[27] In coho salmon alevins exposed for 24 h to water-borne toluene, NPH, or 2-methylNPH, the BCF for 2-methylNPH and NPH were about 20 and 4 times, respectively, higher than that for toluene.[28] In these studies, the tissue concentration of total hydrocarbon (i.e., parent compound plus its metabolites) was measured using radiolabeled compounds. As discussed in other chapters, as well as in a later section of this chapter, the ability of marine organisms to metabolize PAH can have a marked effect on accumulation of PAH. For example, the BCF in fish for a nitrogen-containing analog of BaA, benz(a)acridine, is reported to be nearly an order of magnitude less than that predicted from their octanol-water partition coefficients.[29] Metabolism was estimated to reduce the BCF by 50 to 90% from the hypothetical case in which metabolism did not occur.

Factors other than molecular size also affect the accumulation of PAH by fish. In a study by Solbakken and co-workers, Baltic flounder (*Platichthys flesus*) were exposed to comparable doses of radiolabeled NPH, PHN, or BaP intragastrically,[30] Data presented in Figure 2 show the concentration of total radioactivity of the three PAH in tissues and fluids over time. Based on the lipophilicity of the compounds, one would expect that the concentration in liver would follow the order BaP>PHN>NPH; however, the concentration of PHN-derived radioactivity in liver and bile exceeded that of BaP. Additionally, the concentration of PHN-derived radioactivity exceeded that of BaP in plasma and muscle at the early time points. This study did not include an analysis of the level of metabolites of each PAH in tissues and fluids, which would have been useful in interpreting the results. However, the concentration of PHN-derived radioactivity in all tissues and fluids, except urine, was greater than either BaP or NPH, thus suggesting a greater absorption of orally administered PHN than of BaP or NPH. The much lower concentrations of NPH and its metabolites in tissues and fluids are at least partially related to the ability of fish to excrete low molecular weight PAH, such as NPH, via gills and skin (as discussed later). Furthermore, recent studies suggest that uptake of highly lipophilic compounds having octanol-water partition coefficients (K_{ow} = [concentration of PAH in *n*-octanol]/[concentration of PAH in water]) greater than 10^6, such as BaP (K_{ow} = $10^{6.5}$) from water, is less than predicted because of stereochemical factors that decrease the rate of transport across biological membranes.[31] Whether such considerations apply to absorption by the GI tract remain to be determined.

The extent and route of excretion of PAH from fish are also dependent on molecular size. Low molecular weight PAH — benzene and NPH — are extensively excreted by gills, and to a lesser degree by skin and mucus.[25,32] In contrast, higher molecular weight compounds apparently need to be metabolized prior to excretion.[18,25,30,33-36] Although the pathways of excretion of PAH in fish have generally been identified, the relative importance of each route for PAH of different molecular size has only been examined to a limited degree. In one study, Dolly Varden char (*Salvelinus malma*) were held in a split-box chamber that separated gill excretions from cloacal excretions, and were exposed to radiolabeled toluene, NPH, ANH, or BaP intragastrically.[25] Seawater in both chambers was sampled periodically for 24 h post-exposure, whereas tissues and fluids were sampled at the end of the 24 h exposure. Results of the cumulative excretion showed that as the molecular weight of the PAH increased, the percent dose found in the gill chamber decreased (Table 5). The excretion of ANH and BaP by the gills was very low and did not appear to contribute significantly

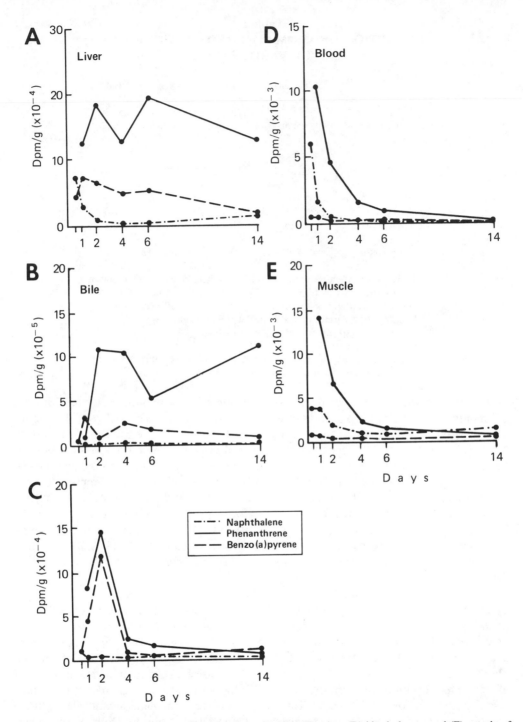

FIGURE 2. Time course of radioactivity in (A) liver, (B) bile, (C) urine, (D) blood plasma, and (E) muscle of flounder *(Platichthys flesus)* after oral administration of 9.25 μg/kg body weight ^{14}C-labeled NPH, PHN or BaP. Each point represents the mean (± S.E.M.) of five fish. The dpm/g values for NPH and BaP have been normalized to the specific activity for PHN. (Adapted from Solbakken, J. E., Solberg, M., and Palmork, K. H., *FiskeriDir. Skr. Ser. Havunders.*, 17, 473, 1984. With permission.)

Table 5
EXCRETION OF AROMATIC HYDROCARBONS BY
DOLLY VARDEN CHAR[a]

	Toluene	NPH	ANH	BaP
Molecular weight	92.13	128.16	178.22	252.30
Partition coefficient[b] (log K_{ow})	2.58	3.33	4.45	6.50[c]
		% Administered dose		
Excreted[d]				
Gill chamber	27.5	10.4	1.9	1.0
Cloacal chamber	2.6	0.4	1.3	0.2
Total excreted	30.1	10.8	3.2	1.2

[a] Table adapted from Reference 25. Fish were held in a split-box chamber that isolated the gills from the rest of the body. Dose (p.o.) of the ^{14}C-aromatic hydrocarbons: toluene 26.1 μmol/kg b.w., NPH 25.7 μmol/kg b.w., ANH 38.7 μmol/kg b.w., and BaP 7.93 μmol/kg b.w.

[b] Log K_{ow} = log{[hydrocarbon]$_{octanol}$/[hydrocarbon]$_{water}$}

[c] From Reference 40.

[d] Values are expressed as percent of administered dose at 24 h after intragastric administration of ^{14}C-hydrocarbons.

to excretion of these PAH. In addition, elimination by way of urine and feces did not apparently contribute to the excretion of NPH, ANH, or BaP, since ≤1.3% of the administered dose of these PAH was found in the cloacal chamber (Table 5). The apparent lack of excretion of NPH, ANH, and BaP via feces and urine may be related to the following factors: (1) the Dolly Varden char were not fed before or during the exposure and, hence, bile production and release may be low, and (2) poor absorption of the administered PAH by the fish may have contributed to the low levels of [14C]NPH, [14C]ANH, and [14C]BaP in the cloacal chamber; in fact, 72, 77, and 82%, respectively, of the radioactivity in the body were present in the GI tract, of which only 10, 6.1, and 28% were metabolites. Because of the low level of radioactivity recovered in the cloacal chamber for NPH, ANH, and BaP, correlations between the extent of excretion via urine and feces and molecular size of PAH cannot be made.

In the study by Thomas and Rice, the excretion of orally administered phenol and cresol by Dolly Varden char was also examined (data not shown).[25] Equal amounts of each compound were excreted by gills (28.3 and 29.1% administered dose, respectively), and these values were similar to the amount of toluene excreted by the gills (Table 5). Further, Nagel and Urich showed that unconjugated phenols (e.g., phenol, 3-nitrophenol, and 3,5-diethyl-phenol) are predominantly excreted via gills, and that urine and bile are the major routes of excretion for the conjugated forms.[37] The log K_{ow} for phenol (1.66) is an order of magnitude less than that for toluene (Table 5). Thus, the size of the compound and not the partition coefficient appears to be a major regulator of branchial excretion by Dolly Varden char. Moreover, a previous study with Dolly Varden char exposed to toluene and NPH in the split-box chamber, showed that the majority of the toluene- and NPH-derived radioactivity in the gill chamber was present mainly as unmetabolized PAH (79.7 and 80.8% of total radioactivity, respectively), indicating branchial excretion of parent PAH.[38] The facile excretion of NPH by the gills of fish certainly contributes significantly to the faster rate of excretion of NPH than NPH metabolites by fish (Table 4).

Assessing the effect of molecular size on the metabolism and disposition of PAH in fish is difficult because of the diversity of experimental parameters used in studies reported in the literature. There are, however, examples which suggest that as the molecular weight of a PAH increases, the amount of the PAH and its metabolites in extrahepatic tissues decreases. For example, Varanasi and Gmur reported that when English sole were placed for 1 week on a marine sediment to which [³H]BaP and [¹⁴C]NPH dissolved in Prudhoe Bay crude oil (PBCO) were added, the ratio of concentrations of radioactivity (PAH plus metabolites) in liver to that in muscle was 40 for BaP and 20 for NPH.[33] A notable example demonstrating a relationship between molecular size and accumulation of PAH in extrahepatic tissue was obtained with gravid female English sole exposed intragastrically to radiolabeled BaP or NPH and sampled 24 h later.[39,40] In these fish, the ratio of administered dose in ovary to that in liver for NPH (6.7) is 20 times the ratio for BaP (0.33). Because PAH administered intragastrically are first transported to the liver by the portal vein, one may speculate that BaP is more efficiently cleared by the liver and more efficiently metabolized than is NPH. Therefore, greater amounts of NPH may reach systemic circulation for deposition in extrahepatic tissues. This hypothesis must be considered preliminary until detailed toxicokinetic studies are conducted to quantitatively assess hepatic clearance of NPH and BaP in fish. For example, in Dolly Varden char exposed orally to NPH, ANH, and BaP, the ratios of percent administered dose in muscle to that in liver (2.1, 3.1, and 2.3, respectively) were not substantially different, thereby suggesting that such an inverse relationship between molecular weight of a PAH and concentration in extrahepatic tissues may not exist.[25] However, in Dolly Varden char, BaP was not metabolized as efficiently (44% unconverted BaP in liver at 24 h after administration) as in English sole (<1% unconverted BaP in liver), and, therefore, more BaP was apparently available for accumulation by extrahepatic tissues of Dolly Varden char. Hence, the conversion of lipophilic PAH to more polar metabolites plays an important role in controlling their distribution to extrahepatic tissues.

C. Environmental Factors

Environmental temperature is known to regulate a variety of biochemical and metabolic functions, including MFO activity, in poikilotherms.[41,42] Temperature also affects the persistence of low molecular weight PAH in water. The lower the water temperature, the greater the persistence of PAH due to lower volatilization and biodegradation. A combination of such physiological/biochemical and physicochemical factors apparently increases the toxicity of toluene towards pink salmon fry (*Oncorhynchus gorbuscha*) at low water temperature.[43,44] A partial explanation for the increased mortality of pink salmon fry exposed to toluene at lower water temperature is the increased retention of low molecular weight PAH (i.e., benzenes, NPH) by fish acclimated to lower temperatures. For example, the results in Table 6 show that decreased temperature (4 vs. 10 or 12°C) increases the retention of NPH-derived radioactivity in coho salmon and starry flounder exposed to [¹⁴C]NPH intragastrically.[45,46] Additionally, the same effect was observed when Dolly Varden char were exposed (p.o.) to radiolabeled NPH and toluene.[47] In the study by Varanasi et al., the effects of temperature on the toxicodynamics of both unmetabolized NPH and NPH metabolites in starry flounder were assessed.[46] The results showed that a decrease in water temperature resulted in significantly higher concentrations of NPH in tissues of starry flounder held at 4°C than in fish held at 12°C. In contrast, the concentration of NPH metabolites in tissues of starry flounder held at 4°C was generally not significantly different from corresponding values for fish held at 12°C (Table 6). These results suggest that the rate of excretion of NPH is more affected by temperature than is either the extent of metabolism of NPH or the excretion of NPH metabolites. This conclusion is further supported by the observations of an apparent compensatory effect on *in vitro* MFO activities as the environmental temperature is lowered, in that microsomal MFO activity is constant when measured at the environmental temperatures, which may indicate a temperature-independent capacity for *in vivo* metabolism of PAH.[48,49]

Table 6
EFFECT OF TEMPERATURE ON CONCENTRATION OF NPH AND NPH METABOLITES IN FISH EXPOSED INTRAGASTRICALLY TO ^{14}C-NPH[a]

| | Total ^{14}C (parent compounds and metabolites) | | | | NPH[b] | | NPH metabolites | |
| | Coho salmon ($C_4°/C_{10}°$)[c] | | Starry flounder ($C_4°/C_{12}°$) | | Starry flounder ($C_4°/C_{12}°$) | | Starry flounder ($C_4°/C_{12}°$) | |
Tissue	8 h[d] 4°C (n = 3)	16 h 10°C (n = 2)	24 h (n = 6)	168 h (n = 6)	24 h (n = 6)	168 h (n = 6)	24 h (n = 6)	168 h (n = 6)
Liver	3.0	5.8e	6.8	8.0	7.8e	34e	1.4	1.6
Muscle								
Dark	1.5	1.5	4.3	9.9	5.6e	26e	2.1	3.6e
Light	1.4	3.0	4.3	9.9	5.6e	26e	2.1	3.6
Bile	2.3	2.0	2.4	0.71	3.3	1.3	2.2	0.7
Stomach	—	—	7.3	0.24	9.8e	4.2e	2.9	0.1
Intestine	—	—	10	0.57	15e	3.4e	2.7	0.3
Skin	—	—	3.0	4.3	3.6e	10e	1.6	2.0
Brain	1.4	2.2e	2.1	3.7	2.2	5.6e	1.1	1.1
Blood	1.4	4.2e	2.1	6.2	1.9	9.6e	2.5	5.6e
Kidney	0.94	4.0	1.5	2.1	1.6	4.8e	1.3	1.6
Gills	—	—	3.5	3.2	3.7e	7.0e	3.0e	1.9

a   Adapted from Reference 45, coho salmon; and from Reference 46, starry flounder.

b   Starry flounder tissues and fluids were analyzed for unmetabolized NPH and for NPH metabolites; coho salmon tissues and fluids were analyzed for total NPH — parent compound plus metabolites.

c   Concentration ratios for tissues and fluids of fish held at 4 and 10 or 12°C.

d   Length of time after p.o. exposure to radiolabeled NPH (coho salmon — 0.9 mg ^{14}C-NPH/kg fish, starry flounder — 0.45 mg ^{3}H-NPH/kg).

e   Concentration value at 4°C was significantly different from value measured at 12°C.

Salinity also affects toxicity and disposition of PAH in fish species such as salmonids during smoltification, and in euryhaline species, such as Dolly Varden char and killifish. The sensitivity of killifish to NPH increases as the salinity of the exposure water increases.[50] Studies have shown that outmigrant salmonids are more sensitive to benzene, NPH, and water-borne crude oil when exposed in saline (10 to 30‰) water than in fresh water.[51,52] During smoltification, salmonids undergo physiological and morphological changes when migrating from fresh water to seawater. Thus, the dual stresses of physiological changes during smoltification and exposure to PAH may be responsible for the decreased tolerance of young salmonids to low - molecular weight PAH in saline water. The studies by Stickle et al. suggest, however, that exposure of coho salmon to toluene and NPH did not directly interfere with the ion-regulating capabilities of the smolts.[52] Rather, the observed alterations in these capabilities were a secondary effect of the PAH toxicity, because alterations in serum concentrations of sodium, potassium, and chloride were only observed at lethal concentrations of toluene and NPH.

It appears that the decreased tolerance of salmonids exposed to toluene and NPH in seawater is related to altered disposition of these PAH.[38,47] Although there were no overall differences in either total [^{14}C]toluene or [^{14}C]NPH excreted, or in percent metabolites of total recovered [^{14}C]toluene or [^{14}C]NPH between Dolly Varden char held in seawater or fresh water, it was observed that for fish held in seawater the neural tissues generally accumulated greater amounts of both PAH and retained more parent hydrocarbon than metabolites.[38] Several studies suggest that neural tissue of fish is a target tissue for low-molecular-weight PAH.[51,53,54] Thus, even though the reasons for the altered disposition of toluene and NPH in euryhaline fish in seawater are not evinced, the higher concentrations of parent hydrocarbon in brain and spinal cord may partially explain the increased sensitivity of fish exposed to water-borne PAH under saline conditions.

D. Species, Age, and Sex Differences
1. Species
The major factors affecting the disposition of PAH in different species of fish appear to be their ability to metabolize PAH and the percent lipid content of tissues. As discussed previously, broad-based comparisons between species is difficult, because no single PAH has been used universally in studies with fish; however, Solbakken and co-workers have exposed several species of fish to [^{14}C]PHN intragastrically and measured levels of PHN-derived radioactivity in tissues and fluids at common sampling times (Table 7).[35,36,55] In the teleosts, rainbow trout, and coalfish (*Pollachius virens*), the gall bladder contained >10% of the dose at maximum accumulation, whereas in the spiny dogfish (*Squalus acanthias*), an elasmobranch, gall bladder contained 1.2% of the dose at the sampling time showing maximum accumulation. The difference in the accumulation of PHN metabolites between teleosts and elasmobranchs is apparently related to the low hepatic MFO activities in spiny dogfish compared to teleosts.[56] The low level of accumulation of PHN metabolites in bile of spiny dogfish is not specific to PHN. Studies have shown high accumulation of NPH in liver (22% of the dose) and only small amounts in gall bladder (0.03% dose) of spiny dogfish.[57]

The influence of lipid content of organs on accumulation of PHN-derived radioactivity can help explain the differences in the results shown in Table 7. The coalfish is a lean fish having major lipid reserves in liver, whereas the rainbow trout is a fat fish having lipid distributed throughout the body. Correspondingly, the ratio of the the percent dose in liver to that in muscle is markedly greater for coalfish (11) than for trout (0.28). Further, the livers of the grunt (*Haemiilon sciurus*) and spiny dogfish have higher hepatic lipid contents (20 to 50%, respectively) than trout (10%) and also accumulate high levels of PHN in liver compared to muscle. The results in Figures 1A and 1B also show that in English sole, a

Table 7
DISPOSITION OF PHN IN FISH SPECIES[a]

	Spiny dogfish		Rainbow trout		Grunts		Coalfish	
	24 h	(max)[b]	24 h	(max)	24 h	(max)	24 h	(max)
Liver	21	74 (168 h)	1.8	1.8 (24 h)	22	22 (24 h)	47	72 (17 h)
Bile	0.1	1.2 (96 h)	4.9	14.0 (72 h)	N.R.[c]	N.R.	3.3	13 (36 h)
Muscle	3.3	20 (96 h)	6.4	6.4 (24 h)	3.5	6.5 (72 h)	3.9	6.4 (17 h)

[a]   Adapted from References 35, 36, and 55. Values represent mean percent dose of PHN and metabolites in
     fish exposed to ^{14}C-PHN intragastrically.
[b]   Values in parentheses represent time of maximum dose in tissue or bile.
[c]   N.R., not reported.

lean fish, exposed to sediment-associated BaP or to BaP intragastrically, liver and bile had
the highest concentrations of BaP and its metabolites. Thus, because the liver of fish is an
important organ for the storage of lipid and because it is the tissue that usually has the
highest levels of phase I and II enzyme activities, it is not surprising that the hepatobiliary
system accumulates high levels of PAH and their metabolites.

2. Age and Sex

There are marked differences in the rates of uptake and elimination of PAH by egg and
larval stages of fish. In general, laboratory studies show that the rates of accumulation and
elimination of a PAH by the egg are much slower than for the larval stages (Figures 3,4,
and 5).[28,58] The differences in the rates of uptake between egg and larval stages appear to
be related to the slower rate of transport of PAH across the chorion compared to the rate
of transport across the gill epithelium of larvae (Figure 3). The slow rate of uptake and
concomitant slow elimination by the egg (Figure 4) results in the ratio of the rate of uptake
to the rate of elimination of ^{14}C]NPH for eggs to be similar to the ratio for larvae. Thus,
since at equilibrium the BCF is proportional to the ratio of the rate of uptake to rate of
elimination, the bioconcentration of ^{14}C]NPH by coho salmon eggs and larvae is similar
(Figure 3).

Studies with early-life stages of fish also show that the rate of elimination of highly
lipophilic PAH is dependent on developmental stage. For example, in contrast to cod larvae
(*Gadus morhua*) no elimination of BaP was observed from cod eggs (Figure 5). Since BaP
is efficiently metabolized by fish species, the greater elimination of BaP by cod larvae is
most probably linked to the higher activity of xenobiotic metabolizing enzymes in fish larvae
than in eggs, as has been shown in the killifish.[59]

It is generally accepted that PAH are most toxic to the embryonic and larval stages of
fish.[60] Embryonic stages are, however, more tolerant to the toxic effects of PAH than the
larval stages.[28] This difference in toxicity may be due to the facts that (1) PAH are primarily
accumulated in the lipid-rich yolk sac of eggs, the contents of which are not used during
embryonic development; (2) activities of xenobiotic metabolizing enzymes are higher in
larval stages than in embryonic stages; and (3) the yolk is utilized during larval development
and, thus, PAH sequestered in the yolk are mobilized and transported to sensitive organs
where metabolism to toxic reactive intermediates may occur.

In addition to the exposure of pelagic embryos and larvae to water-borne PAH (e.g., from
oil spills), the transfer of PAH and other xenobiotics from parental fish to developing gametes
represents an important route by which early-life stages can be exposed to PAH. Furthermore,
exposure via the water column favors uptake of relatively more water-soluble PAH (e.g.,
benzenes, NPH), whereas parental fish living in contaminated sediment or feeding on benthic

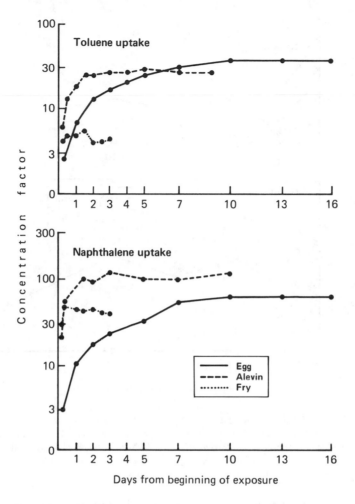

FIGURE 3.   Uptake of toluene and NPH by coho salmon *(Oncorhynchus kisutch)* eggs, mid-alevins, and fry. Concentration of toxicants in water (1.8 mg toluene/l, 108 mg NPH/l) were maintained within 10% of initial value for up to 16 days by changing solutions when needed. Uptake is represented as concentration factor (the mean tissue concentration of seven individuals divided by the mean water concentration of toxicant). Vertical bars are 95% confidence intervals. (Adapted from Korn, S. and Rice, S., *Rapp. P.-V. Reun., Cons. Int. Explor. Mer,* 178, 87, 1981.)

invertebrates are exposed to high concentrations of 3- to 5-benzenoid ring PAH (e.g., BaP) that are associated with the sediments. Thus, an important aspect of the fate of PAH in fish is the accumulation of potentially toxic, teratogenic or carcinogenic PAH in developing gonads and their subsequent disposition.

The accumulation of PAH in gonadal tissue is dependent on the extent that the PAH is metabolized. In our laboratories, we have examined the metabolism and disposition of NPH and BaP in gravid female and male English sole exposed to these PAH when gonadal development is nearly complete (approximately 2 weeks before spawning).[39,40] The results from these studies with female fish showed that ovary accumulated larger amounts of NPH and its metabolites compared to the amount accumulated in liver, while the converse was observed for BaP and its metabolites (Figure 6). The results also clearly show that BaP was present predominantly as metabolites, while NPH was present mainly as parent compound in both liver and ovary (Figure 6). The lower accumulation of BaP and its metabolites by

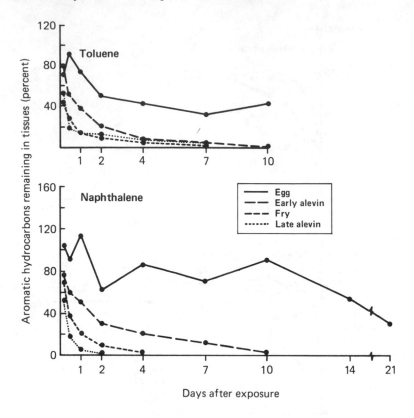

FIGURE 4.   Elimination of toluene and NPH by coho salmon *(Oncorhynchus kisutch)* eggs, mid-alevins, and fry. For details on exposure to toluene and NPH see Figure 3. Elimination is represented as percent of aromatic hydrocarbon remaining in tissues. (Adapted from Korn, S. and Rice, S., *Rapp. P.-V. Reun., Cons. Int. Explor. Mer,* 178, 87, 1981.)

ovary would suggest efficient uptake and metabolism by liver (i.e., first-pass effect), with smaller amounts of unmetabolized BaP than NPH being released to systemic circulation. Although smaller amounts of total BaP than NPH were present in gonads of gravid sole, when these values were compared to juvenile fish which do not accumulate detectable amounts of parent BaP in extrahepatic tissues, ovary and testes of gravid fish had a substantial proportion of unconverted BaP indicating an altered disposition of PAH during gonadal development. The high accumulation of NPH and relatively high percentage of unmetabolized BaP in ovary compared to other tissues indicate that the deposition of lipid in the oocyte strongly favors accumulation of the lipophilic parent compounds. However, the finding that testes of gravid male sole also contained high proportions (33%) of unconverted BaP suggests that other factors, such as altered blood flow, may also strongly favor the deposition of PAH and their metabolites in gonads of sexually mature fish. The hypothesis that the disposition of PAH is significantly altered during gonadal development in fish is supported by the results from studies on the disposition of 2,2′,5,5′-tetrachlorobiphenyl (TCB) in female trout that show that TCB is redistributed from fat depots to developing eggs.[61] Thus, as a consequence of this redistribution, spawning increased the whole-body elimination of TCB. A later study showed that larvae of feral lake trout *(Salvelinum namaycush)* from Lake Michigan and Green Bay had higher hepatic aryl hydrocarbon hydroxylase (AHH) activity than did hatchery controls.[62] These results plus additional supporting biochemical data provided strong evidence that contamination of the feral lake trout gametes by polychlorinated biphenyls (PCB) and possibly other xenobiotics caused induction of hepatic AHH

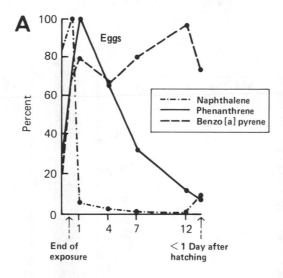

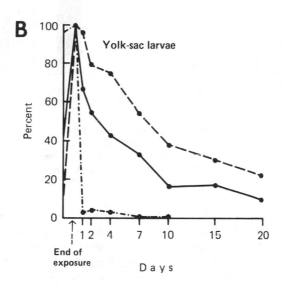

FIGURE 5. Uptake and elimination of ^{14}C-labeled NPH, PHN, and BaP in (A) eggs and (B) yolk-sac larvae of the cod *Gadus morhua* exposed to these compounds dissolved in seawater. Results expressed as percentages of maximum contents of radioactivity for each compound. Arrows: end of 24 h exposure; numbers 1 to 20: days in clean water. (Adapted from Solbakken, J. E., Tilseth, S., and Palmork, K. H., *Gadus morhua, Mar. Ecol. Prog. Ser.*, 16, 297, 1984. With permission.)

activity in the larvae. Thus, these studies serve to illustrate that parents may be an a important vector for contamination of early-life stages of fish with potentially teratogenic PAH.

E. Effects of Exposure to Other Xenobiotics

Aquatic species are exposed to a myriad of contaminants simultaneously and the issue of interactions between xenobiotics on their respective disposition, and interactive effects (synergistic, additive, antagonistic, or potentiating) on biochemical and physiological parameters are of great concern. One area that has received considerable attention is the induction of

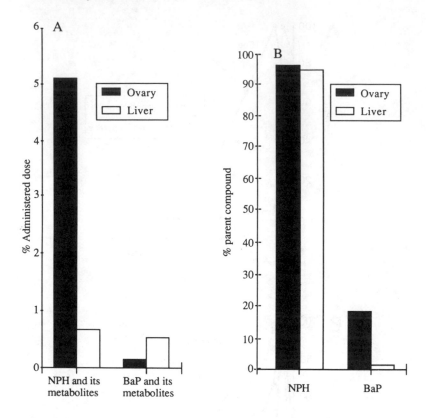

FIGURE 6.   Percent administered dose (A) and percent parent compound (B) in liver and ovary of gravid English sole *(Parophrys vetulus)* exposed intragastrically to either [³H]-NPH or [³H]-BaP). (Adapted from Varanasi, U., Stein, J. E., and Hom, T., *Biochem. Biophys. Res. Commun.*, 103, 780, 1981, and Reichert, W. L. and Varanasi, U., *Environm. Res.*, 27, 316, 1982.)

phase I and II xenobiotic-metabolizing enzymes, because of their obvious importance in influencing the disposition of PAH in aquatic species. In this chapter we will cover the effect of inducers on PAH metabolism and disposition *in vivo,* and in Chapter 5 the effects of inducers on the activities of enzymes involved in the metabolism of PAH are discussed.

The effects of environmental contaminants on the disposition of PAH have received relatively little attention. The studies to date have concentrated on the effects of MFO inducers on tissue and fluid levels of a number of PAH but, in particular, NPH and its alkylated derivatives. The preexposure of salmonids to MFO inducers (e.g., BaA) β-naphthoflavone, and PCB) followed by exposure to radiolabeled NPH or an alkylated NPH resulted in consistently increased levels of radioactivity in bile of induced fish compared to controls.[24,63,64] Changes in hepatic concentrations of radioactivity or percent metabolites, however, were not consistent among these studies. For example, in coho salmon pretreated with various doses of PCB and subsequently given an i.p. injection of [¹⁴C]2,6-dimethylNPH, neither the concentrations nor proportions of the parent PAH or its metabolites in liver were significantly different from that found in control fish.[64] In contrast, pretreatment of rainbow trout with BaA resulted in significantly increased concentrations of 2-methylNPH-derived radioactivity in liver at 6 h post-exposure, whereas pretreatment of trout with β-naphthoflavone resulted in an apparent decrease in the concentration of 2-methylNPH-derived radioactivity in liver and an increase in the percent metabolites in liver.[24,29] Studies by Stein et al. in which English sole were placed on (1) a reference sediment containing BaP and PCB, either singly or together, or (2) sediments from a contaminated site and a reference

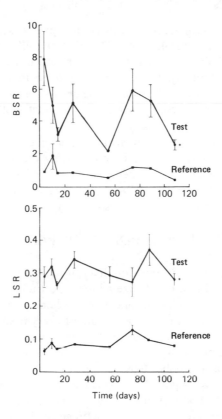

FIGURE 7.   The ratios of the concentration of BaP-derived radioactivity in bile and liver to that in sediment (BSR [pmol BaP equiv./g bile wt]/[pmole BaP/g sediment wet wt]; LSR [pmol BaP equiv./g liver wet wt]/[pmole BaP/g sediment wet wt]) for English sole exposed to test (●) and reference (■) sediments for up to 108 days. Data are shown as X ± S.E.M. (n = 6 to 16). *Indicates significant effect due to exposure of sole to test sediment relative to exposure to reference sediment. The BSR values were significantly affected by length of exposure, but the LSR values were not. (Reprinted from Stein, J. E., Hom, T., Casillas, E., Friedman, A., and Varanasi, U., *Mar. Environ. Res.*, 22, 123, 1987.)

site showed significantly greater accumulation of BaP and its metabolites in liver and bile of English sole exposed to BaP in the presence of other xenobiotics relative to that observed in fish exposed to BaP alone (Figure 7).[10,13] The cumulative results from the above studies on the effects of exposure to MFO inducers on *in vivo* PAH metabolism are that the concentrations of PAH metabolites in bile of fish are increased and that the effect on levels in liver is unpredictable because of the influence of a number of factors, which may include the route of exposure, specific inducer used, and structure of the PAH whose disposition is being examined.

The effect of exposure to xenobiotics on phase I metabolism was the main emphasis of the above studies, while less emphasis was directed towards effects on phase II metabolism or physiological mechanisms of excretion of PAH metabolites. However, alteration in phase II metabolism and mechanisms of excretion may substantially alter the disposition of PAH in fish, and thus give results contrary to that predicted only from induction of enzymes involved in phase I metabolism. For example, pre-exposure (48 h) of Dolly Varden char to water-borne NPH, a weak or noninducer of hepatic AHH in fish, at 75% of its 96-h LC_{50} followed by an exposure (p.o.) to [^{14}C]NPH, resulted in (1) a decrease in percent administered dose of NPH-derived radioactivity in liver, brain, and muscle, (2) no change in percent administered dose in bile, and (3) an increase in percent NPH metabolites in bile, suggesting that the major effect from pretreatment with NPH was altered excretion of unmetabolized

NPH.[65] It is worth noting that mussels (*Mytilus edulis*) preexposed to unlabeled NPH had a significantly higher rate of elimination of [^{14}C]NPH from gills and kidneys than mussels not pre-exposed to NPH.[66] Further, it has been demonstrated that the rate of excretion of PAH metabolites can be inhibited by other substrates of secretory transport systems. For example, the renal excretion of BaP-7,8-diol conjugates by southern flounder (*Paralichthys lethostigma*) was shown to be markedly retarded by pretreatment with the herbicide 2,4-dichlorophenoxyacetic acid.[67]

In conclusion, these studies serve to demonsrate that interactions between xenobiotics can affect both biochemical and physiological systems to alter the disposition of PAH in fish. Moreover, detailed information on the interactive effects of xenobiotics on pathways of absorption, metabolism, and excretion of PAH is necessary to accurately predict effects of coexposure to xenobiotics on the toxicokinetics of PAH.

III. METABOLISM OF PAH

Early studies with subcellular fractions from liver of freshwater and marine fishes clearly demonstrated the ability of fish to metabolize a broad spectrum of organic xenobiotics (see Chapter 5). Subsequent studies with post-mitochondrial homogenates and microsomes from extrahepatic tissues of fish species revealed that almost every tissue or organ examined had detectable MFO activity towards PAH, although the maximum activity was generally found in liver.[68,69] Information from such *in vitro* studies with piscine and mammalian species demonstrated a sequence of metabolic steps, involving oxidation and conjugation reactions that can occur *in vivo* to convert lipophilic organic chemicals into water-soluble metabolites that are easily excreted into bile and urine. Briefly, the first step in oxidation of a PAH involves insertion of an oxygen atom to form an arene oxide, a reaction mediated by the cytochrome P-450-dependent MFO system. Arene oxides of PAH are relatively unstable and can undergo (1) spontaneous rearrangement to yield phenols which can be further oxidized to quinones, (2) epoxide hydrolase-catalyzed hydration to yield dihydrodihydroxy compounds (diols), or (3) a conjugation reaction with glutathione (GSH) mediated by glutathione-S-transferases (GST). Diols and phenols of PAH are substrates for several conjugation reactions mediated by enzymes that catalyze glucuronidation, glycosylation, and sulfation.[70] The diols and phenols can also undergo a second oxidation reaction to form diol-epoxides and phenol-oxides, some of which are believed to be the ulitmate carcinogenic intermediates of PAH and are known to interact covalently with DNA, RNA, and protein.[71] Thus, the balance between activation and detoxication pathways of PAH would partially determine their potential toxic action in a given biological system. In addition to P-450 oxidation, studies with mammalian systems also reveal the existence of oxidative systems involving lipid peroxidation and prostoglandin synthesis.[72] Our current knowledge of such systems in aquatic organisms is essentially nonexistent. Although data on the levels of activities of enzymes catalyzing various metabolic steps are useful in predicting the ability of an organism to metabolize a chemical, studies of the *in vivo* metabolism of xenobiotics are essential in obtaining accurate information on how an organism processes these compounds. The scientific literature reveals that most of the *in vivo* studies of PAH metabolism with fish are conducted primarily with three PAH, namely, NPH, PHN, and BaP, although a few reports of the metabolism of benzene and alkylated NPH are available. Because the general pattern of metabolism of most PAH appears to be similar, a review of the data on the metabolism of NPH, PHN, and BaP in various fish species will provide a comprehensive view of the metabolism of PAH in general.

A. Naphthalene

The pathways for the formation of primary NPH metabolites by rat *in vivo* have been elucidated, providing a basis for comparison of the types of phase I NPH metabolites formed

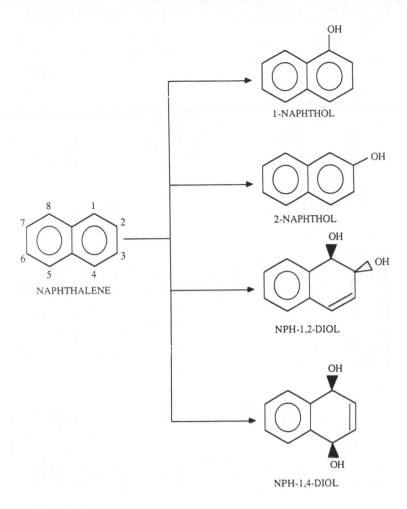

FIGURE 8. Primary metabolites of NPH identified in the rat. (Adapted from Horning, M. G., Stillwell, W. G., Griffin, G. W., and Tsang, W.-S., *Drug Metab. Disposit.*, 8, 404, 1980.)

by fish and mammals. The major route of metabolism in rodents appears to be the formation of NPH-1,2-oxide, which gives rise to phenols and diols (Figure 8). The detection of NPH-1,4-diol suggests formation of a NPH-1,4-oxirane ring, although this is speculative.[73] In addition, several metabolites appear to be substrates for secondary oxidation, giving rise to di-, tri-, and tetrahydroxy derivatives. Extensive studies by Sims and colleagues with rodents have delineated the formation of conjugated metabolites such as sulfates, glucuronides, and mercapturic acid derivatives of phase I metabolites of NPH.[74-76]

As discussed earlier, marine and freshwater fish absorb NPH when exposed to this PAH via water, sediment, or when given intragastrically or intraperitoneally.[18,33,38,77] Data in Table 8 show a variety of metabolites isolated from liver, bile, heart, brain, and muscle of salmonids and pleuronectid fishes exposed to NPH. In most of these studies, fish were exposed to radiolabeled NPH and the metabolites were separated by thin-layer chromatography (TLC) and identified by comparing R_f values of standards, such as the 1-naphthol, 1,2-dihydro-1,2-dihydroxyNPH (NPH-1,2-diol), and the sulfate, mercapturic acid, glycoside, and glucuronide conjugates. As shown in Table 8, profiles of metabolites isolated from body fluids and tissues were similar. Data revealed that fish liver contains mainly 1-naphthol and NPH-1,2-diol and their glucuronide conjugates, whereas gall bladder contains mainly

Table 8
METABOLITES OF NPH IN VARIOUS TISSUES OF FISH SPECIES EXPOSED TO NPH[a]

Species	Dose (mg/kg)	Route of exposure (temp)	Time (h)	Tissue/fluid	Naphthols	NPH-1,2-diol	Glucuronides	Mercapturic acids	Sulfates	Ref.
								% of total NPH metabolites		
Rock sole	0.44	p.o.(12°C)	24	Liver	16	39	13	14	4	18
				Bile	0.8	3.5	81	11	1.1	
				Skin		60	9	6	4	
		p.o. (12°C)	168	Liver	29	7	27	13	14	
				Bile	0.7	1.4	86	8.5	1.8	
			168	Liver	12	12	17	11	35	
				Bile	2.9	3.0	82	8.8	0.9	
Starry flounder	0.44	p.o. (4°C)	24	Liver	29	48	16	11	N.D.	46
				Bile	2.8	7	74	11	1.8	
				Muscle	1.8	77	4.6	3.2	3.6	
			168	Liver	6.2	30	32	11.7	10	
				Bile	0.6	2.5	85	7.7	1.8	
				Muscle	6.9	47	13	7.2	9	
		p.o. (12°C)	24	Liver	11	40	32	10.4	1.9	46
				Bile	1.5	4.0	82	8.9	1.4	
				Muscle	2.4	81	5.4	3	1.5	
			168	Liver	11	10	17	11	35	
				Bile	2.9	3.0	82	8.8	0.9	
				Muscle	22	24	24	8.0	14	
		i.p. (12°C)	24	Liver	21	41	8.3	N.D.	15	
				Bile	0.1	1.8	89	7.0	1.2	
			168	Liver	21	29	9.7	N.D.	24	
				Bile	0.6	0.8	90	6.7	1.3	
English sole		Via sediment exposure[b]	168	Bile	N.D.	0.3	88	2	3	33
Rainbow trout	0.081	p.o. (8—10°C)	16	Brain	N.D.	100	N.D.[c]	N.D.	N.D.	78
				Liver	0.2	65	N.D.	N.D.	33	
				Blood	N.D.	79	0.1	N.D.	20	

Coho salmon (fin-ger-lings)	10.5[d]	i.p. (14°C)	24	Brain	32	5	35	11	17	27
				Liver	40	11	25	22	2	
				Bile	7	7	75	9	2	
				Muscle	47	7	14	6	26	
Coho salmon (adult)	0.86	p.o. (10°C)	16	Liver	1.2	67	1.2	N.D.	1.6	45
				Bile	10	68	6.8	N.D.	8.8	

a Fish were exposed intragastrically to [³H]NPH dissolved in salmon oil except for the study with adult coho salmon,[49] which were exposed to unlabeled NPH. Metabolites were identified by co-chromatography with authentic NPH metabolite standards using thin-layer chromatography; the sulfate and glycoside conjugates standards co-chromatographed in this solvent system, however. the presence of glycoside conjugates was not confirmed by enzymatic hydrolysis. Whereas the presence of glucuronide and sulfate conjugates was confirmed by subsequent treatment with β-glucuronidase and arylsulfatase, respectively.

b Radiolabeled NPH was mixed with marine sediment containing 1% Prudhoe Bay crude oil.

c N.D., not detected.

d 95% Ethanol was used as the solvent carrier.

glucuronide conjugates. An exception was reported with coho salmon (Table 8), where bile contained high proportions (72%) of unconjugated NPH-1,2-diol.[45] However, subsequently it was found that this apparent high level of unconjugated diol was due to inefficient extraction of conjugates by the method used in this study.[78,79] When the glucuronide conjugates from bile of fish are isolated and hydrolyzed with β-glucuronidase, the major unconjugated metabolite detected is the NPH-1,2-diol. Sulfate conjugates do not constitute a major metabolite class in the hepatobiliary systems of either salmonid or pleuronectid fish (Table 8). Glycoside conjugates of PAH are generally not present in significant amounts in hepatobiliary systems of fish, except for one report of high concentrations of glycoside conjugates in bile of starry flounder exposed to 2,6-methylNPH.[81]

The low levels of unconjugated NPH-1,2-diol in bile of most species suggest that conjugation greatly facilitates excretion of metabolites from liver into bile. Marked differences observed in the distribution of types of conjugate classes between liver and bile appear to be from different steps of metabolism being rate limiting.[82] For example, in the case of NPH glucuronides, their formation appears to be the rate-limiting step, whereas for mercapturic acid derivatives the actual excretion, rather than formation, appears to be rate limiting. This hypothesis is supported by the results showing that liver relative to bile contains high proportions of mercapturic acid conjugates when expressed as percent of total metabolites, whereas bile contains predominantly glucuronides (Table 8). Further support for the preferential release of glucuronide conjugates of PAH into bile is demonstrated in a study of the disposition of radiolabeled BaP-4,5-oxide in rat.[83] It was shown that the majority of the radioactivity in the bile was present as the glucuronide conjugate of BaP-4,5-diol (the diol is the hydrolysis product of BaP-4,5-oxide), and that the majority of radioactivity in liver was due to the BaP-4,5-oxide-GSH conjugate, the precursor to the mercapturic acid conjugate.

A recent study has shown that both sulfate and glucuronide conjugates are formed by 12 different freshwater fish species exposed to water-borne 1-naphthol (Table 9).[84] The aquarium water contained a high proportion of sulfate conjugates, relative to the 1-naphthyl-glucuronide. In contrast, analysis of bile showed that the 1-naphthyl-glucuronide was the major metabolite, whereas sulfate conjugates made up a small fraction of the total conjugates in all fish species (Table 9). No GSH conjugates of 1-naphthol metabolites (e.g., phenol oxides) were reported.[84] The presence of a large proportion of 1-naphthyl-sulfate in aquaria water suggests that removal via kidney may be the major route of excretion for sulfate conjugates of 1-naphthol, whereas glucuronide conjugates appear to be excreted via bile.[37] However, because the concentrations of conjugates in bile and in aquaria water were not given, the ratio of the total amount of sulfate to glucuronide conjugates cannot be calculated. *In vitro* studies of the metabolism of PAH suggest that sulfation, relative to glucuronidation, appears generally to be low in fish as seen by higher proportion of glucuronide than sulfate conjugates formed by plaice (*Pleuronectes platessa*) and brown bullhead (*Ictalurus nebulosus*) hepatocytes and cell lines developed from fry of bluegill sunfish (*Lepomis macrochirus*) (BF-2 cells).[85,86]

An early study by Krahn et al. gives some indication of the relative importance of sulfation vs. glucuronidation of naphthols in fish.[87] In bile of rainbow trout exposed to unlabeled NPH at a dose of 7.3 mg PAH per kilogram body weight (b.w.), the concentration of the sulfate conjugate of 1-naphthol was higher (7.4 μg/g, bile) than that for the glucuronide conjugate (4.4 μg/g, bile) as determined by reversed-phase high performance liquid chromatography with fluorescence detection. At ten times the dose (73 mg NPH per kilogram b.w.), however, the concentration of the sulfate decreased to 3.6 μg/g, bile, whereas the glucuronide concentration increased to 10.2 μg/g, bile. This difference in the ratio of sulfate to glucuronide conjugates of 1-naphthol at different doses may be due to the apparent large

Table 9
PROPORTION OF CONJUGATES FORMED BY 12 FISH SPECIES AFTER 48-H EXPOSURE TO WATER-BORNE ^{14}C-1-NAPHTHOL[a]

	Bitterling	Bream	Carp	Goldfish	Gudgeon	Guppy	Minnow	Perch	Roach	Rudd	Stickleback	Tench
No. of fish tested	7	5	3	15	12	45	15	8	12	12	10	12
Exposure concentration of 1-naphthol (mg/l)	3	3	5	5	4	2	2	4	3	3	2	3
% Naphthol-derived radioactivity[b]												
Aquarium water												
Naphthyl sulfate	65	70	76	77	72	55	61	63	76	76	66	70
Naphthyl glucuronide	20	13	8	9	9	2	13	23	16	18	16	12
Total conjugates	85	83	84	86	81	57	74	86	92	94	82	82
Bile												
Naphthyl sulfate	6	6	4	6	5	—c	10	12	11	6	11	12
Naphthyl glucuronide	82	80	91	91	93	—	84	80	87	85	77	79
Total conjugates	88	86	95	97	98	—	94	92	98	91	88	91

a   Adapted from Reference 84.

b   Values represent % of ^{14}C recovered from water or detected in bile. Metabolites were analyzed by TLC.

c   Guppies were too small for bile sampling.

difference in the K_m between UDPGT ($10^{-4} M$) and sulfotransferase ($10^{-6} M$) for phenols.[88] At lower doses, sulfation would be the preferred pathway because of the lower K_m value for sulfotransferase, whereas glucuronidation would predominate at higher dose levels. This is further supported by the studies of Nagel showing that as the phenol concentration in water is increased, the fraction of phenol glucuronide is increased and the fraction of phenol sulfates is decreased in goldfish (*Carassius auratus*).[88] An alternative explanation for the relative decrease in sulfation at higher doses could be that sulfate conjugation of phenol by sulfotransferase is limited by the supply of the cofactor 3'-phosphoadenosine-5'-phospho-sulfate, which may be depleted when conjugation demand is high, as has been suggested.[89-91]

Age of fish appears to influence the disposition of NPH metabolites in specific organs. For example, analyses of metabolites of radiolabeled NPH in extrahepatic tissues of fingerling coho salmon showed that brain contained significant proportions of unconjugated metabolites such as diols (Table 8). Interestingly, brain of young fish also contained significant proportions of conjugates, especially the 1-naphthylglucuronide.[27] However, a later study by Collier et al. showed that brain of adult rainbow trout contained mainly the NPH-1,2-diol and that no conjugates were detected.[78] The presence of conjugates in brain of fingerling salmonids was attributed to the possible underdevelopment of the blood-brain barrier in juvenile fish, thus, allowing conjugated NPH metabolites to enter the brain.[27] Whether such differences in accumulation of NPH metabolites in neural tissues results in differential toxicity in adult and young fish is unknown.

Reichert and Varanasi exposed sexually mature English sole to NPH to study the effect of gonadal maturation on metabolism and disposition.[40] No marked sex-related differences were observed in concentrations or proportions of metabolites in liver and blood. However, the results show that while ovary contained mainly unconjugated metabolites, testes contained both nonconjugates and conjugates in nearly equal concentrations (Table 10). This difference in metabolite profile between male and female gonads may be due to either differences in the absorption of glucuronide conjugates by gonads from blood, or higher levels of UDPGT activity in testes relative to ovary. Concentrations of glucuronides in liver and blood were similar for both male and female sole (Table 10), indicating that the higher level of glucuronides in testes was not due to a concentration effect. However, differences in types of cells between ovary, a lipid-rich organ, and testes may affect the permeability of glucuronides into certain cell types of each organ.

In the above study, both TLC and reversed-phase HPLC were used to identify metabolites of NPH in liver and gonads (Figure 9).[40] HPLC analyses showed that NPH-1,2-diol was the major metabolite in liver, with smaller but appreciable amounts of NPH-derived radioactivity coeluting with naphthol and naphthylsulfate standards. However, subsequent enzymatic hydrolysis showed that the radioactivity cochromatographing with the 1-naphthylsulfate standard on HPLC was due to the NPH-1,2-diol-glucuronide conjugate and that sulfate conjugates were not present in significant amounts in either liver or gonads of sole. These results serve as a caution not to rely solely on a single chromatographic technique to characterize PAH metabolites.

Consequences of preferential retention or release of different classes of conjugates of PAH to the health of fish are not known. In fact, there is virtually no information on toxicity of various NPH metabolites in fish. It is known, however, that exposure of rainbow trout embryos and larvae to increasing concentrations of NPH is accompanied by decreased hatchability and survival of the fish.[92] In addition, 2-naphthol had a slightly lower LC_{50} with rainbow trout embryos (0.07 mg/l) than did NPH (0.11 mg/l), whereas the opposite was observed with largemouth bass (*Micropterus salmoides*) embryos (2-naphthol LC_{50} = 1.77, NPH LC_{50} = 0.51 mg/l).[93] Sublethal concentrations of NPH in water have also been shown to decrease the growth rate of juvenile pink salmon and fathead minnow (*Pimephales*

Table 10
DISTRIBUTION OF METABOLITES IN ENGLISH SOLE GIVEN
³H-NPH p.o.[a]

Metabolites	Female 24 h	48 h	168 h	Male 24 h	48 h	168 h
Liver						
Total conjugates	35	38	>9.6	31	17	7.2
Glucuronides	33	16	>9.6	28	17	>6.6
Total nonconjugates	27	6.2	N.D.	30	6.7	3.8
NPH-1,2-diol	21	4.8	N.D.	23	6.6	3.2
1- and 2-naphthols	2.1	N.D.	N.D.	4.3	<0.12	N.D.
Unknown	5.6	2.9	N.D.	5.7	0.05	N.D.
Blood						
Total conjugates	23	11	>8.4	16	29	>14
Glucuronides	20	10	>8.4	14	25	>14
Total nonconjugates	33	6.8	N.D.	23	12	N.D.
NPH-1,2-diol	29	5.0	N.D.	17	12	N.D.
1- and 2-Naphthols	N.D.	N.D.	N.D.	N.D.	N.D.	N.D.
Unknown	7.3	2.9	N.D.	7.8	4.5	N.D.
Ovary / Testes						
Total conjugates	1.3	0.68	N.D.	10.8	8.2	N.D.
Glucuronides	0.52	Trace	Trace	9.5	8.0	N.D.
Total nonconjugates	19	5.3	N.D.	15	7.8	N.D.
NPH-1,2 diol	17	4.5	Trace	11	6.9	N.D.
1- and 2-Naphthols	1.0	N.D.	N.D.	<0.52	N.D.	N.D.
Unknown	1.5	0.83	N.D.	3.9	2.7	N.D.

Note: N.D., not detected.

[a] Fish were force-fed ³H-NPH dissolved in corn oil (0.3 μmol/kg, b.w.) and sampled at 24, 48, and 168 h after exposure metabolites were separated by TLC. Values are given as picamoles NPH-equivalents per milligram tissue or fluid (dry weight). Adapted from Reference 40.

promelas).[94,95] Without a larger data set on the acute toxicity of NPH and its metabolites, the relative toxicity of parent compound vs. metabolites cannot be assessed. Moreover, results of chronic toxicity tests may be important when attempting to determine the effect of metabolism of NPH on toxicity of this PAH.

B. Phenanthrene

Phenanthrene (PHN) is a component of crude oil and has been shown to be absorbed by fish after intragastric or water exposure.[35,36,58] The major route of metabolism in rat is the formation of several arene oxides (Figure 10). Specifically, oxidation at the 9,10-bond of PHN leading to the PHN-9,10-diol is the major pathway of metabolism in rodents. In addition to the formation of phenols and diols, the glucuronide, sulfate, and GSH conjugates of PHN metabolites are tentatively identified in rodent studies.[96]

Metabolism of PHN by fish has been studied primarily by Solbakken and colleagues using the following teleost species: flounder, rainbow trout, coalfish, Atlantic cod and an elasmobranch species, the spiny dogfish.[97-99] The types and distribution of PHN metabolites in

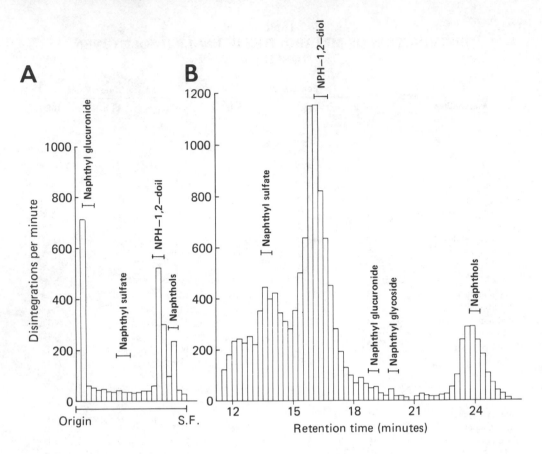

FIGURE 9.   Analysis of NPH and its metabolites present in liver of English sole exposed orally to 0.1 mg
[³H]NPH/kg body weight. NPH and its metabolites were extracted from liver and analyzed by (A) thin-layer
chromatography, using a *n*-butanol:ammonium hydroxide:water (40/10/0.5,v/v/v) solvent system, and (B) reversed-
phase HPLC using a linear gradient from 0.5% acetic acid in water to 100% methanol. (Adapted from Reichert,
W. L. and Varanasi, U., *Environ. Res.*, 27, 316, 1982.)

bile, urine, and contents of stomach and intestine of the various fish species were determined
by gas chromatographic-mass spectrometry (GC-MS) after hydrolysis of conjugated metabolites with β-glucuronidase and arylsulfatase. The metabolites identified in various tissues
and fluids of fish were the PHN-1,2-diol, PHN-3,4-diol, PHN-9,10-diol, 1-hydroxyPHN,
2-hydroxyPHN, 3-hydroxyPHN, and 4-hydroxyPHN (Table 11). In contrast to the results
with rodents and spiny dogfish, in which PHN-9,10-diol, PHN-3,4-diol, and 3-hydroxyPHN
are the major metabolites detected, in all teleost species the glucuronide conjugate of PHN-
1,2-diol was the major metabolite detected in bile. Moreover, in flounder and coalfish, urine
also contained the glucuronide conjugate of PHN-1,2-diol as the major metabolite (>90%).
In addition to the bile and urine, intestine of flounder had a high percentage of PHN-1,2-
diol; however, in trout intestine PHN-9,10-diol was the major metabolite.[97] The ratio of the
proportion of PHN-1,2-diol to PHN-9,10-diol was 46 in bile and 0.25 in intestine of trout.
The relatively high proportion of PHN-9,10-diol in intestine of trout may be due to (1)
differences in the types of cytochrome P-450 isozymes in the intestinal enterocytes of trout,
giving rise to more PHN-9,10-diol than PHN-1,2-diol, (2) preferential removal of PHN-
1,2-diol from intestine through enterohepatic recirculation, thus, increasing the percent of
PHN-9,10-diol in the digestive tract, or (3) metabolism of PHN by gut flora. Recent *in vivo*
and *in vitro* studies of the metabolism of PHN by Atlantic cod, show, however, that the

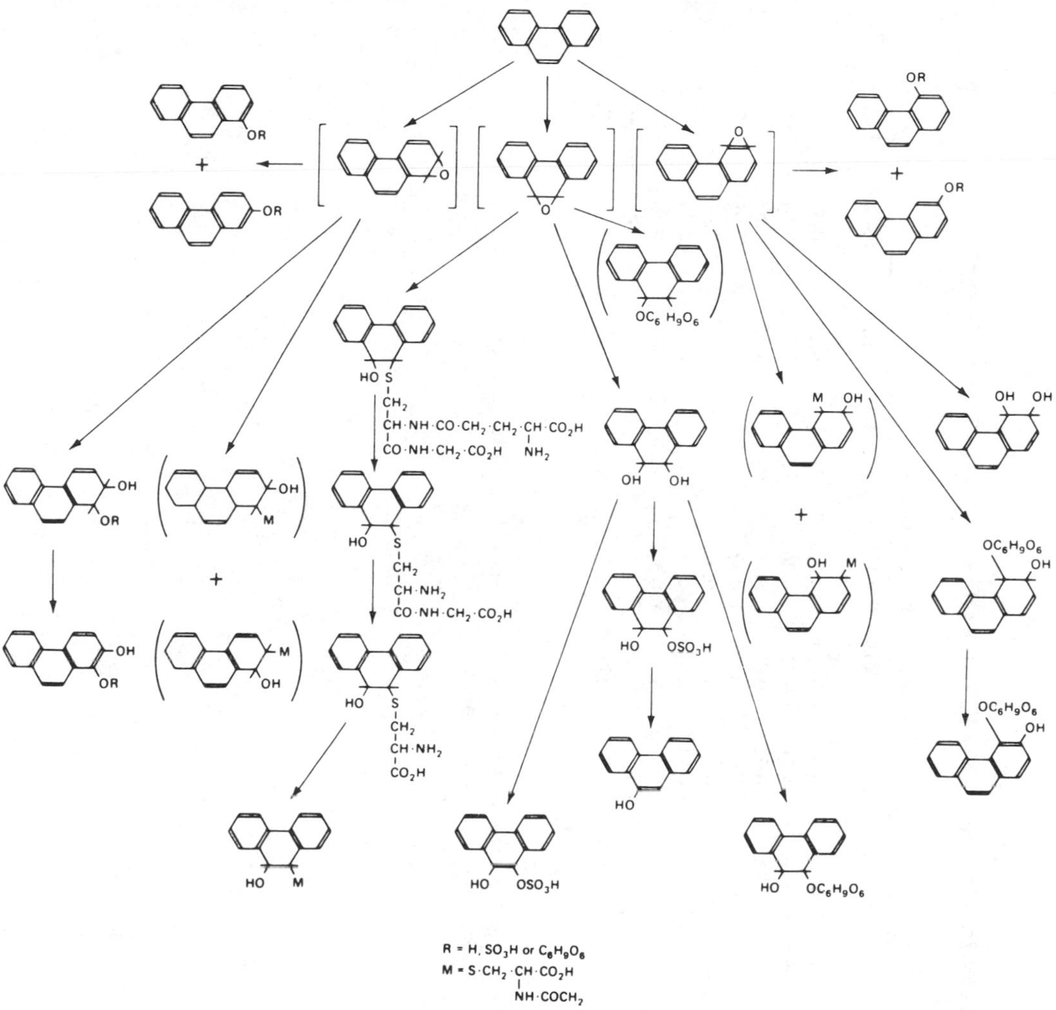

FIGURE 10. Metabolic pathways of PHN in the rat. $C_6H_9O_6$, glucuronic acid. (Adapted from Boyland, E. and Sims, P., *Biochem. J.*, 84, 571, 1965.)

regulation of the metabolism of PHN *in vivo* is complex and cannot be explained by either the level of specific hepatic cytochrome P-450 enzymes or their regioselectivity.[99] In fact, studies with hepatic microsomes from cod revealed that PHN-9,10-diol was a major (34 to 69%) metabolite of PHN, whereas in bile of PHN-exposed cod the glucuronide conjugate of PHN-1,2-diol was the major (91%) metabolite detected (Table 11). These results suggest that differences in conjugation as well as disposition of the two diols may occur *in vivo*. Thus, PHN metabolism in fish species must be conducted *in vivo* and *in vitro* to help identify the important pathways controlling the fate and effects of this PAH, which so far is the only PAH known to undergo substantial K-region metabolism (i.e., metabolism at the 9,10-bond) in fish.

Solbakken et al. reported that in addition to PHN-1,2-diol, trout intestine also contained 2-hydroxyPHN.[97] The detection of 2-hydroxyPHN is of interest because phenolic compounds of PHN are generally detected only in low levels in the hepatobiliary system of fish, and 2-hydroxyPHN is a thermodynamically unfavorable metabolite of the PHN-1,2-oxide.[100] Rearrangement of the epoxide should give predominantly 1-hydroxyPHN, since the carbonium ion formed from the epoxide opening at C-2 has the greatest resonance stabilization.

Table 11
PHASE I METABOLITES OF PHN IN TISSUES AND FLUIDS OF VARIOUS FISH SPECIES[a]

Species	Dose (mg/kg)	Tissue	1-OH[b]	2-OH	3-OH	4-OH	9-OH	1,2-Diol	3,4-Diol	9,10-Diol	Ref.
								% of total metabolites			
Flounder	25	Bile	0.16	0.038	0.058	0.054	0.012	98.6	0.29	0.77	
		Stomach and intestine	0.26	N.D.	N.D.	N.D.	0.26	92.9	1.91	4.65	97
		Urine	0.97	0.46	0.32	0.97	0.28	90.6	1.85	4.57	97
Rainbow trout	25	Bile	3.74	0.44	0.32	0.38	0.13	92.7	0.32	2.03	97
		Intestine	2.86	19.1	0.95	N.D.	0.95	15.2	N.D.	60.9	
Coalfish	75	Bile	1.12	0.29	0.51	0.23	0.16	96.4	N.R.	1.29	98
		Urine	1.83	0.39	0.52	0.92	0.52	93.7	N.R.	2.16	
Atlantic cod	50	Bile	3	2	3	0.6	N.D.	91	N.R.	N.D.	99
Spiny dogfish	25	Bile	N.D.	1.18	33.8	0.68	N.D.	N.D.	18.6	45.7	97

Note: N.D., not detected; N.R., not reported.

[a] All fish were exposed to PHN intragastrically and sampled 48 h after exposure. All samples were treated with glusulase to hydrolyze glucuronide and sulfate conjugates, derivatized, and then analyzed by gas chromatography. Values represent the mean.

[b] OH, hydroxyPHN; Diol, dihydroxydihydroPHN (see Figure 10).

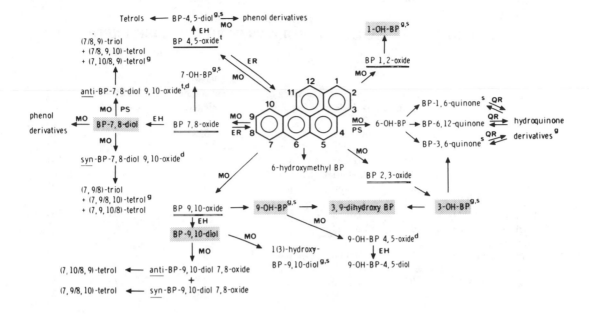

FIGURE 11. Metabolic pathways of BaP in the rat. Metabolites enclosed by a box have been identified in fish. Abbreviations: MO, mono-oxygenase; EH, epoxide hydrolase; ER, epoxide reductase; QR, quinone reductase; PS, prostaglandin synthetase. The superscripts indicate metabolites that: [t], can be converted into a glutathione conjugate;[s], can be converted into a sulfuric acid conjugate,[g] can be converted into a glucuronic acid conjugate; and [d], can contribute to covalent binding of hydrocarbon to nucleic acids in cells or tissues that have been treated with BaP. (Adapted from Cooper, C. S., Grover, P. L., Sims, P., *Prog. Drug Metab.*, 7, 295, 1983.)

Moreover, a recent study has shown that four cytochrome P-450 enzymes isolated from β-naphthoflavone-treated cod form mainly 1-hydroxyPHN and little 2-hydroxyPHN when reconstituted with phospholipid and NADPH-cytochrome P-450 reductase from rat.[99]

Formation of the PHN-1,2-diol in various fish species is of considerable interest because it is the precursor of the PHN-1,2-diol-3,4-epoxide which is the most mutagenic metabolite of PHN, although PHN does not appear to possess carcinogenic activity in mammals.[101] Acute-toxicity testing has shown a relationship between increasing concentration of PHN and reduced survival of rainbow trout embryos and larvae. In addition, the LC_{50} value for PHN was calculated to be 0.04 mg/l for rainbow trout and 0.18 mg/l for largemouth bass.[92] At present it is not known whether the ability to metabolize PHN is important in the differential toxicity of PHN in these fish species.

C. Benzo(a)pyrene

The ubiquitous nature of this pollutant and its demonstrated carcinogenicity in both rodents and fish make it an excellent model compound to study activation and detoxication of environmental carcinogens in biological systems.[71,102] As depicted in Figure 11, BaP is metabolized by rodents to several isomeric phenols, diols, quinones, phenol-oxides, and diol-epoxides.[103] Moreover, several of these metabolites have been identified *in vitro* using fish hepatic microsomes and cell lines (see Chapter 6).

The findings described in the previous section of this chapter demonstrated that 3- to 5-benzenoid-ring PAH, including BaP, that are present in sediments from industrialized water-ways, are bioavailable to benthic fishes and are extensively metabolized.[10,11,13,16] Detailed studies on BaP metabolism and disposition show that regardless of the route of BaP exposure,

Table 12
**DISTRIBUTION OF PHASE II METABOLITES IN BILE OF FISH EXPOSED TO
BaP[a]**

Species	Dose (mg/kg)	Vehicle	Mode of exposure	% Glucuronide and sulfate conjugates	% Thio-ether conjugates	Ref.
English sole	2	Acetone	i.p.	53	47	19
Starry flounder	2	Acetone	i.p.	58	42	19
Rock sole	2	Acetone	i.p.	51	49	104
Rainbow trout	2	Acetone	i.p.	54	46	104
English sole (juvenile)	0.1	Corn oil	p.o.	43	57	105
English sole (adult)	0.1	Corn oil	p.o.	32	68	105
English sole (juvenile)	2	Corn oil	p.o.	56	44	19
Starry flounder (juvenile)	2	Corn oil	p.o.	43	57	19
English sole (gravid)	0.1	Corn oil	p.o.	27[b]	73	38
				23[c]	77	38

[a] Proportions of glucuronide and sulfate conjugates were determined by enzymatic hydrolysis according to methods
     described in Reference 19. Values are given as the mean.
[b] Gravid female English sole.
[c] Gravid male English sole.

the hepatobiliary system of fish accumulates the major proportion of BaP metabolites,
although extrahepatic tissues, such as gonads, also contained low, but detectable levels of
metabolites.[12,13,33] Because of the lipophilicity of BaP and its metabolites, isolation of these
compounds from lipid-rich tissues, such as liver, muscle, and ovary, presents a problem.
Therefore, in most studies identification of BaP metabolites is limited to the water-soluble
conjugates present in bile.

1. Conjugated Metabolites

Our early studies with pleuronectid fishes exposed to radiolabeled BaP demonstrated that
a large portion of the BaP-derived radioactivity (42 to 70%) remained in the aqueous phase
after the bile was treated with β-glucuronidase and arylsulfatase, suggesting the presence
of thio-ether conjugates, including GSH conjugates (Table 12). Because conjugation with
GSH is a major detoxication pathway for reactive intermediates of PAH, such as arene
oxides, epoxides, and quinones, further identification of water-soluble metabolites in the
hydrolyzed bile of BaP-exposed fish was undertaken.[106] The use of alumina oxide column
chromatography and silica-gel TLC to separate BaP conjugates in bile was partially successful
in showing that, indeed, a major proportion of the BaP-derived radioactivity in bile cochromatographed with the 7,8-diol-9,10-epoxide-7,8,9,10-tetrahydroBaP (BPDE)-GSH conjugate standard.[106,107] However, glucuronide and thio-ether conjugates from fish liver or bile
are not well separated by alumina oxide chromatography.[106,108] Recent development of an
ion-pair reversed-phase HPLC method by Plakunov and co-workers has allowed better
separation of glucuronide, sulfate, and thio-ether conjugates present in fish bile.[86,107] Moreover, in this method intact or untreated bile is analyzed, thus, eliminating problems of
incomplete hydrolysis or extraction of conjugates as in other methods. Our data in Figure
12 show that the major metabolites in bile from BaP-exposed English sole and starry flounder
are the glucuronide conjugates of BaP phenols and BaP-7,8-diol, while very little radioactivity coeluted with the 3-hydroxyBaP-sulfate standard. The detection of low levels of

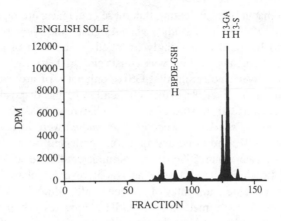

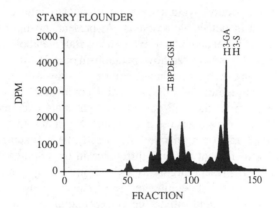

FIGURE 12. Ion-pair reversed-phase HPLC analyses of un-treated bile from (A) English sole *(Parophrys vetulus)*, and (B) starry flounder *(Platichthys stellatus)* exposed intraperitoneally to [³H] BaP (2 mg/kg body weight). The bars show where the standards of *anti*-7,8-diol-9,10-epoxy-7,8,9,10-tetrahydroBaP *(anti*-BPDE)-GSH, 3-hydroxyBaP-glucuronide, and 3-hydroxyBaP-sulfate chromatograph.

sulfates in bile using ion-pair reversed-phase HPLC is in agreement with earlier results showing that sulfate conjugates are minor metabolites detected in the hepatobiliary system of fish.[33] Studies with fish hepatocytes suggest that sulfation is a minor metabolic pathway *in vivo*; however, whether pathways other than biliary excretion (i.e., renal) play an important role in removal of these conjugates needs to be determined.[84] Data in Figure 12 also reveal the presence of several thio-ether conjugates in fish bile. Although the identity of these conjugates is not yet confirmed, the difference in types and proportions of these metabolites suggests qualitative and quantitative differences in the metabolic pathways leading to thio-ether conjugates in English sole and starry flounder. Recent studies by Varanasi et al. have shown that hepatic GST activity in starry flounder is two to three times higher than in English sole.[109] It, thus, appears that higher hepatic GST activity is accompanied by higher thio-ether conjugates in bile of BaP-exposed starry flounder when compared to English sole (Figure 12). Moreover, there are indications that the patterns of hepatic GST enzymes in these two species are quite different.[110]

The level of glucuronide conjugates in liver of BaP-exposed fish is generally very low, which is in contrast to that in bile, indicating that these conjugates are rapidly excreted from liver into gall bladder.[10,33,83] In one study, when English sole were placed on sediment containing radiolabeled BaP and PCB, singly or together, the ratio of proportions of glucuronide to thio-ether conjugates in bile was consistently 10- to 15-fold higher than the corresponding ratios for liver, indicating that GSH conjugates of aromatic compounds are not rapidly released from the liver.[10] Further elucidation of the disposition of conjugated xenobiotics in fish was gained from studies using [^{14}C]styrene oxide-GSH.[111] In these experiments, winter flounder (*Pseudopleuronectes americanus*) were injected intramuscularly with [^{14}C]styrene oxide-GSH, and urine and bile were collected over a 24-h period. Urine was found to contain greater than 60% of the administered dose at 24 h, whereas bile contained only 2% of the administered dose. Analysis by reversed-phase HPLC showed that urine contained mainly cysteine conjugates of styrene oxide, which is an intermediate in the formation of mercapturic acids, a metabolite of GSH conjugates. The actual level of mercapturic acid conjugates in urine was low, however, suggesting a slow N-acetylation of the cysteine conjugate in flounder.[111] In contrast to the urine, bile contained mainly intact GSH conjugates, although the cysteinylglycine, cysteine, and mercapturic acid conjugates of styrene oxide were present in detectable amounts. At present, no information is available on the relative importance of renal and biliary excretion of BaP metabolites in fish. However, Pritchard and co-workers[67,112] investigated how metabolism of BaP affects the renal excretion of conjugated metabolites in southern flounder. Figure 13 shows the efficacy of excretion of BaP, 7-hydroxyBaP conjugates, and BaP-7,8-diol conjugates. Briefly, the lipophilic hydrocarbon, BaP is reabsorbed, the BaP-phenol conjugates are excreted but slightly, and the conjugates of BaP-7,8-diol are excreted at a rate ten times that for 7-hydroxyBaP conjugates. This difference in excretion may not directly be a function of the degree of oxygenation of the BaP moiety, but rather a difference in the conjugation of the two metabolites. 7-HydroxyBaP was present largely as a glucuronide conjugate, whereas BaP-7,8-diol was present largely as the sulfate conjugate. The results also showed that the BaP-7,8-diol-sulfate was excreted more rapidly than the corresponding glucuronide. These results serve to demonstrate the need to obtain detailed information on the excretion and recirculation of individual metabolites to accurately evaluate the disposition and toxicity of PAH in fish.

2. Unconjugated Metabolites

A number of chromatographic and spectrometric techniques have been used in the analysis and characterization of the phase I metabolites of BaP isolated from bile of fish species.[113,114] The major phase I metabolites released after enzymatic hydrolysis of glucuronide conjugates in bile of various fish species are 3- and 1-hydroxyBaP and BaP-7,8-diol; smaller proportions of BaP-9,10-diol and 9-hydroxyBaP are also detected (Table 13).

Studies of BaP metabolism with rodents have demonstrated the formation of 1-,3-,7-, and 9-hydroxyBaP, with 3- and 9-hydroxyBaP being the major metabolites formed *in vitro* and *in vivo*.[103,118-120] As with rodents, 3-hydroxyBaP is generally the major phenolic metabolite detected in pleuronectids and salmonids; however, 7- and 9-hydroxyBaP are formed in very low proportions in these fish both *in vitro* and *in vivo*.[19,114] These results indicate an efficient conversion of BaP-7,8-oxide and BaP-9,10-oxide by fish hepatic epoxide hydrolase (EH) to the corresponding diols. Because bay-region diols can be further activated to form highly carcinogenic diol-epoxides, the preferential formation of these diols in teleost livers may have important toxicological consequences.

In addition to 3-hydroxyBaP, 1-hydroxyBaP is also a major metabolite detected in fish liver. These two phenols are known to be more mutagenic than 9-hydroxyBaP.[121] Unlike rodents, fish do not appear to form K-region metabolites of BaP (e.g., BaP-4,5-diol) to a significant extent, as evidenced by analysis of biliary metabolites (Table 13) as well as metabolites formed by hepatic microsomes (see Chapter 6). Hence, the fish species examined

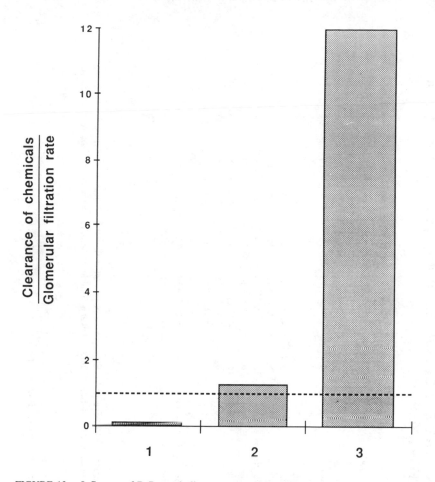

FIGURE 13.   Influence of BaP metabolite structure on the efficacy of excretion by kidney of southern flounder *(Paralichthys lethostigma)*. Data expressed as the ratio of the clearance of the chemical (C_x) to the glomerular filtration rate (GFR). A ratio of 1 indicates no net tubular transport. A ratio of less than 1 denotes removal of the compound from the tubular lumen (i.e., reabsorption). A ratio of greater than one signifies addition of the compound to the tubular lumen (i.e., secretion). (1) BaP, (2) BaP-7-phenol conjugates (largely glucuronide), and (3) BaP-7,8-diol conjugates (largely sulfates, initially). (Adapted from Pritchard, J. B. and Renfro, J. L., in *Aquatic Toxicology*, Vol. 2, Weber, L. J., Ed., Raven Press, New York, 1984, 51.)

thus far show a marked tendency to convert BaP into highly mutagenic and carcinogenic compounds; however, at present, there are only limited reports of toxicity and carcinogenicity of BaP in fish.[122-124]

As shown in Figure 14, BaP-3,6- and BaP-1,6-quinones are the two major quinones detected in bile of English sole and starry flounder. This finding, along with the results showing that 1- and 3-hydroxyBaP are the two major phenols reported to be formed in liver of those fish both *in vivo* and *in vitro*, suggests the possibility that these phenols, rather than 6-hydroxyBaP, are the precursors of the two quinones detected. At present, it is not clear whether in English sole or in any other teleost fish, oxidation via 6-oxyBaP is a significant metabolic pathway. In addition to phenols and quinones, English sole also produce significant amounts of 3,9-dihydroxyBaP (Figure 14). This metabolite was detected in liver, bile, and muscle and its identity was confirmed by UV and mass spectrometry.[113] Although the metabolic pathway leading to the formation of 3,9-dihydroxyBaP *in vivo* is not known,

Table 13

GLUCURONIDE AND SULFATE CONJUGATES OF PHASE I METABOLITES OF BaP IN TISSUES AND BILE OF FISH AND RAT[a]

Species	Vehicle	Route of exposure	Site	Time[a] (h)	% of total identified metabolites							Ref.
					9,10-D[b]	4,5-D	7,8-D	Q	9-OH	1-OH	3-OH	
English sole	Acetone	i.p.[c]	Bile	24	7.3	4.0[d]	24	24	3.1	12	24	19
Starry flounder	Acetone	i.p.	Bile	24	4.4	3.3[d]	18	18	3.1	16	37	19
Rock sole	Acetone	i.p.	Bile	24	4.5	8.3[d]	22	30	5.2	11	18	97
Rainbow trout	Acetone	i.p.	Bile	24	5.6	5.1[d]	22	26	3.2	7.5	30	97
Brown bullhead	Acetone	i.p.	Bile	24	3.9	4.1[d]	23	23	3.5	13	29	97
English sole	Corn oil	p.o.	Bile	24	5.4	3.6[d]	19	12	2.4	11	46	19
		p.o.	Muscle	24	11	N.D.	23	6.8	3.7	4.8	51	115
Starry flounder	Corn oil	p.o.	Bile	24	6.5	4.7[d]	14	19	3.8	9.9	42	19
Rock sole	Corn oil	p.o.	Bile	24	11	4.9[d]	13	17	2.7	18	33	97
English sole	Acetone:emulphor (1:1)	i.m.	Bile	24	9.6	3.9[d]	23	22	7.4	14	42	116
			Liver	24	2.6	2.1[d]	14	9.9	6.4	7.4	58	
			Ovary	24	2.3	4.3[d]	30	14	2.3	9.9	36	
	—	Sediment	Bile	72—168	7.7	5.8[d]	33	12	N.D.	7.7	35	15
Rat	Acetone	i.v.	Bile	22	N.D.	33	11	2.8	<2.8	N.R.	50	117
Sprague-Dawley rat	N.R.	i.v.	Bile	1-6	N.D.	11	4.3	65	3.5	N.R.	16	136

Note: N.D., not detected; N.R., not reported.

[a] All fish were exposed to a dose of 2 mg [³H]BaP/ kg b.w. and sampled at 24 h, except for English sole placed on sediment BaP. Rats were exposed to 0.88 mg BaP/ kg b.w. Phase I metabolites released after treatment with β-glucuronidase and arylsufatase of water-soluble metabolites were analyzed by reversed-phase HPLC. 7-HydroxyBaP was not present in detectable amounts.

[b] D, dihydrodiol; Q, BaP 1,3–,3,6–, and 6,12-quinones; OH, hydroxy.

[c] i.p., intraperitoneal injection; p.o., per os ;i.m., intramuscular injection.

[d] Radioactivity co-chromatographing with BaP-4,5-diol standard in reversed-phase HPLC did not co-chromatograph with BaP-4,5-diol standard in normal-phase HPLC, except for Sprague-Dawley rat.

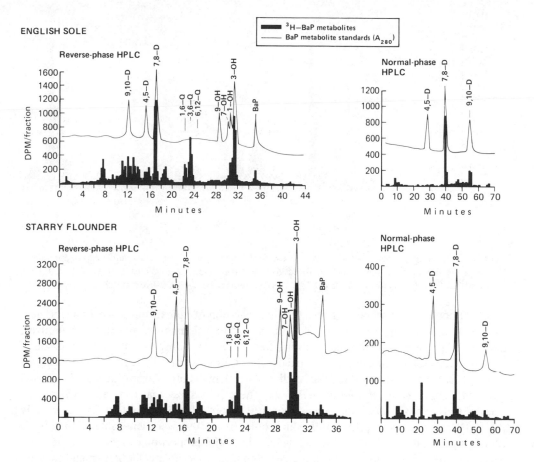

FIGURE 14. Reversed-phase and normal phase HPLC analyses of BaP metabolites released after β-glucuronidase and arylsulfatase treatment of bile from English sole and starry flounder exposed (orally) to 2 mg [³H]-/kg body weight for 24 h. The fractions from 12 to 23 min were isolated from a second reversed-phase HPLC and analyzed further by normal-phase HPLC. BaP metabolite standards were added to the samples before HPLC analysis. The solid line is absorbance at 280 nm and the solid bars represent BaP-derived radioactivity. Abbreviations: DPM; integrations per minute; 9,10-D, BaP-9,10-Diol; 7,8-D, BaP-7,8-Diol; Q, quinones (BaP-1,6-, 3,6-, and 6,12-quinones); 9-OH, 9-hydroxyBaP; 7-OH, 7-hydroxyBaP; 3-OH, 3-hydroxyBaP; and 1-OH, 1-hydroxyBaP. (Reprinted from Varanasi, U., Nishimoto, M., Reichert, W. L., and Eberhart, B.-T. L., *Cancer Res.*, 46, 3817, 1986.)

studies with rat hepatic microsomes show that both 3- and 9-hydroxyBaP are precursors to 3,9-dihydroxyBaP.[118]

As discussed earlier, diols and phenols can undergo further oxidation; however, significant amounts of triols and tetrols are not isolated from bile of BaP-exposed fish. This may be due, at least in part, to low levels of phenol oxides and diol oxides formed by these animals *in vivo*. In addition, the relatively high reactivity of phenoloxides and diolepoxides indicates that reaction with macromolecules and GSH may be responsible for the absence of detectable levels of triols and tetrols among the excretion products.[125]

Although several studies report in detail the disposition of total BaP metabolites in liver and extrahepatic tissues, the information on individual metabolites in hepatic and extrahepatic tissues is meager. Low concentrations of lipophilic BaP metabolites in lipid-rich tissues of fish have hampered attempts to separate and characterize BaP metabolites. However, the importance of such information for liver, muscle, or gonadal tissues cannot be overstressed. In our laboratory, we first used two-dimensional TLC to separate BaP metabolites from liver

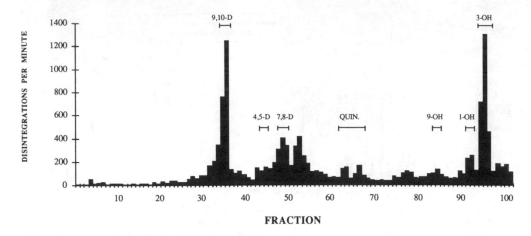

FIGURE 15.   Unconjugated metabolites of BaP in ovary of English sole (*Parophrys vetulus*) exposed intramuscularly to 2 mg BaP/kg body weight and sacrificed 24 h post exposure. Abbreviations: 9,10-D, BaP-9,10-Diol; 7,8-D, BaP-7,8-Diol; Q, quinones (BaP-1,6-, 3,6-, and 6,12-quinones); 3-OH, 3-hydroxyBaP.

of fish. Although this method was effective in separation of lipids from phase I metabolites, it was not effective in the resolution of monohydroxy derivatives of BaP.[33] Moreover, oxidation of phenols was substantial during the procedure. Subsequently, column chromatography with either a silica gel or Sephadex® LH-20 was used to remove lipids and unconverted BaP before HPLC analysis of the phase I metabolites from muscle, liver, and gonads of English sole exposed to BaP.[115,116] The results show that liver and extrahepatic tissues of BaP-exposed sole contain predominantly water-soluble metabolites (approximately 60% of BaP-derived radioactivity) of which only 10 to 20% were amenable to hydrolysis with β-glucuronidase and arylsulfatase. Hence, the major proportion of metabolites in these tissues was present presumably as thio-ether conjugates and as metabolites bound to macromolecules (discussed later).

Profiles of phase I metabolites released after hydrolysis of glucuronide and sulfate conjugates in liver and extrahepatic tissues of English sole exposed to BaP were similar to those for bile, with 1- and 3-hydroxyBaP and BaP-7,8-diol as the major metabolites (Table 13). Interestingly, profiles of unconjugated metabolites present in extrahepatic tissues were dissimilar to those present in liver. For example, 3-hydroxyBaP, quinones, and BaP-7,8-diol were present in liver, while 3-hydroxyBaP and BaP-9,10-diol were the major metabolites present in ovary of the same fish (Figure 15). This difference in disposition of unconjugated metabolites can be partially explained in terms of the dynamics of BaP metabolites observed in rodent cells. It has been reported that BaP-9,10-diol is rapidly released into extracellular fluid of rat hepatocytes and organ cultures from hamster lung exposed to BaP.[126,127] Moreover, BaP-7,8-diol and 3-hydroxyBaP are detected in extracellular fluid mainly as glucuronide and sulfate conjugates. Preliminary work with rainbow trout hepatocytes exposed to [³H]BaP shows that BaP-9,10-diol is the major unconjugated metabolite present in extracellular media.[107] Based on this information, it can be assumed that once formed in fish liver, BaP-9,10-diol is rapidly released and, hence, may be available for systemic circulation and deposition in extrahepatic tissues. On the other hand, BaP-7,8-diol and 3-hydroxyBaP may not be released from liver prior to conjugation, and the presence of these metabolites in ovary may be the result of metabolism of BaP by ovarian MFO. Recent studies have shown that extrahepatic tissues have low, but significant levels of cytochrome P-450 enzymes.[128] Hence, in addition to the transport of BaP metabolites from liver into extrahepatic tissues, *in situ* metabolism of BaP and its metabolites should also influence disposition of potentially toxic metabolites in critical sites such as gonads of fish just prior to spawning. It was reported that both gonads and blood of gravid English sole had a significantly higher proportion of unconverted BaP than did liver.[39] Because lipophilic PAH and their metabolites are deposited in extrahepatic tissues, the exposure of gametes and early-life stages (e.g., eggs, embryo) to contaminants via parental route may be very important in benthic fishes that reside in areas having high levels of contaminants in sediments. At present, no information is available on the importance of the water column or sea-surface microlayer as a source of PAH in pelagic eggs of benthic fishes. A number of factors such as salinity at the microlayer and temperature and weather conditions at spawning sites may strongly affect the resident time of embryos in these two compartments. Moreover, the importance of sediment as a source of PAH to demersal eggs needs to be evaluated.

3. Macromolecular Binding

Investigation of phase I and II metabolism provides useful information on the ability of fish species to activate and detoxicate carcinogenic PAH. However, a more direct measure of metabolic activation can be obtained by measuring levels of metabolites bound to tissue macromolecules, especially DNA. The binding of xenobiotics to DNA has been used as a molecular dosimeter to determine their cacinogenic potential in rodents.[102] The first step in tumor initiation is believed to be the nonrandom binding of ultimate carcinogens to target sites in DNA.[129-133] Fixation of a biochemical change by cell proliferation is considered the second step in initiation.[133,134] For an understanding of the process of initiation, it is important, therefore, to measure the biologically effective dose by determining the extent of binding of a PAH to DNA rather than simply measuring tissue concentrations of the PAH and its metabolites. Measurement of the level of PAH metabolites bound to DNA allows for an estimate of the degree of activation of a PAH to reactive metabolites. Various factors such as types of DNA adducts, levels of adducts, rate of DNA repair, mitotic indices of target tissues, and persistence of critical DNA adducts may be relevant to initiation. Therefore, a study of these factors in fish exposed to the well-studied carcinogenic PAH,BaP, may give an insight into species differences in susceptibilities to PAH-induced liver cancer. At present, however, limited studies with a few fish species are reported in which covalent modification of DNA is determined. In most of those studies, the level of binding of BaP metabolites to cellular protein is also reported which could be used as an additional molecular dosimeter of the amount of reactive intermediates formed by a tissue.

Using English sole, a fish species having a high prevalence of liver lesions when sampled from contaminated estuaries,[6] we were the first to show that BaP is metabolized in fish to intermediates that covalently bind to hepatic DNA (Table 14) and protein *in vivo*.[20] The level of modification of hepatic DNA in juvenile sole exposed to BaP orally was an order of magnitude higher than that reported for rodents exposed to BaP orally.[102] To ensure that variability in conditions of exposure or in methods of isolation of DNA did not contribute to the substantial difference observed in the binding values for sole[20] and rodents,[102,136] in a subsequent study, juvenile English sole and Sprague-Dawley rats were exposed to equimolar concentrations of BaP intraperitoneally and hepatic DNA was isolated using identical procedures.[19] The results confirmed the earlier study in which sole were fed BaP[20] and demonstrated that the level of modification of hepatic DNA in sole exposed to BaP intraperitoneally was substantially (about 100 times) higher than that for rat (i.p. exposure) at 24 h after exposure. This difference in binding could be due, in part, to the differences in activation of BaP by the two species under investigation. For example, both *in vivo* and *in vitro* studies of BaP metabolism show that sole liver converts considerably higher proportions of BaP into BaP-7,8-diol, the penultimate carcinogen of BaP, compared to rat liver.[19] Moreover, differences in rates of excision repair of BaP-modified DNA may also contribute significantly to the observed differences in binding levels between English sole and Sprague-Dawley rats. Our results show that the level of binding of BaP to hepatic DNA in juvenile English sole

Table 14
SPECIES DIFFERENCES IN THE COVALENT BINDING OF BaP METABOLITES TO HEPATIC DNA

Species[a]	Dose (mg/kg)	Time (h)	Vehicle	Mode of exposure	% administered dose in liver	fmol BaP equiv./ mg protein	fmol BaP equiv. / mg DNA	CBI[b]	Ref.
Aquatic									
English sole	0.1	24	Corn oil	p.o.	0.8	90	51	45	20
	2.0	24	Corn oil	p.o.	0.43	1,800	2,100	85	19
	2.0	24	Corn oil	i.p.	0.78	N.R.	250	9.8	107
	2.0	48	Corn oil	i.p.	1.59	N.R.	880	31	107
	2.0	24	Acetone	i.p.	60	N.R.	28,000	1,050	19
	2.0	24	Ethanol	i.p.	N.R.	N.R.	42,000	1,600	107
Starry flounder	2.0	24	Corn oil	p.o.	0.6	600	500	21	19
	2.0	24	Acetone	i.p.	3.5	N.R.	14,000	540	19
California killifish	0.6	24	Corn oil	i.p.	N.R.	740	8.9	1.2	135
Speckled sandab	0.6	24	Corn oil	i.p.	N.R.	1,400	14	1.8	135
Terrestrial									
Sprague-Dawley rat	2.0	24	Acetone	i.p.	1	N.D.	300	13	19
Sprague-Dawley rat	2.5	1	N.R.	i.v.	N.R.	N.R.	2,500	9.8	136
C57B1/6J mouse	0.6	24	Corn oil	i.p.	N.R.	430	72	9.3	135

Note: Values are given as mean. Abbreviations: p.o., per os; i.p., intraperitoneal; i.v., intravenous; N.R., not reported.

[a]

[b] Covalent binding index (CBI) = ([μmol BaP equiv./mol nucleotides]/[mmol BaP administered/kg b.w.]).

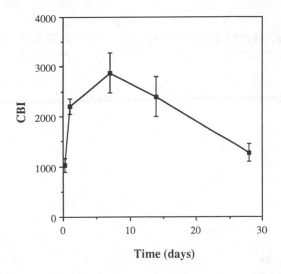

FIGURE 16.   Persistence of BaP-DNA adducts in liver of English sole *(Parophrys vetulus)* exposed to [³H]BaP (2 mg/kg body weight) by i.p. injection. Hepatic DNA was isolated from liver and radioactivity associated with purified DNA was determined by liquid scintillation spectrometry.

does not change significantly until several weeks after a single exposure to BaP (Figure 16),[20,109] whereas for rats a significant decline in the level of binding is observed over a 4-week period, most probably due to the repair of the DNA lesions.[137] Studies with fish and rodent cells show that, in general, fish cells have a much slower rate of repair of the DNA lesions than rodent cells.[138-140] Additionally, the observed differences in level of DNA modification by BaP metabolites between sole and rodent liver may also be due to difference in detoxication of reactive intermediates of BaP by conjugation, for example, with GSH. Although no data are available to compare the ability of fish and rodents to detoxicate putative carcinogenic intermediates, studies with rodents show that DNA binding is inversely proportional to both GST activity as well as cellular level of GSH.[141-143] Because glucuronidation is another major detoxification pathway in both fish and rodents, differences in the hepatic UDPGT activities between these species may also play a role in the detoxification of the BaP-7,8-diol, the precursor of *anti*-BaPDE. Moreover, BaP-7,8-diol may be converted to the sulfate conjugate in mammals,[136] but in teleosts, sulfation does not appear to be significant in liver. Thus, differences in conjugation of reactive intermediates or their precursors may explain the differences in the levels of binding of BaP metabolites to hepatic DNA in fish and rodents.

Exposure of fish to BaP is generally confined to either p.o. or i.p. administration when *in vivo* binding studies have been conducted. However, because of the differences in absorption of BaP from the intestine and the peritoneal cavity, and the possibility of metabolism of BaP by intestinal tissue,[23,144] the degree of activation of BaP to reactive metabolites in liver may be dependent on the route of administration. Hence, to determine the effect of route of exposure, juvenile English sole were exposed, either orally or intraperitoneally, to BaP in different solvent carriers (Table 14). At 24 h after the exposure of fish to BaP dissolved in corn oil, DNA binding levels for fish given BaP orally were approximately ten times greater than for fish given BaP via an i.p. injection. However, the level of binding in fish given BaP via i.p. injection increased fourfold by 48 h after exposure (Table 14). In contrast, no increase in binding at 48 h was seen for fish exposed p.o. to BaP dissolved in corn oil (data not shown). Data in Table 15 from experiments with gravid fish also show

Table 15

COVALENT MODIFICATION OF HEPATIC AND GONADAL DNA BY BaP METABOLITES IN GRAVID ENGLISH SOLE

Tissue	Time (h)	Sex	fmol BaP equiv. /mg protein	fmol BaP equiv. /mg DNA	CBI
Liver	24	Male	26	13	98
	48		28	8.1	57
	168		1.9	0.47	2.8
	24	Female	8.6	1.9	16
	48		4.6	4.4	37
	168		11.0	0.91	7.8
Gonads	24	Male	0.48	1.6	9.8
	48		0.19	0.06	0.36
	168		0.02	0.28	1.5
	24	Female	0.145	N.R.[a]	N.R.
	48		0.23	N.R.	N.R.
	168		0.16	N.R.	N.R.

Note: Fish were exposed p.o. to BaP at a dose of 0.1 mg [^{3}H]BaP/kg b.w. Values are given as mean. Covalent Binding Index (CBI) = ([μmol BaP equiv./mol nucleotides]/[mmol BaP administered/kg b.w.])

[a] N.R., not reported.

Table adapted from Reference 39.

that in fish given BaP dissolved in corn oil (p.o.), the DNA binding levels do not increase from 24 to 168 h. Thus, the maximum binding of BaP intermediates to hepatic DNA occurs within 48 h after oral administration of BaP dissolved in corn oil. The increase in DNA binding over time for fish given BaP intraperitoneally could be attributed to slow but continual absorption of BaP from the peritoneal cavity. This is supported by the observations that droplets of corn oil were still present in the peritoneal cavity at 48 h after exposure, and that for fish given BaP orally no corn oil was seen in the intestinal tract of these fish at any of the sampling times, indicating removal of the source of BaP from these fish. Further comparison of the effect of carrier solvents for BaP on the level of binding after exposure via i.p. injections is given in Table 14. When English sole were given equimolar doses of BaP, the level of DNA binding at 24 h was 20-fold higher when BaP was dissolved in either acetone or ethanol rather than corn oil. Levels of DNA binding in fish given BaP in corn oil p.o. were not measured beyond 48 h and, hence, no information is available about the time and the level of maximum binding for these fish. Nevertheless, the data in Table 14 demonstrate clearly that when comparing different studies, it is important to bear in mind that carrier solvents and route of exposure profoundly affect both the level and the time of maximum DNA binding. On the other hand, neither of these factors seems to affect the type of BaP metabolites detected in bile (Table 13), thus, the difference in levels of binding appears to be mainly due to the differences in effective dose reaching the liver.

Limited information is available on the effect of age and sex on chemical modification of DNA in PAH-exposed fish. In one study, it was reported that young English sole (<2 years) had higher binding value (40 ± 2.3 fmol BaP equiv./mg DNA) of BaP to hepatic DNA than adult (>6 years) fish (5.0 ± 1.9 fmol BaP equiv./mg DNA) when exposed (p.o.) to equimolar concentrations of BaP and sampled 24 h later.[106] Such differences in binding may be related to either the activation of BaP to DNA-binding intermediates, or the removal of these intermediates by detoxication pathways, such as GSH conjugation. For example, in the above study, the adult sole showed a greater proportion of thio-ether conjugates in bile than juvenile sole (Table 12). Conjugation with GSH is competitive with EH for primary oxides (e.g., BaP-7,8-oxide) and with DNA for other reactive BaP intermediates (e.g., phenol-oxides, diol-epoxides). In another study,[39] gravid English sole were exposed to BaP intragastrically to determine the level of covalent binding of BaP to macromolecules of liver and gonads (Table 15). Substantial variation in binding values for individual fish precluded any assessment of sex-related differences in the level of binding of BaP metabolites to hepatic DNA. The finding that low but detectable levels of BaP metabolites were bound to gonadal macromolecules (testicular DNA and protein, and ovarian protein) of ripe fish exposed to BaP is significant in view of the fact that BaP exhibits teratogenic effects in rodents.[145] Moreover, exposure of sexually mature female flathead sole (*Hippoglossoides elassodon*) to BaP was accompanied by a significantly lowered hatching success of eggs that were fertilized with milt from control males.[146] In addition, exposure of flatfish eggs to BaP caused a decline in hatching success and an increase in developmental abnormalities (e.g., twinning, overgrowth of tissues, arrested development).[147] These few studies suggest that BaP or its metabolites exert teratogenic effects in various fish species, although the mechanisms of these effects are not known.

Epizootological evidence demonstrates certain species-specific differences in the prevalence of hepatic neoplasms exhibited by fishes residing in industrialized waterways in the U.S. (see Chapter 8). Such species-specific differences in susceptibility to liver cancer in PAH-contaminated areas provide an excellent opportunity to compare biochemical processes involved in activation and detoxication of carcinogenic PAH. Hence, we have initiated detailed studies to measure both the *in vivo* and *in vitro* metabolism of BaP by pleuronectid fishes that show marked differences in the prevalences of hepatic neoplasms when sampled from the same contaminated sites.[19] As stated previously, English sole residing in urban waterways in Puget Sound, Washington exhibit a high prevalence of hepatic neoplasms that is statistically correlated with the concentrations of PAH in the sediments, whereas a closely related species, starry flounder, exhibits a very low prevalence of hepatic neoplasms. However, the levels of fluorescent aromatic compounds in bile of English sole and starry flounder are comparable when sampled from urban or industrialized areas, indicating similar exposure to xenobiotics such as PAH.[148] When these fish species were sampled from reference (non-urban) areas and given equimolar doses of BaP either orally or intraperitoneally, the level of binding of BaP metabolites to hepatic DNA was two to four times higher in English sole than in starry flounder, regardless of the route of exposure (Figure 17 and Table 14). Comparison of phase I metabolites released after enzymatic hydrolysis of bile shows that sole contained significantly higher proportions of BaP-7,8-diol than did flounder (Figure 17). These findings, along with the results showing that both English sole and starry flounder liver microsomes metabolize BaP essentially to a single DNA adduct, namely, the (+)-*anti*-BPDE-dG,[149,150] suggest that the higher binding of BaP intermediates to hepatic DNA in sole may most probably be due to more BPDE available in English sole than in starry flounder.[19] Although binding of other reactive intermediates (e.g., free-radicals, phenol-oxides) to DNA *in vivo* cannot be excluded, no consistent difference was observed in the proportions of phenols or quinones released after enzymatic hydrolyses of bile from the two species (Table 13). In a separate study, starry flounder and English sole were sampled from several urban and nonurban sites in Puget Sound, and the comparison of activities of hepatic AHH and GST activities showed consistently that, hepatic AHH activity was significantly higher and GST activity was significantly lower in English sole.[148] These field investigations lend support to the above laboratory study showing lower binding of BaP intermediates to hepatic DNA in starry flounder than English sole (Table 14).[19,109] Moreover, as described

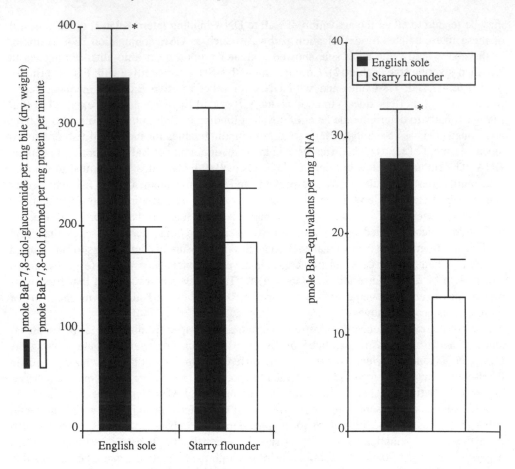

FIGURE 17.   Metabolism and DNA binding of BaP by English sole *(Parophrys vetulus)* and starry flounder *(Platichthys stellatus)* exposed to [³H]BaP intraperitoneally at a dose of 2.0 mg BaP/kg body weight. Biochemical parameters include (a) the concentration of BaP-7,8-diol-glucuronide conjugate in bile of fish, (b) rates of formation of BaP-7,8-diol by hepatic microsomes of fish previously exposed to BaP, and (c) level of hepatic DNA binding in fish exposed to BaP. (Adapted from Varanasi, U., Nishimoto, M., Reichert, W. L., and Eberhart, B.-T. L., *Cancer Res.*, 46, 3817, 1986.)

earlier, analyses of conjugated metabolites in bile of flounder and sole (Table 12, Figure 12) show that, consistent with the above findings, starry flounder bile had lower proportions of BaP 7,8-diol glucuroride and higher proportions of thio-ether conjugates. These results serve to illustrate that the use of several biochemical parameters (i.e., phase I and II metabolites, binding to DNA) in concert is necessary to better understand species-specific responses to toxicants in the aquatic environment.

A recent study by von Hofe and Puffer allows further comparisons of species differences in binding of BaP to DNA. Their studies showed that both the California killifish (*Fundulus parvipinnis*) and speckled sanddab (*Ictharichthys stigmaeous*) activated [³H]BaP *in vivo* to intermediates that became bound to hepatic DNA (Table 14).[135] By comparing covalent binding indices [CBI = (μmol BaP equiv./mol nucleotide)/(mmol BaP administered/kg fish)] that normalize for different doses, the results in Table 14 show that the CBI for both the killifish (8.9) and sanddab (14) given BaP dissolved in corn oil (p.o.) had significantly lower levels of binding at 24 h than the binding level in English sole (250) under similar experimental conditions. The differences in CBI for fish species were most probably not due to variability in experimental techniques used in the two studies, but represented species-specific differences, because rats and mice used as positive controls in these two studies[19,35] had CBI values comparable to literature values (Table 14). It is apparent from the above discussion that factors, such as differences in solvent vehicles, differences in water temperature, differential rates of formation and detoxication of reactive intermediates, and repair of BaP-DNA adducts, may play important roles in the toxicokinetics of BaP and ultimately in determining the level of DNA modification in fish species.

4. Characterization of BaP-DNA Adducts

An important area of research that has just begun to yield useful information is the characterization of the types of adducts between PAH metabolites and DNA. Recent studies have shown that for carcinogenic PAH only certain adducts are correlated with toxic effects, indicating that not all of the adducts formed may be involved in the development of subsequent deleterious events. For example, total binding of carcinogens to DNA and their persistence do not always correlate with tumor susceptibility of the tissue.[151,152] Therefore, identification of individual DNA adducts is important if DNA damage is to be correlated with the initiation of carcinogenesis.

A recent study with bluegill sunfish (*Lepomis macrochirus*) exposed to a single dose of BaP intraperitoneally shows that acid hydrolysis of hepatic DNA or hemoglobin at 72 h after exposure released BaP tetrols which are products from the hydrolysis of adducts between DNA and *anti*- and *syn*-BPDE.[153] The major tetrol released from liver DNA was *trans*-2-tetrol (85 to 88%), indicating that *anti*-BPDE was formed in fish liver. The proportion of *anti*-BPDE bound to hemoglobin, as evidenced by the presence of isomeric tetrols from *syn*- and *anti*-BPDE released on acid hydrolysis of hemoglobin, was slightly lower (75 to 76%) than the proportion of *anti*-BPDE bound to hepatic DNA (85 to 88%); however, the small sample size (two determinations) precludes valid statistical comparisons between proportion of *syn*-BPDE adducts with DNA and hemoglobin. Lowering of water temperature (from 20 to 13°C) at which fish were maintained resulted in substantial decrease in the level of adducts bound to liver DNA or hemoglobin, which may be due to lower absorption of administered BaP at lower water temperature.[46]

The study of Shugart et al.[153] provided indirect proof for the presence of BPDE-DNA adducts in fish exposed to BaP. The advent of sensitive techniques, such as ^{32}P-postlabeling of DNA, makes it increasingly feasible to characterize PAH-DNA adducts formed *in vivo*.[154-156] Recent results from our laboratory, using ^{32}P-postlabeling analysis of hepatic DNA isolated from BaP-exposed English sole, show that the major adduct (85 to 97% of labeled adducts) had similar chromatographic properties as the standard *anti*-BPDE bound to the exocyclic amino group of deoxyguanosine-3′,5′-bis-phosphate (Figure 18). Additionally, acid hydrolysis of the DNA showed the presence of tetrols derived from BPDE, also indicating that BaP was converted into *anti*-BPDE and bound to liver DNA of English sole.[157,158] This finding *in vivo* is in agreement with the results on metabolism of BaP catalyzed by sole hepatic microsomes showing the formation of a single *anti*-BPDE-dG adduct.[149,150] Thus, several lines of evidence strongly suggest that BaP metabolism *in vivo* proceeds mainly via activation of BaP-7,8-diol into diol epoxides that bind to hepatic DNA of English sole. However, the question of the stability of PAH-DNA adducts during isolation of DNA and subsequent hydrolysis of the DNA raises the posssibility that adducts other than *anti*-BPDE-dG may have been formed, but were lost during workup. For example, the BPDE-N^7-dG adduct is unstable.[159] Moreover, at present very little information is available to show if continuous exposure to xenobiotics in a contaminated environment alters the ability of feral fishes to activate and detoxicate environmental carcinogens, and, hence, results in formation of adducts other than those reported above with fish sampled from relatively uncontaminated areas.[100]

D. PAH Metabolites and DNA Adducts in Environmental Samples

The major emphasis of investigations on the metabolism of PAH by fish has been with

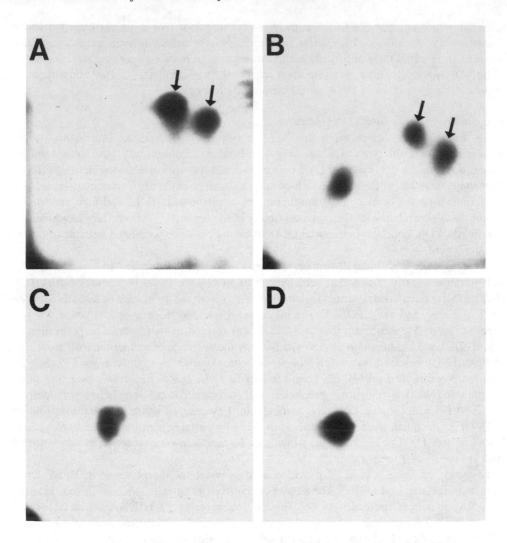

FIGURE 18.   Autoradiographs of PEI-cellulose thin-layer chromatograms of ^{32}P-labeled digests of hepatic DNA from English sole treated intraperitoneally with (A) acetone carrier only, (B) 15 mg BaP/kg body weight, (C) 100 mg BaP/kg body weight, and of salmon sperm DNA incubated with *anti*-BaP-7,8-diol-9,10-epoxide standard (D). The spots indicated by arrows are present at low levels in nearly all DNA samples and can be considered background. Due to high level of adducts in autoradiographs C and D only a short exposure was required during autoradiography and thus the two spots were not evident.

single compounds. However, because numerous xenobiotics are present in the aquatic environment, studies with mixtures of PAH and other xenobiotics are needed to determine the metabolism and disposition of PAH under environmentally relevant conditions. Studies with rodents have shown that PAH can have both antagonistic and synergistic effects on the metabolism and DNA binding of carcinogenic PAH.[161-164] Thus, a very complex problem of studying metabolism of multiple PAH present in urban areas becomes even more confounding. It is obvious that increasingly sophisticated analytical techniques are needed to identify metabolites and DNA adducts of these xenobiotics.

Recently, Krahn et al. have used reversed-phase HPLC-fluorescence detection and GC/MS to characterize PAH metabolites in bile of fish sampled from urban estuaries.[165,166] These results show the presence of hydroxylated derivatives of several PAH, including NPH, fluorene, biphenyl, PHN, anthracene, fluoranthene, pyrene, and BaP. The metabolites of

PHN, fluoranthene, and pyrene were most prominent and were consistently present in bile of English sole sampled from industrialized areas in Puget Sound. The presence of metabolites of 5-ring PAH was not confirmed by GC/MS, although HPLC-fluorescence measurement of the bile samples suggests the presence of BaP metabolites, such as BaP-7,8-diol and 3- and 1-hydroxyBaP.[165] If the bile contained high proportions of thio-ether conjugates, such as GSH conjugates of 4- to 5-ring PAH, these compounds would not have been detected by the techniques used, because only those metabolites conjugated with glucuronic acid were characterized. It should be noted that high proportions of GSH conjugates of BaP are present in bile of English sole exposed to a low level of this PAH added to sediment.[12] Moreover, the level of BaP metabolites bound to GSH or macromolecules in sole liver increases when these fish are exposed to BaP in the presence of other xenobiotics.[12] Hence, chronic exposure of fish to a myriad of chemicals present in contaminated areas may have significant effects on their ability to metabolize PAH.

The identification of PAH metabolites in bile of fish from contaminated areas shows that PAH present in contaminated areas are bioavailable and may be activated into potentially toxic intermediates that bind to DNA in liver. Analysis by ^{32}P-postlabeling of hepatic DNA isolated from English sole treated with PAH extracted from urban sediments showed the presence of several adducts having chromatographic properties similar to those for metabolites of 4- to 5-ring PAH (chrysene, dibenz(a)nthracene, BaP bound to DNA.[157] The profiles of adducts obtained with English sole exposed in the laboratory to urban-sediment extracts appeared similar to the profiles obtained from ^{32}P-postlabcling analyses of DNA from liver of lesion-free juvenile English sole caught from contaminated areas of Puget Sound.[157] Thus, the combined use of techniques such as fluorescence-detection/GC-MS identification of biliary metabolites, and ^{32}P-postlabeling analyses of PAH-DNA adducts from tissues of young fish caught in PAH-contaminated areas may provide an early indication of exposure of these fish to genotoxic PAH and allow better interpretation of relationships between contaminants and biological effects in the aquatic environment.

IV. CONCLUDING REMARKS

The laboratory studies to date on the metabolism and disposition of individual PAH in fish have not only broadened our overall knowledge of the mechanisms of activation and detoxication of xenobiotics, but have led to generic information useful for assessing the impact of PAH pollution on the quality of aquatic environments. For example, new techniques (e.g., measurements of aromatic compounds in fish bile and DNA-xenobiotic adducts in fish liver) for monitoring studies to assess the level of exposure of fish to PAH in contaminated areas were developed based on information from laboratory studies of PAH metabolism. Further investigations are needed, nevertheless, to strengthen our understanding of the metabolism and toxic actions of these environmental pollutants. Throughout this chapter we have identified several promising avenues of research, and here only a few areas are discussed for added emphasis.

Most of the studies on PAH metabolism are done with fish that are either sampled from relatively pristine areas or that have been cultured. Virtually no information is available on the effect of chronic exposure to a contaminated environment on the *in vivo* metabolism and disposition of PAH in fish, especially with emphasis on interaction of metabolites with DNA. Fish sampled from contaminated areas show significant liver necrosis and other pathological diseases.[167] Because factors such as cytotoxicity and mitotic indices of the target cells influence the process of carcinogenesis in rodent models,[133] information on certain parameters (e.g., cell cycle) in fish chronically exposed to pollutants is vital in our attempts to delineate causal relationships between PAH and the prevalence of diseases in fish residing in PAH-contaminated areas. Any significant alteration in crucial biochemical systems (e.g., xenobiotic-metabolizing enzymes, antioxidants, membrane lipids) involved in processing of xenobiotics may profoundly influence the activation and detoxication of PAH and, hence, their subsequent deleterious effects on fish residing in contaminated areas. The fact that fish residing in urban areas often suffer from diseases, which may influence their ability to metabolize PAH, is a confounding issue. Thus, caution must be exercised in designing and interpreting field studies to evaluate whether long-term residence in a contaminated environment alters uptake, metabolism, or disposition of PAH in these fish.

Additionally, studies are needed to evaluate PAH metabolism in the entire life-cycle of fishes, including spawning fish, and to correlate the information with toxicity data. Such studies will provide valuable insights into why certain life-stages are particularly susceptible to specific PAH or their metabolites. Moreover, another area of research that deserves further investigation is species-specific differences in the activation and detoxication of PAH, because epizootological studies show a strong correlation between sediment levels of PAH and prevalences of liver neoplasms in certain bottom-dwelling fish species residing in urban areas, as well as show marked differences in the prevalence of liver neoplasms among closely related species. Furthermore, preliminary results from studies with pleuronectids show differences in the levels of activation and detoxication enzymes among these species. Thus, an excellent opportunity exists to conduct comparative studies on PAH metabolism with closely related fish species to better understand the role of PAH in tumor induction in feral fish. Such comparative studies would be immensely valuable in providing fundamental information to broaden our knowledge of toxic actions of PAH in aquatic species, and may help identify steps in the metabolism of PAH that play key roles in governing the susceptibility of a species, aquatic or mammalian, to PAH-induced carcinogenesis.

NOTE ADDED IN PROOF

Several recent studies[168-170] have addressed the issue of DNA damage in livers of fish residing in industrialized waterways. For example, our investigations show that hepatic DNA from English sole sampled from the Duwamish Waterway and Eagle Harbor, Puget Sound, WA, and winter flounder sampled from Boston Harbor, MA, contained a suite of adducts indicative of exposure to genotoxic compounds.[168] Autoradiograms of TLC of ^{32}P-labeled hepatic DNA digests from English sole from these contaminated sites exhibited up to three diagonal radioactive zones (DRZ). These DRZ contained several distinct spots as well as what appeared to be multiple overlapping adduct spots. Furthermore, all autoradiograms of DNA from English sole and winter flounder from these sites exhibited a DRZ designated as DRZ-2 in Figure 19, where DNA adducts of chrysene, benzo(a)pyrene and dibenz(a,h)anthracene, formed *in vitro* using English sole hepatic microsomes, were shown to chromatograph (Figure 20).[169] Hepatic DNA of English sole sampled from a site (Useless Bay, WA) of very low contamination in sediment did not reveal the presence of any adducts that chromatographed in the regions of the three DRZs demonstrating the ability of the ^{32}P-postlabeling assay to distinguish between fish sampled from areas of high and low contamination. Recent analysis[170] of ^{32}P-labeled hepatic DNA digests from brown bullhead *(Ictalurus nebulosus)* sampled from the Buffalo and Detroit Rivers, also revealed elevated levels of DNA adducts in the area of the autoradiogram we have designated as DRZ-2,[169] compared to the levels of adducts for control fish which were raised in aquaria. Sediments from both of these rivers, as well as from the three industrialized sites in our study, are contaminated with high levels of PAH. When English sole sampled from Useless Bay were treated with organic-solvent extracts of the sediments from the two contaminated sites in Puget Sound, WA, hepatic DNA adduct profiles were generally similar to those for English sole sampled from the respective contaminated sites. The chromatographic characteristics of the adducts and the similarities in adduct profiles between field-sampled English sole and those treated

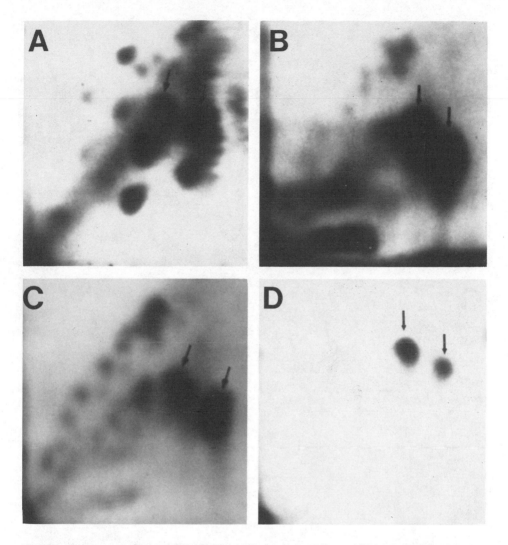

FIGURE 19.   Autoradiograms of PEI-cellulose thin-layer chromatograms of ^{32}P-labeled digests of hepatic DNA from English sole and winter flounder. English sole were caught from contaminated (A: Duwamish Waterway; B: Eagle Harbor) and reference (D: Useless Bay) sites of Puget Sound, Washington, whereas, winter flounder were caught from a contaminated (C: Boston Harbor) site off Massachusetts.[168] The spots indicated by arrows are present in nearly all DNA samples and can be considered background. The origin is located at the bottom left-hand corner of each TLC map.

with contaminated sediment extracts suggested that hydrophobic aromatic compounds of anthropogenic origin were adducted to hepatic DNA of sole from contaminated sites, but not in sole from the reference site. In addition to analysis of hepatic DNA, the ^{32}P-postlabeling assay should allow determination of the level of DNA adducts in extrahepatic tissues such as blood and gonads. Such information may be useful in studies of impaired reproductive processes reported in fish from PAH-contaminated areas.[171]

ACKNOWLEDGMENTS

The authors are grateful to Drs. Tracy Collier and William L. Reichert (National Marine Fisheries Service/National Oceanic and Atmospheric Administration [NOAA]) for critical review of this chapter, and to Bich-Thuy Le Eberhart and Lisa Wirick for their excellent

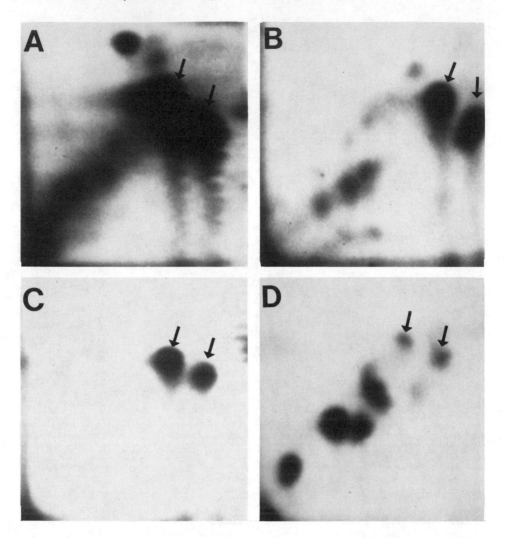

FIGURE 20.   Autoradiograms of PEI-cellulose TLC of ^{32}P-labeled digests of hepatic DNA from English sole exposed to extracts of sediments from Duwamish Waterway (A) and Eagle Harbor (B). Control fish from reference area were given vehicle only (C). A composite autoradiogram of DNA adducts formed from the metabolism by English sole hepatic microsomes of BaP, chrysene and dibenz[a,h]anthracene (D) is given to show the chromatographic properties of ^{32}P-labeled DNA adducts of 4-5 ring aromatic hydrocarbons.[169] The spots indicated by arrows are present in nearly all DNA samples and can be considered background. The origin is located at the bottom left-hand corner of each TLC map.

help in the preparation of the manuscript. The authors' work described in this chapter was supported, in part, by the Office of Oceanography and Marine Association-NOAA and through a National Cancer Institute-NOAA Interagency Agreement (Y01-CP-40507).

REFERENCES

1. **Adamson, R. H. and Sieber, S. M.,** The disposition of xenobiotics in fishes, in *Survival in Toxic Environments,* Khan, M. A. Q. and Bederka, J. P., Jr., Eds., Academic Press, New York, 1974, 203.

2. **Buhler, D. R. and Rasmusson, M. E.,** The oxidation of drugs by fishes, *Comp. Biochem. Physiol.,* 25, 223, 1968.

3. **Neff, J. M.,** *Polycyclic Aromatic Hydrocarbons in the Aquatic Environment,* Applied Science Publishers, London, England, 1979.

4. **Baumann, P. C., Smith, W. D., and Ribick, M.,** Hepatic tumor rates and polynuclear aromatic hydrocarbon levels in two populations of brown bullhead *(Ictalurus nebulosus),* in *Polynuclear Aromatic Hydrocarbons: 6th Int. Symp. on Physical and Biological Chemistry,* Cooke, M., Dennis, A. J., and Fischer, G. L., Eds., Battelle Press, Columbus, Ohio 1982, 93.

5. **Black, J. J.,** Field and laboratory studies of environmental carcinogenesis in Niagara River fish, *J. Great Lakes Res.,* 9, 326, 1983.

6. **Malins, D. C., McCain, B. B., Brown, D. W., Chan, S.-L., Myers, M. S., Landahl, J. T., Prohaska, P. G., Friedman, A. J., Rhodes, L. D., Burrows, D. G., Gronlund, W. D., and Hodgins, H. O.,** Chemical pollutants in sediments and diseases in bottom-dwelling fish in Puget Sound, Washington, *Environ. Sci. Technol.,* 18, 705, 1984.

7. **Chan, G. L.,** A study of the effects of the San Francisco oil spill on marine life, in *Proc. 1975 Conf. on Prevention and Control of Oil Pollution,* American Petroleum Institute, Washington, D.C., 1975, 457.

8. **Wolfe, D. A., Ed.,** *Fate and Effects of Petroleum Hydrocarbons in Marine Organisms and Ecosystems,* Pergamon Press, New York, 1977.

9. **Clark, R. C., Jr. and MacLeod, W. D., Jr.,** Inputs, transport mechanisms, and observed concentrations of petroleum in the marine environment, in *Effect of Petroleum on Arctic and Subarctic Marine Environments and Organisms, Vol. 1, Nature and Fate of Petroleum,* Malins, D. C., Ed., Academic Press, New York, 1977, 1.

10. **Stein, J. E., Hom, T., Casillas, E., Friedman, A., and Varanasi, U.,** Simultaneous exposure of English sole *(Parophrys vetulus)* to sediment-associated xenobiotics. II. Chronic exposure to an urban estuarine sediment, with added ^{3}H-benzo(a)pyrene and ^{14}C-polychlorinated biphenyls, *Mar. Environ. Res.,* 22, 123, 1987.

11. **Varanasi, U., Reichert, W. L., Stein, J. E., Brown, D. W., and Sanborn, H. R.,** Bioavailability and biotransformation of aromatic hydrocarbons in benthic organisms exposed to sediment from an urban estuary, *Environ. Sci. Technol.,* 19, 836, 1985.

12. **Varanasi, U. and Gmur, D. J.,** unpublished observations, 1980.

13. **Stein, J. E., Hom, T., and Varanasi, U.,** Simultaneous exposure of English sole *(Parophrys vetulus)* to sediment-associated xenobiotics. I. Uptake and disposition of ^{14}C-polychlorinated biphenyls and ^{3}H-benzo[a]pyrene, *Mar. Environ. Res.,* 13, 97, 1984.

14. **Balk, L., Meijer, J., DePierre, J. W., and Appelgren, L.-E.,** The uptake and distribution of [^{3}H]benzo(a)pyrene in the Northern pike *(Esox lucius).* Examination by whole-body autoradiography and scintillation counting, *Toxicol. Appl. Pharmacol.,* 74, 430, 1984.

15. **Plesha, P. P.,** unpublished observations, 1986.

16. **Pizza, J. C. and O'Connor, J. M.,** PCB dynamics in Hudson River striped bass. II. Accumulation from dietary sources, *Aquat. Toxicol.,* 3, 313, 1983.

17. **Thomann, R. V. and Connolly, J. P.,** Model of PCB in the Lake Michigan lake trout food chain, *Environ. Sci. Technol.,* 18, 65, 1984.

18. **Varanasi, U., Gmur, D. J., and Treseler, P. A.,** Influence of time and mode of exposure on biotransformation of naphthalene by juvenile starry flounder *(Platichthys stellatus)* and rock sole *(Lepidopsetta bilineata), Arch. Environ. Contam. Toxicol.,* 8, 673, 1979.

19. **Varanasi, U., Nishimoto, M., Reichert, W. L., and Eberhart, B.-T. L.,** Comparative metabolism of benzo(a)pyrene and covalent binding to hepatic DNA in English sole, starry flounder, and rat, *Cancer Res.,* 46, 3817, 1986.

20. **Varanasi, U., Stein, J. E., and Hom, T.,** Convalent binding of benzo(a)pyrene to DNA in fish liver, *Biochem. Biophys. Res. Commun.,* 103, 780, 1981.

21. **James, M. O. and Bend, J. R.,** Polycyclic aromatic hydrocarbon induction of cytochrome P-450-dependent mixed-function oxidases in marine fish, *Toxicol. Appl. Pharmacol.,* 54, 117, 1980.

22. **Williams, W. M., Chen, T. S. T., and Huang, K. C.,** Renal handing of aromatic amino acids, sugar, and standard glomerular markers in winter flounder, *Am. J. Physiol.,* 227, 1380, 1974.

23. **Vetter, R. D., Carey, M. C., and Patton, J. S.,** Coassimilation of dietary fat and benzo(a)pyrene in the small intestine: an absorption model using the killifish, *J. Lipid Res.,* 26, 428, 1985.

24. **Melancon, M. J., Jr. and Lech, J. J.,** Uptake, biotransformation, disposition, and elimination of 2-methylnaphthalene and naphthalene in several fish species, in *Aquatic Toxicology,* ASTM STP 667, Marking, L. L. and Kimerle, R. A., Eds., Cleveland, OH, 1979, 5.

25. **Thomas, R. E. and Rice, S. D.,** Metabolism and clearance of phenolic and mono-, di- and polynuclear aromatic hydrocarbons by Dolly Varden char, in *Physiological Mechanisms of Marine Pollutant Toxicity,* Vernberg, J., Calabrese, A., Thurberg, T. P., and Thurberg, W. B., Eds., Academic Press, New York, 1982, 161.

26. **Roubal, W. T., Stranahan, S. J., and Malins, D. C.,** The accumulation of low molecular weight aromatic hydrocarbons of crude oil by coho salmon *(Oncorhynchus kisutch)* and starry flounder *(Platichthys stellatus), Arch. Environ. Contam. Toxicol.,* 7, 237, 1978.

27. **Roubal, W. T., Collier, T. K., and Malins, D. C.,** Accumulation and metabolism of carbon-14 labeled benzene, naphthalene, and anthracene by young coho salmon, *(Oncorhynchus kisutch), Arch. Environ. Contam. Toxicol.,* 5, 513, 1977.

28. **Korn, S. and Rice, S.,** Sensitivity to, and accumulation and depuration of, aromatic petroleum components by early life stages of coho salmon *(Oncorhyncus kisutch), Rapp. P. V. Reun. Cons. Int. Explor. Mer,* 178, 87, 1981.

29. **Southworth, G. R., Keffer, C. C., and Beauchamp, J. J.,** The accumulation and disposition of benz(a)acridine in the fathead minnow, *Pimephales promelas, Arch. Environ. Contam. Toxicol.,* 10, 561, 1981.

30. **Solbakken, J. E., Solberg, M., and Palmork, K. H.,** A comparative study on the disposition of three aromatic hydrocarbons in flounder *(Platichthys flesus), FiskDir. Skr. Ser. Havunders.,* 17, 473, 1984.

31. **Hawker, D. W. and Connell, D. W.,** Relationships between partition coefficient, clearance rate constant, and time to equilibrium for bioaccumulation, *Chemosphere,* 14, 1205, 1985.

32. **Varanasi, U., Uhler, M., and Stranahan, S.,** Uptake and release of naphthalene and its metabolites in skin and epidermal mucus of samonids, *Toxicol. Appl. Pharmacol.,* 44, 277, 1978.

33. **Varanasi, U. and Gmur, D. J.,** Hydrocarbons and metabolites in English sole *(Parophrys vetulus)* exposed simultaneously to [³H]benzo(a)pyrene and [¹⁴C]naphthalene in oil-contaminated sediment, *Aquat. Toxicol.,* 1, 49, 1981.

34. **Lee, R. F., Sauerheber, R., and Dobbs, G. H.,** Uptake, metabolism and discharge of polycyclic aromatic hydrocarbons by marine fish, *Mar. Biol.,* 17, 201, 1972.

35. **Solbakken, J. E., Palmork, K. H., Neppelberg, T., and Scheline, R. R.,** Distribution of radioactivity in coalfish *(Pollachius virens)* following intragastric administration of [9-¹⁴C]phenanthrene, *Bull. Environ. Contam. Toxicol.,* 23, 100, 1979.

36. **Solbakken, J. E. and Palmork, K. H.,** Distribution of radioactivity in the chondrichthyes Squalus acanthias and the osteichthyes Salmo gairdneri following intragastric administration of (9-¹⁴C)phenanthrene, *Bull. Environ. Contam. Toxicol.,* 25, 902, 1980.

37. **Nagel, R. and Urich, K.,** Kinetic studies on the elimination of different substituted phenols by goldfish *(Carassius auratus), Bull. Environ. Contam. Toxicol.,* 24, 374, 1980.

38. **Thomas, R. E. and Rice, S. D.,** Excretion of aromatic hydrocarbons and their metabolites by freshwater and seawater Dolly Varden char, in *Biological Monitoring of Marine Pollutants,* Vernberg, J., Calabrese, A., Thurberg, F. P., and Vernberg, W. B., Eds., Academic Press, New York, 1981, 425.

39. **Varanasi, U., Nishimoto, M., Reichert, W. L., and Stein, J. E.,** Metabolism and subsequent covalent binding of benzo(a)pyrene to macromolecules in gonads and liver of ripe English sole *(Parophrys vetulus), Xenobiotica,* 12, 417, 1982.

40. **Reichert, W. L. and Varanasi, U.,** Metabolism of orally administered naphthalene in spawning English sole *(Parophyrs vetulus), Environ. Res.,* 27, 316, 1982.

41. **Fry, F. E. J.,** Temperature compensation, *Annu. Rev. Physiol.,* 20, 202, 1958.

42. **Hazel, J. R. and Prosser, C. L.,** Molecular mechanisms of temperature compensation in poikilotherms, *Physiol. Rev.,* 54, 620, 1974.

43. **Korn, S., Moles, D. A., and Rice, S. D.,** Effects of temperature on the median tolerance limit of pink salmon and shrimp exposed to toluene, naphthalene, and Cook Inlet crude oil, *Bull. Environ. Contam. Toxicol.,* 21, 521, 1979.

44. **Thomas, R. E. and Rice, S. D.,** The effect of exposure temperatures on oxygen consumption and opercular breathing rates of pink salmon fry exposed to toluene, naphthalene, and water-soluble fractions of Cook Inlet crude oil and no. 2 fuel oil, in *Marine Pollution: Functional Responses,* Vernberg, W. B., Calabrese, A., Thurberg, F. P., and Vernberg, F. J., Eds., Academic Press, New York, 1979, 39.

45. **Collier, T. K., Thomas, L. C., and Malins, D. C.,** Influence of environmental temperature on disposition of dietary naphthalene in coho salmon *(Onchorhynchus kisutch):* isolation and identification of individual metabolites, *Comp. Biochem. Physiol.,* 61C, 23, 1978.

46. **Varanasi, U., Gmur, D. J., and Reichert, W. L.,** Effect of environmental temperature on naphthalene metabolism by juvenile starry flounder *(Platichthys stellatus), Arch. Environ. Contam. Toxicol.,* 10, 203, 1981.

47. **Thomas, R. E. and Rice, S. D.,** Effect of temperature on uptake and metabolism of toluene and naphthalene by Dolly Varden char, *Salvelinus malma, Comp. Biochem. Physiol.,* 84C, 83, 1986.

48. **Egaas, E. and Varanasi, U.,** Effects of polychlorinated biphenyls and environmental temperature on *in vitro* formation of benzo(a)pyrene metabolites by liver of trout *(Salmo gairdneri), Biochem. Pharmacol.,* 31, 561, 1982.

49. **Koivusaari, U., Harri, M., and Hanninen, O.,** Seasonal variation of hepatic biotransformation in female and male rainbow trout *(Salmo gairdneri), Comp. Biochem. Physiol.,* 70C, 149, 1981.

50. **Levitan, W. M. and Taylor, M. H.,** Physiology of salinity-dependent naphthalene toxicity in *Fundulus heteroclitus, J. Fish. Res. Board Can.,* 36, 615, 1979.

51. **Moles, O. A., Rice, S. D., and Korn, S.,** Sensitivity of Alaskan freshwater and anadromus fishes to Prudhoe Bay crude oil and benzene, *Trans. Am. Fish. Soc.,* 108, 408, 1979.

52. **Stickle, W. B., Sabourin, T. D., and Rice, S. D.,** Sensitivity and osmoregulation of coho salmon, *Oncorhynchus kisutch,* exposed to toluene and naphthalene at different salinities, in *Physiological Mechanisms of Marine Pollutant Toxicity,* Vernberg, W. B., Calabrese, A., Thurberg, F. P., and Vernberg, J. F., Eds., Academic Press, New York, 1982, 331.

53. **DiMichele, L. and Taylor, M. H.,** Histopathological and physiological responses of *Fundulus heteroclitus* to naphthalene exposure, *J. Fish. Res. Board Can.,* 35, 1060, 1978.

54. **Dixit, D. and Anderson, J. W.,** Distribution of naphthalenes within exposed *Fundulus simulus* and correlations with stress behavior, in *Proc. 1977 Oil Spill Conf. (Prevention Behavior, Control, Cleanup),* American Petroleum Institute, Washington, D.C., 1977, 633.

55. **Solbakken, J. E., Knap, A. H., and Palmork, K. H.,** Disposition of (9-^{14}C) phenanthrene in a subtropical marine teleost *(Haemulon sciurus), Bull. Environ. Contam. Toxicol.,* 28, 285, 1982.

56. **Pohl, R. J., Bend, J. R., Guarino, A. M., and Fouts, J. R.,** Hepatic microsomal mixed-function oxidase activity of several marine species from coastal Maine, *Drug Metab. Dispos.,* 2, 545, 1974.

57. **Guarino, A. M., Briley, P. M., Anderson, J. B., Kinter, M. A., Schneiderman, S., Klipp, L. D., and Adamson, R. H.,** Renal and hepatic excretion of foreign chemicals by *Squalus acanthias, Bull. Mt. Desert Isl. Biol. Lab.,* 12, 41, 1972.

58. **Solbakken, J. E., Tilseth, S., and Palmork, K. H.,** Uptake and elimination of aromatic hydrocarbons and a chlorinated biphenyl in eggs and larvae of cod *Gadus morhua, Mar. Ecol. Prog. Ser.,* 16, 297, 1984.

59. **Binder, R. L. and Stegeman, J. J.,** Microsomal electron transport and xenobiotic monooxygenase activities during the embryonic period of development in the killifish, *Fundulus heteroclitus, Toxicol. Appl. Pharmacol.,* 73, 432, 1984.

60. **Rosenthal, H. and Alderdice, D. F.,** Sublethal effects of environmental stressors, natural and pollutional, on marine fish eggs and larvae, *J. Fish. Res. Board Can.,* 33, 2047, 1976.

61. **Guiney, P. D., Melancon, M. J., Jr., Lech, J. J., and Peterson, R. E.,** Effects of egg and sperm maturation and spawning on the distribution and elimination of a polychlorinated biphenyl in rainbow trout *(Salmo gairdneri), Toxicol. Appl. Pharmacol.,* 47, 261, 1979.

62. **Binder, R. L. and Lech, J. J.,** Xenobiotics in gametes of Lake Michigan lake trout *(Salvelinus namaycush)* induce hepatic monooxygenase activity in their offspring, *Fundam. Appl. Toxicol.,* 4, 1042, 1984.

63. **Statham, C. N., Elcombe, C. R., Szyjka, S. P., and Lech, J. J.,** Effect of polycyclic aromatic hydrocarbons on hepatic microsomal enzymes and disposition on methylnaphthalene in rainbow trout *in vivo, Xenobiotica,* 8, 65, 1978.

64. **Collier, T. K., Gruger, E. H., Jr., and Varanasi, U.,** Effect of Aroclor 1254 on the biological fate of 2,6-dimethylnaphthalene in coho salmon, *(Oncorhynchus kisutch), Bull. Environ. Contam. Toxicol.,* 34, 114, 1985.

65. **Thomas, R. E. and Rice, S. D.,** Effect of pretreatment exposure to toluene and naphthalene on the subsequent metabolism of dietary toluene and naphthalene by Dolly Varden, *Salvelinus malma,* in *Marine Pollution and Physiology: Recent advances,* Vernberg, F. J., Thurberg, F. P., Calabrese, A., and Vernberg, W., Eds., University of South Carolina Press, Columbia, 1985, 505.

66. **Widdows, J., Moore, S. L., Clarke, K. R., and Donkin, P.,** Uptake, tissue distribution and elimination of [1-^{14}C]naphthalene in the mussel *Mytilus edulis, Mar. Biol.,* 76, 109, 1983.

67. **Pritchard, J. B. and Bend, J. R.,** Mechanisms controlling the renal excretion of xenobiotics in fish: effects of chemical structure, *Drug Metab. Rev.,* 15, 655, 1984.

68. **Bend, J. R. and James, M. O.,** Xenobiotic metabolism in marine and freshwater species, in *Biochemical and Biophysical Perspectives in Marine Biology,* Vol. 4, Malins, D. C. and Sargent, J. R., Eds., Academic Press, New York, 1978, 125.

69. **Stegeman, J. J.,** Polynuclear aromatic hydrocarbons and their metabolism in the marine environment, in *Polycyclic Hydrocarbons and Cancer,* Vol. 3, Gelboin, H. V. and Ts'o, P. O. P., Eds., Academic Press, New York, 1981, 1.

70. **James, M. O.,** Xenobiotic conjugation in fish and other aquatic species, in *Xenobiotic Conjugation Chemistry,* Paulson, G. D., Caldwell, J., Hutson, D. H., and Menn, J. J., Eds., American Chemical Society, Washington, D.C., 1986, 29.

71. **Langenbach, R., Newnow, S., and Rice, J. M., Eds.,** *Organ and Species Specificity in Chemical Carcinogenesis,* Plenum Press, New York, 1983.

72. **Sims, P., Grover, P. L., Swaisland, A., Pal, K., and Hewer, A.,** Metabolic activation of benzo(a)pyrene proceeds by a diol-epoxide, *Nature (London),* 252, 326, 1974.

73. **Horning, M. G., Stillwell, W. G., Griffin, G. W., and Tsang, W.-S.,** Epoxide intermediates in the metabolism of naphthalene by the rat, *Drug Metab. Dispos.,* 8, 404, 1980.

74. **Boyland, E. and Sims, P.,** Metabolism of polycyclic compounds. XII. An acid-labile precursor of 1-naphthylmercapturic acid and naphthol: an N-acetyl-S-(1:2-dihydrohydroxynaphthyl)-L-cysteine, *Biochem. J.,* 68, 440, 1958.

75. **Sims, P.,** Metabolism of polycyclic compounds. XIV. The conversion of naphthalene into compounds related to *trans*-1:2-dihydroxynaphthalene by rabbits, *Biochem. J.,* 73, 389, 1959.

76. **Boyland, E., Ramsey, G. S., and Sims, P.,** Metabolism of polycyclic compounds. XVIII. The secretion of metabolites of naphthalene, 1:2-dihydronaphthalene and 1:2-epoxy-1:2:3:4-tetrahydronaphthalene in rat bile, *Biochem. J.,* 79, 376, 1961.

77. **McCain, B. B., Hodgins, H. O., Gronlund, W. D., Hawkes, J. W., Brown, D. W., Myers, M. S., and Vandermeulen, J. H.,** Bioavailability of crude oil from experimentally oiled sediments to English sole *(Parophrys vetulus),* and pathological consequences, *J. Fish. Res. Board Can.,* 35, 657, 1978.

78. **Collier, T. K., Krahn, M. M., and Malins, D. C.,** The disposition of naphthalene and its metabolites in the brain of Rainbow trout *(Salmo gairdneri), Environ. Res.,* 23, 35, 1980.

79. **Thomas, L. C., MacLeod, W. D., and Malins, D. C.,** Extraction and quantitation of metabolites of naphthalene in marine organisms by high-pressure liquid chromatography, in *Proc. 9th Materials Research Symp. on Trace Organic Analysis,* National Bureau of Standards Spec. Publ. 519, National Bureau of Standards, 1978, 79.

80. **Collier, T. K.,** personal communication, 1987.

81. **Gruger, E. H., Jr., Schnell, J. V., Fraser, P. S., Brown, D. W., and Malins, D. C.,** Metabolism of 2,6-dimethylnaphthalene in starry flounder *(Platichthys stellatus)* exposed to naphthalene and p-cresol, *Aquat. Toxicol.,* 1, 37, 1981.

82. **Klaasen, C. and Watkins, J. B., III,** Mechanisms of bile formation, hepatic uptake, and biliary excretion, *Pharmacol. Rev.,* 36, 1, 1984.

83. **Plummer, J. L., Smith, B. R., Ball, L. M., and Bend, J. R.,** Metabolism and biliary excretion of benzo[a]pyrene 4,5-oxide in the rat, *Drug Metab. Dispos.,* 8, 68, 1980.

84. **Layiwola, P. J., Linnecar, D. F. C., and Knights, B.,** The biotransformation of three ^{14}C-labelled phenolic compounds in twelve species of freshwater fish, *Xenobiotica,* 13, 107, 1983.

85. **Morrison, H., Young, P., and George, S.,** Conjugation of organic compounds in isolated hepatocytes from a marine fish, the plaice, *Pleuronectes platessa, Biochem. Pharmacol.,* 34, 3944, 1985.

86. **Plakunov, I., Smolarek, T. A., Fischer, D. L., Wiley, J. C., Jr., and Baird, W. M.,** Separation by ion-pair high-performance liquid chromatography of the glucuronide, sulfate and glutathione conjugates formed from benzo(a)pyrene in cell cultures from rodents, fish and humans, *Carcinogenesis (London),* 8, 59, 1987.

87. **Krahn, M. M., Brown, D. W., Collier, T. K., Friedman, A. J., Jenkins, R. G., and Malins, D. C.,** Rapid analysis of naphthalene and its metabolites in biological systems: determination by high-performance liquid chromatography/fluorescence detection and by plasma desorption/chemical ionization mass spectrometry, *J. Biochem. Biophys. Methods,* 2, 233, 1980.

88. **Nagel, R.,** Species differences, influence of dose and application on biotransformation of phenol in fish, *Xenobiotica,* 13, 101, 1983.

89. **Bock, K. W.,** Dual role of glucuronyl- and sulfotransferases converting xenobiotics into reactive or biologically inactive and easily excretable compounds, *Arch. Toxicol.,* 39, 77, 1977.

90. **Mulder, G. J. and Keulemans, K.,** Metabolism of inorganic sulphate in the isolated perfused rat liver: effect of sulphate concentration on the rate of sulphation by phenol sulphotransferase, *Biochem. J.,* 176, 959, 1978.

91. **Mulder, G. J. and Scholtens, E.,** The availability of inorganic sulphate in blood for sulphate conjugation of drugs in rat liver *in vivo, Biochem. J.,* 172, 247, 1978.

92. **Black, J. A., Birge, W. J., Westerman, A. G., and Francis, P. C.,** Comparative aquatic toxicology of aromatic hydrocarbons, *Fundam. Appl. Toxicol.,* 3, 353, 1983.

93. **Milleman, R. E., Birge, W. J., Black, J. A., Cushman, R. M., Daniels, K. L., Franco, P. J., Giddings, J. M., McCarthy, J. F., and Stewart A. J.,** Comparative acute toxicity to aquatic organisms of components of coal-derived synthetic fuels, *Trans. Am. Fish. Soc.,* 113, 74, 1984.

94. **Moles, A. and Rice, S. D.,** Effects of crude oil and naphthalene on growth, caloric content, and fat content of pink salmon juveniles in seawater, *Trans. Am. Fish. Soc.,* 112, 205, 1983.

95. **DeGraeve, G. M., Elder, R. C., Woods, D. C., and Bergman, H. L.,** Effects of naphthalene and benzene on fathead minnows and rainbow trout, *Arch. Environ. Contam. Toxicol.,* 11, 487, 1982.

96. **Boyland, E. and Sims, P.,** Metabolism of polycyclic compounds: the metabolism of phenanthrene in rabbits and rats: dihydrodihydroxy compounds and related glucosiduronic acids, *Biochem. J.,* 84, 571, 1965.

147

97. **Solbakken, J. E. and Palmork, K. H.,** Metabolism of phenanthrene in various marine animals, *Comp. Biochem. Physiol.,* 70C, 21, 1981.
98. **Solbakken, J. E., Palmork, K. H., Neppelberg, T., and Scheline, R. R.,** Urinary and biliary metabolites of phenanthrene in the coalfish *(Pollachius virens), Acta Pharmacol. Toxicol.,* 46, 127, 1980.
99. **Goksoyr, A., Solbakken, J. E., and Klungsoyr, J.,** Regioselective metabolism of phenanthrene in Atlantic cod *(Gadus morhua):* studies on the effects of monooxygenase inducers and role of cytochrome P-450, *Chem. Biol. Interact.,* 60, 247, 1986.
100. **Dipple, A., Moschel, R. C., and Bigger, C. A. H.,** Polynuclear aromatic hydrocarbons, in *Chemical Carcinogens,* Vol. 1, Searle, C. E., Ed., American Chemical Society, Washington, D. C., 1984, 41.
101. **Conney, A. H.,** Induction of microsomal enzymes by foreign chemicals and carcinogenesis by polycyclic aromatic hydrocarbons: G. H. A. Clowes Memorial Lecture, *Cancer Res.,* 42, 4875, 1982.
102. **Lutz, W. K.,** *In vivo* covalent binding of organic chemicals to DNA as a quantitative indicator in the process of chemical carcinogenesis, *Mutat. Res.,* 65, 289, 1979.
103. **Cooper, C. S., Grover, P. L., and Sims, P.,** The metabolism and activation of benzo(a)pyrene, *Prog. Drug Metab.,* 7, 295, 1983.
104. **Varanasi, U. et al.,** unpublished observation, 1986.
105. **Mannervik, B.,** The isozymes of glutathione transferase, in *Advances in Enzymology,* Vol. 57, Meister, A., Ed., John Wiley & Sons, New York, 1985, 357.
106. **Varanasi, U., Nishimoto, M., and Stover, J.,** Analyses of biliary conjugates and hepatic DNA binding in benzo(a)pyrene-exposed English sole, in *Polynuclear Aromatic Hydrocarbons: 8th Int. Symp. on Mechanisms, Methods and Metabolism,* Cooke, M. W. and Dennis, A. J., Eds., Battelle Press, Columbus, Ohio 1983, 1315.
107. **Nishimoto, M. and Varanasi, U.,** unpublished observations, 1987.
108. **Balk, L., Knall, A., and DePierre, J. W.,** Separation of the different classes of conjugates formed by metabolism of benzo(a)pyrene in the northern pike *(Esox lucius), Acta Chem. Scand.,* B36, 403, 1982.
109. **Varanasi, U., Stein, J. E., Nishimoto, M., Reichert, W. L., and Collier, T. K.,** Chemical carcinogenesis in feral fish: uptake, activation, and detoxication of organic xenobiotics, *Environ. Health Perspect.,* 71, 155, 1987.
110. **Awasthi, Y. and Singh, S.,** personal communication, 1987.
111. **Yagen, B., Foureman, G. L., Ben-Zvi, Z., Ryan, A. J., Hernandez, O., Cox, R. H., and Bend, J. R.,** The metabolism and excretion of ^{14}C-styrene oxide-glutathione adducts administered to the winter flounder, *Pseudopleuronectes americanus,* a marine teleost, *Drug Metab. Dispos.,* 12, 389, 1984.
112. **Pritchard, J. B. and Renfro, J. L.,** Interactions of xenobiotics with teleost renal function, in *Aquatic Toxicology,* Vol. 2, Weber, L. J., Ed., Raven Press, New York, 1984, 51.
113. **Gmur, D. J. and Varanasi, U.,** Characterization of benzo(a)pyrene metabolites isolated from muscle, liver, and bile of a juvenile flatfish, *Carcinogenesis (London),* 3, 1397, 1982.
114. **Krahn, M. M., Schnell, J. V., Uyeda, M. Y., and MacLeod, W. D., Jr.,** Determination of mixtures of benzo(a)pyrene, 2,6-dimethylnaphthalene and their metabolites by high-performance liquid chromatography with fluorescence detection, *Anal. Biochem.,* 113, 27, 1981.
115. **Varanasi, U., Stein, J. E., Nishimoto, M., and Hom, T.,** Benzo(a)pyrene metabolites in liver, muscle, gonads and bile of adult English sole *(Parophrys vetulus),* in *Polynuclear Aromatic Hydrocarbons: 7th Int. Symp. on Formation, Metabolism and Measurement,* Cooke, M. W. and Dennis, A. J., Eds., Battelle Press, Columbus, OH, 1982, 1221.
116. **Stein, J. E. et al.,** unpublished observations, 1986.
117. **Chipman, J. K., Frost, G. S., Hirom, P. C., and Millburn, P.,** Biliary excretion, systemic availability and reactivity of metabolites following intraportal infusion of [^{3}H]benzo(a)pyrene in the rat, *Carcinogenesis (London),* 2, 741, 1981.
118. **Prough, R. A., Saeki, Y., and Capdevilla, J.,** The metabolism of benzo(a)pyrene phenols by rat liver microsomal fractions, *Arch. Biochem. Biophys.,* 212, 136, 1981.
119. **Selkirk, J. K., Croy, R. G., and Gelboin, H. V.,** High-pressure liquid chromatographic separation of 10 benzo(a)pyrene phenols and the identification of 1-phenol and 7-phenol as new metabolites, *Cancer Res.,* 36, 922, 1976.
120. **Yang, S. K., Deutsch, J., and Gelboin, H. V.,** Benzo(a)pyrene metabolism: activation and detoxification, in *Polycyclic Hydrocarbons and Cancer, Vol. 1,* Gelboin, H. V. and Ts'o, P. O. P., Eds., Academic Press, New York, 1978, 205.
121. **Owens, I. S., Koteen, G. M., and Legraverend, C.,** Mutagenesis of certain benzo[a]pyrene phenols *in vitro* following further metabolism by mouse liver, *Biochem. Pharmacol.,* 28, 1615, 1979.
122. **Hoover, K. L., Ed.,** Use of Small Fish Species in Carcinogenicity Testing, NCI Monogr. 65, NIH Publ. No. 84-2653, National Institutes of Health, Bethesda, MD, 1984, 1.
123. **Black, J. J., Maccubbin, A. E., and Schiffert, M.,** A reliable efficient, microinjection apparatus and methodology for the *in vivo* exposure of rainbow trout and salmon embryos to chemical carcinogens, *J. Natl. Cancer Inst.,* 75, 1123, 1985.

124. **Hendricks, J. D., Meyers, T. R., Shelton, D. W., Casteel, J. L., and Bailey, G. S.,** Hepatocarcinogenicity of benzo[a]pyrene to rainbow trout by dietary exposure and intraperitoneal injection, *J. Natl. Cancer Inst.,* 74, 839, 1985.

125. **Harvey, R. G.,** Polycyclic hydrocarbons and cancer, *Am. Sci.,* 70, 386, 1982.

126. **Jones, C. A., Moore, B. P., Cohen, G. M., Fry, J. R., and Bridges, J. W.,** Studies on the metabolism and excretion of benzo(a)pyrene in isolated adult rat hepatocytes, *Biochem. Pharmacol.,* 27, 693, 1978.

127. **Cohen, G. M. and Moore, B. P.,** The metabolism of benzo(a)pyrene, 7,8-dihydro-7,8-dihydroxybenzo(a)pyrene and 9,10-dihydro-9,10-dihydroxybenzo(a)pyrene by short-term organ cultures of hamster lung, *Biochem. Pharmacol.,* 26, 1481, 1977.

128. **Schenkman, J. B. and Kupfer, D., Eds.,** *International Encyclopedia of Pharmacology and Therapeutics, Section 108: Hepatic Cytochrome P-450 Monooxygenase System,* Pergamon Press, Oxford, England, 1982.

129. **Goodman, J. I., Vorce, R. L., and Baranyi-Furlong, B. L.,** Genetic toxicology: chemical carcinogens modify DNA in a nonrandom fashion, *Trends Pharmacol. Sci.,* 7, 354, 1986.

130. **Harvey, R. G., Ed.,** *Polycyclic Hydrocarbons and Carcinogenesis,* ACS Symp. Ser. 283, American Chemical Society, Washington, D.C., 1985.

131. **Weinstein, I. B., Gattoni-Celli, S., Kirschmeier, P., Hsiao, W., Horowitz, A., and Jeffrey, A.,** Cellular targets and host genes in multistage carcinogenesis, *Fed. Proc.,* 43, 2287, 1984.

132. **Weinstein, I. B.,** Current concepts and controversies in chemical carcinogenesis, *J. Supramol. Struct. Cell. Biochem.,* 17, 99, 1981.

133. **Farber, E. and Sarma, D. S. R.,** Chemical carcinogenesis: the liver as a model, *Pathol. Immunopathol. Res.,* 5, 1, 1986.

134. **Goldsworthy, T. L., Hanigan, M. H., and Pitot, H. C.,** Models of hepatocarcinogenesis in the rat—contrasts and comparisons, *CRC Crit. Rev. Toxicol.,* 17, 61, 1987.

135. **von Hofe, E. and Puffer, H. W.,** *In vitro* metabolism and *in vivo* binding of benzo(a)pyrene in the California killifish *(Fundulus parvipinnis)* and speckled sanddab *(Citaricthys stigmaeous), Arch. Environ. Contam. Toxicol.,* 15, 251, 1986.

136. **Boroujerdi, M., Kung, H.-C., Wilson, A. G. E., and Anderson, M. W.,** Metabolism and DNA binding of benzo(a)pyrene *in vivo* in the rat, *Cancer Res.,* 41, 951, 1981.

137. **Kleihues, P., Doejer, G., Ehret, M., and Guzman, J.,** Reaction of benzo(a)pyrene and 7,12-dimethylbenz(a)anthracene with DNA of various rat tissues *in vivo, Quant. Aspects Risk Assess. in Chem. Carcinogenesis, Arch. Toxicol.,* 3, 237, 1980.

138. **Scovassi, A. I., Wicker, R., and Bertazzoni, U.,** A phylogenetic study on vertebrate mitochondrial DNA polymerase, *Eur. J. Biochem.,* 100, 491, 1979.

139. **Kelly, J. J. and Maddock, M. B.,** *In vitro* induction of unscheduled DNA synthesis by genotoxic carcinogens in the hepatocytes of the oyster toadfish *(Opsanus tau), Arch. Environ. Contam. Toxicol.,* 14, 555, 1985.

140. **Walton, D. G., Acton, A. B., and Stich, H. F.,** DNA repair synthesis in cultured mammalaian and fish cells following exposure to chemical mutagens, *Mutat. Res.,* 124, 153, 1983.

141. **Shen, A. L., Fahl, W. E., and Jefcoate, C. R.,** Metabolism of benzo(a)pyrene by isolated hepatocytes and factors affecting covalent binding of benzo(a)pyrene metabolites to DNA in hepatocyte and microsomal systems, *Arch. Biochem. Biophys.,* 204, 511, 1980.

142. **Hesse, S. and Jernstrom, B.,** Role of glutathione-S-transferases: detoxification of reactive metabolites of benzo(a)pyrene-7,8-dihydrodiol by conjugation with glutathione, in *Biochemical Basis of Chemical Carcinogenesis,* Greim, H., Jung, R., Kramer, M., Marquardt, H., and Oesch, F., Eds., Raven Press, New York, 1984.

143. **Jernstrom, B., Martinez, M., Meyer, D. J., and Ketterer, B.,** Glutathione conjugation of the carcinogenic and mutagenic electrophile (±)-7,8-dihydroxy-9,10-oxy-7,8,9,10-tetrahydrobenzo(a)pyrene catalyzed by purified rat liver glutathione transferase, *Carcinogenesis (London),* 6, 85, 1985.

144. **Van Veld, P., Vetter, R. D., Lee, R. F., and Patton, J. S.,** Metabolism of triglyceride-solubilized benzo(a)pyrene by the intestine of fish: influence of triglyceride concentration and cytosolic glutathione transferases, in press.

145. **Nebert, D. W., Thorgeirsson, S. S., and Lambert, G.,** Genetic aspects of toxicity during development, *Environ. Health,* 18, 35, 1976.

146. **Hose, J. E., Hannah, J. B., Landolt, M. L., Miller, B. S., Felton, S. P., and Iwaoka, W. T.,** Uptake of benzo[a]pyrene by gonadal tissue of flatfish (family *Pleuronectidae*) and its effects on subsequent egg development, *J. Toxicol. Environ. Health,* 7, 990, 1981.

147. **Hose, J. E., Hannah, J. B., DiJulio, D., Landolt, M. L., Miller, B. S., Iwaoka, W. T., and Felton, S. P.,** Effects of benzo(a)pyrene on early development of flatfish, *Arch. Environ. Contam. Toxicol.,* 11, 167, 1982.

148. **Collier, T. K. and Varanasi, U.,** Xenobiotic metabolizing enzymes in two species of benthic fish showing different prevalences of hepatic neoplasms in contaminated areas, *Mar. Environ. Res.,* 24, 113, 1988.

149. **Nishimoto, M. and Varanasi, U.**, Metabolism and DNA adduct formation of benzo[a]pyrene and the 7,8-dihydrodiol of benzo[a]pyrene by fish liver enzymes, in *Polycyclic Aromatic Hydrocarbons: 9th Int. Symp. on Chemistry, Characterization and Carcinogenesis,* Cooke, M. and Dennis, A. J., Eds., Battelle Press, Columbus, OH, 1986, 685.

150. **Nishimoto, M. and Varanasi, U.**, Benzo(a)pyrene metabolism and DNA adduct formation mediated by English sole liver enzymes, *Biochem. Pharmacol.,* 34, 263, 1985.

151. **Lucier, G. W. and Hook, E. E. R., Eds.**, DNA adducts: dosimeters to monitor human exposure to environmental mutagens and carcinogens, in *Environmental Health Perspectives, Vol.* 62, Research Triangle Park, North Carolina, 1985.

152. **Singer, B. and Grunberger, D.**, *Molecular Biology of Mutagens and Carcinogens,* Plenum Press, New York, 1983.

153. **Shugart, L., McCarthy, J., Jimenez, B., and Daniels, J.**, Analysis of adduct formation in the bluegill sunfish *(Lepomis macrochirus)* between benzo(a)pyrene and DNA of the liver and hemoglobin of the erythrocyte, *Aquat. Toxicol.,* 9, 319, 1987.

154. **Randerath, K., Reddy, M. V., and Gupta, R. C.**, ^{32}P-labeling test for DNA damage, *Proc. Natl. Acad. Sci. U.S.A.,* 78, 6126, 1981.

155. **Gupta, R. C., Reddy, M. V., and Randerath, K.**, ^{32}P-postlabeling analysis of non-radioactive aromatic carcinogen-DNA adducts, *Carcinogenesis (London),* 3, 1081, 1982.

156. **Reddy, M. V., Gupta, R. C., Randerath, E., and Randerath, K.**, ^{32}P-postlabeling test for covalent DNA binding of chemicals *in vivo;* application to a variety of aromatic carcinogens and methylating agents, *Carcinogenesis (London),* 5, 231, 1984.

157. **Reichert, W. L. and Varanasi, U.**, Detection of damage by ^{32}P-analysis of hepatic DNA in English sole from an urban area and in fish exposed to chemicals extracted from the urban sediment, *Proc. Am. Assoc. Cancer Res.,* 28, 95, 1987.

158. **Varanasi, U. and Reichert, W. L.**, ^{32}P-postlabeling of DNA adducts of PAHs extracted from an urban sediment, benzo(a)pyrene, benzo(b)fluoranthene, and dibenz(a,h)anthracene formed by English sole liver *in vivo* and *in vitro, Fed. Proc.,* 46, 743, 1987.

159. **Osborne, M. R., Jacobs, S., Harvey, R. G., and Brookes, P.**, Minor products from the reaction of (+) and (−) benzo[a]pyrene-*anti*-diolepoxide with DNA, *Carcinogenesis (London),* 2, 553, 1981.

160. **Nishimoto, M. and Varanasi, U.**, Formation of reactive intermediates of benzo[a]pyrene by hepatic microsomes of English sole *(Parophrys vetulus),* sampled from reference and contaminated areas of Puget Sound, Washington, *Aquat. Toxicol.,* 11, 419, 1988.

161. **Van Duuren, B. L. and Goldschmidt, B. M.**, Carcinogenic and tumor-promoting agents in tobacco carcinogenesis, *J. Natl. Cancer Inst.,* 56, 1237, 1976.

162. **Slaga, T. J., Jecker, L., Bracken, W. M., and Weeks, C. E.**, The effects of weak or non-carcinogenic polycyclic hydrocarbons on 7,12-dimethylbenz(a)anthracene and benzo(a)pyrene skin tumor-initiation, *Cancer Lett.,* 7, 51, 1979.

163. **DiGiovanni, J., Rymer, J., Slaga, T. J., and Boutwell, R. K.**, Anticarcinogenic and cocarcinogenic effects of benzo(e)pyrene and dibenz(a,c)anthracene on skin tumor initiation by polycyclic hydrocarbons, *Carcinogenesis (London),* 3, 371, 1982.

164. **Rice, J. E. Hosted, T. J., and LaVoie, E.**, Fluoranthene and pyrene enhanced benzo(a)pyrene-DNA adduct formation *in vivo* in mouse skin, *Cancer Lett.,* 24, 327, 1984.

165. **Krahn, M. M., Burrows, D. G., MacLeod, W. D., Jr., and Malins, D. C.**, Determination of individual metabolites of aromatic compounds in hydrolyzed bile of English sole *(Parophrys vetulus)* from polluted sites in Puget Sound, Washington, *Arch. Envrion. Contam. Toxicol.,* 16, 511, 1987.

166. **Krahn, M. M., Myers, M. S., Burrows, D. G., and Malins, D. C.**, Determination of metabolites of xenobiotics in bile of fish from polluted waterways, *Xenobiotica,* 14, 633, 1984.

167. **Myers, M. S., Rhodes, L. D., and McCain, B. B.**, Pathologic anatomy and patterns of occurrrence of hepatic neoplasms, putative preneoplastic lesions, and other idiopathic hepatic conditions in English sole *(Parophrys vetulus)* from Puget Sound, Washington, *J. Natl. Cancer Inst.,* 78, 333, 1987.

168. **Varanasi, U., Stein, J. E. and Reichert, W. L.**, ^{32}P-Postlabeling of DNA adducts in liver of wild English sole *(Parophrys vetulus)* and winter flounder *(Pseudopleuronectes americanus), Cancer Res.,* 1988 (accepted).

169. **Varansi, U., Reichert, W. L., Eberhart, B. -T. L., and Stein, J. E.**, Formation and persistence of benzo(a)pyrene-diolepoxide-DNA adducts in liver of English sole *(Parophrys vetulus), Chem.-Biol. Interact.,* 1988.

170. **Dunn, B. P., Black, J.J., and Maccubbin, A.**, ^{32}P-postlabeling analysis of aromatic DNA adducts in fish from polluted areas, *Cancer Res.,* 47, 6543, 1987.

171. **Johnson, L. L., Casillas, E., Collier, T. K., McCain, B. B., and Varanasi, U.**, Contaminants effects on ovarian development in English sole *(Parophrys vetulus)* from Puget Sound, Washington, *Canad. J. Fish. Aqua. Sci.,* 1988.

Chapter 5, Part I

ENZYMES INVOLVED IN METABOLISM OF PAH BY FISHES AND OTHER AQUATIC ANIMALS: OXIDATIVE ENZYMES (OR PHASE I ENZYMES)

Donald R. Buhler and David E. Williams

TABLE OF CONTENTS

I. INTRODUCTION

Freshwater and coastal marine environments act as sinks for the deposition of numerous chemicals (xenobiotics) of natural and anthropogenic origin.[1-4] Many of these environmental contaminants are toxic to both terrestrial and aquatic species. Xenobiotics such as the polycyclic aromatic hydrocarbons (PAH) and the polychlorinated biphenyls (PCB) resist chemical and biological degradation and, hence, tend to persist in the aquatic environment. Because of their high lipophilicity, such chemicals are readily bioaccumulated by aquatic organisms in proportion to their octanol:water partition coefficients.[5]

Depending on the chemical and its exposure concentration, this bioconcentration can result in toxicity or death, or in delayed effects such as the development of tumors, reduction of reproductive success, teratogenicity, etc.[2,3,6-8] Chemical residues in the tissues of fish or other aquatic species may also pose a potential threat to humans who utilize these animals for food. Enzymes responsible for the biotransformation of xenobiotics, however, are generally ubiquitous among vertebrates and other eukaryotes. Aquatic animals, therefore, are capable of enzymatically biotransforming many xenobiotics. Usually, such biotransformation reactions result in the formation of products that are less toxic than the parent chemical. Thus, metabolism of most xenobiotics by aquatic organisms results in their detoxification through conversion to more water-soluble and readily excretable products. However, some xenobiotics such as the PAH are enzymatically transformed to metabolites that are more toxic than the parent compound. This type of metabolic activation is the basis for many toxicities, including chemical carcinogenesis.[4,7] In recent years, a number of natural fish populations have been found to show high incidences of various tumors, particularly liver tumors.[2,3,7-15] Postulations that this high frequency of neoplasia is related to environmental exposure to PAH[2,13,16,17] have been strengthened by the finding that benzo(a)pyrene (BaP) is hepatocarcinogenic in laboratory studies.[18] The frequent occurrence of such "cancer epidemics" is of concern in that it may be predictive of a quantitative and qualitative decline of an important world food source and may, as well, be indicative of a direct health risk for humans.

Because of these well-publicized cancer epidemics or epizootics among fish populations, as well as the increasing use of fish as animal models for toxicity and carcinogenicity testing,[19] there has been an expanding interest in the study of basic mechanisms for chemical carcinogenesis in fish and other aquatic animals. Since metabolic activation of carcinogenic chemicals is generally a prerequisite for the initiation of neoplasia, this, in turn, has focused attention on the metabolic pathways in aquatic species that are involved in the biotransformation of PAH and related chemicals.

Biotransformation reactions are usually classified into two categories, phase I and phase II reactions. Phase I reactions introduce polar groups into the xenobiotic molecule through oxidative, hydrolytic, or reductive processes. Phase II reactions involve conjugation of xenobiotics or their phase I metabolites with polar cellular constituents such as glucuronic acid, sulfate, or glutathione to form highly water-soluble conjugates, easily excreted via bile, kidney, or gill. Although not as thoroughly studied as in mammals,[20,21] the comparative metabolism of xenobiotics by fish and other aquatic animals has been the subject of several previous reviews.[4,22-28] The present chapter will focus primarily on the enzymes involved in the oxidative metabolism of PAH by fishes, crustaceans, mollusks, and other aquatic animals. Special emphasis will be given to the cytochrome P-450-dependent mixed-function oxidase (MFO) system.

II. CYTOCHROME P-450

A. Background

Oxidative metabolism of xenobiotics is primarily catalyzed by the cytochrome P-450-

dependent MFO system which is localized in the endoplasmic reticulum of the liver and other tissues. Fish were originally thought to be devoid of MFO activity,[29] but subsequent investigations[30-34] showed that such enzymes were, indeed, present in these species. Although similar in many respects to their mammalian counterparts, the MFO system from fish and other aquatic organisms exhibits a number of unique properties that distinguish it from the mammalian enzymes. For example, the MFO system from aquatic animals generally has a lower temperature optimum than the mammalian system.

A number of reports have appeared during the past 20 years describing the cytochrome P-450-dependent oxidation of PAH and other aromatic chemicals by fish and other aquatic organisms. Such studies have demonstrated that the constitutive fish MFO system is very active towards PAH (reviewed in References 4, 27, and 35), with activities, in many cases, much greater than those seen in untreated mammals. The fish system is also very responsive to induction by PAH and related inducing agents, resulting in a many-fold increase in activity towards xenobiotics such as BaP (reviewed in References 36 and 37).

In mammals, MFO substrates include a diverse spectrum of chemicals of both natural and anthropogenic origin, as well as endogenous compounds such as steroids, fatty acids, and prostaglandins. The types of reactions catalyzed by the MFO are also quite varied and include aliphatic and aromatic hydroxylation, o- and n-dealkylation, epoxidation, s-oxidation, etc.[20,21] The MFO system in fish functions in a similar manner. Thus, various species of fish have been shown to exhibit cytochrome P-450-dependent activity towards numerous xenobiotics, including PAH as well as endogenous compounds such as sex steroids and fatty acids.[38-42]

Over the past 20 years, numerous studies have been performed to characterize the physical and catalytic properties of microsomal cytochrome P-450 from a number of fish species. The fish MFO system has many characteristics similar to mammals, including comparable substrate specificities and responsiveness to PAH-type inducers, including BaP, 3-methylcholanthrene (3-MC), β-naphthoflavone (BNF), and PCB.[43-49] Fish MFO also exhibit some unique properties, including: (1) a very wide range of catalytic turnovers and specific contents* which are generally lower than most mammalian species;[4,27,35] (2) the absence of induction by phenobarbital (PB)-type inducers;[31,36,37,50] (3) the absence in some species of a shift from 450 nm to 447 to 448 nm in the λ_{max} of the CO-reduced difference spectrum;[43,44] and (4) lower temperature optima of 20 to 30°C for P-450-catalyzed MFO reactions for most aquatic species, with some warmwater fishes and marine elasmobranchs exhibiting temperature optima closer to the 37 to 45°C range common to mammals. These lower temperature optima are consistent with the poikilothermic nature of these aquatic animals, yet it is of interest to note that the optima observed are usually higher than the environmental temperature to which the fish have been acclimated.[51-56]

Of particular concern for this review is the observation that fish cytochromes P-450 exhibit a high degree of regioselectivity for oxygenation of PAH such as BaP at the non-K or bay region. Such oxidation seems to occur whether the cytochrome P-450 appear to be constitutive, or are the result of induction by PAH or other inducing chemicals in the laboratory, or through environmental exposure. This BaP-hydroxylase (aryl hydrocarbon hydroxylase [AHH]) activity results in the production of metabolites with mutagenic and carcinogenic properties such as the BaP-7,8-dihydrodiol-9,10-epoxides. Studies performed with liver microsomes of rainbow trout[43,57] and other fish species[58-62] demonstrate significant biotransformation of BP metabolites that covalently bind to DNA.

A number of laboratories have succeeded in the purification and partial characterization of cytochromes P-450 from at least two marine[63,64] and one freshwater[45,47] species of fish, as well as from lobster[42] and crab.[65,66] Studies with these purified cytochromes P-450 from

---

* Nanomoles of cytochrome P-450 per milligram of microsomal protein.

aquatic animals have confirmed previous studies with whole microsomes; that is, that the PAH-inducible form(s) of cytochrome P-450 are similar, in many respects, to the PAH-inducible cytochrome P-450 or "P-448" gene family[67] in mammals.

In addition to cytochrome P-450, other components of the MFO system include the flavoprotein, NADPH-cytochrome P-450 reductase, and the phosolipid of the endoplasmic reticulum membrane. Cytochrome b_5 appears to play a role in the mammalian MFO system, but little is known of the possible contribution of this hemoprotein to the corresponding fish system. Klotz et al.[68] have recently reported that cytochrome b_5 stimulated MFO activity with only one of the purified cytochrome P-450 isozymes from the marine fish, scup.

B. Occurrence

1. Species Distribution

Numerous studies have demonstrated the occurrence of BaP-hydroxylase in a variety of freshwater (Table 1) and marine (Table 2) species. Although there appear to be marked interspecies differences in basal BaP-hydroxylase activities in aquatic animals, it is important to realize that these data were obtained using different analytical procedures and/or standards and different fluorometric, radiometric, or HPLC detection techniques. Some of the observed variability could, therefore, be due to the methodologies employed. Other differences may reflect the contribution of environmental, genetic, or physiological factors (see below).

Nevertheless, one can draw some general conclusions from the available data. For example, there do not appear to be any significant differences in hepatic BaP-hydroxylase activities between freshwater and marine fishes. Considerable interspecies variability in hepatic MFO activity was observed (Tables 1 and 2) with, in some cases, hepatic BaP-hydroxylase levels in fishes exceeding those found in mammals. Kinetic analysis of liver microsomal BaP-hydroxylase from rainbow trout indicated that the V_{max} value was higher than that observed with the rat (0.14 vs. 0.027 nmol/min/mg microsomal protein), and the apparent K_m value for BaP in trout microsomes was lower than rat (8 vs. 20 μM).[57] Comparable V_{max} and K_m values for BaP-hydroxylase with hepatic microsomes from coho salmon were 0.12 nmol/min/mg protein and 2 μM, respectively,[82] whereas an apparent K_m value of 3 μM was observed for BaP-hydroxylase in rainbow trout hepatic microsomes by Pedersen et al.[98] Stegeman and Kaplan[91] observed V_{max} and K_m values of 0.008 nmol/min/mg protein and 0.7 μM, respectively, for the BaP-hydroxylase system in digestive gland microsomes from the barnacle *Balinus eburneus*. Various investigators have also demonstrated in aquatic animals a high capacity for *in vivo* and *in vitro* oxygenation of BaP to carcinogenic and mutagenic metabolites.[4,58-61,63,74,75,99,100]

In general, teleosts have higher BaP-hydroxylase activities than elasmobranchs, while crustaceans, mollusks, and other marine animals have fairly low activities. However, the large differences in BaP-hydroxylase found among aquatic species are not matched by equally large variations in the activities of other MFO systems.[4] Although BaP-hydroxylase is low in crustaceans and mollusks, cytochrome P-450 levels in these species are remarkably high (Table 2). Moreover, hepatopancreas and digestive gland preparations from crustaceans and mollusks also show significant activities towards other MFO substrates.[4,25,101] These results suggest that some of the nonteleost species may be deficient in the P-450 isozyme primarily responsible for BaP-hydroxylation.[4]

There does, however, appear to be some controversy regarding the occurrence of BaP-hydroxylase activity in marine invertebrates. While several groups have reported such activities in digestive gland and other tissues of various marine molluskan species,[4,25,91-95] other investigators[96] failed to detect any activity. Furthermore, in a series of recent reports, Kurelec et al. have shown that the postmitochondrial fraction of the digestive gland from the marine mussels *Mytilus galloprovincialis*[102,103] and *M. edulis*[104] were devoid of cytochrome P-450-dependent BaP-hydroxylase activities. Such preparations also failed to activate

Table 1

**HEPATIC MICROSOMAL BaP-HYDROXYLASE ACTIVITIES OF
FRESHWATER SPECIES**

Species	Cytochrome P-450 (nmol/mg protein)	BaP-hydroxylase (nmol/min/mg protein)	Ref.
Teleosts			
Lake trout	—	0.306	57
—		0.800	58
Rainbow trout	—	0.210	52
	0.22	0.140	57
	—	0.091	69
	—	0.059	70
	0.11	0.020	71
Northern pike	0.13	0.066	72
Goldfish	—	0.031	73
Perch	0.05	0.020	69
Bluegill	—	0.020—0.061	55
Vendace	0.07	0.0084	71
Brown bullhead	—	0.005	74
Roach	0.09	0.004—0.014	58, 71
Carp	—	0.001	75
Crustacea			
Crayfish[a] (Astacus astacus)	0.31	ND[b]	76

[a] Hepatopancreas.
[b] ND, not detected.

Table 2

**HEPATIC MICROSOMAL BaP-HYDROXYLASE ACTIVITIES OF MARINE
SPECIES**

Species	Cytochrome P-450 (nmol/mg protein)	BaP-hydroxylase (nmol/min/mg protein)	(FU[a]/ min/mg protein)	Ref.
Teleosts				
Scup	0.62	0.69	—	77, 78
	0.61	1.23	—	79
	0.27	3.0-3.8	3.1	80
Sheepshead	—	0.28	—	81
Coho salmon	—	0.13	—	48
	—	0.12	—	82
	—	0.027	—	59
Mullet	0.047	—	2.9	81
	—	0.053	—	83
Starry flounder	—	0.040	—	59
Mangrove snapper	0.025	—	6.3	80
Pigfish	—	—	5.1	80
Mummichog	—	—	4.1	84
Sea bass	—	—	3.6	80
Winter flounder	0.17	—	2.54	84
	0.12—0.60	—	0.7—6.4	85
Sculpin	—	0.90	—	86
Black drum	0.15	—	0.53	80
Codfish	—	—	0.51	86

Table 2 (continued)
HEPATIC MICROSOMAL BaP-HYDROXYLASE ACTIVITIES OF MARINE SPECIES

Species	Cytochrome P-450 (nmol/mg protein)	BaP-hydroxylase (nmol/min/mg protein)	BaP-hydroxylase (FU[a]/ min/mg protein)	Ref.
Southern flounder	0.11	—	0.25	80
Eel	—	0.21	—	84
Mackerel	—	—	<0.07	84
King of Norway	—	—	0.004	84
Elasmobranchs				
Nurse shark	0.47	—	1.4	80
Atlantic stingray	0.50	—	0.77	80
Large skate	0.29—0.36	—	0.30	84
Little skate	0.32	—	0.17	80
Bluntnose ray	0.30	—	0.15	80
Thorny skate	—	—	0.12	84
Dogfish shark	0.23—0.29	—	0.07	84
Southern stingray	0.31	—	ND[b]	80
Crustacea[c]				
Crabs				
Uca pugnax	0.14—0.23	—	0.133—0.517	88
Callinectes sapidus	0.04—0.19	—	0.018—0.127	88
	0.18	—	0.008	80
	—	0.057 (F)[d]	—	89
	—	0.00075 (M)[d]	—	89
Menippe mercenaria	0.20—1.00	—	0.008	87
Libinia sp.	0.36—0.56	—	0.002—0.011	87
Sesarma cinerum	0.31—0.51	—	ND—0.003	87
Una minax	0.06—0.14	—	ND—0.001	87
U. pugilator	0.09—0.16	—	ND	87
Lobsters				
Homarus americanus	—	—	0.025—0.065	90
	—	—	n.d.—0.02	80
	—	—	<0.01	7
Spiny lobster				
Panulirus argus	0.91	—	0.03	80
Mollusca[e]				
Barnacle				
Balanus eburneus	0.11	0.043	—	91
Mussel				
Mytilus galloprorincialis	0.047	0.024	—	92
M. edulis	0.047	0.054	—	4
	0.134	0.019—0.031	—	25
Periwinkle				
Littorina littorea	—	0.046	—	93
Snail				
Tegula funebralis	—	0.073	—	94
Thais haemastoma	—	0.001—0.013	—	25, 95
Softshell clam				
Mya arenaria	—	—	ND	96
European oyster				
Ostrea edulis	—	—	ND	96
Other phyla				
Starfish[e]				
Asterias sp.	—	—	0.08	88

Table 2 (continued)
HEPATIC MICROSOMAL BaP-HYDROXYLASE ACTIVITIES OF MARINE SPECIES

Species	Cytochrome P-450 (nmol/mg protein)	BaP-hydroxylase		Ref.
		(nmol/min/mg protein)	(FU[a]/ min/mg protein)	
Sea urchin				
Strongylocentrotus sp.[e]	—	—	0.08	88
S. purpuratus[f]	—	—	0.040	97
Lugworm[f]				
Arenicola sp.	—	—	ND	90

[a] Activity expressed in fluorescence units (FU) defined somewhat differently by different authors; for example, 1 FU is the fluorescent intensity of hydroxylated BaP metabolites at excitation wavelength 400 nm and emission wavelength 525 nm that is equal in fluorescent intensity to a solution of 3 μg of quinine sulfate per milliliter in 0.1 N H_2SO_4.[79]

[b] ND, not detected.

[c] Hepatopancreas.

[d] Male (M) and female (F).

[e] Digestive gland.

[f] Larvae.

BaP to *Salmonella typhimurium* TA98 mutagens. The mussel digestive gland preparations, however, possessed significant flavin-containing monooxygenase activity and were capable of activating various carcinogenic aromatic amines to mutagenic metabolites. Similar findings have been reported[105] for four species of marine sponges.

Recent studies with purified trout, scup, and cod P-450 isozymes have established unequivocally that the high degree of BaP-hydroxylase *in vivo* and *in vitro* displayed by these species is due to the presence of a specific P-448 type of isozyme with high activity towards PAH and regioselectivity towards the production of mutagenic and carcinogenic bay region metabolites.[47,63,64] The observed variabilities in BaP-hydroxylase activities (Tables 1 and 2), not paralleled by similar variations in the activities of other MFO,[4] therefore, presumably reflects differences in the relative proportions of this P-448 isozyme in relation to the other cytochromes P-450.

Microsomal preparations from the livers of several noninduced teleost species, including rainbow trout, mullet, winter flounder, English sole, starry flounder, scup, and coho salmon, metabolized BaP, forming significant quantities (40 to 60% of total metabolites) of the bay region BaP-7,8-dihydrodiol and BaP-9,10-dihydrodiol metabolites.[61,62,106,107] Other metabolites produced by the fish microsomal preparations included 3-hydroxy-BaP, 9-hydroxy-BaP, BaP-4,5-dihydrodiol, and BaP-quinones. Hepatic microsomes from the elasmobranch, little skate, showed similar regiospecificity in BaP metabolism, although the activity was considerably lower than that seen in teleosts. By contrast, metabolism of BaP to phenolic metabolites and quinones predominated in liver microsomes from untreated rats and formation of the bay region BaP metabolites (BaP-7,8- and 9,10-dihydrodiols), was relatively low.

In addition to BaP-hydroxylase, other P-450-dependent MFO enzyme activities are frequently measured in tissue preparations from aquatic species. Activities commonly determined include aminopyrine *N*-demethylase (APD), aniline hydroxylase, benzphetamine *N*-demethylase (BND), 7-ethoxycoumarin *O*-deethylase (ECOD), and 7-ethoxyresorufin *O*-deethylase (EROD).[4,22-27]

In general, BaP-hydroxylase and EROD are indicative of a P-448-type isozyme, responsive

Table 3
STRAIN DIFFERENCES IN RAINBOW TROUT LIVER MICROSOMAL BaP-HYDROXYLASE[a]

	nmol/min/mg protein	
Strain	Untreated	Treated
Chambers Creek	2.28	6.73
Spokane	1.51	5.39
Mt. Whitney	1.34	3.36
Hagerman	0.65	—
Donaldson	0.55	1.27
Chester Morse	0.04	1.91
Mt. Shasta[b]	0.091	—
Sweden[c]	0.059	

[a]   Data from Reference 98.
[b]   Reference 69.
[c]   Reference 70.

to 3-MC-type inducers. The high correlation between EROD and BaP-hydroxylase activities, together with the high sensitivity and relative ease and safety involved with the EROD assay, has prompted several researchers to employ the latter assay when studying BaP-hydroxylase (AHH) induction in fish and mammals. For the purposes of this review, however, we will only present data from studies in which BaP or another PAH was used as the enzymatic substrate.

2. Strain Differences

In addition to large variations in BaP-hydroxylase or other MFO activities among aquatic animals, large strain differences in MFO activity may occur in a particular species. Such differences have been observed in mammals,[108] but evidence for such variation in aquatic species is less firm. Pedersen et al.[98] obtained 1- to 3-year-old fish from six geographically and genetically distinct strains of rainbow trout. Fish were acclimated in the same water under the same photoperiod and fed the same diet for at least 1 month prior to analysis. A greater than 50-fold difference in hepatic BaP-hydroxylase was seen among the six trout strains examined (Table 3). This marked variability in MFO activity among trout strains appeared to be relatively specific for BaP-hydroxylase, since only slight differences in activity were noted with another MFO activity, aniline hydroxylase. For comparison, values of BaP-hydroxylase in other rainbow trout strains (Mt. Shasta, Sweden) are also presented in Table 3.

The findings of Pedersen et al.[98] are based on the assumption that cytochrome P-450 induction does not persist for a prolonged period after exposure to the inducing agent has ceased, and that a 1-month acclimation period is sufficient to return elevated BaP-hydroxylase activities to "background" levels. These assumptions, however, may be invalid, since Forlin has recently observed[109] that EROD levels were still elevated threefold at 59 d following injection of a single 50-mg/kg i.p. dose of BNF. The 1-month acclimation period used by Pedersen et al.[98] therefore, may not have been sufficient, particularly if persistent inducers such as the PCB were present in the fish. The high variability between trout strains seen by Pedersen et al. was not due to some defect in the ability to synthesize the cytochrome P-450 isozyme active towards BaP, since following treatment with 3-MC, all strains exhibited induced BaP-hydroxylase activities ranging from 1.2 to 6.7 nmol/min/mg.

Table 4
EXTRAHEPATIC BaP-HYDROXYLASE ACTIVITIES

Species	Tissue	Percent of liver activity	Ref.
Little skate	Kidney	18	110
	Gill	12	
	Spleen	18	
	Spiral valve	<2	
	Stomach	12	
	Pancreas	18	
	Heart	12	
Sheepshead	Kidney	18	80
	Intestine	7	
	Gill, ovary, testes	<1	
Flounder	Kidney	46	80
Scup	Kidney	33	78
	Testes	9	
	Foregut	2	
	Gill	2	
	Heart	1	
Rainbow trout	Kidney	67	111
	Kidney	15	71
	Intestine	13	
	Heart	8	
	Gill	2	
Vendace	Kidney	22	71
	Intestine	4	
	Heart	3	
	Gill	12	
Perch	Kidney	—	71
	Intestine	5	
	Heart	6	
	Gill	14	
Roach	Kidney	52	71
	Intestine	30	
	Heart	6	
	Gill	55	

3. Hepatic and Extrahepatic Distribution

As is the case with mammals, the liver is the organ in fishes which displays the highest levels of P-450-dependent BaP-hydroxylase (Tables 1 and 2), although activity has been observed in most extrahepatic tissues (Table 4) including the kidney, heart, gill, gut, spleen, testis, ovaries, and muscle.[4,31,71,78,79,111-117] Most of the metabolism of xenobiotics in aquatic animals, however, is concentrated in the liver of teleosts and elasmobranchs, the hepatopancreas of crustacea, and the digestive gland of mollusks. Comparison of the regioselectivity for BaP metabolism in liver, kidney, and gill microsomes from the scup showed similar BaP metabolite distribution patterns.[107]

In most cases, extrahepatic microsomal BaP-hydroxylase activities in aquatic species are less than 50% of those seen for liver (Table 4). However, the kidney is second to the liver with respect to the rate of P-450-dependent oxygenation of PAH, and under some conditions, the levels of total kidney P-450 and activity towards certain substrates can be higher in kidney than in liver. One such substrate is lauric acid, which is hydroxylated at the (ω-1) position in rainbow trout kidney microsomes at a rate which equals or exceeds that found in liver.[41,117] In sexually mature male rainbow trout, the kidney microsomal P-450 isozymes displayed little activity towards BaP and, therefore, may not have a significant role in metabolizing PAH in the "uninduced" fish.[117] Rainbow trout kidney appears to be a target organ for benzene toxicity, yet little metabolism of this xenobiotic occurs in this organ compared to liver.[118]

Trout kidney is capable of responding to "PAH-type" inducers and it appears that the induction of a P-448 type of isozyme, active towards 7-ethoxyresorufin, can suppress the level of a constitutive P-450, active towards fatty acids.[41] Therefore, exposure to PAH could adversely affect the function of constitutive cytochromes P-450 and their metabolism of endogenous substrates.

The kidney is an important organ for further study for a number of reasons. The structures of fish kidneys are quite varied. In rainbow trout, the kidney is located in a dorsal retroperitoneal position and is composed of a connecting head kidney, which is the major hematopoietic organ, and the trunk kidney which contains the majority of nephrons and ducts. The head kidney also contains interrenal and chromaffin tissue which corresponds physiologically to the mammalian adrenal cortex (steroid synthesis) and medulla (catecholamine synthesis), respectively. The trout kidney has a much higher relative rate of blood perfusion (almost the entire cardiac output) than the mammalian kidney and possesses an active xenobiotic excretion system for compounds such as BaP and its metabolites.[119]

A high P-450-specific content has been reported for cardiac microsomes from the marine fish, scup.[78,79] In these fish, the average specific content values for P-450 in microsomes were 0.61 nmol/mg protein for liver and 0.18 and 0.25 nmol/mg for ventricle and atrium, respectively.[79] Whereas P-450 levels were quite high in heart tissue, BaP-hydroxylase, ECOD, and APD activities were 10 to 30 times lower in cardiac microsomes than in hepatic microsomes. Examination of the regiospecificity of BaP metabolism in uninduced scup cardiac microsomes showed that similar levels of the BaP 7,8- and 9,10-dihydrodiols were being formed, accounting for 50 to 70% of total BaP metabolism.

BaP-hydroxylase and other MFO activities are localized primarily in the hepatopancreas and green gland of crustaceans;[89-91] however, significant activities have also been observed in the intestine[91] and stomach.[89,120] Lower MFO activities are associated with the gills and reproductive tissues[89,90] of these species. In the case of mollusks, BaP-hydroxylase is concentrated in the digestive gland and other tissues show very low or nondetectable activities.[25]

4. Cellular Organelles

In mammals, cytochrome P-450 exists in cellular organelles other than the endoplasmic reticulum, especially the mitochondria and the nuclear envelope. Mammalian mitochondrial cytochromes P-450 catalyze the rate-limiting step in the synthesis of sex steroids, glucocorticoids, and mineral corticoids from cholesterol and are present in high amounts in steroidogenic tissues such as the adrenals and gonads. In addition, mammalian mitochondrial P-450 is active towards xenobiotics, including BaP.[121]

In the case of teleosts and elasmobranchs, hepatic and extrahepatic MFO activities, including BaP-hydroxylase, are primarily associated with the microsomal fraction of the endoplasmic reticulum.[4,31,43,78,84,122] The BaP-hydroxylase system is also concentrated in the microsomal fraction of hepatopancreas preparations from crustaceans.[25,123] Similar findings are noted for microsomes from the digestive gland of mollusks.[91] Considerable variability in MFO activities has been observed, however, particularly in hepatopancreas from crustaceans.[80] Various explanations have been advanced for the low or variable activities of such preparations,[84,91] including the presence of proteolytic enzymes, endogenous inhibitors, and/ or low NADPH-cytochrome P-450 reductase activities.

With respect to aquatic species, very little is currently known about the existence and properties of cytochromes P-450 in organelles other than the endoplasmic reticulum. Presumably, fish or other aquatic animals possess mitochondrial cytochromes P-450 which function in the biosynthesis of steroids. If these enzymes can also oxidize PAH in a manner similar to that observed in mammals,[121] this may have significance with respect to effects of environmental PAH on development and reproduction in aquatic populations. Reduced reproductive success of starry flounder in San Francisco Bay has been correlated with pollutant exposure and resulting elevation in hepatic microsomal BaP-hydroxylase levels.[124]

C. Effect of Inducers

Numerous studies have appeared over the last 15 to 20 years demonstrating that many elasmobranchs and fishes are very responsive to the PAH class of P-450 inducers.[4,27,35-37,125] The results summarized in Table 5 are not complete, but give a representative sampling. Compounds which have been demonstrated to be capable of inducing fish P-450-dependent BaP-hydroxylase include 3-MC, BNF, BaP, 2,3,7,8-tetrachlorodibenzo-p-dioxin (TCDD), 1,2,3,4-dibenzanthracene (DBA), PCB, polybrominated biphenyls (PBB), crude oil fractions, municipal waste water, and bleached Kraft mill effluent[49,70,87,126-154] (Table 5). In all cases, except stingray, there appear to be no significant differences in the effectiveness of induction between various routes of exposure. Other agents, known to produce induction of the mammalian P-450 system, are ineffective in fish including PB, dichlorodiphenyltrichloroethane (DDT), phenylbutazone, mirex, kepone, and non-coplanar isomers of PCB and PBB.[31,36,37,50] Sex steroids such as testosterone, or synthetic steroids such as 16α-cyanopregnenolone, and compounds such as isosafrole, which are also inducers of mammalian cytochromes P-450, appear to induce the fish MFO activity towards certain drugs, but not towards PAH.[37,49,132] Little information on the effectiveness of other classes of inducers, such as ethanol, on induction of fish P-450, is available. The limited data available suggest that cytochromes P-450 in extrahepatic tissues such as gill[137] and kidney[110] respond to inducing chemicals in the same manner as liver. However, Payne et al.[149] found marked induction of both P-450 and BaP-hydroxylase in kidney, but not in liver tissues of winter flounder collected at an oil spill site in Newfoundland.

The characteristics of the induction response to "PAH-type" inducers is similar in fish and mammals, except that, in many cases, fish do not exhibit the marked hypsochromic shift from 450 nm to 447 to 448 nm in the CO-difference spectrum, which is diagnostic for mammalian induction. The relative change in P-450-specific content in fish following induction is variable, ranging from a 40% decrease to a 340% increase (Table 5). Such variability could be due to other factors to be discussed later. The degree of BaP-hydroxylase induction is also variable, probably due to the large variation in "constitutive" or "background" activity between species, strains, or individuals of the same species and strain. In most cases, the relative increase in BaP-hydroxylase far exceeds that of the total increase in P-450. This same observation has been made with mammals and recent applications of immunochemical techniques have demonstrated that this is due to a many-fold increase in the transcription of genes producing mRNA coding for "P_1-448" or "P-450"* which have high activity towards BaP.[155,156] These PAH-inducible isozymes of P-450, are very low or absent in nontreated animals. Induction of P-448 or P_1-450, in many cases, represses the levels of constitutive cytochromes P-450 such that the relative percentage of total P-450 as P-448 increases from 3 to 83%.[157] Cases where the P-450 levels and BaP-hydroxylase activities are already fairly high and are only increased a few-fold by induction, could be

---

* The BaP-inducible cytochrome P-450 in mammals has been designated as "cytochrome P_1-450" by Nebert.[67]

Table 5
INDUCTION OF HEPATIC P-450 AND BaP-HYDROXYLASE[a]

Species	Compound	Dose (route)	(Fold induction)		Ref.
			P-450	BaP-hydroxylase	
Sheepshead	3-MC	2 × 20 mg/kg (i.p.)	1.9	8	87
	DBA	2 × 10 mg/kg (i.p.)	—	12	27
	BNF	50 mg/kg (i.p.)	—	5	
	Aroclor 1254	100 mg/kg (i.p.)	—	6	87
Southern flounder	3-MC	2 × 10 mg/kg (i.p.)	1.6	15	87
Stingray	3-MC	3 × 20 mg/kg (i.p.)	0.9	2	87
		2 × 15 mg/kg (i.m.)	1.2	10	
Little skate	3-MC	2 × 50 mg/kg (i.p.)	—	8	126, 127
	TCDD	2 × 0.005 mg/kg (i.p.)	—	19	110
	DBA	3 × 10 mg/kg (i.p.)	0.6	44	87
Dogfish shark	DBA	3 × 10 mg/kg (i.p.)	-	6	87
	3-MC	10—20 mg/kg (i.p.)	2.1	4	
Croaker	BaP	15 mg/kg (i.p.)	2.9	16	128
Northern pike	3-MC	3 × 20 mg/kg (i.p.)	0.9	34	72
Coho salmon	PCB	1 mg/kg (diet)	—	2	48
Rainbow trout	BNF	100 mg/kg (i.p.)	1.5—1.6	16—41	44, 112, 129, 130
		500 ppm (diet)	1.6	11	130
	3-MC	20 mg/kg (i.p.)	2.2	10	131
		10—20 mg/kg (i.p.)	1.8	19	128
		20 mg/kg (i.p.)	1.5	22	129
		2 × 20 mg/kg (i.p.)	3.4	18	132
	Aroclor 1242	150 mg/kg (i.p.)	1.2	11	133, 134
	Aroclor 1254	200 mg/kg (i.p.)	—	30	112
		150 mg/kg (i.p.)	1.3	13	133, 134
		30 mg/kg (i.p.)	1.7	3	135
		100 mg/kg (diet)	1.6	47	136

Clophen A50	2 × 500 mg/kg (i.p.)	1.2—2.0	5—22	132
Firemaster BP6	150 mg/kg (i.p.)	1.3	7	132, 133
BaA-treated municipal waste water	10 mg/kg (i.p.)	1.6	7	129
		1.0	9	70

[a] Abbreviations: 3-MC, 3-methylcholanthrene; DBA, (1,2,3,4-dibenzanthracene; BNF, (β-naphthoflavone); Aroclor 1242 and 1254, polychlorinated biphenyl mixtures; TCDD, (2,3,7,8-tetrachlorodibenzo-*p*-dioxin); BaP, benzo[a]pyrene; Firemaster BP6, a polybrominated biphenyl mixture; and BaA benz[a]anthracene.

due to "environmental" exposure to PAH. The absence of induction of cytochromes P-450 in fish by PB-type inducers,[37] including phenylbutazone, DDT, and certain PCB and PBB isomers, has been the object of study at several laboratories.[31,36,37] Nonresponsiveness at the translational level has been demonstrated by the absence of any enhanced incorporation of [^{35}S]-methionine into hepatic microsomal protein by PB-pretreated rainbow trout.[158] Recent studies by Kleinow and co-workers[159] also could not detect any PB stimulation of P-450 mRNA in trout livers. These findings support the conclusion that, contrary to mammals, no *de novo* synthesis of cytochromes P-450 occurs in fish following exposure to PB-type compounds. The reason for the observed lack of response of fish to PB-type inducers remains to be elucidated.

Reports on the induction of MFO in aquatic invertebrates are relatively scarce. Lee et al.[160] observed an elevation in intestinal BaP-hydroxylase activity for over three generations following exposure of the marine polychaete *Capitella capitata* to crude petroleum or BaA. Various invertebrates (snails, sea urchins, starfish, and lugworms) that were exposed to oil-saturated sediments over a 1-week period failed to show any increase in BaP-hydroxylase activity.[90] Male and female lobsters similarly did not yield any increase in BaP-hydroxylase after exposure to a surface slick of crude oil for a period of 5 months.[90] More definitive findings were reported recently by Livingstone et al.,[93] who found that short-term (1 to 8 d) or long-term (4 to 16 months) exposure of mussel (*Mytilus edulis*) and periwinkle (*Littorina littorea*) to 29 or 123 ppb of diesel oil resulted in significant increases in digestive gland microsomal P-450 levels. Corresponding increases, however, did not occur in BaP-hydroxylase activities. There appears to be little evidence for BaP-hydroxylase induction in marine invertebrates. Furthermore, there are data that even question the presence of this MFO in some marine invertebrates.[96,102-105] An intriguing recent finding also suggests that PB may be able to induce MFO in aquatic invertebrates, since the frequency of sister chromatid exchanges produced by cyclophosphamide, a chemical requiring metabolic activation for mutagenicity, was significantly increased in adult and larval mussels (*M. edulis*) pretreated with this inducer.[161]

The problem of exactly what changes in P-450 isozyme composition occur during induction in fish or other aquatic animals leads into a consideration of the existence, properties, and regulation of multiple forms of P-450 in fish. Until recently, researchers studying induction in fish have addressed the question of the existence of multiple forms of P-450 in untreated and induced fish liver microsomes by examining properties such as substrate specificity and regioselectivity, spectral properties, molecular weight, and effects of *in vitro* inhibitors and activators. PAH induction in fish results in an increased oxygenation of PAH without affecting activity towards other substrates such as aminopyrine, benzphetamine, or ethylmorphine. Activity towards endogenous substrates such as steroids and fatty acids is usually unchanged or decreased.

Induction can also alter the regioselective metabolism of a particular substrate. For instance, it has previously been shown that prior or coadministration of BNF or PCB protected rainbow trout from the hepatocarcinogenic effects of aflatoxin B_1 (AFB$_1$).[162,163] Subsequent studies *in vitro* with liver microsomes demonstrated that BNF induction increases the P-448-catalyzed 4-hydroxylation of AFB$_1$ to yield aflatoxin M_1 (AFM$_1$) (detoxication), while reducing the formation of the ultimate toxic and carcinogenic metabolite AFB$_1$-2,3-epoxide.[164]

Untreated and induced fish display few differences in the regioselective metabolism of BaP.[106,107] Furthermore, there is no change in the K_m for BaP upon induction. These observations suggest that PAH induction is increasing the amount of BaP-metabolizing cytochromes P-450 found in uninduced animals (which usually comprise only a small percentage of the total liver microsomal P-450).

In vitro inhibitors of MFO activity, such as SKF-525A, metyrapone, and α-naphthoflavone

(ANF) have proved useful in distinguishing multiple P-450 forms in mammals. In general, SKF-525A can be thought of as a fairly nondiscriminating inhibitor. Metyrapone appears to be more effective at inhibiting activities catalyzed by the PB-inducible family of cytochromes P-450 and ANF is fairly specific for inhibiting the activity of PAH-inducible cytochromes P-450. In mammals, ANF effectively inhibits microsomal BaP-hydroxylase in PAH-induced animals, but in many cases actually stimulates BaP-hydroxylase in untreated animals. Such results have been interpreted as demonstrating that the majority of BaP-hydroxylase in uninduced mammals is catalyzed by cytochromes P-450, which are distinct from those that are induced by PAH. The results of such experiments in fish have been equivocal. In some fish, ANF inhibits BaP-hydroxylase effectively in untreated and induced fish (reviewed in References 4 and 125). Studies with wild populations of fish have demonstrated that fish with high "constitutive" BaP-hydroxylase activity are inhibited by ANF, whereas individuals of the same species with low BaP-hydroxylase activity are not subject to ANF inhibition. Both groups display similar sensitivity upon treatment with PAH, suggesting that the differences may be environmental and not genetic (as with Ah-positive and Ah-negative strains of mice).

A more direct method to examine the changes in fish P-450 forms upon PAH induction involves sodium dodecyl sulfate-polyacrylamide gel electrophoresis (SDS-PAGE) (separation by molecular weight) of liver microsomal proteins from untreated and induced fish. Following treatment with inducer, rainbow trout and flounder both display large increases in the relative amounts of liver microsomal proteins with a calculated molecular weight of 57,000 Da. PAH-induction in scup leads to a similar increase in a band with a slightly lower molecular weight (54,000). The relative intensity of this band positively correlated with the BaP-hydroxylase displayed by different fish.[4,37] Furthermore, studies in which rainbow trout were dosed with [^{35}S]-methionine upon induction, incorporated 60-fold higher specific activity into a protein with a molecular weight on SDS-PAGE of 57,000.[158]

D. Influence of Other Factors

1. Age and Development

The expression of genes coding for various isozymes of mammalian cytochromes P-450, including forms active towards PAH, has been known for a number of years to be under developmental control (reviewed in Reference 165). In most mammals, the levels of P-450 are low in the fetus and increase rapidly following birth, reaching a peak during early adulthood and then gradually decreasing with senescence. Immunochemical quantitation and molecular biology studies have shown that different P-450 isozymes are regulated independently.

Binder and Stegeman have followed the developmental patterns of BaP-hydroxylase during embryonic development in the killifish[166,167] and brook trout.[168] BaP-hydroxylase, apparently P-450 dependent, was readily detectable as early as 4 d postfertilization, prior to the appearance of the liver. Activity was localized in the microsomal fraction from whole embryos and eleutheroembryos. Levels of BaP-hydroxylase in whole embryos at developmental stages prior to the appearance of the liver rudiment were comparable to those found in the livers of nonspawning adult fish. At hatching, however, BaP-hydroxylase activity increased three- to ninefold. BaP-hydroxylase activity has also been observed in sea urchin larvae.[97]

Fish embryos and eleutheroembryos were sensitive to induction of BaP-hydroxylase at all developmental stages.[166,168,169] For example, BaP-hydroxylase activity was induced following immersion exposure of near-hatching killifish embryos to PCB or fuel oil,[166] and of late brook trout embryos to PCB.[168] PCB also induced BaP-hydroxylase activities in both liver and extrahepatic tissues of similarly exposed yolk-sac fry. Lake trout embryos and fry derived from females environmentally exposed to PCB showed significant elevation of P-450 levels and the activities of BaP-hydroxylase, APD and EROD.[169] Similar induction was seen in lake trout fry given immersion exposure to PCB. A new cytochrome P-450 band (molecular weight 58,000) was seen upon SDS-PAGE analysis of hepatic microsomes from the resulting swim-up fry. This newly induced P-450 isozyme was similar in properties[47] to, and was immunologically indistinguishable[170] from, the major P-450 isozyme from BNF-induced rainbow trout.

The mechanism for the large increase in BaP-hydroxylase activity posthatching is unknown. Binder and Stegeman[168] found that neither age nor developmental stage was a factor, but rather, it was the process of hatching that appeared to be critical.

The developmental regulation of BaP-hydroxylase activity through later stages of the life cycle has not been systematically investigated. Furthermore, no studies have, as yet, been performed examining the developmental appearance of distinct P-450 isozymes. The acquisiton of specific polyclonal and monoclonal antibodies and the production of cDNA probes (discussed in a later section), will provide the tools necessary to carry out such studies.

2. Sex

Certain species of mammals, most notably the rat, display a marked sex difference in the P-450-dependent metabolism of certain drugs, xenobiotics, and/or steroids (reviewed in Reference 171). This sex difference can be neonatally imprinted, requires an intact hypothalamic-pituitary-liver axis, and is probably mediated by growth hormone.[172,173] Recently, sex-specific isozymes of P-450 have been purified from liver microsomes of both mouse and rat.[174-176]

Some fish appear to display a similar sex difference and, as in rat, males typically exhibit higher MFO activities than females.[117,132,177,178] It is, therefore, not surprising that administration of testosterone induces fish P-450, whereas estrogens are inhibitory.[40,179,180]

Interestingly, trout appear to express this sex difference by a mechanism distinct from rats, as the former do not require an intact pituitary for the sex difference to be expressed. A greater than 20-fold sex difference in the amounts of a particular P-450 isozyme (LM$_2$) has recently been documented in rainbow trout kidney microsomes.[117] These same fish displayed little or no sex-associated differences in BaP-hydroxylase activity. Similar but less dramatic differences were seen in liver microsomes from these fish.

3. Environmental

Interest in the environmental induction of MFO in fish and other aquatic species has been catalyzed by two considerations: the effect of such induction on the health and survival of aquatic populations and the potential use of this phenomenon as an environmental monitoring tool. In this section, we consider the influence of two environmental variables, temperature and chemical exposure, which have been found to affect BaP-hydroxylase activity in fish. Studies with bluegills demonstrated[55] that, following acclimation at 10, 20, or 30°C, liver/body weight ratios and hepatic P-450 levels decreased with increasing acclimation temperatures. Liver microsomal BP-hydroxylase activity measured *in vitro* at incubation temperatures of 10, 20, and 30°C showed significant differences in K_m, but no differences in V_{max}.

Numerous studies have demonstrated seasonal fluctuations in fish MFO activity[181-183] which are thought to result primarily from compensation by the fish to changes in water temperature.[54-56] Temperature-derived alterations in membrane lipids[184] and other changes that maintain enzyme activities at a relatively constant level in spite of water temperature changes are well-established mechanisms for temperature compensation by fishes. The interpretation of the results from such studies requires a thorough knowledge of the effect of other factors on the species examined, such as developmental stage, sex, and diet. Acclimation temperature has also been demonstrated to have a marked influence on the degree of BaP-hydroxylase induction in fish.[51,53-55]

The second factor to be considered is the induction of BaP-hydroxylase in fish populations by exposure to PAH or other pollutants in the environment. It was proposed by Payne and Penrose[137] that BaP-hydroxylase activities of hepatic microsomal preparations from natural fish populations, particularly marine species, could serve as environmental monitors for aquatic pollution by petroleum or its refined products. Subsequently, many investigators have confirmed that petroleum or certain of its constituents, through laboratory or field exposure, could increase hepatic BaP-hydroxylase activities in various freshwater and marine fish species.[48,138,139,141,142,147,149,154,185-188] However, studies with marine fish, such as the scup, indicate that a significant portion of the population may already be in an "environmentally induced" state.[4]

Nevertheless, the concept of utilizing BaP-hydroxylase activity as an environmental monitoring tool has merit. However, numerous factors can influence this MFO system (discussed in this and previous sections), greatly complicating interpretation of results. For example, while petroleum and its various aromatic constituents including the PAH are good inducers of the fish BaP-hydroxylase system, other environmental contaminants such as PCB and various other halogenated aromatic chemicals also stimulate the same MFO system.[4,27,36,125] This nonspecificity detracts from the practical utilization of BaP-hydroxylase activity as an environmental monitoring tool of PAH levels. Factors such as interindividual variability, strain differences, degree of sexual maturation, and migratory movement of the species being examined all influence enzyme activity. Another difficulty arises in that fish exposed to high levels of PAH could still display low BaP-hydroxylase activity if the general health of the animal was poor or if specific BaP-hydroxylase inhibitors were present (e.g., BaP itself as a competitive inhibitor). Exposing trout to 50 ppm of acrylamide in the water selectively represses EROD activity and reduces the trout P-450 LM_{4b} isozyme to virtually undetectable levels.[189] Some of the inherent pitfalls in utilizing BaP-hydroxylase to monitor PAH exposure could be circumvented by developing sensitive immunoassay techniques to detect specific P-450 isozymes sensitive to induction by PAH.

4. Dietary

In mammals, BP-hydroxylase activity can be modulated by dietary factors such as the ratio of protein to carbohydrate, the levels and degree of polyunsaturation of dietary fat, and intake of endogenous dietary MFO inhibitors or inducers.

Unfortunately, little is known regarding the effects of dietary manipulations on BP-hydroxylase or other MFO activities in fish. Starvation for periods of weeks is not unusual during certain life stages of aquatic species, and, therefore, it is not surprising that starvation for 6 to 12 weeks has relatively little effect on P-450 levels or BaP-hydroxylase activity in rainbow trout.[111]

Numerous studies have shown that fish respond to dietary induction of BaP-hydroxylase by PAH and other inducing chemicals (Table 5), but relatively little is known regarding the occurrence or effectiveness of such dietary constituents or the influence of other normal dietary factors on P-450 levels or BaP-hydroxylase activities in a particular aquatic species.

E. Isolation, Purification, and Characterization

The cytochromes P-450 are a group of closely related isozymes with differing and overlapping catalytic capacities. The relative concentrations of these enzymes are influenced by a myriad of species, developmental, dietary, physiological, and environmental factors. Metabolism of different substrates, and the regioselectivity or stereospecificity of oxidative metabolism of chemicals catalyzed by cytochromes P-450, can also vary tremendously depending on the particular isozymes involved. Metabolism of endogenous chemicals and xenobiotics *in vitro* and *in vivo* is, therefore, controlled to a large degree by the relative proportion of cytochromes P-450 present in a particular cell, organ, tissue, strain, species, or animal. Because of the obvious complexities arising from the presence of a mixture of P-450 isozymes present in the endoplasmic reticulum, it has proven difficult to arrive at a clear understanding of the role of each isozyme in a given metabolic pathway through the use of microsomal preparations. Much progress has been made towards elucidating the contributions of the various isozymes, however, following their purification and characterization.

1. Marine Fish

The first report of the successful solubilization, separation, and reconstitution of fish MFO constituents appeared in 1977 from the laboratory of Bend.[127] Liver microsomes from the marine elasmobranch, little skate, were solubilized with sodium cholate and purified by DEAE-cellulose and hydroxylapatite chromatography. Following detergent removal, the final P-450 fraction had a specific content of 3.0 nmol/mg protein, a λ_{max} in the CO-difference spectrum of 450 nm, and, when reconstituted with lipid and partially purified little skate reductase, a turnover number with 7-ethoxycoumarin of about 5 min^{-1}. Subsequently, this same laboratory resolved two separate cytochromes P-450, P-451 and P-448, from liver microsomes of DBA-treated little skate.[190] The P-448 fraction was purified about five- or sixfold relative to the microsomes, had a λ_{max} in the CO-difference spectrum of 448 nm, and metabolized BaP with a turnover number of 0.7 to 0.8 min^{-1}. The yield of P-448, relative to P-451, was greatly increased by pretreatment with DBA, consistent with the observed increase in BaP-hydroxylase.

The first studies in which P-450 was purified to homogeneity from a marine fish were reported in 1983 by Klotz et al.[63] Solubilization of liver microsomes from an untreated marine teleost, scup, and purification by polyethylene glycol fractionation and resolution on DE-52, yielded five distinct P-450 fractions. Scup P-450E (so named because it was the last to elute from the DEAE column) was obtained in highest yield and was subsequently purified to a high specific content (11.7 nmol/mg) by hydroxylapatite chromatography. Scup P-450E appeared homogenous on SDS-PAGE with a minimum molecular weight of 54,500 Da. The N-terminal amino acid sequence of the first nine residues of scup P-450E was unlike any previously characterized mammalian P-450. The absolute spectrum displayed a Soret band maximum at 418 nm, indicative of a low spin cytochrome and a λ_{max} in the CO-difference spectrum of 447 nm. Reconstitution with purified NADPH-cytochrome P-450 reductase, from either scup or rat, resulted in a turnover number with BaP of 0.56 min^{-1}, which was inhibited 70 to 80% by 0.1 mM ANF. Upon addition of purified rat hepatic epoxide hydrolase, scup P-450E displayed a high regioselectivity for oxygenation of BaP at the 7,8- (34% BaP-7,8-dihydrodiol) and the 9,10- (42% BaP-9,10-dihydrodiol) positions with no detectable activity towards the 4,5-position. All of these characteristics (molecular weight, spectral, and catalytic properties) are entirely consistent with the hypothesis that scup P-450E is the major P-450 isozyme induced by PAH in this species. The high yield of P-450E from the livers of natural populations of ''untreated'' scup is consistent with the high BaP-hydroxylase activity displayed by these fish, and raises the possibility that fish populations are being induced by environmental exposure to PAH.[4]

In a recent report, Klotz et al.[68] described the isolation and purification of two additional P-450 isozymes from the scup, designated P-450A and P-450B. The P-450A was localized in the first effluent peak from the initial DEAE-cellulose column and was isolated by the use of a second DEAE-cellulose column together with hydroxylapatite and CM-cellulose columns. P-450B, from the second peak in the initial DEAE-cellulose column followed by a hydroxylapatite column step. P-450A exhibited a molecular weight of 52,500 Da and a λ_{max} at 447.5 nm in the CO-difference spectrum, whereas P-450B yielded a molecular weight of 45,900 and a λ_{max} at 449.5 nm. Reconstitution of the P-450A isozyme with scup or rat NADPH-cytochrome P-450 reductase gave modest BaP-hydroxylase (0.1 min^{-1}), ECOD

(0.42 min^{-1}), and testosterone 6β-hydroxylase (0.8 min^{-1}) activities. The reconstituted P-450B isozyme oxidized testosterone at several positions including the 15α-position (0.07 min^{-1}).

Multiple forms of P-450 have also been isolated from liver microsomes of BNF-induced Atlantic cod.[64,191] The major form obtained, P-450C, had a high specific content (13.4 nmol/mg) and was homogeneous on SDS-PAGE with a minimum molecular weight of 58,000 Da. Cod P-450c was obtained in low spin form (416.5 nm) and had a λ_{max} of 448 nm in the CO-difference spectrum. Reconstitution of P-450c gave a turnover number with 7-ethoxyresorufin of 1.07 min^{-1}.[191]

A recent report describing the partial purification (tenfold) of P-450 from coho salmon has appeared.[192] Preliminary evidence indicates that the major P-450 in four other salmon species (sockeye, chinook, chum, and pink) have different properties. Of particular interest was the observation that, in spawning female coho salmon, up to 30% of the total microsomal protein was P-450.

2. Freshwater Fish

Multiple forms of P-450 have been purified from liver microsomes of BNF-treated rainbow trout by Williams and Buhler.[45,47] We have found that a procedure employing 3-[(3-chol-amidopropyl)dimethylammonio]-1-propane sulfonate (CHAPS) detergent to solubilize the microsomal P-450, followed by chromatography on tryptamine-Sepharose®, DEAE-Sepharose®, and hydroxylapatite, yields the best results.

The major form purified from BNF-treated rainbow trout was originally designated trout P-450F[45] or DEcHA$_2$,[46] but is now referred to as trout P-450 LM$_{4b}$.[47] Purified trout P-450 LM$_{4b}$ has a minimum molecular weight of 58,000 Da, a specific content of 11 to 12 nmol/mg, and a λ_{max} in the CO-difference spectrum of 447 to 448 nm.

The physical and catalytic properties of trout P-450 LM$_{4b}$ were demonstrated to be much more like those exhibited by the major form of P-450 induced by PAH in rats (P-448 or BNF-B using the nomenclature of Guengerich) compared to the major form from PB-treated rats (PB-B).[46]

Trout P-450 LM$_{4b}$ regioselectively oxygenates BaP at the bay or non-K-region, and, as is the case with scup P-450E and rat P-450 BNF-B, in the presence of added epoxide hydrolase, primarily produces the 7,8- and 9,10-dihydrodiols.[47]

A minor form of P-450, designated LM$_{4a}$[47] was also isolated from liver microsomes of BNF-treated rainbow trout. Trout P-450 LM$_{4a}$ exhibited many properties which were indistinguishable from trout LM$_{4b}$, including molecular weight on SDS-PAGE, spectral properties, substrate specificity, regioselectivity towards BaP, and peptide profiles following limited proteolysis. Slight differences in amino acid composition and sensitivity to inhibition by ANF were observed. Immunochemically, trout LM$_{4a}$ and LM$_{4b}$ are not distinguishable by Ouchterlony double precipitation analysis, but differ slightly in sensitivity to inhibition of BaP-hydroxylase by rabbit-anti-LM$_{4a}$ or anti-LM$_{4b}$-IgG (see section on immunological studies). Trout cytochromes P-450 LM$_{4a}$ and LM$_{4b}$ were resolved by DEAE-Sepharose.® It is possible that trout cytochromes P-450 LM$_{4a}$ and LM$_{4b}$ are similar enzymes of a trout P-448 gene family, analogous to the situation with rat P-450c and P-450d, rabbit LM$_4$ and LM$_6$, and mouse P-448 and P$_1$-450.[165]

A third form of P-450 (LM$_2$) was highly purified from BNF-induced trout which displayed physical and catalytic properties markedly different from trout LM$_{4a}$-LM$_{4b}$.[47] Trout LM$_2$ appears to be a major constitutive form of P-450 in trout liver and kidney (see section on immunological studies) and displays high activity towards AFB$_1$[164] and lauric acid,[41] but does not metabolize BP.

Recently, five cytochrome P-450 isozymes have been purified from the livers of untreated male rainbow trout.[193] While one isozyme was immunologically and biochemically indistinguishable from trout P-450 LM_2, the other cytochromes P-450 were dissimilar from previously characterized forms. These isozymes, designated LM_{C1}, LM_{C3}, LM_{C4}, and LM_C had negligible BaP- hydroxylase activity, but oxidized various steroid and MFO substrates.

The only other PAH tested to date as a substrate for purified trout cytochromes P-450 is 2-methylnaphthalene (2-MeN). Trout P-450 LM_{4a} and LM_{4b}, in the presence of added epoxide hydrolase, yielded the 3,4-, 5,6-, and 7,8-dihydrodiols of 2-MeN in a ratio of about 1:4:10 with no activity towards the 2-methyl position.[194] This regioselectivity is similar to that displayed by rat P-448 (BNF-B). By contrast, trout LM_2 produced 2-hydroxymethylnaphthalene as the major metabolite, similar to rat P-450 PB-B.

3. Marine Invertebrates

In addition to fish, P-450 has been partially purified (specific content 1.8 nmol/mg) from the hepatopancreas of the male blue crab.[195] In another study, three P-450 forms were resolved from the hepatopancreas of four species of marine crabs.[65] Specific contents of the purified cytochromes P-450 ranged from 0.5 to 5 nmol/mg. Interestingly, the relative yields of the three forms differed markedly from crabs collected in the spring or fall and suggest a possible relationship to the reproductive cycle. Recently, a single form of P-450 has been purified from the spiny crab.[66]

A form of P-450, highly effective in the hydroxylation (at the 16α-position) of progesterone and testosterone, has been purified to apparent homogeneity (specific content 12 to 14 nmol/mg) from the hepatopancreas of the spiny lobster by James and Shiverick.[42] The properties of the P-450 responsible for BaP-hydroxylase activity in lobster appear quite different compared to fish. Total P-450, solubilized and partially purified from hepatopancreas of spiny lobster, exhibited a high turnover towards BaP (1 to 3 min^{-1}), but not ethoxyresorufin (<0.05 min^{-1}). This activity was not inhibited by ANF and the relative amount of BaP-4,5-dihydrodiol produced was as great as for the benzo-ring (7,8- and 9,10)-dihydrodiols.[42,196] These results suggest that lobster and other invertebrate cytochromes P-450, active towards PAH, may differ substantially from the fish family of isozymes immunochemically related to trout LM_{4b}. The effectiveness of PAH as inducers of BaP-hydroxylase in marine invertebrates is still uncertain.[90,93,195-198]

4. Immunological Studies

The acquisition of highly purified fish cytochromes P-450 has now opened the door for the utilization of antibodies and immunochemical techniques for immunoquantitation, probing the catalytic and chemical properties of fish cytochromes P-450 and for future molecular biology studies.

Polyclonal antibodies to trout cytochromes P-450 LM_{4a}, LM_{4b}, and LM_2 have been raised in rabbit.[47] Studies with these antibodies have provided much useful information including:

1.   Trout LM_{4a} and LM_{4b} appear to be minor and major forms, respectively, of cytochromes P-448 (active towards BaP) in liver microsomes of BNF-treated rainbow trout which are not distinguishable immunochemically.
2.   Immunoquantitation via Western blotting[200] demonstrates that trout LM_{4a}/LM_{4b} is induced many-fold following treatment with BNF or PCB, but is not affected by PB.[201]
3.   Trout P-450 LM_2 is a major constitutive isozyme in liver and kidney and is not inducible by BNF, PCB, or PB.[117,201]
4.   All species of fish tested to date possess a liver microsomal P-450 which immunochemically cross-reacts with the antibody to trout LM_{4b}. Scup, gulf killifish, brook trout, flounder, cod, and English sole, induced either in the laboratory or through environmental exposure to PAH, display BaP-hydroxylase activity which is effectively inhibited by rabbit anti-trout LM_{4b}-I_{gG} and/or is immunochemically recognized by the

antibody on Western blots.[170,202] Recently, this antibody has been utilized by Varanasi's group and this author's group, to demonstrate an excellent correlation (r = +0.96) between BaP-hydroxylase activity and levels of English sole P-448 (immunochemically cross-reacting with trout LM_{4b}).[202]

5. Sexually mature rainbow trout display a marked sex difference in the levels of LM_2 in kidney, with males expressing an approximately 20-fold higher level of immunochemically determined LM_2.[117] Consistent with the relatively higher levels of LM_2, males demonstrated 15- to 20-fold higher activity towards lauric acid and AFB_1, compounds previously demonstrated to be substrates for trout LM_2. A similar sex difference is observed in liver, but the level in males is only 2-fold higher rather than the 20-fold difference seen in kidney. No significant difference was observed in kidney or liver with respect to immunochemically determined levels of LM_{4b} or activity towards BaP.

Stegeman's laboratory has produced both monoclonal and polyclonal antibodies to P-450E, the major isozyme from BNF-induced scup.[63,203] Polyclonal antibodies to scup P-450E do not cross-react with four other scup (A through D) P-450 isozymes. Titration with polyclonal anti-P-450E inhibited greater than 90% of the BaP-hydroxylase and EROD activity of scup liver microsomes. A monoclonal antibody to scup P-450E was isolated (Mab 1-12-3) which was more potent than the polyclonal antibody at inhibiting EROD activity by scup hepatic microsomes, yet did not affect BaP-hydroxylase activity, suggesting that these activities may be catalyzed by two distinct, yet very similar isozymes.[203] Stegeman et al.,[204] in an exciting application of these antibodies, demonstrated that even deep sea fish *(Coryphaenoides armatus)* appear to be environmentally induced. The degree of induction was much higher in fish taken from an area high in chlorinated hydrocarbon inducers (Hudson Canyon, off New York) than in fish taken from a relatively cleaner environment (Carson Canyon, off Newfoundland). As is the case with trout, the antibody to the major BNF-inducible form in scup cross-reacts strongly with other species. Recently, polyclonal antibodies have been prepared to the major P-450 isozyme purified from BNF-treated cod, P-450c, and utilized in immunoquantitation (along with antibodies to trout LM_4 and scup P-450E) assays (Western blotting) and microsomal inhibition studies to demonstrate increased levels of P-448 following BNF-induction in cod, trout, scup, hagfish, herring, Northern pike, perch, and plaice.[205]

III. CYTOCHROME P-450 REDUCTASE

A. Background

NADPH-cytochrome *c* (P-450) reductase is a microsomal flavoprotein which transfers electrons (in two separate, one-electron reductions) from NADPH to cytochrome P-450 (reviewed in Reference 206). The mammalian enzyme has a monomeric molecular weight of 74,000 to 80,000 Da and is anchored in the microsomal membrane by a hydrophobic tail. The large hydrophilic portion of the enzyme can be removed by treatment with lipase or protease, in which case activity towards cytochrome *c* remains, but the ability to reduce P-450 is lost. The intact reductase can be successfully solubilized with detergent and purified utilizing affinity chromatograpahy on 2′,5′-ADP-agarose as the crucial step.[207] It is of interest that reductase purified from a particular species is capable of reducing P-450 from other species which may be phylogenetically quite distant. As is the case for total P-450 levels in fish, hepatic microsomal NADPH-cytochrome *c* reductase activities are typically 10 to 50% that found in most mammals.

B. Isolation, Purification, and Characterization

The techniques described for purifying the mammalian enzyme, especially the use of affinity chromatography as developed by Yasukochi and Masters,[207] have been successfully modified to purify the enzyme from fish. Hepatic microsomal NADPH-cytochrome P-450 reductase was purified from little skate by Pohl et al.[208]

The enzyme from little skate differed in molecular weight (74,000 Da on SDS-PAGE) compared to the rabbit liver enzyme. The little skate reductase was also more thermolabile than the rabbit enzyme and both reductases yielded quite distinct peptide patterns on SDS-PAGE following limited proteolysis. The flavin content of both enzymes was comparable.

Immunochemically, antibody to the rabbit reductase did not cross-react with the flavoprotein from little skate when analyzed by double immunodiffusion on Ouchterlony plates, but was capable of inhibiting the activity of the little skate enzyme, although at higher levels of antibody than was required to inhibit the rabbit enzyme.

The little skate reductase also differed from the rabbit enzyme in that the former was isolated in a partially reduced, semiquinone state. These results suggested that elasmobranch NADPH-cytochrome P-450 reductase differed somewhat from the mammalian enzyme and the greater thermolability of the little skate enzyme was responsible for the lower temperature optimum typically observed in fish MFO reactions.

The reductase has also been isolated from hepatic microsomes of rainbow trout by this author's laboratory.[209] As with the previous study, differences in primary structure and properties of the fish and mammalian enzymes were observed. The molecular weight (77,000 Da) of the purified trout enzyme (as determined by SDS-PAGE) was about 2000 Da higher than the rat enzyme (75,000). The flavin composition (approximately 1 mol of FAD and 1 mol of FMN/mole of enzyme) of both enzymes was comparable, but significant differences were observed in amino acid compositions and peptide patterns following limited proteolysis. No immunochemical cross-reactivity was observed between the trout enzyme and antibody to rat reductase on Ouchterlony plates; however, the antibody was quite effective at inhibiting the reduction of cytochrome c by both the rat and trout enzymes. Unlike the case with little skate reductase, trout reductase did not appear to be any more thermolabile than the rat enzyme. Further studies on the temperature-sensitive components of the trout MFO system demonstrated that when any combination of rat and trout P-448 (trout LM_{4b} and rat BNF-B), NADPH-cytochrome P-450 reductase, and microsomal lipid was reconstituted into micelles, a temperature optimum (30 to 40°C) was observed which resembled that of rat microsomes.[209] The lower temperature optimum of trout microsomes could only be mimicked when all three trout MFO components were reconstituted into liposomes. These results, and recent findings from Mannering's laboratory,[210] suggest that there is some structural or functional organization within the trout microsomal membrane which is responsible for the lower temperature optima of MFO reactions observed with many fish species.

NADPH-cytochrome P-450 reductase has also been isolated from the marine teleost scup by Klotz et al.[63] As with trout and little skate, the purified scup reductase displayed a higher molecular weight on SDS-PAGE (82,000 Da) than the mammalian (rat) enzyme (77,700 Da). Also, as with little skate, the flavoprotein was isolated as the one-electron reduced semiquinone. The flavin content was 1.0 mol of FAD and 0.8 mol of FMN/mol of enzyme with a specific activity of 45 to 60 U/mg.

NADPH-cytochrome c (P-450) reductase has, therefore, been purified to homogeneity from a marine elasmobranch, a marine teleost, and a freshwater teleost. In all cases, the minimum molecular weight of the fish reductase was higher than the mammalian (rat or rabbit) enzyme. Fish and mammalian reductase both contain a single mole of FAD and FMN, exhibit the same turnover with cytochrome c as electron receptor, and can both function in P-450 reconstitution systems, reducing mammalian or fish cytochromes P-450. Fish NADPH-cytochrome P-450 reductase has readily discernible differences compared to mammals, as determined by amino acid composition and peptide profiles following limited proteolysis. Immunochemical analysis suggests that fish and mammalian reductase share some antigenically related structural features.

IV. COOXIDATION OF PAH

In addition to the P-450-dependent MFO system, mammals can oxygenate PAH, such as BaP or their oxidation products, to electrophilic, toxic metabolites by mechanisms involving cooxidation during nonenzymatic lipid peroxidation (reviewed in References 211 and 212) or during prostaglandin biosynthesis with arachidonic acid or other polyunsaturated fatty acid substrates (reviewed in Reference 213). In mammals, this pathway is thought to play a significant role in the metabolic activation of xenobiotics, especially in extrahepatic tissues which are high in prostaglandin synthase and low in P-450.

Relatively little is known regarding the potential for cooxidation to play a significant role in the activation of PAH in aquatic animals. Poikilothermic species, especially those acclimated to colder temperatures, have large amounts of polyunsaturated fatty acids in cell membranes,[184] and, therefore, the potential exists for significant cooxidation of PAH to occur during lipid peroxidation. Furthermore, prostaglandins are produced in fish[214-218] and it is, therefore, possible that prostaglandin synthase-catalyzed cooxidation of PAH may contribute to the overall biotransformation of PAH in fish. As in mammals, the high activity of prostaglandin synthase may result in this pathway being more significant than the P-450-MFO system in metabolizing PAH in extrahepatic tissues.[216-218] Recent results in this author's laboratory suggest that arachidonic acid-dependent cooxidation could be a significant pathway for xenobiotic metabolism in fish as evidenced by the N-oxidation of 2-aminofluorene by rainbow trout.[219] If this system is important in fish, changes associated with increased prostaglandin production (related to reproduction) may influence the metabolism and/or toxicity of PAH.[220]

V. FLAVIN-CONTAINING MONOOXYGENASE

Mammalian flavin-containing monooxygenase (FMO) catalyzes the N-oxidation of various tertiary, secondary, and primary amines as well as the oxidation of certain sulfur-, phosphorus-, and selenium-containing compounds (reviewed in Reference 221). The FMO, like the P-450 MFO system, is localized primarily in the liver endoplasmic reticulum. The relative contributions of FMO and cytochromes P-450 to the oxidation of a given substrate can be approximated through the use of antibodies toward P-450-reductase, selective inhibitors, and the higher pH optimum and greater temperature sensitivity of the FMO system.[222]

Fish[223,224] and marine invertebrates such as *Mytilus edulis*,[104] *M. galloprovincialis*,[102,103] and four species of marine sponges[105] have been shown to contain a microsomal enzyme with properties similar to those of mammalian FMO. The enzyme from mussel digestive gland,[104] for example, oxidized several aromatic amines, including *N,N*-dimethylaniline, 2-aminofluorene, and 2-aminoanthracene, to yield mutagenic metabolites detectable via the Ames-*Salmonella* assay. Kurelec and colleagues[102-105] failed to detect cytochrome P-450-dependent BaP-hydroxylase activity in marine invertebrates. These investigators suggest, instead, that aromatic amine rather than PAH activation may be a more important mechanism for mutagenic, clastogenic, and carcinogenic end-points found in these animals. A number of nitrogen- and sulfur-containing xenobiotics, including some PAH have become significant environmental pollutants in aquatic ecosystems.[8,9] If the properties of the FMO associated with aquatic animals are similar to the mammalian enzyme, this enzyme could play a role in the oxidative metabolism and possible metabolic activation of such compounds.

The FMO system in mussels is not inducible[103] and may have important physiological functions in marine animals.[102] In fish, especially marine species, this enzyme may function in the N-oxidation of trimethylamine to trimethylamine-N-oxide, a physiologically important end-product of nitrogen metabolism.

VI. SUMMARY AND PROSPECTUS

PAH are an important class of environmental pollutants which serve as inducers and substrates of the microsomal MFO system. These lipophilic compounds are oxygenated by cytochromes(s) P-450 to produce either detoxication metabolites, usually in conjunction with the phase II conjugation enzymes, or to activated electrophilic metabolites which can covalently react with macromolecules producing a variety of toxic effects, including carcinogenesis.

The metabolic activation of PAH, such as BaP, has been extensively studied in mammals. Acquisition of the purified P-450 isozymes involved and a combined utilization of metabolism and molecular biology studies have resulted in a more complete understanding of the properties, function, and regulation of this important pathway.

A good deal of effort by a number of laboratories has established that the fish P-450-dependent MFO system is similar in many respects to the mammalian one and, in fact, may be a simpler model for studying this phylogenetically primitive P-450 gene family.

The purification of several cytochrome P-450 isozymes from teleosts and the preparation of polyclonal and monoclonal antibodies to these proteins have made possible molecular biology studies which will more fully elucidate the structural properties and regulation of PAH-inducible cytochromes P-450 in fish. This, in turn, will allow us to better understand their role in the metabolic disposition and effects of PAH.

NOTE ADDED IN PROOF

Cytochrome P-450[225] and FMO[226] activities have been characterized in the gumboot chiton (*Cryptochiton stelleri*). Recently, the major cytochrome P_1-450 (LM_{4b}) cDNA has been isolated and sequenced from liver of 3-MC-treated rainbow trout.[227] Using an immunohistochemical technique, the cellular localization of cytochrome P-450 LM_2 and LM_{4b} has been demonstrated in the kidneys of male and female rainbow trout.[228] The P-450 LM_2 was concentrated in the cytoplasm of cells in the second portion (P_2) of the proximal tubules, while low staining for P-450 LM_{4b} appeared in the cytoplasm of cells in both the first portion of the proximal tubules and in the P_2 section. Moderate staining for P-450 LM_{4b} also occurred in the interrenal cells of the trout head kidney.

ACKNOWLEDGMENTS

The authors very much appreciate the helpful suggestions and comments by Drs. J. R. Bend and L. Forlin. This work was supported by Grants ES0040, ES00210, and ES03850 from the National Institute of Environmental Health Sciences.

REFERENCES

1. **Eglinton, G., Simoneit, B. R. T., and Zoro, J. A.,** The recognition of organic pollutants in aquatic sediments, *Proc. R. Soc. London,* 189B, 415, 1975.
2. **Malins, D. C., McCain, B. B., Brown, D. W., Chan, S.-L., Myers, M. S., Landahl, J. T., Prohaska, P. G., Friedman, A. J., Rhodes, L. D., Burrows, D. G., Gronlund, W. D., and Hodgins, H. O.,** Chemical pollutants in sediments and diseases of bottom-dwelling fish in Puget Sound, Washington, *Environ. Sci. Technol.,* 18, 705, 1984.

3. **Malins, D. C., Krahn, M. M., Myers, M. S., Rhodes, L. D., Brown, D. W., Krone, C. A., McCain, B. B., and Chan, S.-L.,** Toxic chemicals in sediments and biota from a creosote-polluted harbor: relationships with hepatic neoplasms and other hepatic lesions in English sole *(Parophrys vetulus), Carcinogenesis,* 6, 1463, 1985.

4. **Stegeman, J. J.,** Polynuclear aromatic hydrocarbons and their metabolism in the marine environment, in *Polycyclic Hydrocarbons and Cancer,* Vol. 3, Gelboin, H. V. and Ts'o, P. O. P., Eds., Academic Press, New York, 1981, 1.

5. **Veith, D. G., Defoe, D. L., and Bergstedt, B. V.,** Measuring and estimating the bioconcentration factor of chemicals in fish, *J. Fish. Res. Board Can.,* 36, 1040, 1979.

6. **Anderson, P. D. and D'Apollonia, S.,** Aquatic animals, in *Principles of Ecotoxicology,* Butler, G. C., Ed., John Wiley & Sons, New York, 1978, 187.

7. **Malins, D. C.,** Alterations in the cellular and subcellular structure of marine teleosts and invertebrates exposed to petroleum in the laboratory and field: a critical review, *Can. J. Fish. Aquat. Sci.,* 39, 877, 1982.

8. **Malins, D. C., Krahn, M. M., Brown, D. W., Rhodes, L. D., Myers, M. S., McCain, B. B., and Chan, S.-L.,** Toxic chemicals in marine sediment and biota from Mukilteo, Washington: relationships with hepatic neoplasms and other hepatic lesions in English sole *(Parophyrs vetulus), J. Natl. Cancer Inst.,* 74, 487, 1985.

9. **Krahn, M. M., Rhodes, L. D., Myers, M. S., Moore, L. K., MacLeod, W. D., Jr., and Malins, D. C.,** Associations between metabolites of aromatic compounds in bile and the occurrence of hepatic lesions in English sole *(Parophrys vetulus)* from Puget Sound, Washington, *Arch. Environ. Contam. Toxicol.,* 15, 61, 1986.

10. **Dawe, C. J., Stanton, M. F., and Schwartz, F. J.,** Hepatic neoplasms in native bottom-feeding fish of Deep Creek Lake, Maryland, *Cancer Res.,* 24, 1194, 1964.

11. **Brown, E. R., Hazdra, J. J., Keith, L., Greenspan, I., Kapinski, J. B. G., and Beamer, P.,** Frequency of fish tumors in a polluted watershed as compared to nonpolluted Canadian waters, *Cancer Res.,* 33, 189, 1973.

12. **Sonstegard, R. A.,** Environmental carcinogenesis studies in fishes of the Great Lakes of North America, *Ann. N.Y. Acad. Sci.,* 298, 261, 1977.

13. **Black, J. J.,** Field and laboratory studies of environmental carcinogenesis in Niagara River fish, *J. Great Lakes Res.,* 9, 326, 1983.

14. **Baumann, P. C. and Harshbarger, J. C.,** Frequencies of liver neoplasia in a feral fish population and associated carcinogens, *Mar. Environ. Res.,* 17, 324, 1985.

15. **Murchelano, R. A. and Wolke, R. E.,** Epizootic carcinoma in the winter flounder, *Pseudopleuronectes americanus, Science,* 228, 587, 1985.

16. **Couch, J. A. and Harshburger, J. C.,** Effect of carcinogenic agents on aquatic animals: an environmental and experimental overview, *Environ. Carcinogenesis Rev.,* 3, 63, 1985.

17. **Schultz, M. E. and Schultz, R. J.,** Induction of hepatic tumors with 7,12-dimethylbenz[a]anthracene in two species of viviparous fishes (Genus *Poeciliopsis), Environ. Res.,* 27, 337, 1982.

18. **Hendricks, J. D., Meyers, T. R., Shelton, D. W., Casteel, J. R., and Bailey, G. S.,** Hepatocarcinogenicity of benzo(a)pyrene to rainbow trout *(Salmo gairdneri)* by dietary exposure and intraperitoneal injection, *J. Natl. Cancer Inst.,* 74, 839, 1985.

19. **Bailey, G. S., Hendricks, J. D., Nixon, J. E., and Pawloski, N. E.,** The sensitivity of rainbow trout and other fish to carcinogens, *Drug Metab. Rev.,* 15, 725, 1984.

20. **Caldwell, J. and Jakoby, W. B.,** *Biological Basis of Detoxication,* Academic Press, New York, 1983, 1.

21. **Jakoby, W. B., Bend, J. R., and Caldwell, J.,** Metabolic basis of detoxication, in *Metabolism of Functional Groups,* Academic Press, New York, 1982, 1.

22. **Franklin, R. B., Elcombe, C. R., Vodicnik, M. J., and Lech, J. J.,** Comparative aspects of the disposition and metabolism of xenobiotics in fish and mammals, *Fed. Proc.,* 39, 3144, 1980.

23. **Chambers, J. E. and Yarbrough, J. D.,** Xenobiotic biotransformation systems in fishes, *Comp. Biochem. Physiol.,* 55C, 77, 1976.

24. **James, M. O., Fouts, J. R., and Bend, J. R.,** Xenobiotic metabolizing enzymes in marine fish, in *Pesticides in the Aquatic Environments,* Khan, M. A. Q., Ed., Plenum Press, New York, 1977, 171.

25. **Livingstone, D. R.,** Responses of the detoxication/toxication enzyme systems of molluscs to organic pollutants and xenobiotics, *Mar. Pollut. Bull.,* 16, 158, 1985.

26. **Lech, J. J. and Vodicnik, M. J.,** Biotransformation, in *Fundamentals of Aquatic Toxicology. Methods and Applications,* Rand, G. M. and Petrocelli, S. R., Eds., Hemisphere Publishing, New York, 1985, 526.

27. **Bend, J. R. and James, M. O.,** Xenobiotic metabolism in marine and freshwater species, in *Biochemical and Biophysical Perspectives in Marine Biology,* Vol. 4, Malins, D. C. and Sargent, J. R., Eds., Academic Press, New York, 1978, 125.

28. **Tan, B. and Melius, P.,** Polynuclear aromatic hydrocarbon metabolism in fishes, *Comp. Biochem. Physiol.,* 83C, 217, 1986.

29. **Brodie, B. B. and Maickel, R. P.,** Comparative biochemistry of drug metabolism, in *Proc. 1st Int. Pharmacology Meeting,* Brodie, B. B. and Erdos, F. G., Eds., Macmillan, New York, 1962, 299.

30. **Buhler, D. R.,** Hepatic drug metabolism in fishes, *Fed. Proc.,* 25, 343, 1966.

31. **Buhler, D. R. and Rasmusson, M. E.,** The oxidation of drugs by fishes, *Comp. Biochem. Physiol.,* 25, 223, 1968.

32. **Creaven, P. J., Parke, D. V., and Williams, R. T.,** A fluorometric study of the hydroxylation of biphenyl *in vitro* by liver preparations of various species, *Biochem. J.,* 96, 879, 1965.

33. **Dewaide, J. H. and Henderson, P. T.,** Hepatic N-demethylation of aminopyrine in rat and trout, *Biochem. Pharmacol.,* 17, 1901, 1968.

34. **Adamson, R. H.,** Drug metabolism in marine vertebrates, *Fed. Proc.,* 26, 1047, 1967.

35. **Lech, J. J. and Bend, J. R.,** Relationship between biotransformation and the toxicity and fate of xenobiotic chemicals in fish, *Environ. Health Perspect.,* 34, 115, 1980.

36. **Lech, J. J., Vodicnik, M. J., and Elcombe, C. R.,** Induction of monooxygenase activity in fish, in *Aquatic Toxicology,* Weber, L. J., Ed., Raven Press, New York, 1982, 107.

37. **Binder, R. L., Melancon, M. J., and Lech, J. J.,** Factors influencing the persistence and metabolism of chemicals in fish, *Drug Metab. Rev.,* 15, 697, 1984.

38. **Hansson, T.,** Androgenic regulation of hepatic metabolism of 4-androstene-3,17-dione in the rainbow trout, *Salmo gairdneri, J. Endocrinol.,* 92, 409, 1982.

39. **Hansson, T. and Rafter, J.,** *In vitro* metabolism of estradiol-17β by liver microsomes from juvenile rainbow trout, *Salmo gairdneri, Gen. Comp. Endocrinol.,* 49, 490, 1983.

40. **Stegeman, J. J., Pajor, A. M., and Thomas, P.,** Influence of estradiol and testosterone on cytochrome P-450 and monooxygenase activity in immature brook trout, *Salvelinus fontinulis, Biochem. Pharmacol.,* 31, 3979, 1982.

41. **Williams, D. E., Okita, R. T., Buhler, D. R., and Masters, B. S. S.,** Regiospecific hydroxylation of lauric acid at the (ω-1) position by hepatic and kidney microsomal cytochromes P-450 from rainbow trout, *Arch. Biochem. Biophys.,* 231, 503, 1984.

42. **James, M. O. and Shiverick, K. T.,** Cytochrome P-450-dependent oxidation of progesterone, testosterone and ecdysone in the spiny lobster, *Panulirus argus, Arch. Biochem. Biophys.,* 233, 1, 1984.

43. **Ahokas, J. T., Pelkonen, O., and Karki, N.,** Characterization of benzo[a]pyrene hydroxylase of trout liver, *Cancer Res.,* 37, 3737, 1977.

44. **Elcombe, C. R. and Lech, J. J.,** Induction and characterization of hemoprotein(s) P-450 and monooxygenation in rainbow trout *(Salmo gairdneri), Toxicol. Appl. Pharmacol.,* 49, 437, 1979.

45. **Williams, D. E. and Buhler, D. R.,** Purification of cytochromes P-448 from β-naphthoflavone treated rainbow trout, *Biochim. Biophys. Acta,* 717, 398, 1982.

46. **Williams, D. E. and Buhler, D. R.,** Comparative properties of purified cytochrome P-448 from β-naphthoflavone treated rats and rainbow trout, *Comp. Biochem. Physiol.,* 75C, 25, 1983.

47. **Williams, D. E. and Buhler, D. R.,** Benzo(a)pyrene-hydroxylase catalyzed by purified isozymes of cytochrome P-450 from β-naphthoflavone-fed rainbow trout, *Biochem. Pharmacol.,* 33, 3743, 1984.

48. **Gruger, E. H., Jr., Wekell, M. M., Numoto, P. T., and Craddock, D. R.,** Induction of hepatic aryl hydrocarbon hydroxylase in salmon exposed to petroleum dissolved in seawater and to petroleum and polychlorinated biphenyls, separate and together, in food, *Bull. Environ. Contam. Toxicol.,* 17, 512, 1977.

49. **Vodicnik, M. J., Elcombe, C. R., and Lech, J. J.,** The effect of various types of inducing agents on hepatic microsomal monooxygenase activity in rainbow trout, *Toxicol. Appl. Pharmacol.,* 59, 364, 1981.

50. **Addison, R. F., Zincke, M. E., and Willis, D. E.,** Mixed function oxidase enzymes in trout *(Salvelinus fontinalis)* liver: absence of induction following feeding of p,p'-DDT or p,p'-DDE, *Comp. Biochem. Physiol.,* 57C, 39, 1977.

51. **Stegeman, J. J.,** Temperature influence on basal activity and induction of mixed function oxidase activity in *Fundulus heteroclitus, J. Fish. Res. Board Can.,* 36, 1400, 1979.

52. **Egaas, E. and Varanasi, U.,** Effects of polychlorinated biphenyls and environmental temperature on *in vitro* formation of benzo[a]pyrene metabolites by liver of trout *(Salmo gairdneri), Biochem. Pharmacol.,* 31, 561, 1982.

53. **Andersson, T. and Koivusaari, U.,** Influence of environmental temperature on the induction of xenobiotic metabolism by β-naphthoflavone in rainbow trout, *Salmo gairdneri, Toxicol. Appl. Pharmacol.,* 80, 43, 1985.

54. **Koivusaari, U. and Andersson, T.,** Partial temperature compensation of hepatic biotransformation enzymes in juvenile rainbow trout *(Salmo gairdneri)* during the warming of water in spring, *Comp. Biochem. Physiol.,* 78B, 223, 1984.

55. **Karr, S. W., Reinert, R. E., and Wade, A. E.,** The effects of temperature on the cytochrome P-450 system of thermally acclimated bluegill, *Comp. Biochem. Physiol.,* 80C, 135, 1985.

56. **Ankley, G. T., Reinert, R. E., Wade, A. E., and White, R. A.,** Temperature compensation in the hepatic mixed-function oxidase system of bluegill, *Comp. Biochem. Physiol.,* 81C, 125, 1985.

57. **Ahokas, J. T., Pelkonen, O., and Karki, N. T.,** Metabolism of polycyclic hydrocarbons by a highly active aryl hydrocarbon hydroxylase system in the liver of a trout species, *Biochem. Biophys. Res. Commun.,* 63, 635, 1975.

58. **Ahokas, J. T., Saarni, H., Nebert, D. W., and Pelkonen, O.,** The *in vitro* metabolism and covalent binding of benzo[a]pyrene to DNA catalyzed by trout liver microsomes, *Chem. Biol. Interact.,* 25, 103, 1979.

59. **Varanasi, U. and Gmur, D. J.,** Metabolic activation and covalent binding of benzo[a]pyrene to deoxyribonucleic acid catalyzed by liver enzymes of marine fish, *Biochem. Pharmacol.,* 29, 753, 1980.

60. **Varanasi, U., Stein, J. E., and Hom, T.,** Covalent binding of benzo[a]pyrene to DNA in fish liver, *Biochem. Biophys. Res. Commun.,* 103, 780, 1981.

61. **Nishimoto, M. and Varanasi, U.,** Benzo[a]pyrene metabolism and DNA adduct formation mediated by English sole liver enzymes, *Biochem. Pharmacol.,* 34, 263, 1985.

62. **Varanasi, U., Nishimoto, M., Reichert, W. L., and Eberhart, B.-T.L.,** Comparative metabolism of benzo(a)pyrene and covalent binding to hepatic DNA in English sole, starry flounder and rat, *Cancer Res.,* 46, 3817, 1986.

63. **Klotz, A. V., Stegeman, J. J., and Walsh, C.,** An aryl hydrocarbon hydroxylating hepatic cytochrome P-450 from the marine fish *Stenotomus chrysops, Arch. Biochem. Biophys.,* 226, 578, 1983.

64. **Goksoyr, A.,** Purification of hepatic microsomal cytochromes P-450 from β-naphthoflavone-treated Atlantic cod *(Gadus morhua),* a marine teleost fish, *Biochim. Biophys. Acta,* 840, 409, 1985.

65. **Quattrochi, L. C. and Lee, R. F.,** Microsomal cytochromes P-450 from marine crabs, *Comp. Biochem. Physiol.,* 79C, 171, 1984.

66. **Batel, R., Bihari, N., and Zahn, R. K.,** Purification and characterization of a single form of cytochrome P-450 from the spiny crab *Maja crispata, Comp. Biochem. Physiol.,* 83C, 165, 1986.

67. **Nebert, D. W., Eisen, H. J., Negishi, M., Lang, M. A., Hjelmeland, L. M., and Okey, A. B.,** Genetic mechanisms controlling the induction of polysubstrate monooxygenase (P-450) activities, *Annu. Rev. Pharmacol. Toxicol.,* 21, 431, 1981.

68. **Klotz, A. V., Stegeman, J. J., Woodin, B. R., Snowberger, E. A., Thomas, P. E., and Walsh, C.,** Cytochrome P-450 isozymes from the marine teleost *Stenotomus chrysops:* their roles in steroid hydroxylation and the influence of cytochrome b₅, *Arch. Biochem. Biophys.,* 249, 326, 1986.

69. **Eisele, T. A., Coulombe, R. A., Williams, J. L., Shelton, D. W., and Nixon, J. E.,** Time and dose dependent effects of dietary cyclopropenoid fatty acids on the hepatic microsomal mixed function oxidase system of rainbow trout, *Aquat. Toxicol.,* 4, 139, 1983.

70. **Forlin, L. and Hansson, T.,** Effects of treated municipal wastewater on the hepatic, xenobiotic, and steroid metabolism in trout, *Ecotoxicol. Environ. Saf.,* 6, 41, 1982.

71. **Lindstrom-Seppa, P., Koivusaari, U., and Hanninen, O.,** Extrahepatic xenobiotic metabolism in North-European freshwater fish, *Comp. Biochem. Physiol.,* 69C, 259, 1981.

72. **Balk, L., Meijer, J., Seidegard, J., Morgenstein, R., and DePierre, J. W.,** Initial characterization of drug-metabolizing systems in the liver of the Northern pike, *Esox lucius, Drug Metab. Dispos.,* 8, 98, 1980.

73. **Maemura, S. and Omura, T.,** Drug-oxidizing monooxygenase system in liver microsomes of goldfish *(Carassius auratus), Comp. Biochem. Physiol.,* 76C, 45, 1983.

74. **Swain, L. and Melius, P.,** Characterization of benzo[a]pyrene metabolites formed by 3-methylcholanthrene-induced goldfish, black bullhead and brown bullhead, *Comp. Biochem. Physiol.,* 79C, 151, 1984.

75. **Protic'-Sabljic', M. and Kurelec, B.,** High mutagenic potency of several polycyclic aromatic hydrocarbons induced by liver postmitochondrial fractions from control and xenobiotic-treated immature carp, *Mutat. Res.,* 118, 177, 1983.

76. **Lindstrom-Seppa, P., Koivusaari, U., and Hanninen, O.,** Cytochrome P-450 in the hepatopancreas of freshwater crayfish, *Astacus astacus* L., in *Cytochrome P-450, Biochemistry, Biophysics and Environmental Implications,* Hietanen, E., Latinen, M., and Hanninen, O., Eds., Elsevier, New York, 1982, 251.

77. **Stegeman, J. J. and Binder, R. L.,** High benzo[a]pyrene hydroxylase activity in the marine fish *Stenatomus versicolor, Biochem. Pharmacol.,* 28, 1686, 1979.

78. **Stegeman, J. J., Binder, R. L., and Orren, A.,** Hepatic and extrahepatic microsomal electron transport components and mixed-function oxygenases in the marine fish *Stenotomus versicolor, Biochem. Pharmacol.,* 28, 3431, 1979.

79. **Stegeman, J. J., Woodin, B. R., Klotz, A. V., Wolke, R. E., and Orme-Johnson, N. R.,** Cytochrome P-450 and monooxygenase activity in cardiac microsomes from the fish *Stenotomus chrysops, Mol. Pharmacol.,* 21, 517, 1982.

80. **James, M. O., Khan, M. A. Q., and Bend, J. R.,** Hepatic microsomal mixed-function oxidase activities in several marine species common to coastal Florida, *Comp. Biochem. Physiol.,* 62C, 155, 1979.

81. **Little, P. J., James, M. O., Bend, J. R., and Ryan, A. J.,** Imidazole derivatives as inhibitors of cytochrome P-450-dependent oxidation and activators of epoxide hydrolase in hepatic microsomes from a marine fish, *Biochem. Pharmacol.,* 30, 2876, 1981.
82. **Schnell, J. V., Gruger, E. H., Jr., and Malins, D. C.,** Monooxygenase activities of coho salmon *(Oncorhynchus kisutch)* liver microsomes using three polycyclic aromatic hydrocarbon substrates, *Xenobiotica,* 10, 229, 1980.
83. **Schoor, W. P. and Srivastava, M.,** Position-specific induction of benzo[a]pyrene metabolism by 3-methylcholanthrene and phenobarbital in mullet *(Mugil cephalus),* a marine fish, *Comp. Biochem. Physiol.,* 78C, 391, 1984.
84. **Pohl, R. J., Bend, J. R., Guarino, A. M., and Fouts, J. R.,** Hepatic microsomal mixed-function oxidase activity of several marine species from coastal Maine, *Drug Metab. Dispos.,* 2, 545, 1974.
85. **Bend, J. R., Foureman, G. L., Ben-Zvi, Z., and Albro, P. W.,** Heterogeneity of hepatic aryl hydrocarbon hydroxylase activity in feral winter flounder: relevance to carcinogenicity testing, in Use of Small Fish Species in Carcinogenicity Testing, Hoover, K. L., Ed., National Cancer Institute Monogr. No. 65, U.S. Government Printing Office, Washington, D.C., 1984, 359.
86. **Payne, J. F. and Fance, L. L.,** Effect of long term exposure to petroleum on mixed function oxygenases in fish: further support for use of the enzyme system in biological monitoring, *Chemosphere,* 11, 207, 1982.
87. **James, M. O. and Bend, J. R.,** Polycyclic aromatic hydrocarbon induction of cytochrome P-450 dependent mixed-function oxidases in marine fish, *Toxicol. Appl. Pharmacol.,* 54, 117, 1980.
88. **Lee, R. F., Conner, J. W., Page, D., Ray, L. E., and Giam, C. S.,** Cytochrome P-450 dependent mixed-function oxygenase systems in marsh crabs, in *Physiological Mechanisms of Marine Pollutant Toxicity,* Vernberg, W. B., Calabrese, A., Thurbing, F. P., and Vemberg, F. J., Eds., Academic Press, New York, 1982, 145.
89. **Singer, S. C. and Lee, R. F.,** Mixed function oxygenase activity in the blue crab, *Callinectes sapidus:* tissue distribution and correlation with changes during molting and development, *Biol. Bull.,* 153, 377, 1977.
90. **Payne, J. F. and May, N.,** Further studies on the effect of petroleum hydrocarbons on mixed-function oxidases in marine organisms, in *Pesticides and Xenobiotic Metabolism in Aquatic Organisms,* Khan, M. A., Lech, J. J., and Menn, J. J., Eds., American Chemical Society, Washington, D.C., 1979, 339.
91. **Stegeman, J. J. and Kaplan, H. B.,** Mixed-function oxygenase activity and benzo[a]pyrene metabolism in the barnacle *Balanus eburneus* (Crustacea: Cirripedia), *Comp. Biochem. Physiol.,* 68C, 55, 1981.
92. **Ade, P., Banchelli Soldaini, M. G., Castelli, M. G., Chiesara, E., Clementi, F., Fanelli, R., Funari, E., Ignesti, G., Marabini, A., Orunesu, M., Palmero, S., Pirisino, R., Ramundo Orlando, A., Rizzi, R., Silano, V., Viarengo, A., and Vittozzi, L.,** Comparative biochemical and morphological characterization of microsomal preparations from rat, quail, trout, mussel and *Daphnia magna,* in *Cytochrome P-450 Biochemistry, Biophysics and Environmental Implications,* Hietanen, E., Latinen, M., and Hanninen, O., Eds., Elsevier, New York, 1982, 387.
93. **Livingstone, D. R., Moore, M. N., Lowe, D. M., Nasci, C., and Farrar, S. V.,** Responses of the cytochrome P-450 monooxygenase system to diesel oil in the common mussel, *Mytilus edulis* L., and the periwinkle, *Littorina littorea* L., *Aquat. Toxicol.,* 7, 79, 1985.
94. **Williams, D. E., Caldwell, R. S., and Buhler, D. R.,** Analysis of microsomal and cytosolic AHH activity in control and oil-exposed *Tegula funebralis,* unpublished observations.
95. **Livingstone, D. R., Stickle, W. B., Kapper, M., and Wang, S.,** Microsomal detoxication enzyme responses of the marine snail, *Thais haemastoma,* to laboratory oil exposure, *Bull. Environ. Contam. Toxicol.,* 36, 843, 1986.
96. **Vandermeulen, J. H. and Penrose, W. E.,** Absence of aryl hydrocarbon hydroxylase (AHH) in three marine bivalves, *J. Fish. Res. Board Can.,* 35, 643, 1978.
97. **McKim, J. M., Caldwell, R. S., and Buhler, D. R.,** A preliminary determination of cytochrome P-450 levels and mixed-function oxidase activity in sea urchin larvae, in press.
98. **Pedersen, M. G., Harshbarger, W. K., Zachariah, P. K., and Juchau, M. R.,** Hepatic biotransformation of environmental xenobiotics in six strains of rainbow trout *(Salmo gairdneri), J. Fish. Res. Board Can.,* 33, 666, 1976.
99. **Miyauchi, M.,** Conversion of procarcinogens to mutagens by the S-9 fraction from the liver of rainbow trout *(Salmo gairdneri):* inducibility with PCB, 3-methylcholanthrene and phenobarbital and inhibition by metyrapone and α-naphthoflavone, *Comp. Biochem. Physiol.,* 79C, 363, 1984.
100. **Milling, D. M. and Maddock, M. B.,** Activation of polycyclic aromatic hydrocarbons by hepatic S-9 from a marine fish, *Bull. Environ. Contam. Toxicol.,* 35, 301, 1985.
101. **Khan, M. A. Q., Coello, W., Khan, A. A., and Pinto, H.,** Some characteristics of the microsomal mixed-function oxidase in freshwater crayfish, *Cambarus, Life Sci.,* 11, 405, 1972.
102. **Kurelec, B., Britvic, S., Krea, S., and Zahn, R. K.,** Metabolic fate of aromatic amines in the mussel *Mytilus galloprovincialis, Mar. Biol.,* 91, 523, 1986.

103. **Britvic, S. and Kurelec, B.,** Selective activation of carcinogenic aromatic amines to bacterial mutagens in the marine mussel *Mytilus galloprovincialis, Comp. Biochem. Physiol.,* 85C, 111, 1986.

104. **Kurelec, B.,** Exclusive activation of aromatic amines in the marine mussel *Mytilus edulis* by FAD-containing monooxygenase, *Biochem. Biophys. Res. Commun.,* 127, 773, 1985.

105. **Kurelec, B., Britvic, S., Krca, S., Muller, W. E. G., and Zahn, R. K.,** Metabolism of some carcinogenic aromatic amines in four species of marine sponges, *Comp. Biochem. Physiol.,* 86C, 17, 1987.

106. **Melius, P.,** Comparative benzo[a]pyrene metabolite patterns in fish and rodents, in Use of Small Fish Species in Carcinogenicity Testing, Hoover, K. L., Ed., National Cancer Institute Monogr. N. 65, U.S. Government Printing Office, Washington, D.C., 1984, 387.

107. **Stegeman, J. J., Woodin, B. R., and Binder, R. L.,** Patterns of benzo[a]pyrene metabolism by varied species, organs and developmental stages of fish, in *Use of Small Fish Species in Carcinogenicity Testing,* Hoover, K. L., Ed., National Cancer Institute Monogr. No. 65, U.S. Government Printing Office, Washington, D.C., 1984, 371.

108. **Baty, J. D.,** Species, strain, and sex differences in metabolism, *Foreign Compd. Metab. Mamm.,* 6, 133, 1981.

109. **Forlin, L.,** personal communication, 1987.

110. **Pohl, R. J., Fouts, J. R., and Bend, J. R.,** Response of hepatic microsomal MFO in the little skate and the winter flounder to pretreatment with TCDD or DBA, *Bull. Mt. Desert Isl. Biol. Lab.,* 15, 64, 1975.

111. **Andersson, T., Koivusaari, U., and Forlin, L.,** Xenobiotic biotransformation in the rainbow trout liver and kidney following starvation, *Comp. Biochem. Physiol.,* 82C, 221, 1985.

112. **Melancon, M. J., Elcombe, C. R., Vodicnik, M. J., and Lech, J. J.,** Induction of cytochromes P-450 and mixed-function oxidase activity by polychlorinated biphenyls and β-naphthoflavone in carp *(Cyprinus carpio), Comp. Biochem. Physiol.,* 69C, 219, 1981.

113. **Lindstrom-Seppa, P., Koivusaari, U., and Hanninen, O.,** Metabolism of xenobiotics by vendace *(Coregonus albula), Comp. Biochem. Physiol.,* 68C, 121, 1981.

114. **Balk, L., Maner, S., Bergstrand, A., and DePierre, J. W.,** Preparation and characterization of subcellular fractions suitable for studies of drug metabolism from the trunk kidney of the Northern pike *(Esox lucius)* and assay of certain enzymes of xenobiotic metabolism in these subfractions, *Biochem. Pharmacol.,* 33, 2447, 1984.

115. **Balk, L., Maner, S., Bergstrand, A., Birberg, W., Pilotti, A., and DePierre, J. W.,** Preparation and characterization of subcellular fractions from the head kidney of the Northern pike *(Esox lucius)*, with particular emphasis on xenobiotic-metabolizing enzymes, *Biochem. Pharmacol.,* 34, 789, 1985.

116. **Balk, L., Maner, S., Meijer, J., Bergstrand, A., and DePierre, J. W.,** Preparation and characterization of subcellular fractions from the intestinal mucosa of the Northern pike *(Esox lucius)*, with special emphasis on enzymes involved in xenobiotic metabolism, *Biochim. Biophys. Acta,* 838, 277, 1985.

117. **Williams, D. E., Masters, B. S. S., Lech, J. J., and Buhler, D. R.,** Sex differences in cytochrome P-450 isozyme composition and activity in kidney microsomes of mature rainbow trout, *Biochem. Pharmacol.,* 35, 2017, 1986.

118. **Cooper, K. R., Kindt, V., and Snyder, R.,** Correlation of benzene metabolism and histological lesions in rainbow trout *(Salmo gairdneri), Drug. Metab. Rev.,* 15, 673, 1984.

119. **Pritchard, J. B. and Bend, J. R.,** Mechanisms controlling the renal excretion of xenobiotics in fish: effects of chemical structure, *Drug Metab. Rev.,* 15, 655, 1984.

120. **Singer, S. C., March, P. E., Gonsoulin, F., and Lee, R. F.,** Mixed function oxygenase activity in the blue crab, *Callinectes sapidus:* characterization of enzyme activity from stomach tissue, *Comp. Biochem. Physiol.,* 65C, 129, 1980.

121. **Niranjan, B. G., Avadhani, N. G., and DiGiovanni, J.,** Formation of benzo[a]pyrene metabolites and DNA adducts catalyzed by a rat liver mitochondrial monooxygenase system, *Biochem. Biophys. Res. Commun.,* 131, 935, 1985.

122. **Balk, L., Meijer, J., Bergstrand, A., Astrom, A., Morgenstern, R., Seidegard, J., and DePierre, J. W.,** Preparation and characterization of subcellular fractions from the liver of the Northern pike, *Esox lucius, Biochem. Pharmacol.,* 31, 1491, 1982.

123. **Elmamlouk, T. H., Gessner, T., and Brownie, A. C.,** Occurrence of cytochrome P-450 in hepatopancreas of *Homarus americanus, Comp. Biochem. Physiol.,* 48B, 419, 1974.

124. **Spies, R. B., Rice, D. W., Jr., Montagna, P. A., and Ireland, R. R.,** Reproductive success, xenobiotic contaminants and hepatic mixed-function oxidase (MFO) activity in *Platichthys stellatus* populations from San Francisco Bay, *Mar. Environ. Res.,* 17, 117, 1985.

125. **Melancon, M. J., Binder, R. L., and Lech, J. J.,** Environmental induction of monooxygenase activity in fish, in *Toxic Contaminants and Ecosystem Health: A Great Lakes Focus,* Evans, M. S., Ed., John Wiley & Sons, New York, 1988, 215.

126. **Bend, J. R., Pohl, R. J., Davidson, N. P., and Fouts, J. R.,** Response of hepatic and renal mixed-function oxidase systems in little skate, *Raja erinacea,* to pretreatment with 3-methylcholanthrene or TCDD (2,3,7,8-tetrachlorodibenzo-p-dioxin), *Bull. Mt. Desert Isl. Biol. Lab.,* 14, 7, 1974.

127. **Bend, J. R., Pohl, R. J., Arinc, E., and Philpot, R. M.,** Hepatic microsomal and solubilized mixed-function oxidase systems from the little skate, *Raja erinacea,* a marine elasmobranch, in *Microsomes and Drug Oxidations,* Ullrich, V., Rocts, I., Hildebrant, A., Estabrook, R. W., and Conney, A. H., Eds., Pergamon Press, Oxford, 1977, 160.

128. **Stegeman, J. J., Klotz, A. V., Woodin, B. R., and Pajor, A. M.,** Induction of hepatic cytochrome P-450 in fish and the indication of environmental induction in scup *(Stenotomus chrysops), Aquat. Toxicol.,* 1, 197, 1981.

129. **Statham, C. N., Elcombe, C. R., Szyjka, S. P., and Lech, J. J.,** Effect of polycyclic aromatic hydrocarbons on hepatic microsomal enzymes and disposition of methylnaphthalene in rainbow trout *in vivo, Xenobiotica,* 8, 65, 1978.

130. **Eisele, T. A., Coulombe, R. A., Pawlowski, N. E., and Nixon, J. E.,** The effects of route of exposure and combined exposure of mixed function oxidase inducers and suppressors on hepatic parameters in rainbow trout *(Salmo gairdneri), Aquat. Toxicol.,* 5, 211, 1984.

131. **Schwen, R. J. and Mannering, G. J.,** Hepatic cytochrome P-450-dependent monooxygenase systems of the trout, frog and snake. III. Induction, *Comp. Biochem. Physiol.,* 71B, 445, 1982.

132. **Forlin, L.,** Effects of Clophen A50, 3-methylcholanthrene, pregnenolone-16α-carbonitrile and phenobarbital on the hepatic microsomal cytochrome P-450-dependent monooxygenase system in rainbow trout, *Salmo gairdneri,* of different age and sex, *Toxicol. Appl. Pharmacol.,* 54, 420, 1980.

133. **Elcombe, C. R., Franklin, R. B., and Lech, J. J.,** Induction of hepatic microsomal enzymes in rainbow trout, in *Pesticide and Xenobiotic Metabolism in Aquatic Organisms,* Khan, M. A. Q., Lech, J. J., and Menn, J. J., Eds., American Chemical Society, Washington, D.C., 1979, 319.

134. **Elcombe, C. R., Franklin, R. B., and Lech, J. J.,** Induction of microsomal hemoprotein(s) P-450 in the rat and rainbow trout by polyhalogenated biphenyls, *Ann. N.Y. Acad. Sci.,* 320, 193, 1979.

135. **Gerhart, E. H. and Carlson, R. M.,** Hepatic mixed-function oxidase activity in rainbow trout exposed to several polycyclic aromatic compounds, *Environ. Res.,* 17, 284, 1978.

136. **Voss, S. D., Shelton, D. W., and Hendricks, J. D.,** Effects of dietary Aroclor 1254 and cyclopropene fatty acids on hepatic enzymes in rainbow trout, *Arch. Environ. Contam. Toxicol.,* 11, 87, 1982.

137. **Payne, J. F. and Penrose, W. R.,** Induction of aryl hydrocarbon benzo[a]pyrene hydroxylase in fish by petroleum, *Bull. Environ. Contam. Toxicol.,* 14, 112, 1975.

138. **Payne, J. F.,** Field evaluation of benzo[a]pyrene hydroxylase induction as a monitor for marine petroleum pollution, *Science,* 191, 945, 1976.

139. **Kurelec, B., Britvic, S., Rijayec, M., Muller, W. E. G., and Zahn, R. K.,** Benzo[a]pyrene monooxygenase induction in marine fish-molecular response to oil pollution, *Mar. Biol.,* 44, 211, 1977.

140. **Forlin, L. and Lidman, U.,** Effects of Clophen A50®, 4-, 2,5,2′,5′-tetra- and 2,4,5,2′,4′,5′-hexachlorobiphenyl on the mixed-function oxidase system of rainbow trout *(Salmo gairdneri* Rich.) liver, *Comp. Biochem. Physiol.,* 60C, 193, 1978.

141. **Walton, D. G., Penrose, W. R., and Green, J. M.,** The petroleum-inducible mixed-function oxidase of cunner *(Tautogolabrus adspersus* Walbaum 17): some characteristics relevant to hydrocarbon monitoring, *J. Fish. Res. Board Can.,* 35, 1547, 1978.

142. **Stegeman, J. J.,** Influence of environmental contamination on cytochrome P-450 mixed-function oxygenases in fish: implications for recovery in the wild harbor marsh, *J. Fish. Res. Board Can.,* 35, 668, 1978.

143. **Kurelec, B., Matijasevic, Z., Rijavec, M., Alacevic, M., Britvic, S., Muller, W. E. G., and Zahn, R. K.,** Induction of benzo[a]pyrene monooxygenase in fish and the Salmonella test as a tool for detecting mutagenic/carcinogenic xenobiotics in the aquatic environment, *Bull. Environ. Contam. Toxicol.,* 21, 799, 1979.

144. **Forlin, L. and Lidman, U.,** Effects of Clophen A50® and 3-methylcholanthrene on the hepatic mixed function oxidase system in female rainbow trout, *Salmo gairdneri, Comp. Biochem. Physiol.,* 70C, 297, 1981.

145. **Addison, R. F., Zinck, M. E., Willis, D. E., and Wrench, J. J.,** Induction of hepatic mixed function oxidase activity in trout *(Salvelinus fontinalis)* by Aroclor 1254® and some aromatic hydrocarbon PCB replacements, *Toxicol. Appl. Pharmacol.,* 63, 166, 1982.

146. **Vodicnik, M. J. and Lech, J. J.,** Characterization of the induction response of the hepatic microsomal monooxygenase (MO) system of fishes, in *Cytochrome P-450, Biochemistry, Biophysics and Environmental Implications,* Hietanen, E., Laitinen, M., and Hanninen, O., Eds., Elsevier, Biomedical Press New York, 1982, 225.

147. **Ridlington, J. W., Chapman, D. E., Boese, B. L., and Johnson, V. G.,** Petroleum refinery wastewater induction of the hepatic mixed-function oxidase system in Pacific Staghorn sculpin, *Arch. Environ. Contam. Toxicol.,* 11, 123, 1982.

148. **Collodi, P., Stekoll, M. S., and Rice, S. D.,** Hepatic aryl hydrocarbon hydroxylase activities in coho salmon *(Oncorhynchus kisutch)* exposed to petroleum hydrocarbons, *Comp. Biochem. Physiol.,* 79C, 337, 1984.

149. **Payne, J. F., Bauld, C., Dey, A. C., Kiceniuk, J. W., and Williams, U.,** Selectivity of mixed-function oxygenase enzyme induction in flounder *(Pseudopleuronectes americanus)* collected at the site of the Baie Verte, Newfoundland oil spill, *Comp. Biochem. Physiol.,* 79C, 15, 1984.

150. **Payne, J. F.,** Mixed-function oxygenases in biological monitoring programs: review of potential usage in different phyla of aquatic animals, in *Ecotoxicological Testing for Marine Environment,* Vol. 1, Personne, G., Jaspers, E., and Clans, C., Eds., State University of Ghent, Belgium, 1984, 652.

151. **Lindstrom-Seppa, P., Koivusaari, U., Hanninen, O., and Pyysalo, H.,** Cytochrome P-450 and mono-oxygenase activities in the monitoring of aquatic environment, *Pharmazie,* 40, 232, 1985.

152. **Collier, T. K., Gruger, E. H., Jr., and Varanasi, U.,** Effect of Aroclor 1254® on the biological fate of 2,6-dimethylnaphthalene in coho salmon *(Oncorhynchus kisutch), Bull. Environ. Contam. Toxicol.,* 34, 114, 1985.

153. **Andersson, T., Pesonen, M., and Johansson, C.,** Differential induction of cytochrome P-450-dependent monooxygenase, epoxide hydrolase, glutathione transferase and UDP glucuronosyl transferase activities in the liver of the rainbow trout by β-naphthoflavone or Clophen A50,® *Biochem. Pharmacol.,* 34, 3309, 1985.

154. **Collier, T. K., Stein, J. E., Wallace, R. J., and Varanasi, U.,** Xenobiotic metabolizing enzymes in spawning English sole *(Parophyrs vetulus)* exposed to organic-solvent extracts of marine sediments from contaminated and reference areas, *Comp. Biochem. Physiol.,* 84C, 297, 1986.

155. **Negishi, M. and Nebert, D. W.,** Structural gene products of the Ah complex. Increases in large mRNA from mouse liver associated with cytochrome P_1-450 induction by 3-methylcholanthrene, *J. Biol. Chem.,* 256, 3085, 1981.

156. **Foldes, R. L., Hines, R. N., Ho, K.-L., Shen, M.-L., Nagel, K. B., and Bresnick, E.,** 3-Methyl-cholanthrene-induced expression of the cytochrome P-450c gene, *Arch. Biochem. Biophys.,* 239, 137, 1985.

157. **Guengerich, F. P., Wang, P., and Davidson, N. K.,** Estimation of isozymes of microsomal cytochrome P-450 in rats, rabbits, and humans using immunochemical staining coupled with sodium dodecyl sulfate-polyacrylamide gel electrophoresis, *Biochemistry,* 21, 1698, 1982.

158. **Vodicnik, M. J., Rau, L. A., and Lech, J. J.,** The effect of monooxygenase inducing agents on the incorporation of [^{35}S]-methionine into hepatic microsomal protein of rainbow trout *(Salmo gairdneri), Comp. Biochem. Physiol.,* 79C, 271, 1984.

159. **Kleinow, K. M., Melancon, M. J., and Lech, J. J.,** Implications for toxicity, bioaccumulation and monitoring of environmental xenobiotics in fish, *Environ. Health Persp.,* 71, 105, 1988.

160. **Lee, R. F., Singer, S. C., Tenore, K. R., Gardner, W. S., and Philpot, R. M.,** Detoxification system in polychaete worms: importance in the degradation of sediment hydrocarbons, in *Marine Pollution. Functional Responses,* Vernberg, W. B., Thurberg, F. P., Calabrese, A., and Vernberg, F. J., Eds., Academic Press, New York, 1979, 23.

161. **Dixon, D. R., Jones, I. M., and Harrison, F. L.,** Cytogenetic evidence of inducible processes linked with metabolism of a xenobiotic chemical in adult and larval *Mytilus edulis, Sci. Total Environ.,* 46, 1, 1985.

162. **Hendricks, J. D., Putnam, T. P., Bills, D. D., and Sinnhuber, R. O.,** Inhibitory effect of polychlorinated biphenyl (Aroclor 1254®) on aflatoxin B_1 carcinogenesis in rainbow trout *(Salmo gairdneri), J. Natl. Cancer Inst.,* 59, 1545, 1977.

163. **Nixon, J. E., Hendricks, J. D., Pawlowski, N. E., Pereira, C. B., Sinnhuber, R. O., and Bailey, G. S.,** Inhibition of aflatoxin B_1 carcinogenesis in rainbow trout by flavone and indole compounds, *Carcinogenesis,* 5, 615, 1984.

164. **Williams, D. E. and Buhler, D. R.,** Purified form of cytochrome P-450 from rainbow trout with high activity toward conversion of aflatoxin B_1 to aflatoxin B_1-2,3-epoxide, *Cancer Res.,* 43, 4752, 1983.

165. **Adesnik, M. and Atchison, M.,** Genes for cytochrome P-450 and their regulation, *CRC Crit. Rev. Biochem.,* 19, 247, 1986.

166. **Binder, R. L. and Stegeman, J. J.,** Induction of aryl hydrocarbon hydroxylase activity in embryos of an estuarine fish, *Biochem. Pharmacol.,* 29, 949, 1980.

167. **Binder, R. L. and Stegeman, J. J.,** Microsomal electron transport and xenobiotic monooxygenase activities during the embryonic period of development in the killifish, *Fundulus heteroclitus, Toxicol. Appl. Pharmacol.,* 73, 432, 1984.

168. **Binder, R. L. and Stegeman, J. J.,** Basal levels and induction of hepatic aryl hydrocarbon hydroxylase activity during embryonic period of development in brook trout, *Biochem. Pharmacol.,* 32, 1324, 1983.

169. **Binder, R. L. and Lech, J. J.,** Xenobiotics in gametes of Lake Michigan lake trout *(Salvelinus namaycush)* induce hepatic monooxygenase activity in their offspring, *Fundam. Appl. Toxicol.,* 4, 1042, 1984.

170. **Williams, D. E., Stegeman, J. J., Lech, J. J., Melancon, M., Goksoyr, A., Schoor, P., and Buhler, D. R.,** unpublished observations.

171. **Kato, R.,** Sex-related differences in drug metabolism, *Drug Metab. Rev.,* 3, 1, 1974.

172. **Levin, W., Ryan, D., Kuntzman, R., and Conney, A. H.,** Neonatal imprinting and the turnover of microsomal cytochrome P-450 in rat liver, *Mol. Pharmacol.,* 11, 190, 1975.

173. **Mode, A., Gustafsson, J. A., Jannson, J. O., Eden, S., and Isaksson, O.,** Association between plasma level of growth hormone and sex differentiation of hepatic steroid metabolism of the rat, *Endocrinology,* 111, 169, 1982.

174. **Waxman, D. J.,** Rat hepatic cytochrome P-450 isoenzyme 2c. Identification as a male-specific developmentally induced steroid 16α-hydroxylase and comparison to a female-specific cytochrome P-450 isoenzyme, *J. Biol. Chem.,* 259, 15481, 1984.

175. **Harada, N. and Negishi, M.,** Mouse liver testosterone 15α-hydroxylase (cytochrome P-450$_{15\alpha}$), purification, regioselectivity, stereospecificity and sex-dependent expression, *J. Biol. Chem.,* 259, 1265, 1984.

176. **Noshiro, M., Serabjit-Singh, C. J., Bend, J. R., and Negishi, M.,** Female-predominant expression of testosterone 16α-hydroxylase (''I''-P-450 16α) and its repression in strain 129/J, *Arch. Biochem. Biophys.,* 244, 857, 1986.

177. **Hansson, T. and Gustafsson, J.-A.,** Sex differences in the hepatic *in vitro* metabolism of 4-androstene-3,17-dione in rainbow trout, *Salmo gairdneri, Gen. Comp. Endocrinol.,* 44, 181, 1981.

178. **Stegeman, J. J. and Chevion, M.,** Sex differences in cytochrome P-450 and mixed-function oxygenase activity in gonadally mature trout, *Biochem. Pharmacol.,* 29, 553, 1980.

179. **Forlin, L. and Hansson, T.,** Effects of oestradiol-17β and hypophysectomy on hepatic mixed function oxidases in rainbow trout, *J. Endocrinol.,* 95, 245, 1982.

180. **Vodicnik, M. J. and Lech, J. J.,** The effect of sex steroids and pregnenolone-16α-carbonitrile on the hepatic microsomal monooxygenase system of rainbow trout *(Salmo gairdneri), J. Steroid Biochem.,* 18, 323, 1983.

181. **Koivusaari, U., Harri, M., and Hanninen, O.,** Seasonal variation of hepatic biotransformation in female and male rainbow trout *(Salmo gairdneri), Comp. Biochem. Physiol.,* 70C, 149, 1981.

182. **Hanninen, O., Koivusaari, U., and Lindstrom-Seppa, P.,** Seasonal variation of cytochrome P-450 and monooxygenase activities in several freshwater fish, in *Cytochrome P-450. Biochemistry, Biophysics and Environmental Implications,* Hietanen, E., Laitinen, M., and Hanninen, O., Eds., Elsevier, New York, 1982, 209.

183. **Walton, D. G., Fancey, L. L., Green, J. M., Kiceniuk, J. W., and Penrose, W. R.,** Seasonal changes in aryl hydrocarbon hydroxylase activity of a marine fish *Tautogo labrus adspersus* (Walbaum) with and without petroleum exposure, *Comp. Biochem. Physiol.,* 76C, 247, 1983.

184. **Hazel, J. R.,** Influence of thermal acclimation on membrane lipid composition of rainbow trout liver, *Am. J. Physiol.,* 236, R91, 1979.

185. **Burns, K. A.,** Microsomal mixed function oxidases in an estuarine fish, *Fundulus heteroclitus,* and their induction as a result of environmental contamination, *Comp. Biochem. Physiol.,* 53B, 443, 1976.

186. **Chambers, J. E.,** Induction of microsomal mixed-function oxidase system components in striped mullet by short-term exposure to crude oil, *Toxicol. Lett.,* 4, 227, 1979.

187. **Spies, R. B., Felton, J. S., and Dillard, L.,** Hepatic mixed-function oxidases in California flatfishes are increased in contaminated environments and by oil and PCB ingestion, *Mar. Environ. Res.,* 14, 412, 1984.

188. **Yarbrough, J. D. and Chambers, J. E.,** Crude oil effects on microsomal mixed-function oxidase system components in the striped mullet *(Mugil cephalus), Life Sci.,* 21, 1095, 1977.

189. **Petersen, D. W. and Lech, J. J.,** Hepatic effects of acrylamide in rainbow trout, *Toxicol. Appl. Pharm.,* 89, 249, 1987.

190. **Bend, J. R., Ball, L. M., Elmamlouk, T. H., James, M. O., and Philpot, R. M.,** Microsomal mixed-function oxidation in untreated and polycyclic aromatic hydrocarbon-treated fish, in *Pesticide and Xenobiotic Metabolism in Aquatic Organisms,* Khan, M. A. Q., Lech, J. J., and Menn, J. J., Eds., American Chemical Society, Washington, D.C., 1979, 297.

191. **Goksoyr, A., Klungsoyr, J., Solberg, T., and Solbakken, J. E.,** Cytochromes P-450 in cod *(Gadus morhua):* effects of inducers on regiospecificity of phenanthrene metabolism and immunochemical properties, *Mar. Environ. Res.,* 17, 87, 1985.

192. **French, J. S. and Kennish, J. M.,** Cytochromes P-450 in liver microsomes of five Pacific salmon species, *Mar. Environ. Res.,* 17, 149, 1985.

193. **Miranda, C. L., Wang, J.-L., Henderson, M. C., and Buhler, D. R.,** Purification and characterization of hepatic steroid hydroxylases cytochrome P-450 isozymes from untreated rainbow trout, *Toxicologist, Arch. Biochem. Biophys.,* in press.

194. **Melancon, M. J., Williams, D. E., Buhler, D. R., and Lech, J. J.,** Metabolism of 2-methylnaphthalene by rat and rainbow trout hepatic microsomes and purified cytochrome P-450, *Drug Metab. Dispos.,* 13, 542, 1985.

195. **Conner, J. W. and Singer, S. C.,** Purification scheme for cytochrome P-450 of blue crab *(Callinectes sapidus* Rathbun), *Aquat. Toxicol.,* 1, 271, 1981.

196. **James, M. O., Sherman, B., Fisher, S. A., and Bend, J. R.,** Benzo(a)pyrene metabolism in reconstituted monooxygenase systems containing cytochrome P-450 from lobster *(Homarus americanus)* hepatopancreas fractions and NADPH cytochrome P-450 reductase from pig liver, *Bull. Mt. Desert Isl. Biol. Lab.,* 22, 37, 1982.

197. **Lee, R. F., Singer, S. C., and Page, D. S.,** Response of cytochrome P-450 systems in marine crab and polychaetes to organic pollutants, *Aquat. Toxicol.,* 1, 355, 1981.

198. **Bihari, N., Batel, R., Kurelec, B., and Zahn, R. K.,** Tissue distribution, seasonal variation and induction of benzo[a]pyrene monooxygenase activity in the crab *Maja crispata, Sci. Total Environ.,* 35, 41, 1984.

199. **James, M. O. and Little, P. J.,** 3-Methylcholanthrene does not induce *in vitro* xenobiotic metabolism in spiny lobster hepatopancreas, or affect *in vivo* disposition of benzo[a]pyrene, *Comp. Biochem. Physiol.,* 78C, 241, 1984.

200. **Burnette, W. N.,** "Western blotting": electrophoretic transfer of proteins from sodium dodecylsulfate-polyacrylamide gels to unmodified nitrocellulose and radiographic detection with antibody and radioiodinated protein A, *Anal. Biochem.,* 112, 195, 1981.

201. **Williams, D. E., Bender, R. C., Morrissey, M. T., Selivonchick, D. P., and Buhler, D. R.,** Cytochrome P-450 isozymes in salmonids determined with antibodies to purified forms of P-450 from rainbow trout, *Mar. Environ. Res.,* 14, 13, 1984.

202. **Varanasi, U., Collier, T. K., Williams, D. E., and Buhler, D. R.,** Hepatic cytochrome P-450 isozymes and aryl hydrocarbon hydroxylase in English sole *(Parophrys vetulus), Biochem. Pharmacol.,* 35, 2967, 1986.

203. **Park, S. S., Miller, H., Klotz, A. V., Koepper-Sams, P. J., Stegeman, J. J., and Gelboin, H. V.,** Monoclonal antibodies to liver microsomal cytochrome P-450E of the marine fish *Stenotomus chrysops* (scup): cross-reactivity with 3-methylcholanthrene induced rat cytochrome P-450, *Arch. Biochem. Biophys.,* 249, 339, 1986.

204. **Stegeman, J. J., Kloepper-Sams, P. J., and Farrington, J. W.,** Monooxygenase induction and chlorobiphenyls in the deep-sea fish *Coryphaenoides armatus, Science,* 231, 1287, 1986.

205. **Goksoyr, A., Andersson, T., Buhler, D. R., Stegeman, J. J., Williams, D. E., and Forlin, L.,** An immunological comparison of β-naphthoflavone-inducible microsomal cytochrome P-450 (P-450I) in different fish species and rat, *Arch. Biochem. Biophys.,* in press.

206. **Masters, B. S. S. and Okita, R. T.,** The history, properties and function of NADPH-cytochrome P-450 reductase, *Pharmacol. Ther.,* 9, 227, 1980.

207. **Yasukochi, Y., and Masters, B. S. S.,** Some properties of a detergent-solubilized NADPH-cytochrome *c* (cytochrome P-450) reductase purified by biospecific affinity chromatography, *J. Biol. Chem.,* 251, 5337, 1976.

208. **Pohl, R. J., Serabjit-Singh, C. J., Slaughter, S. R., Albro, P. W., Fouts, J. R., and Philpot, R. M.,** Hepatic microsomal NADPH-cytochrome P-450 reductase from little skate, *Raja erinacea,* comparison of thermolability and other molecular properties with a mammalian enzyme, *Chem. Biol. Interact.,* 45, 283, 1983.

209. **Williams, D. E., Becker, R. R., Potter, D. W., Guengerich, F. P., and Buhler, D. R.,** Purification and comparative properties of NADPH-cytochrome P-450 reductase from rat and rainbow trout: differences in temperature optima between reconstituted and microsomal trout enzymes, *Arch. Biochem. Biophys.,* 225, 55, 1983.

210. **Gurumurthy, P. and Mannering, G. J.,** Membrane bound cytochrome P-450 determines the optimal temperatures of NADPH-cytochrome P-450 reductase and cytochrome P-450-linked monooxygenase reactions in rat and trout hepatic microsomes, *Biochem. Biophys. Res. Commun.,* 127, 571, 1985.

211. **Dix, T. A. and Marnett, L. J.,** Metabolism of polycyclic aromatic hydrocarbon derivatives to ultimate carcinogens during lipid peroxidation, *Science,* 221, 77, 1983.

212. **McNeil, J. M., Gower, J. D., and Wills, E. D.,** The formation of the ultimate carcinogen of benzo(a)pyrene during non-enzyme lipid peroxidation, *Biochem. Pharmacol.,* 34, 4066, 1985.

213. **Eling, T., Boyd, J., Reed, G., Mason, R., and Sivarajah, K.,** Xenobiotic metabolism by prostaglandin endoperoxide synthetase, *Drug Metab. Rev.,* 14, 1023, 1983.

214. **Bandhyopadhyay, G. K., Dutta, J., and Ghosh, S.,** Synthesis of diene prostaglandins in freshwater fish, *Lipids,* 17, 755, 1982.

215. **Srivastava, K. C. and Mustafa, T.,** Arachidonic acid metabolism and prostaglandins in lower animals, *Mol. Physiol.,* 5, 53, 1984.

216. **Anderson, A. A., Fletcher, T. C., and Smith, G. M.,** Prostaglandin biosynthesis in the skin of the plaice, *Pleuronectes platessa, Comp. Biochem. Physiol.,* 70C, 195, 1981.

217. **Mai, J., Goswami, S. K., Bruckner, G., and Kinsella, J. E.,** A new prostaglandin, $C22-PGF_{4\alpha}$, synthesized from docosahexaneoic acid (C_{22}:6n3) by trout gill, *Prostaglandins,* 21, 691, 1981.

218. **Herman, C. A., Zimmerman, P. R., and Doolittle, K.,** Prostaglandin synthesis in goldfish heart, *Carassius auratus, Gen. Comp. Endocrinol.,* 54, 478, 1984.

219. **Williams, D. E.,** unpublished observations, 1986.

220. **Cetta, F. and Goetz, F. W. M.,** Ovarian and plasma prostaglandin E and F levels in brook trout *Salvelinus fontinalis* during pituitary-induced ovulation, *Biol. Reprod.,* 27, 1216, 1982.

221. **Ziegler, D. M.,** Microsomal flavin-containing monooxygenase: oxygenation of nucleophilic nitrogen and sulfur compounds, in *Enzymatic Basis of Detox. tion,* Vol. 1, Jakoby, W. B., Ed., Academic Press, New York, 1980, 201.

222. **Tynes, R. E. and Hodgson, E.,** The measurement of FAD-containing monooxygenase activity in microsomes containing cytochrome P-450, *Xenobiotica,* 14, 515, 1984.

223. **Baker, J. R., Struempler, A., and Chaykin, S.,** A comparative study of trimethylamine-N-oxide biosynthesis, *Biochim. Biophys. Acta,* 71, 58, 1963.

224. **Agustsson, I. and Strom, A. R.,** Biosynthesis and turnover of trimethylamine oxide in the teleost cod, *Gadus morhua, J. Biol. Chem.,* 256, 8045, 1981.

225. **Schlenk, D. and Buhler, D. R.,** Cytochrome P-450 and phase II activities in the gumboot chiton *(Cryptochiton stelleri), Aquatic Toxicol.,* 13, 167, 1988.

226. **Schlenk, D. and Buhler, D. R.,** Flavin-containing monooxygenase activity in the gumboot chiton *(Cryptochiton stelleri)., Marine Biol.,* submitted.

227. **Heilmann, L. J., Sheen, Y. -Y., Bigelow, S. W. and Nebert, D. W.,** Trout P450IA1:cDNA and deduced protein sequence, expression in liver, and evolutionary significance. *DNA* 7, 379, 1988.

228. **Lorenzana, R. M., Hedstrom, O. R., and Buhler, D. R.,** Localization of cytochrome P-450 in the head and trunk kidney of rainbow trout *(Salmo gairdneri), Tox. Appl. Pharm.,* in press.

Chapter 5, Part II

ENZYMES INVOLVED IN METABOLISM OF PAH BY FISHES AND OTHER AQUATIC ANIMALS: HYDROLYSIS AND CONJUGATION ENZYMES (OR PHASE II ENZYMES)

Gary L. Foureman

TABLE OF CONTENTS

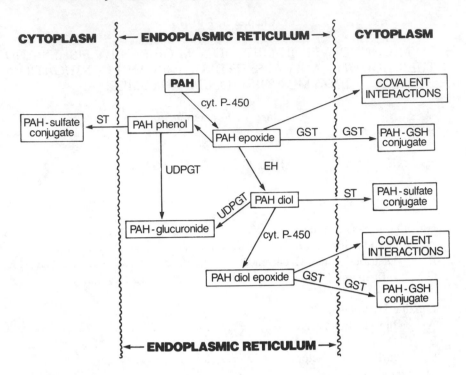

A diagrammatic representation of the possible metabolic fates of polycyclic aromatic hydrocarbons (PAH) and their metabolites in a cell. The cytochrome P-450 system (cyt. P-450) is discussed in the prior chapter. Glutathione *S*-transferase (GST), epoxide hydrolase (EH), UDP-glucuronosyl-transferase (UDPGT), and sulfotransferase (ST) will be discussed in this chapter.

I. PREFACE

Subsequent to the oxidative processes described in Part I of this chapter, some polycyclic aromatic hydrocarbons (PAH) may undergo further metabolism by enzymes which are the subject of Part II of the chapter. The enzymes involved in this secondary metabolism include the glutathione *S*-transferases (GST), UDP-glucuronosyltransferases, and sulfotransferases. According to the classical definition of Williams,[1] these are phase II enzymes which function to increase the polarity of the lipophilic xenobiotic (phase I metabolites of PAH) by conjugating it with the endogenous compounds indicated in the names of the enzymes; this increase in polarity makes phase II metabolites more readily excreted by the organism. Although the other enzyme discussed in this chapter, epoxide hydrolase, is usually not considered to be a phase II enzyme, it does fit into William's broad definition (water being quite endogenous), and it is therefore included.

Most studies of PAH metabolism in aquatic species have concentrated on the principal constituent of oxidative (phase I) metabolism, the cytochrome P-450 monooxygenase system. This was largely due to the realization that cytotoxicity, mutagenicity, and carcinogenicity of many PAH were dependent upon metabolic alteration of parent PAH by components within tissues,[2-4] and that, in mammalian species, the cytochrome P-450-dependent oxidation of PAH to PAH oxides was an important toxication reaction.[5] In turn, this realization intensified research on biological pathways capable of metabolizing PAH oxides, such as those mediated by the GST and epoxide hydrolase. Studies on PAH metabolism were redoubled by the experimental evidence[6] and proposal[7,8] that, with some PAH, the cytochrome P-450 system acted in concert with epoxide hydrolase to produce an ultimate carcinogenic species, a diol epoxide. This condensed discussion of PAH metabolism makes evident the importance of investigating the *complete* metabolism, including secondary and tertiary biotransformation of primary metabolites, of any compound implicated as being toxic to a biological system.

There are several recent excellent reviews available on the enzymes involved in conjugation and hydrolysis of PAH metabolites[9-11] and on xenobiotic metabolism in aquatic species.[10-15] In combining these two concerns, this chapter emphasizes recent studies of these enzymes in aquatic species, wherein metabolites of PAH or closely related substrates were employed to show how the ability of an organism to metabolize PAH can be evaluated from a detailed consideration of activation and detoxication enzymes. The last section of the chapter points out a few areas of prospective investigation into the phase II enzymes of aquatic species. This review covers the literature from 1976 up to the summer of 1987.

II. GLUTATHIONE *S*-TRANSFERASES

A. Introduction

The nucleophilic thiol group of the tripeptide glutathione (GSH) represents a source for neutralization of the potent electrophilic species generated from PAH by phase I enzymes. A remarkably wide range of living systems maintain GSH at high levels and provide a group of enzymes, the GST, to catalyze the formation of GSH-xenobiotic conjugates. These enzymes have been studied extensively due to their capability to directly catalyze the metabolism of the activated intermediate of PAH, the epoxide.

GST activity is present in phylogenetically diverse species as has been demonstrated by Stenersen et al.[16] In this exhaustive study, workers found GST activity (towards 1-chloro-2,4-dinitrobenzene; CDNB) in 71 of 72 species representing 9 different phyla. GST is present in prodigious quantities; in the aforementioned study, GST was estimated to represent from 0.3 to 0.7% of the soluble protein present in the homogenates from these various species.[16] Although GST activity is primarily localized in the cytosol, both microsomal[17] and mitochondrial[18] activities have been distinguished. Studied primarily in the liver, GST activity has been found in virtually all mammalian organs. GST activity has been purified from a variety of sources and is shown to consist of several distinct proteins, all of which are apparently dimeric permutations of up to six subunits,[16,19,20] having molecular weights from 21,900 to 29,000. Through the elegant and ingenious work of Armstrong and co-workers,[21-24] studies on the purified enzymes have been extended into the mechanism(s) of catalysis, substrate stereoselectivity towards PAH oxide substrates, and the topology of the hydrophobic- and GSH binding sites.

Conjugation of xenobiotics with GSH is the initial step in formation of the corresponding mercapturic acids, i.e., the *S*-conjugates of *N*-acetylcysteine. Found in the urine, mercapturic acids result from the successive cleavage and covalent alteration of the initial GSH-conjugate in a process involving three other enzymes besides GST (see Tate[25] for a review on mercapturic acid formation).

B. Distribution and Occurrence

The hepatic GST activities of rat and several aquatic species toward the PAH substrate (±)benzo(a)pyrene 4,5-oxide (BaPO) are listed in Table 1. Although not yet proven to be as general a substrate for GST as CDNB, radiolabeled BaPO is an excellent substrate to use as it is quite stable and gives a very low background.[26] These data demonstrate the remarkable phylogenetic ubiquity of this enzyme activity with a PAH substrate.

The presence of GST activity in extrahepatic tissues of an aquatic species was initially reported by Bend et al.,[27] who detected activity (CDNB) in kidney, testes, gill, spiral valve mucosa, spleen, heart, and pancreas of the little skate, *Raja erinacea*. Later, Bauermeister et al.[29] performed a similar study in rainbow trout, *Salmo gairdneri*. Tate and Herf[30] found

Table 1
IN VITRO GST ACTIVITY IN THE LIVER OR HEPATOPANCREAS OF THE RAT AND SEVERAL AQUATIC SPECIES WITH (±) BENZO(A)PYRENE 4,5-OXIDE AS SUBSTRATE

Species		GST activity (nmol/min/mg protein)
Little skate	*Raja erinacea*	131
Sheepshead	*Archosargus probatocephalus*	53
Black drum	*Pogonius cromi*	9
Dogfish shark	*Squalus acanthas*	8
Winter flounder	*Pseudopleuronectes americanus*	5
Eel	*Anguilla rostrata*	3
Lobster	*Homarus americanus*	3
Mangrove snapper	*Lutjanus griseus*	2
Atlantic Stingray	*Dasyatis sabina*	2
Spiny lobster	*Panulirus argus*	<1
Mussel	*Mytilus edulis*	<1
Clam	*Mya arenaria*	<1
Rock crab	*Cancer irroratus*	<1
Rat		24

Data arranged from Bend et al.[27] and Foureman and Bend.[28]

easily measured levels of GST (CDNB) in the gills and excretory gland of the blue crab, *Callinectes sapidus*. In demonstrating GST activity in the clodoceran *Daphnia magna*, LeBlanc and Cochrane[31] segregated activity towards CDNB and ethacrynic acid into two ammonium sulfate fractions, thereby indicating the presence of isozymes of GST. Also, GST activity towards non-PAH substrates has been detected in the green gland, gills, and intestines of the crayfish, *Astacus astacus*.[32] Nimmo and co-workers[33,34] have also made extensive studies of extrahepatic GST activities in rainbow trout and have succeeded in partially purifying a kidney GST.

Hepatic GST have been isolated and purified to varying degrees from little skate,[28] thornyback shark, *Platyrhinoides triserata*,[35] salmon, *Salmo salar*,[36] and rainbow trout.[37-41] These studies, especially those of Nimmo and co-workers,[39-41] have elucidated parallels between the mammalian and fish GST; e.g., both exist as a group of isoenzymes and both can efficiently bind organic anions. In the case of the little skate, there is immunologic and electrophoretic evidence that the various GST, like rat GST, are dimeric, comprised of permutations of two to four subunits, all of which have molecular weights in the range of 21,000 to 26,000 Da.[28] Foureman et al.[42] examined the activity of the principal little skate hepatic cytosolic GST, E4, with several PAH arene oxides and a diol epoxide of BaP (Table 2). The rate of the E4 catalyzed reaction between GSH and BaPO is nearly 150-fold that reported for rat GST "C", the most reactive hepatic transferase with this substrate.[43] The rates observed with the other K-region oxides listed in Table 2 were also high, demonstrating the remarkable capacity of these enzymes to deal with phase I metabolites of PAH. The finding that E4 exhibits reactivity with the 7,8-dihydrodiol-9,10-epoxide of BaP (DIEP) is significant, as diol epoxides are the ultimate carcinogenic species of PAH possessing "bay" regions.[44] As diol epoxides are apparently poor substrates for epoxide hydrolase,[45] the other oxide-metabolizing enzyme present in little skate, and as fish P-450s generate relatively large amounts of bay or non-K region oxides (see previous portion of this chapter), then GST in this species may be of importance in the detoxication of these ultimate carcinogenic compounds.

In further studies, Foureman et al.[42] examined the stereoselectivity of little skate E4 towards

Table 2
THE SPECIFIC ACTIVITY (SA) AND STEREOSELECTIVITY (R:S) OF MALE LITTLE SKATE HEPATIC GST E4 TOWARDS SEVERAL POLYCYCLIC ARENE OXIDES

Name	SA[a]	R:S[b]
(±)-Benzo(a)arthracene 5,6-oxide (BaAO)	47	>99.7: <0.3
(±)-Benzo(a)pyrene 4,5-oxide (BaPO)	14	>99.4: <0.6
Pyrene 4,5-oxide (PO)	13	93.0: 7.0
Phenanthrene 9,10-oxide (Pho)	2	<98.0: <2.0
(±)-7β,8α-Dihydroxy-9α,10α-epoxy-7,8,9,10-tetrahydrobenzo(a)pyrene (DIEP)	0.2	N.D.

Note: N.D., not determined.

[a] μmol GSH conjugated with epoxide/min/mg E4.

[b] Percent of conjugates formed from parent PAH-oxide in which the R- or S-configured oxirane carbon reacted with GSH. All values listed for S, save that of PO, represent the minimum detectable level of the HPLC assay.

several K-region oxides (Table 2). Using high-pressure liquid chromatographic techniques, authentic standards, and radiolabeled BaPO, they demonstrated that E4 catalyzed conjugate formation only at the R-configured carbon of either (+) or (−)BaPO or benzo(a)anthracene 5,6-oxide (BaAO). Of the other two K-region oxides in Table 2, *S*-configured conjugate was detected only with phenanthrene 9,10-oxide (PhO). Armstrong and co-workers have made strikingly similar observations in hepatic GST from rat.[22] To exhibit such stereospecificity, the active site of this elasmobranch enzyme has to be highly ordered, imposing restraints on this xenobiotic substrate (i.e., PAH epoxide) as stringent as those imposed upon endogenous substrates by physiologic enzymes.

C. Influencing Factors

Although sex differences appear to exist in fish cytochrome P-450-dependent monooxygenase activities, there have been no reports of sex differences in GST activities in any aquatic species. As pointed out by Koivusaari et al.,[46] however, sex differences in fish species may exist only at times during and contiguous to spawning. Moreover, there have been few reports comparing activity levels in fetal or neonatal and adult specimens in any aquatic species, although marked differences have been demonstrated between neonate and adult specific activities of GST in hepatic and extrahepatic tissues of rats.[47]

Inducibility of GST activity by PAH and PAH-type compounds in mammals is well characterized, with studies extending our knowledge of inducer effects to the subunit level of the enzyme.[48] Reports of induction of GST activity by PAH-type compounds indicate that some aquatic species are refractory to induction (at least at the dosages studied), while others are not. For example, James and Bend[49] reported that hepatic GST activity in sheepshead, southern flounder (*Paralicthyes lethostigma*), stingray, little skate, and dogfish shark were not significantly increased by 1,2,3,4-dibenzanthracene and Balk[50] observed no effect of 3-methycholanthrene (3-MC) on hepatic GST activity in Northern pike (*Esox lucius*). Aroclor 1254® administered i.p. increased GST activity a mere 20% in catfish (*Ictalurus punctatus*).[51] Recently, though, Andersson and co-workers[52] reported moderate induction (twofold over controls) of hepatic GST activity by β-naphthoflavone in rainbow trout. Preliminary results of Collier and Varanasi[53] have shown that hepatic GST activity of English sole (*Parophrys vetulus*) from a polluted area in Puget Sound, Washington is higher than that of fish taken from a reference area. Subsequent attempts to experimentally induce GST activity in sole from reference areas using BaP, *trans*-stilbene oxide, and an extract of polluted sediment were unsuccessful. GST activity is usually very high in animals in which initiation of carcinogenesis has occurred;[54] the higher GST activity in English sole sampled from polluted areas may be due to this fact rather than simple induction by chemicals. Accordingly, these results demonstrate the need for further work in this area.

A consequence of GSH conjugation to PAH metabolites may be temporary depletion of GSH, a condition which may comprise the capability of the organism to neutralize electrophilic species such as PAH oxides.[55] In rats, for example, hepatic GSH content was decreased to 42% of control after phenanthrene administration.[56] In aquatic species, the toxicological implications of this decrease could be exacerbated by two factors. First, the levels of hepatic GSH have been measured in several species, includinig sea bass, *Centropristis striata,*[57] thorny-back shark,[35] little skate,[58] large skate,[58] and thorny skate,[58] and have been found to be less than half of the concentrations found in rat liver. Second, the K_m values of GSH reported for fish GST are an order of magnitude greater than those of rat GST.[28,35,38,58] Thus, minor decreases in GSH content could possibly impair the conjugating capability and, thus, the capacity of some aquatic species to cope with an acute exposure to PAH as could occur with an oil spill or in an area with a high index of PAH pollution. An initial study in this area, however, indicates that the biochemical systems in some aquatic species are able to avert this consequence. After i.p. administration of an acute dose of 3-MC (10 mg/kg) to plaice, *Pleuronectes platessa*), hepatic levels of GSH were unchanged over a 14-d period.[59]

The synthesis and regulation of GSH in aquatic species require further study, although some results have recently been published.[57,60] Thomas and Wofford[60] exposed mullet, *Mugil cephalus*, to a water-soluble fraction of fuel oil for 10 d, after which the hepatic content of GSH and the incorporation of radiolabeled glycine into GSH were measured. Rather than decreasing, the hepatic GSH levels and the incorporation of radiolabel into the GSH pool were both nearly doubled. The authors suggested that oil and other chemicals may elevate GSH content by increasing the hepatic uptake of the amino acids required in GSH synthesis, although this was not tested.

III. EPOXIDE HYDROLASE

A. Introduction

Epoxides (alkene and arene oxides) generated from phase I metabolism can be conjugated with GSH by GST or hydrated to vicinal *trans*-dihydrodiols by epoxide hydrolase (E.C. 3.3.2.3). In rat and mouse, epoxide hydrolase (EH) activity is found in both microsomal and cytosolic fractions.[61] The microsomal enzyme is especially active with certain oxides of PAH,[26,62] although much EH characterization has been done with styrene 7,8-oxide. The activity of the murine cytosolic EH with PAH oxides is reported to be lower by several orders of magnitude.[63] The dual role of epoxide hydrolase in the metabolism of PAH oxides must be emphasized. In most cases dihydrodiols formed from PAH oxides are not toxic and are even substrates for subsequent conjugative reactions;[64] thus, hydration of these oxides represents metabolic detoxication. Bay region dihydrodiols of PAH[44] such as 7,8-dihydro-7,8-dihydroxybenzo(a)pyrene (7,8-BaP-diol) may recycle through the cytochrome P-450 system producing an ultimate carcinogenic species of PAH, a diol epoxide. Diol epoxides of PAH are apparently not substrates for microsomal EH.[45] The involvement of dihydrodiols in the formation of diol epoxides is reviewed by Sims and Grover.[65]

B. Distribution and Occurrence

The hepatic microsomal EH activities of several aquatic species toward the PAH oxide, BaPO, are listed in Table 3. With BaPO as substrate, EH activity has also been detected in several extrahepatic organs of the lobster, including green gland, egg masses, and gill.[27] To the author's knowledge, EH activity has been detected in every tissue assayed. It therefore appears that in aquatic species, as in mammalian species,[67] microsomal EH activity is

Table 3
IN VITRO MICROSOMAL EH ACTIVITY IN THE LIVER OR
HEPATOPANCREAS OF RAT AND SEVERAL AQUATIC SPECIES
WITH THE PAH SUBSTRATE (±) BENZO(A)PYRENE 4,5-OXIDE

Species		EH activity (nmol/min/mg protein)
Spiny lobster	*Panulirus argus*	19
Lobster	*Homarus americanus*	6
Blue crab	*Cancer sapidus*	4
Winter flounder	*Pseudopleruonectes americanus*	4
Dogfish shark	Squalus acanthas	3
Sheepshead	*Archosargus probatocephalus*	3
Eel	*Anguilla rostrata*	2
Rock crab	*Cancer irroratus*	2
Clam	*Mya arenaria*	2
Atlantic stingray	*Dasyatis sabina*	<1
Little skate	*Raja erinacea*	<1

Data from James et al.[66] and Jerina et al.[26]

ubiquitous; such ubiquity could indicate an endogenous substrate or function for EH, although none has so far been confirmed.

Dihydrodiols were detected as major metabolites of the PAH, BaP in hepatic microsomal preparations from several species. Varanasi and co-workers[68,69] and Ahokas et al.[70] reported that in use of post-mitochondrial fractions from livers of coho salmon, *Oncorhynchus kisutch* starry flounder, *Platichthys stellatus* English sole, *Parophrys vetulus* and rainbow trout, the major BaP diols isolated from the incubation mixtures were the bay region 7,8-BaP-diol, a precursor of the diol epoxides of BaP and 9,10-dihydro-9,10-dihydroxy BaP.[68,69] The presence of isomeric dihydrodiols of BaP in these studies and of 2-methylnaphthalenes in the study of Breger et al.[71] indicates that the microsomal EH in fish species, as in rat,[26] exhibits little regioselectivity in the metabolism of the oxides generated from the cytochrome P-450 system. The author is not aware of any study on cytosolic EH in aquatic species, although Balk and co-workers reported that one third of the total hepatic EH activity in the liver of the Northern pike was present in the microsomal supernatant cystolic fraction.[50]

C. Influencing Factors

Only moderately induced (twofold) in rodent species by some PAH and polychlorinated biphenyls (PCB), EH activity is not appreciably induced in either mammals or any aquatic species by 3-MC-type compounds which so effectively induce the activities of cytochromes P-450 involved in PAH metabolism (see previous portion of chapter).[49-52,72] In light of the dual role of EH activity in PAH metabolism, the consequences of this situation would be equivocal; upon induction the rate of formation of dihydrodiols of PAH *in vivo* would be increased, while the rate of formation of diol epoxides may or may not be increased dependent on the kinetics of the enzymes and their localization. There have been no successful attempts to purify EH in aquatic species and, thus, no meaningful kinetic data have been obtained. Differential induction of EH and cytochrome P-450 activity has been partially achieved in rats using *trans*-stilbene oxide,[73] although preliminary results showed that this compound does not induce EH activity in liver of English sole.[53] It has been established that the cytosolic EH is not induced by any of the classical inducing compounds.[61] Other known modifiers of EH activity in mammals that have not been examined in aquatic species include age and sex-linked differences. Laitinen et al.[74] noted that hepatic EH activity was inhibited 30% in splake (*Salvelinus fontinalis* × *Salvelinus namaycush*) exposed four times to the anesthetic MS-222.

In a series of interesting studies,[75,76] Laitinen and co-workers examined the effects of water acidification on the microsomal EH levels in the liver and several extrahepatic tissues of the whitefish, *Coregonus peled*. After an 8-h exposure at pH 3, the hepatic microsomal EH activity of the fish was increased, whereas the cytochrome P-450-dependent aryl hydrocarbon hydroxylase (AHH) activity was unchanged; in kidney, EH activity was increased and AHH activity was decreased. These studies are among the first to consider the results of acidification (a consequence of "acid rain") on xenobiotic metabolism in aquatic species and demonstrate the need for further research in this area.

IV. UDP-GLUCURONOSYLTRANSFERASES

A. Introduction

Oxidative metabolites of PAH are potential substrates for conjugation to metabolically activated UDP-glycosides such as UDP-glucuronic acid to form a D-glucuronide. This reaction is catalyzed by the UDP-glucuronosyltransferase (UDPGT; E.C. 2.4.1.17), an enzyme that is thought to be at least partly imbedded in the endoplasmic reticulum *in situ*, and, in microsomal preparations, located on the lumenal side of microsomal vesicles. This intramembranal localization is apparently the reason for the latency of the activity of this enzyme; measured activities of UDPGT increase when surfactants which would alter the structure of the membrane are included in the incubation mixtures.[77] The oxidative metabolites of PAH would form *O*-glucuronides with UDPGT, the phenol or the dihydrodiol being the substrate (or aglycone) in the reaction. With naphthalene, for example, glucuronides can form at either the 1- or 2-position of 1,2-dihydro-1,2-dihydroxynaphthalene.[78]

Studies have shown UDPGT activity present in more than one form. In rat liver, evidence exists for at least two forms of UDPGT: G1, inducible by 3-methylcholanthrene, and G2, inducible by phenobarbital.[79-81] UDPGT(G1) preferentially glucuronidates flat (<4.5 Å) aromatic molecules such as PAH, whereas UDPGT(G2) glucuronidates more bulky molecules. Interestingly, recent work with rat liver has shown further heterogeneity of the UDPGT(G1) based on specificity toward PAH-type substrates, because the rate of glucuronidation of 4-phenyl-7-hydroxycoumarin, which is a bulky molecule and therefore a G2 substrate, was induced by 3-MC, which is a G1 inducer.[82]

B. Distribution and Occurrence

Lindstrom-Seppa et al.[83] measured hepatic UDPGT activity (4-nitrophenol) in vendace (*Coregonus albula*) and rainbow trout, showing that under certain conditions the level of this enzyme activity is comparable to that of rat liver (Table 4). In a subsequent study[84] these authors also demonstrated UDPGT activity to be present in extrahepatic tissues of several aquatic species; in two of these species, vendace and roach *(Rutilus rutilus)* the specific UDPGT activity in the gills was twice that found in liver.

UDPGT activity in fish microsomes is activated in the presence of surfactants, but the degree of latency appears to be less in fish than in mammals. Activity in trout liver[52] towards 1-naphthol was activated 3-fold by digitonin as compared to 12-fold in rat liver preparations.[85] Hanninen et al.[86] tested the ability of four different surfactants to activate UDPGT activity in trout liver and found a similar level of latency (about twofold) when they employed the neutral surfactant, Emulgen 911. Even this minor level of latency in UDPGT activity can obscure and confound interpretation of results, as was recently demonstrated by Andersson and co-workers.[52] Following β-naphthoflavone (βNF) treatment of trout, they observed distinct elevations in hepatic UDPGT activity towards 1-naphthol and testosterone only *after* activation with digitonin. To avoid such problems, these workers recommend that UDPGT activity be measured only in its fully activated state, as this is probably a more accurate assessment of the amount of enzyme present.[77]

Table 4
UDPGT ACTIVITY AT 18°C
TOWARDS 4-NITROPHENOL IN THE
HEPATIC MICROSOMAL FRACTION
OF VENDACE, RAINBOW TROUT,
AND RAT

Species	Activity (pmol/mg protein/min)
Vendace (males)	74.3
Rainbow trout	83.2
Rat (males)	58.0

Data from Lindstrom-Seppa et al.[83]

The possibility that isoenzymes of UDPGT are present in aquatic species has been addressed in two studies. Chowdhury and co-workers[87] found that in contrast to mammalian systems,[88] there was no difference in UDPGT activity towards bilirubin in the livers of adult or fetal spiny dogfish, *Squalus acanthas*; this result indicates that if isozymes of UDPGT exist in dogfish, they cannot be distinguished by their ability to conjugate bilirubin. Using rainbow trout, Andersson et al.[52] demonstrated a marked nonparallel pattern of hepatic UDPGT induction with three different substrates. After a single dose of βNF, UDPGT activity towards 1-naphthol and 4-nitrophenol increased for 1 to 2 weeks and remained at maximum levels until the end of the sixth week, whereas UDPGT activity towards testosterone reached peak values at 1 week, but returned to normal values at the end of the second week. The basis of this result could be the existence of one or more isoenzymes of UDPGT in trout liver.

Conjugating activity with UDP-glucose rather than UDP-glucuronic acid was observed in unactivated microsomal preparations from the hepatopancreas of the crayfish, *Astacus astacus* by Hanninen et al.[86] This activity in both fish and mammals appears to be partially latent, as activation of the activity occurs in the presence of anionic surfactants.

Forlin and Andersson[89] have shown UDPGT activity to be stable for up to 1 year in microsomal preparations of rainbow trout stored in 20% glycerol and at −80°C.

C. Influencing Factors

The induction of UDPGT activity achieved in fish with the 3-MC type of inducers is moderate in comparison to increases observed in cytochrome P-450 activities: 1- to 3-fold for UDPGT vs. 172-fold for 7-ethoxyresorufin deethylase activity.[52] In a manner parallel to the cytochrome P-450 system, induced UDPGT catalysis does not only lead to increased metabolism of the inducing agent, but also to increased metabolism of endogenous substrates such as steroids. One consequence of this increase could be reduced levels of steroids and subsequent impairment of reproductive physiology. In testing this possibility, Forlin and Haux[90] observed that biliary excretion of radiolabeled estradiol increased twofold in trout treated with βNF. Also, Sivarajah et al.[91] reported reduced serum levels of steroids concomitant with elevated hepatic UDPGT activities in trout and carp treated with Aroclor 1254®. Thus, PAH could stress aquatic species in an indirect manner by induction of an enzyme having a critical physiologic function.

The effect of environmental temperature on enzyme activities involved in xenobiotic metabolism is an area of special concern in aquatic species. Exposure of fish to PAH may occur year round, especially in the populated temperate regions of the planet. Koivusaari et al.[46] have shown the hepatic cytochrome P-450 system in rainbow trout to have perfect temperature compensation, while UDPGT appears to be more temperature resistant. In a recent study, Andersson et al.[92] examined the effect of temperature on βNF induction of UDPGT activity in livers of rainbow trout. βNF-induced UDPGT activity was measured in liver microsomal preparations of rainbow trout acclimated at 5 and 17°C. The induction was significant in the 5°C-acclimated fish, however, only when activity was measured at the acclimated temperature (5°C), not at the higher one (17°C). The causes for such results may be related to the conformation or composition of the membrane with which the enzyme is associated, although the author is not aware of any studies with aquatic species in this area.

Hanninen[86] observed differences in the latency of hepatic UDPGT activity in rainbow trout during spawning. In both sexes a twofold higher activity was observed in microsomal preparations from spawning as compared to nonspawning fish. Lindstrom-Seppa[93] also observed a similar spawning effect on hepatic UDPGT activity in both male and female vendace.

Hepatic UDPGT activity (unactivated) was inhibited up to 85% of normal values in rainbow trout exposed for 4 d to either trichlorophenol, pentachlorophenol, or dehydroabietic acid, chemicals known to occur in effluent from pulp and paper mills.[94] In another study where fourhorn sculpin, *Myoxocephalus quadricornis*, were exposed for up to 9 months to pulp mill waste water, no change in hepatic UDPGT activity was noted.[95] Definitive reasons for these differences, such as substances in the pulp mill water that could counter an inhibitory effect or low levels of the halophenols present in the pulp mill water, require follow-up studies. It is interesting to note that neither of these studies considered interactions between UDPGT and sulfotransferase (ST) activity as halophenols, are known to inhibit ST activity (see below).

The pH of the water is also capable of affecting UDPGT activity as has been demonstrated by Laitinen et al.[75] After an 8-h exposure to acidic water (pH 3), the hepatic UDPGT activity towards 4-nitrophenol of splake (*Salvelimus fontinalis*) and laveret (*Coregonus lavaretus*) was significantly increased, while the opposite was true for cytochrome P-450-dependent AHH activity.

Starvation, which influences UDPGT activity in mammals, was reported to have no effect on UDP-glucosyltransferase activity in crayfish[86] or UDPGT activity in rainbow trout.[96] Hormones, also known to affect UDPGT activity in mammals,[88] have yet to be evaluated for effects on this enzyme activity in aquatic species.

V. SULFOTRANSFERASES

A. Introduction

The phenolic or hydroxyl groups of an oxidative PAH metabolite may be conjugated with UDP-glucuronic acid (above) or with the metabolically activated 3'-phosphoadenosine 5'-phosphosulfate (PAPS) to form a sulfonate monoester in a reaction catalyzed by the ST (E.C.2.8.2.1; also called aryl ST). The endogenous role of ST in sulfonation of steroids[97] and bile acids[98] has been thoroughly investigated. In rabbit liver cytosol the enzyme exists in at least three forms, two of which transfer the intermediate PAPS to phenols.[99] Nemoto et al.[100,101] reported that the oxidative, phase I metabolites of BaP were substrates for the ST present in rat hepatic cytosol. They also point out that the sulfonate conjugates of PAH are intermediate in polarity between the parent PAH and their corresponding glucuronide conjugates.

B. Distribution and Occurrence

In southern flounder, *Paralicthys lethostigma*, Pritchard and Bend[102] demonstrated the *in vivo* formation of sulfonate conjugates of the penultimate carcinogenic PAH, 7,8-dihydro-7,8-dihydroxybenzo(a)pyrene. They showed the excretion rates of the sulfonate conjugates

Table 5
CONVERSIONS TO CONJUGATES FOLLOWING 48-H EXPOSURE OF
FISH TO [^{14}C]-PHENOL IN AQUARIUM WATER

Species		% ^{14}C found in medium as:	
		Phenyl sulfate	Phenyl glucuronide
Goldfish	Carassius auratus	45.1	— (not detected)
Guppy	Poecilia reticulata	41.5	—
Roach	Rutilus rutilus	34.8	9.6
Bream	Abramis brama	33.8	19.7
Tench	Tinca tinca	28.3	—
Rudd	Scardinius erythropthalamus	22.7	17.0
Minnow	Phoxinus phoxinus	19.4	—
Perch	Perca fluviatilis	5.3	25.5

Data abstracted from Layiwola and Linnecar.[106]

to be greater than those of the glucuronide conjugates which were also identified as metabolites.

Although sulfonate conjugation has been reported to be quite low or absent in rainbow trout,[103-105] Layiwola and Linnecar[106] demonstrated that sulfonate conjugates represented a greater proportion of the metabolites than did glucuronides in seven out of eight freshwater species examined (Table 5) which had been exposed to phenol for 48 h. On the other hand, bile from English sole, *Paraphrys retulus*, which had been administered BaP contained a larger ratio of glucuronide than sulfonate conjugate (4:1).[107] This difference may be related to the difference in substrates (PAH vs. hydroxy derivative of PAH) used in these two studies. Using liver slices from carp, *Cyprinus carpio*, spiny lobster, *Panulirus japonicus*, and goldfish, *Carassius auratus*, Kobayashi et al.[108] also found ST activity towards phenol which, in the goldfish, approached 30% of the activity in similar preparations from rat liver.

C. Influencing Factors

There is no known compound that will induce ST activity towards PAH or any other substrate, although most of the classical-type inducers have been tried in mammals.[109] The compounds 2,6-dichloro-4-nitrophenol[110] and pentachlorophenol,[111] however, both show promise for use as specific *in vivo* inhibitors of sulfonation. Recently, the latter compound was shown to inhibit the ST activity in liver-soluble fraction of goldfish.[112] In mammals, ST activity appears to be under hormonal control,[113] but no data exist on this aspect of ST activity in any aquatic species.

The studies of Pritchard and Bend[102] and of Layiwola and Linnecar[106] indicate that with various PAH metabolites, sulfonation may predominate over glucuronidation in some aquatic species. Reports suggesting sulfonation to be extremely low in aquatic species have been based exclusively on studies in rainbow trout,[103-105] a species having only trace amounts of ST activity when compared to four other aquatic species.[108] In light of these studies, the case may be that metabolism of PAH by ST in aquatic species is underestimated.

VI. OVERVIEW AND PROSPECTIVES

The aquatic species so far examined appear to have a full complement of hepatic and extrahepatic phase II enzymes, albeit with lower activity than the well-characterized mammalian systems. The efficacy of phase II enzymes in aquatic species has been amply demonstrated by recent *in vivo* studies with PAH substrates,[69,102,114-120] as appreciable concentrations of administered PAH were recovered as phase II conjugates. In comparing various hepatic phase II enzymes (EH, GST, UDPGT) in trout and seven terrestrial species, Gregus et al.[121] observed low values for the trout enzymes relative to the terrestrial species, but pointed out large variations in activity towards different substrates. For example, while trout UDPGT activity towards 1-naphthol was the lowest of the eight species tested, trout had the highest activity towards testosterone. While aquatic species have been proposed as models in several toxicity test systems,[122,123] such variability among species points out the difficulties in extrapolating toxicological results from aquatic to mammalian species.

Besides anthropogenic factors, the environment itself imposes on the physiologies of aquatic organisms. The seasonal variations of phase I and II enzyme activities observed by Koivusaari[46] and Lindstrom-Seppa[93] are due to the aquatic organism responding to these impositions. These seasonal variations are of toxicological interest, as phase I and some phase II enzymes, such as UDPGT, are differently affected in temperature acclimitization and during spawning.[46,92,93] In a recent study[93] in vendace, *Coregonus albula*, the level of hepatic GSH, the cosubstrate of GST, was also altered during spawning. The predominance of phase I over phase II enzymes in fish exposed to PAH could result in a buildup of electrophilic PAH metabolites produced by the easily induced cytochrome P-450 system. This possibility should be considered during risk assessment.

The routes of exposure to PAH in aquatic species are a consequence of their behavior and environment. Workers have employed much ingenuity in varying routes of administration in experimental studies with fish. These include force-feeding of encapsulated material,[119] associating the dose with sediment for bottom-dwelling fish,[115,117] and aqueous exposure for free-swimming fish.[106] The few studies that have carefully compared dosing routes have noted differences. James and co-workers[49,66] showed with Atlantic stingray that the levels of induction of cytochrome P-450-dependent monooxygenase activity were higher with intramuscular (i.m.) rather than intraperitoneal (i.p.) injection. In comparing dietary and i.p. dosing in teleosts, Varanasi et al.[119] observed marked differences in the ratios of conjugated to nonconjugated metabolites of naphthalene in liver and bile of the starry flounder. When an equimolar dose of BaP was administered orally to one group of starry flounder and intraperitoneally to another, the former group showed a 14-fold lower level of hepatic DNA modification than did the latter group.[124] As the types of conjugates and the chromatographic profiles of bile hydrolysates from the two groups were similar, the authors suggest that the lower levels of binding of BaP metabolites to hepatic DNA in the former group were due to the lower effective dose that reached the liver, the remainder of the dose being excreted before it reached that organ. Yet another study with rainbow trout[121] has demonstrated a strong correlation between oral and i.p. toxicity of more than 30 organic compounds. These studies emphasize that wide interspecies variability may exist and that extrapolation of results with one species to different fish species could be complicated. In future studies, the route of exposure, then, should be carefully chosen to most closely reflect the habitat, physiology, and behavior of the individual species.

The gradual postnatal development of some phase II enzyme activities in rats and humans can have toxicological implications (e.g., UDPGT activity and fetal jaundice). As mentioned above, the facile induction of phase I enzymes compared to the refractory induction of phase II enzymes could result in an imbalance in the metabolism of PAH producing a higher proportion of toxic, electrophilic intermediates; insufficient phase II activities could place juveniles at high risk for exposure to these intermediates. Investigation into this area is overdue. To this author's knowledge, there exists only a single study addressing this issue in aquatic species, namely, that of Chowdhury et al.[87] with juvenile dogfish sharks.

Another important consideration in the area of phase II enzymes is the ability of these enzymes to metabolically activate, rather than detoxicate, xenobiotics. A recent study shows this consideration applies directly to PAH metabolism. Watabe and co-workers[125] demonstrated covalent binding to DNA by the sulfate derivative of the PAH carcinogen, 7-hydroxymethyl-12-methylbenz(a)anthracene. Another study has demonstrated enterohepatic circulation of PAH metabolites in fish;[126] after dosing goldfish, *Carassius auratus* with phenol, phenylglucuronide was found in the aquarium water only after oral dosing of a β-glucuronidase inhibitor. The hydrolytic activity was apparently associated with the intestinal flora of the goldfish. This cycle (glucuronide hydrolysis and intestinal reabsorption) could reexpose the organism to a chemical species capable of being activated. Probably the most perplexing observation of phase II activation is that of Fahl et al.;[127] in rat liver microsomes incubated with BaP UDPGT was observed to *enhance* the extent of conjugation of diol epoxides to DNA. The recent study by Kari et al.[128] raises the possibility that glucuronides could serve as vehicles for PAH, distributing potentially mutagenic PAH metabolites throughout the body and releasing them in areas of high glucuronidase activity. Paradoxical activation by phase II enzymes is just one area of PAH metabolism where need for research is clear. It is hoped that this review has succeeded in pointing out to the reader the multitude of unexplored pathways in phase II metabolism of PAH that need investigation through rigorous application of ideas and experimentation with aquatic species.

ACKNOWLEDGMENT

The author is grateful for partial support from the National Institute of Environmental Health Sciences No. ES 01978 during the writing of this work.

REFERENCES

1. **Williams, R. T.,** *Detoxication Mechanisms, The Metabolism and Detoxication of Drugs, Toxic Substances, and Other Organic Compounds,* John Wiley & Sons, New York, 1959, 734.
2. **Gelboin, H. V., Huberman, E., and Sachs, L.,** Enzymatic hydroxylation of benzopyrene and its relationship to cytotoxicity, *Proc. Natl. Acad. Sci. U.S.A.,* 64, 1188, 1969.
3. **Huberman, E. and Sachs, L.,** Cell-mediated mutagenesis of mammalian cells with chemical carcinogens, *Intl. J. Cancer,* 13, 326, 1974.
4. **Kinoshita, N. and Gelboin, H. V.,** Aryl hydrocarbon hydroxylase in 7,12-dimethylbenz(a)anthracene skin tumorgenesis: on the mechanism of 7,8-benzoflavone inhibition of tumorgenesis, *Proc. Natl. Acad. Sci. U.S.A.,* 69, 824, 1972.
5. **Jerina, D. J., Daly, J. W., Witkop, B., Zaltzam-Nirenberg, P., and Udenfriend, S.,** The role of arene oxide-oxepin systems in the metabolism of aromatic substrates. III. Formation of 1,2-naphthalene oxide from naphthalene by liver microsomes, *J. Am. Chem. Soc.,* 90, 6525, 1968.
6. **Borgen, A., Darvey, H., Castagnoli, N., Crocker, T. T., Rasmussen, R. E., and Wang, I. Y.,** Metabolic conversion of benzo(a)pyrene by Syrian hamster in liver microsomes and binding of metabolites to DNA, *J. Med. Chem.,* 16, 502, 1973.
7. **Jerina, D. M. and Daly, J. W.,** Arene oxides. A new aspect of drug metabolism, *Science,* 185, 573, 1974.
8. **Sims, P., Grover, P. L., Swaisland, A., and Hewer, A.,** Metabolic activation of benzo(a)pyrene proceeds by a diol-epoxide, *Nature (London),* 252, 326, 1974.
9. **Gelboin, H. V. and Ts'o, P. O. P.,** *Polycyclic Hydrocarbons and Cancer,* Vol. 1, Academic Press, New York, 1978.
10. **Gelboin, H. V. and Ts'o, P. O. P.,** *Polycyclic Hydrocarbons and Cancer,* Vol. 3, Academic Press, New York, 1981.
11. **James, M. O.,** Xenobiotic conjugation in fish and other aquatic species, in *Xenobiotic Conjugation Chemistry,* Paulson, G. D., Caldwell, J., Hutson, D. H., and Menn, J. J., Eds., American Chemical Society, Washington, D.C., 1986, 29.
12. **Bend, J. R. and James, M. O.,** Xenobiotic metabolism in marine and freshwater species, in *Biochemical and Biophysical Perspectives in Marine Biology,* Vol. 4, Malins, D. C. and Sargent, J. R., Eds., Academic Press, New York, 1978, 125.
13. **Lech, J. J. and Bend, J. R.,** Relationship between biotransformation and the toxicity and fate of xenobiotic chemicals in fish, *Environ. Health Perspects.,* 34, 115, 1980.

14. **Neff, J. M.,** *Polycyclic Aromatic Hydrocarbons in the Aquatic Environment,* Applied Science, London, 1979.
15. **Tan, B. and Melius, P.,** Polynuclear aromatic hydrocarbon metabolism in fishes, *Comp. Biochem. Physiol.,* 83C, 217, 1986.
16. **Stenersen, J., Kobro, S., Bjerk, M., and Arenal, U.,** Glutathione transferases in aquatic and terrestrial animals from nine phyla, *Comp. Biochem. Physiol.,* 86C, 73, 1987.
17. **Morgenstern, R. J., DePierre, J. W., and Ernster, L.,** Activation of microsomal glutathione S-transferase activity by sulfhydryl reagents, *Biochem. Biophys. Res. Commun.,* 87, 657, 1979.
18. **Wahllander, A., Soboll, S., and Sies, H.,** Hepatic mitochondrial and cytosolic glutathione content and the subcellular distribution of GSH-S-transferases, *FEBS Lett.,* 97, 138, 1979.
19. **Habig, W. H., Pabst, M. J., and Jakoby, W. B.,** Glutathione S-transferases, the first enzymatic step in mercapturic acid formation, *J. Biol. Chem.,* 249, 7130, 1974.
20. **Bhargava, M. M., Ohmi, N., Listowsky, I., and Arias, I. M.,** Structural, catalytical, binding, and immunological properties associated with each of the two subunits of rat liver ligandin, *J. Biol. Chem.,* 222, 718, 1980.
21. **Armstrong, R. N., Chen, W. J., and deSmidt, P. C.,** Kinetic and spectroscopic studies of stereoisomeric product complexes of glutathione S-transferases, *Fed. Proc.,* 44, 1810, 1985.
22. **Cobb, D., Boehlert, C., Lewis, D., and Armstrong, R. N.,** Stereoselectivity of isozyme *c* of glutathione *S*-transferase toward arene and azaarene oxides, *Biochemistry,* 22, 805, 1983.
23. **Boehlert, C. C. and Armstrong, R. N.,** Investigation of the kinetic and stereochemical recognition of arene and azaarene oxides by isozymes A2 and C2 of glutathione S-transferase, *Biochem. Biophys. Res. Commun.,* 121, 980, 1984.
24. **deSmidt, P. C., McCarrick, M. A., Darnow, J. N., Mervic, M., and Armstrong, R. N.,** Stereoselectivity and enantioselectivity of glutathione S-transferase towards stilbene oxide substrates, *Biochem. Intl.,* 14, 401, 1987.
25. **Tate, S. S.,** Enzymes of mercapturic acid formation, in *Enzymatic Basis of Detoxication,* Vol. 2, Jakoby, W. B., Ed., Academic Press, New York, 1980, chap. 5.
26. **Jerina, D. M., Dansette, P. M., Lu, A. Y. H., and Levin, W.,** Hepatic microsomal epoxide hydrase: a sensitive radiometric assay for hydration of arene oxides of carcinogenic aromatic hydrocarbons, *Mol. Pharmacol.,* 13, 342, 1977.
27. **Bend, J. R., James, M. O., and Dansette, P. M.,** *In vitro* metabolism of xenobiotics in some marine animals, *Ann. N.Y. Acad. Sci.,* 298, 505, 1977.
28. **Foureman, G. L. and Bend, J. R.,** The hepatic glutathione transferases of the male little skate, *Raja erinacea, Chem. Biol. Interact.,* 49, 89, 1984.
29. **Bauermeister, A., Lewenoon, A., Ramage, P. I. N., and Nimmo, I. A.,** Distribution and some properties of the glutathione S-transferase and gamma-glutamyl transpeptidase activities of rainbow trout, *Comp. Biochem. Physiol.,* 74C, 89, 1983.
30. **Tate, L. G. and Herf, D. A.,** Characterization of glutathione S-transferase activity in tissues of the blue crab, *Callinectes sapidus, Comp. Biochem. Physiol.,* 61C, 165, 1978.
31. **LeBlanc, G. S. and Cochrane, B. J.,** Modulation of substrate specific glutathione S-transferase activity in *Daphnia magna* with concomitant effects on toxicity tolerance, *Comp. Biochem. Physiol.,* 82C, 37, 1985.
32. **Lindstrom-Seppa, P., Koivusaari, U., and Hanninen, O.,** Metabolism of foreign compounds in freshwater crayfish (*Astacus astacus* L.) tissues, *Aquat. Toxicol.,* 3, 35, 1983.
33. **Nimmo, I. A.,** The glutathione S-transferase activity in the gills of rainbow trout (*Salmo gairdneri*), *Comp. Biochem. Physiol.,* 80B, 365, 1985.
34. **Nimmo, I. A. and Spalding, C. M.,** The glutathione S-transferase activity in the kidney of rainbow trout (*Salmo gairdneri*), *Comp. Biochem. Physiol.,* 82B, 91, 1985.
35. **Sugiyama, Y., Tadataka, Y., and Kaplowitz, N.,** Glutathione S-transferases in elasmobranch liver, *Biochem. J.,* 199, 749, 1981.
36. **Ramage, P. I. N., Rae, G. H., and Nimmo, I. A.,** Purification and properties of the hepatic glutathione S-transferases of the Atlantic salmon (*Salmo salar*), *Comp. Biochem. Physiol.,* 83B, 23, 1986.
37. **Ramage, P. I. N. and Nimmo, I. A.,** The purification of the hepatic glutathione S-transferases of rainbow trout by glutathione affinity chromatography alters their isoelectric behaviour, *Biochem. J.,* p. 523, 1983.
38. **Dierickx, P. J.,** Hepatic glutathione S-transferases in rainbow trout and their interaction with 2,4-dichlorophenoxyacetic acid and 1,4-benzoquinone, *Comp. Biochem. Physiol.,* 82C, 495, 1985.
39. **Ramage, P. I. N. and Nimmo, I. A.,** The substrate specificities and subunit composition of the hepatic glutathione S-transferases of rainbow trout (*Salmo gairdneri*), *Comp. Biochem. Physiol.,* 78B, 189, 1984.
40. **Nimmo, I. A. and Clapp, J. B.,** A comparison of the glutathione S-transferases of trout and rat liver, *Comp. Biochem. Physiol.,* 63B, 423, 1979.

41. **Nimmo, I. A., Coghill, D. R., Hayes, J. D., and Strange, R. C.,** A comparison of the subcellular distribution, subunit composition and bile acid-binding activity of glutathione S-transferases from trout and rat liver, *Comp. Biochem. Physiol., 68B,* 579, 1981.

42. **Foureman, G. L., Hernandez, O., Bhatia, A., and Bend, J. R.,** The steroselectivity of four hepatic glutathione-S-transferases purified from a marine elasmobranch with several K-region polycyclic arene oxide substrates, *Biochim. Biophys. Acta,* 914, 127, 1987.

43. **Nemoto, N., Gelboin, H. V., Habig, W. H., Ketley, J. N., and Jakoby, W. B.,** K-region benzo(a)pyrene 4,5-oxide is conjugated by homogenous glutathione S-transferases, *Nature,* 255, 512, 1974.

44. **Jerina, D. M., Yagi, H., Lehr, R. E., Thakker, D. R., Schaefer-Ridder, M., Karle, J. M., Levin, W., Wood, A. W., Chang, R. L., and Conney, A. H.,** The bay-region theory of carcinogenesis by polycyclic aromatic hydrocarbons, in *Polycyclic Hydrocarbons and Cancer,* Vol. 1, Gelboin, H. V. and Ts'o, P. O. P., Eds., Academic Press, New York, 1978, 173.

45. **Wood, A. W., Chang, R. L., Levin, W., Ryan, D. E., Thomas, P. E., Lehr, R. E., Kumar, S., Schaefer-Ridder, M., Engelhardt, U., Yagi, H., Jerina, D. M., and Conney, A. H.,** Mutagenicity of diol-epoxides and tetrahydroepoxides of benz(a)acridine and benz(c)acridine in bacteria and in mammalian cells, *Cancer Res.,* 43, 1656, 1983.

46. **Koivusaari, U., Harri, M., and Hanninen, O.,** Seasonal variation of hepatic biotransformation in female and male rainbow trout *(Salmo gairdneri), Comp. Biochem. Physiol., 70C,* 149, 1981.

47. **Gregus, Z., Varga, F., and Schmelas, A.,** Age-development and inducibility of hepatic glutathione S-transferase activities in mice, rats, rabbits, and guinea pigs, *Comp. Biochem. Physiol., 80C,* 85, 1985.

48. **Di Simplicio, P., Jensson, H., and Mannervik, B.,** Identification of the isozymes of glutathione transferase induced by *trans*-stilbene oxide, *Acta Chem. Scand. Ser. B,* 255, 1983.

49. **James, M. O. and Bend, J. R.,** Polycyclic aromatic hydrocarbon induction of cytochrome P-450-dependent mixed-function oxidases in marine fish, *Toxicol. Appl. Pharmacol.,* 54, 117, 1980.

50. **Balk, L., Meijer, J., Seidegard, J., Morgenstern, R., and DePierre, J. W.,** Initial characterization of drug-metabolizing systems in the liver of the Northern pike, *Esox lucius, Drug Metab. Dispos.,* 8, 98, 1980.

51. **Ankley, G. T., Blazer, V. S., Reinert, R. E., and Agosin, M.,** Effects of Arochlor 1254 on cytochrome P-450-dependent monooxygenase, glutathione S-transferase, and UDP-glucuronosyltransferase activities in channel catfish liver, *Aquat. Toxicol.,* 9, 91, 1986.

52. **Andersson, T., Pesonen, M., and Johansson, C.,** Differential induction of cytochromne P-450-dependent monooxygenase, epoxide hydrolase, glutathione transferase and UDP-glucuronosyltransferase activities in the liver of the rainbow trout by β-naphthoflavone or Clophen A50, *Biochem. Pharmacol.,* 34, 3309, 1985.

53. **Collier, T. K. and Varanasi, U.,** Field and laboratory studies of xenobiotic metabolizing enzymes in English sole, *Proc. Pac. Northwest Assoc. Toxicol.,* 1, 16, 1984.

54. **Sato, K., Kitahara, A., Yin, Z., Elina, T., Satoh, K., Hatayama, I., Nishinura, K., Yamazaki, T., Tsuda, H., Ito, N., and Dempo, K.,** Molecular forms of glutathione S-transferase and UDP-glucuronyltransferase as hepatic preneoplastic marker enzymes, *Ann. N.Y. Acad. Sci.,* 417, 213, 1983.

55. **Gillette, J. R., Mitchell, J. R., and Brodie, B. B.,** Biochemical mechanisms of drug toxicity, *Annu. Rev. Pharmacol.,* 14, 271, 1974.

56. **Suga, T., Ohata, I., and Akagi, M.,** Studies on mercapturic acids, *J. Biochem.,* 59, 209, 1966.

57. **Braddon, S. A., McIlvaine, C. M., and Balthrop, J. E.,** Distribution of GSH and GSH cycle enzymes in black sea bass *(Centropristis striata), Comp. Biochem. Physiol., 80B,* 213, 1985.

58. **James, M. O., Fouts, J. R., and Bend, J. R.,** *In vitro* epoxide metabolism in some marine species, *Bull. Mt. Desert Isl. Biol. Lab.,* 14, 41, 1974.

59. **George, S. G. and Young, P.,** The time course of effects of cadmium and 3-methylcholanthrene on activities of enzymes of xenobiotic metabolism and metallothionein levels in the plaice, *Pleuronectes platessa, Comp. Biochem. Physiol., 83C,* 37, 1986.

60. **Thomas, P. and Wofford, H. W.,** Effects of metals and organic compounds on hepatic glutathione, cysteine, and acid-soluble thiol levels in mullet *(Mugil cephalis* L.), *Toxicol. Appl. Pharmacol.,* 76, 172, 1984.

61. **Hammock, B. D. and Ota, K.,** Differential induction of cytosolic epoxide hydrolase, microsomal epoxide hydrolase, and glutathione S-transferase activities, *Toxicol. Appl. Pharmacol.,* 71, 254, 1983.

62. **Lu, A. Y. H., Jerina, D. M., and Levin, W.,** Liver microsomal epoxide hydrase, *J. Biol. Chem.,* 252, 3715, 1977.

63. **Oesch, F. and Golan, M.,** Specificity of mouse liver cytosolic epoxide hydrolase for K-region epoxides derived from polycyclic aromatic hydrocarbons, *Cancer Lett.,* 9, 169, 1980.

64. **Nemoto, N. and Gelboin, H. V.,** Enzymatic conjugation of benzo(a)pyrene oxides, phenols and dihydro-diols with UDP-glucuronic acid, *Biochem. Pharmacol.,* 25, 1221, 1976.

65. **Sims, P. and Grover, P. L.,** Involvement of dihydrodiols and diol epoxides in the metabolic activation of polycyclic hydrocarbons other than benzo(a)pyrene, in *Polycyclic Hydrocarbons and Cancer,* Vol. 3, Gelboin, H. V. and Ts'o, P. O. P., Eds., Academic Press, New York, 1981, 117.

66. **James, M. O., Bowen, E. R., Dansette, P. M., and Bend, J. R.,** Epoxide hydrase and glutathione S-transferase activities with selected alkene and arene oxides in several marine species, *Chem. Biol. Interact.,* 25, 321, 1979.

67. **Oesch, F., Glatt, H., and Schmassmann, H.,** The apparent ubiquity of epoxide hydratase in rat organs, *Biochem. Pharmacol.,* 26, 603, 1977.

68. **Varanasi, U. and Gmur, D. J.,** Metabolic activation and covalent binding of benzo(a)pyrene to deoxyribonucleic acid catalyzed by liver enzymes of marine fish, *Biochem. Pharmacol.,* 29, 753, 1980.

69. **Varanasi, U., Gmur, D. J., and Krahn, M. M.,** Metabolism and subsequent binding of benzo(a)pyrene to DNA in *Pleuronectid* and salmonid fish, in *Polynuclear Aromatic Hydrocarbons: Chemistry and Biological Effects,* Bjoerseth, A. and Dennis, A. J., Eds., Battelle Press, Columbus, Ohio, 1980, 455.

70. **Ahokas, J. T., Saarni, H., Nebert, D. W., and Pelkonen, O.,** The *in vitro* metabolism and covalent binding of benzo(a)pyrene to DNA catalyzed by trout liver microsomes, *Chem. Biol. Interact.,* 25, 103, 1979.

71. **Breger, R. K., Franklin, R. B., and Lech, J. J.,** Metabolism of 2-methylnaphthalene to isomeric dihydrodiols by hepatic microsomes of rat and rainbow trout, *Drug Metab. Dispos.,* 9, 88, 1981.

72. **James, M. O. and Little, P. J.,** Polyhalogenated biphenyls and phenobarbital: evaluation as inducers in the sheepshead, *Archosargus probatocephalus, Chem. Biol. Interact.,* 36, 229, 1981.

73. **Oesch, F. and Schmassmann, H.,** Species and organ specificity of the *trans*-stilbene oxide induced effects on epoxide hydratase and benzo(a)pyrene monooxygenase activity in rodents, *Biochem. Pharmacol.,* 28, 171, 1979.

74. **Laitinen, M., Nieminen, M., Pasanen, P., and Hietanen, E.,** Tricaine (MS-222) induced modification on the metabolism of foreign compounds in the liver and duodenal mucosa of the splake *(Salvelinus fontinalis* × *Salvelinus namaycush), Acta Pharmacol. Toxicol.,* 49, 92, 1981.

75. **Laitinen, M., Nieminen, M., and Hietanen, E.,** The effect of pH changes of water on the hepatic metabolism in fish, *Acta Pharmacol. Toxicol.,* 51, 24, 1982.

76. **Laitinen, M., Hietanen, E., Nieminen, M., and Pasanen, P.,** Acidification of water and extrahepatic biotransformation in fish, *Environ. Pollut. Ser. A,* 35, 271, 1984.

77. **Dutton, G. J.,** *Glucuronidation of Drugs and Other Compounds,* CRC Press, Boca Raton, FL, 1980.

78. **Boyland, E., Ramsay, G. S., and Sims, P.,** The secretion of naphthalene, 1:2-dihydronaphthalene and 1:2-epoxy-1:2:3:4-tetrahydronaphthalene in rat bile, *Biochem. J.,* 78, 376, 1961.

79. **Lilienblum, W., Walli, A. K., and Bock, K. W.,** Differential induction of rat liver microsomal UDP-glucuronosyltransferase activities by various inducing agents, *Biochem. Pharmacol.,* 31, 907, 1982.

80. **Bock, K. W., Lilienblum, W., Pfeil, H., and Eriksson, L. C.,** Increased uridine diphosphate-glucuronyltransferase activity in preneoplastic liver nodules and Morris hepatomas, *Cancer Res.,* 42, 3747, 1982.

81. **Bock, K. W., Burchell, B., Dutton, G. J., Hanninen, O., Mulder, G. J., Owens, I. S., Siest, G., and Tephly, T. R.,** UDP-glucuronosyltransferase activities, *Biochem. Pharmacol.,* 32, 953, 1983.

82. **Boutin, J. A., Thomassin, J., Siest, G., and Cartier, A.,** Heterogeneity of hepatic microsomal UDP-glucuronosyltransferase activities, *Biochem. Pharmacol.,* 34, 2235, 1985.

83. **Lindstrom-Seppa, P.,** Metabolism of xenobiotics by vendace *(Coregnous albula), Comp. Biochem. Physiol.,* 68C, 121, 1981.

84. **Lindstrom-Seppa, P., Koivusaari, U., and Hanninen, O.,** Extrahepatic xenobiotic metabolism in north-European freshwater fish, *Comp. Biochem. Physiol.,* 69C, 259, 1981.

85. **Winsnes, A.,** Studies on the activation *in vitro* of glucuronyltransferase, *Biochim. Biophys. Acta,* 191, 279, 1969.

86. **Hanninen, O., Lindstrom-Seppa, P., Koivusaari, U., Vaisanen, M., Julkunen, A., and Juvonen, R.,** Glucuronidation and glucosidation reactions in aquatic species in boreal regions, *Biochim. Soc. Trans.,* 12, 13, 1984.

87. **Chowdhury, N. R., Chowdhury, J. R., and Arias, I. M.,** Bile pigment composition and hepatic UDP-glucuronyltransferase activity in the fetal and adult dogfish shark, *Squalus acanthus, Comp. Biochem. Physiol.,* 73B, 651, 1982.

88. **Wishart, G. J.,** Functional heterogeneity of UDP-glucuronosyltransferase as indicated by its differential development and inducibility by glucocorticoids, *Biochem. J.,* 174, 485, 1978.

89. **Forlin, L. and Andersson, T.,** Storage conditions of rainbow trout liver cytochrome P-450 and conjugating enzymes, *Comp. Biochem. Physiol.,* 80B, 569, 1985.

90. **Forlin, L. and Haux, C.,** Increased excretion in the bile of 17β-[³H]-estradiol-derived radioactivity in rainbow trout treated with β-naphthoflavone, *Aquat. Toxicol.,* 6, 197, 1985.

91. **Sivarajah, K., Franklin, C. S., and Williams, W. P.,** The effects of polychlorinated biphenyls on plasma steroid levels and hepatic microsomal enzymes in fish, *J. Fish Biol.,* 13, 401, 1978.

92. **Andersson, T. and Koivusaari, U.,** Influence of environmental temperature on the induction of xenobiotic metabolism by β-naphthoflavone in rainbow trout, *Salmo gairdneri, Toxicol. Appl Pharmacol.,* 80, 43, 1985.

93. **Lindstrom-Seppa, P.,** Seasonal variation of the xenobiotic metabolizing enzyme activities in the liver of male and female vendace *(Coregonus albula* L.,*), Aquat. Toxicol.,* 6, 323, 1985.

94. **Castren, M. and Oikari, A.,** Changes of the liver UDP-glucuronosyltransferase activity in trout *(Salmo gairdneri Rich.)* acutely exposed to selected aquatic toxicants, *Comp. Biochem. Physiol.,* 86C, 357, 1987.

95. **Andersson, T., Bengtsson, B.-E., Forlin, L., Hardig, J., and Larsson, A.,** Long-term effects of bleached kraft mill effluent on carbohydrate metabolism and hepatic xenobiotic biotransformation enzymes in fish, *Ecotoxicol. Environ. Safety,* 13, 53, 1987.

96. **Andersson, T., Koivusaari, U., and Forlin, L.,** Xenobiotic biotransformation in the rainbow trout liver and kidney during starvation, *Comp. Biochem. Physiol.,* 82C, 221, 1985.

97. **Miyazaki, M., Yoshizawa, I., and Fishman, J.,** Direct *O*-methylation of estrogen catechol sulfates, *Biochemistry,* 8, 1669, 1969.

98. **Chen, L. J., Bolt, R. J., and Admirand, W. H.,** Enzymatic sulfation of bile salts. Partial purification and characterization of an enzyme from rat liver that catalyzes the sulfation of bile salts, *Biochim. Biophys., Acta,* 480, 219, 1977.

99. **Nose, Y. and Lipmann, F.,** Separation of steroid sulfokinases, *J. Biol. Chem.,* 233, 1348, 1958.

100. **Nemoto, N., Takayama, S., and Gelboin, H. V.,** Enzymatic conversion of benzo(a)pyrene phenols, dihydrodiols, and quinones to sulfate conjugates, *Biochem. Pharmacol.,* 26, 1825, 1977.

101. **Nemoto, N., Takayama, S., and Gelboin, H. V.,** Sulfate conjugation of benzo(a)pyrene metabolites and derivatives, *Chem. Biol. Interact.,* 23, 19, 1978.

102. **Pritchard, J. B. and Bend, J. R.,** Mechanisms controlling the renal excretion of xenobiotics in fish: effects of chemical structure, *Drug Metab. Rev.,* 15, 655, 1984.

103. **Parker, R. S., Morrissey, M. T., Moldeus, P., and Selivonchick, D. P.,** The use of isolated hepatocytes from rainbow trout *(Salmo gairdneri)* in the metabolism of acetaminophen, *Comp. Biochem. Physiol.,* 70B, 631, 1981.

104. **Statham, C. N., Pepple, S. K., and Lech, J. J.,** Biliary excretion products of 1-[1-^{14}C]naphthyl-N-methylcarbamate (carbaryl) in rainbow trout *(Salmo gairdneri), Drug Metab. Dispos.,* 3, 400, 1975.

105. **Andersson, T., Forlin, L., and Hansson, T.,** Biotransformation of 7-ethoxycoumarin in isolated perfused rainbow trout liver, *Drug Metab. Dispos.,* 11, 494, 1983.

106. **Layiwola, P. J., and Linnecar, D. F. C.,** The biotransformation of [^{14}C]phenol in some freshwater fish, *Xenobiotica,* 11, 167, 1981.

107. **Varanasi, U., Stein, J. E., Nishimoto, M., and Hom, T.,** Benzo(a)pyrene metabolites in liver, muscle, gonads, and bile of adult English sole *(Parophrys vetulus),* in *Polynuclear Aromatic Hydrocarbons: 2nd Intl. Symp. on Formation, Metabolism and Measurement,* Cooke, M.W. and Dennis, A. J., Eds., Battelle Press, Columbus, Ohio, 1982, 1221.

108. **Kobayashi, K., Kimura, S., and Akitake, H.,** Studies on the metabolism of chlorophenols in fish. VII. Sulfate conjugation of phenol and PCP by fish livers, *Bull. Jpn. Soc. Sci. Fish.,* 42, 171, 1976.

109. **Mulder, G. J.,** Conjugation of phenols, in *Metabolic Basis of Detoxication,* Jakoby, W. B., Bend, J. R., and Caldwell, J., Eds., Academic Press, New York, 1982, 247.

110. **Mulder, G. J. and Scholtens, E.,** Phenolsulphotransferase and UDP-glucuronyltransferase from rat liver *in vivo* and *in vitro*: 2,6-dichloro-4-nitrophenol as selective inhibitor of sulfation, *Biochem. J.,* 165, 553, 1977.

111. **Kimbrough, R. D. and Linder, R. E.,** The effect of technical and purified pentachlorophenol in the rat liver, *Toxicol. Appl. Pharmacol.,* 46, 151, 1978.

112. **Kobayashi, K., Kimura, S., and Oshima, Y.,** Sulfate conjugation of various phenols by liver-soluble fraction of goldfish, *Bull. Jpn. Soc. Sci. Fish.,* 50, 833, 1984.

113. **Jakoby, W. B., Sekura, R. D., Lyon, E. S., Marcus, C. J., and Wang, J.-L.,** Sulfotransferases, in *Enzymatic Basis of Detoxication,* Vol. 2, Jakoby, W. B., Ed., Academic Press, New York, 1980, chap. 11.

114. **Varanasi, U., Nishimoto, M., Reichert, W. L., and Stein, J. E.,** Metabolism and subsequent covalent binding of benzo(a)pyrene to macromolecules in gonads and liver of ripe English sole, *Xenobiotica,* 12, 417, 1982.

115. **Varanasi, U. and Gmur, D. J.,** Hydrocarbons and metabolites in English sole exposed simultaneously to [^{3}H]benzo(a)pyrene and [^{14}C]naphthalene in oil-contaminated sediment, *Aquat. Toxicol.,* 1, 49, 1981.

116. **Reichert, W. L., Le Eberhart, B.-T., and Varanasi, U.,** Exposure of two deposit-feeding species amphipods to sediment-associated [^{3}H]benzo(a)pyrene: uptake, metabolism and covalent binding to tissue macromolecules, *Aquat. Toxicol.,* 6, 46, 1985.

117. **Stein, J. E., Hom, T., and Varanasi, U.,** Simultaneous exposure of English sole *(Parophrys vetulus)* to sediment-associated xenobiotics. I. Uptake and disposition of ^{14}C-polychlorinated biphenyls and ^{3}H-benzo(a)pyrene, *Mar. Environ. Res.,* 13, 97, 1984.

118. **Collier, T. K., Thomas, L. C., and Malins, D. C.,** Influence of environmental temperature on disposition of dietary naphthalene in coho salmon *(Oncorhyncus kisutch):* isolation and identification of individual metabolites, *Comp. Biochem. Physiol.,* 61C, 23, 1978.

119. **Varanasi, U., Gmur, D. J., and Treseler, P. A.,** Influence of time and mode of exposure on biotransformation of naphthalene by juvenile starry founder *(Platichthys stellatus)* and rock sole *(Lepidopsetta bilineata)*, *Arch. Environ. Contam. Toxicol.*, 8, 673, 1979.

120. **Roubal, W. T., Collier, T. K., and Malins, D. C.,** Accumulation and metabolism of carbon-14 labeled benzene, naphthalene, and anthracene by young coho salmon *(Oncorhynchus kisutch)*, *Arch. Environ. Contam. Toxicol.*, 5, 513, 1977.

121. **Gregus, Z., Watkins, J. B., Thompson, T. N., Harvey, M. J., Rozman, K., and Klaassen, C. D.,** Hepatic phase I and phase II biotransformations in quail and trout: comparison to other species commonly used in toxicity testing, *Toxicol. Appl. Pharmacol.*, 67, 430, 1983.

122. **Hoover, K. L.,** The Use of Small Fish Species for Carcinogenicity Testing, National Cancer Institute Monogr. 65, U. S. Government Printing Office, Washington, D. C., 1984.

123. **Hodson, P. V.,** A comparison of the acute toxicity of chemicals to fish, rats, and mice, *J. Appl. Toxicol.*, 4, 220, 1985.

124. **Varanasi, U., Nishimoto, M., Reichert, W. L., and Le Eberhart, B.-T.,** Comparative metabolism of benzo(a)pyrene and covalent binding to hepatic DNA in English sole, starry flounder, and rat, *Cancer Res.*, 46, 3817, 1986.

125. **Watabe, T., Fujieda, T., Hiratsuka, A., Ishizuka, T., Hakamata, Y., and Ogura, K.,** The carcinogen, 7-hydroxymethyl-12-methylbenz(a)anthracene, is activated and covalently binds to DNA via a sulfate ester, *Biochem. Pharmacol.*, 34, 3002, 1985.

126. **Layiwola, P. J., Linnecar, D. F. C., and Knights, B.,** Hydrolysis of the biliary glucoronic acid conjugate of phenol by the intestinal mucus/flora of goldfish *(Carassius auratus)*, *Xenobiotica*, 13, 27, 1983.

127. **Fahl, W. E., Shen, A. L., and Jefcoate, C. R.,** UDP-glucuronyltransferase and the conjugation of benzo(a)pyrene metabolites to DNA, *Biochem. Biophys. Res. Commun.*, 85, 891, 1978.

128. **Kari, F. W., Kauffman, F. C., and Thurman, R. G.,** Characterization of mutagenic glucuronide formation from benzo(a)pyrene in the nonrecirculating perfused rat liver, *Cancer Res.*, 44, 5073, 1984.

Chapter 6

METABOLIC ACTIVATION OF PAH IN SUBCELLULAR FRACTIONS AND CELL CULTURES FROM AQUATIC AND TERRESTRIAL SPECIES*

Usha Varanasi, Marc Nishimoto, William M. Baird, and Teresa A. Smolarek

TABLE OF CONTENTS

*   This paper is the official writing of employees (U. Varanasi and M. Nishimoto) of the U.S. Government. As such, it resides in the public domain and is not subject to copyright.

I. INTRODUCTION

Cellular metabolism of polycyclic aromatic hydrocarbons (PAH) results in the conversion of these hydrophobic compounds into forms that can be excreted from the cell and ultimately from the organism. Metabolism serves mainly as a pathway of detoxication of these compounds; however, some of the metabolites that are intermediates in this detoxication process have been demonstrated to possess carcinogenic, mutagenic, and cytotoxic activity in rodents.[1,2] Thus, determining both qualitatively and quantitatively how PAH are metabolized by different tissues in various species is essential for an understanding of the potential genotoxic activity of these chemicals. Initial studies with rodents have demonstrated that PAH were metabolized to polar derivatives and that the metabolites present in the bile, urine, and feces resulted from both oxidation and conjugation of the parent PAH.[3,4] Because of the multiple metabolic steps involved and the inability to obtain a quantitative recovery of these metabolites, it is impossible to evaluate the overall pathways of PAH metabolism *in vivo* or to determine the relative importance of various pathways involved in both detoxication and tumor induction. Two major approaches have been developed for the qualitative and quantitative analysis of the metabolites formed from PAH. The first approach involves the study of hydrocarbon metabolites formed in incubations of PAH with subcellular fractions of several tissues from various species in the presence of appropriate cofactors.[1] By the use of radioisotope-containing PAH and various chromatographic techniques, it was possible to determine the proportions of individual metabolites formed.[2] The second major approach involves analyses of the metabolites formed in cell cultures from various species and tissues.[5]

One of the most widely studied PAH is benzo(a)pyrene (BaP), a common environmental contaminant found in food, air, and water.[1] The general pathways of metabolism of BaP are illustrated in Figure 1. Initially, metabolism involves oxidation of the PAH that is usually catalyzed by the microsomal cytochrome P-450-dependent mixed function oxidases (MFO). In most cases, this results in the formation of an epoxide or arene oxide, although oxidation at specific positions (e.g., the C-6 of BaP) probably occurs through other intermediates.[6] The arene oxides can be formed at a number of positions and specific forms of cytochrome P-450 demonstrate stereoselectivity in the proportions of specific isomeric arene oxides formed.[1] The arene oxides are relatively unstable metabolites that undergo three major reactions: (1) spontaneous rearrangement to form phenols, (2) hydration to form *trans* dihydroxydihydro compounds (diols), a reaction catalyzed by the enzyme epoxide hydrolase (EH), and (3) conjugation with glutathione (GSH) involving either catalysis by a family of glutathione-S-transferase (GST) enzymes, or by direct reaction at a slower rate. The GSH conjugates are water-soluble precursors of mercapturic acids and are generally considered to be detoxication products.[2] The diols and phenols are usually metabolized further prior to excretion: two major pathways involve conjugation with either glucuronic acid, mediated by UDP glucuronyl transferase (UDPGT), or sulfation, mediated by sulfotransferases.[1] Both the sulfates and glucuronides are water soluble and are generally considered detoxication products, although certain sulfate conjugates, such as the sulfate ester of the 7-hydroxymethylbenz(a)anthracene, have been shown to be capable of inducing mutations in a bacterial system.[7] Dihydrodiols and phenols may also be metabolized by further oxidation to form multiple hydroxylated derivatives such as triols and tetrols.[1] These metabolites are generally considered detoxication products, although one class of metabolic intermediates in the formation of such compounds, the bay-region diol-epoxides (Figure 2), are ultimate carcinogenic metabolites,[1,2,6,8,9] capable of inducing cytotoxicity and mutation in cells in culture and tumors in animals.[1,2] The formation of such reactive metabolites is generally detected in cells by analysis of the reaction products formed with cellular macromolecules such as protein, RNA, and DNA.[2,6] The analysis of PAH-DNA interaction products will be discussed later in this chapter. Other unsubstituted PAH are metabolized by pathways similar to those

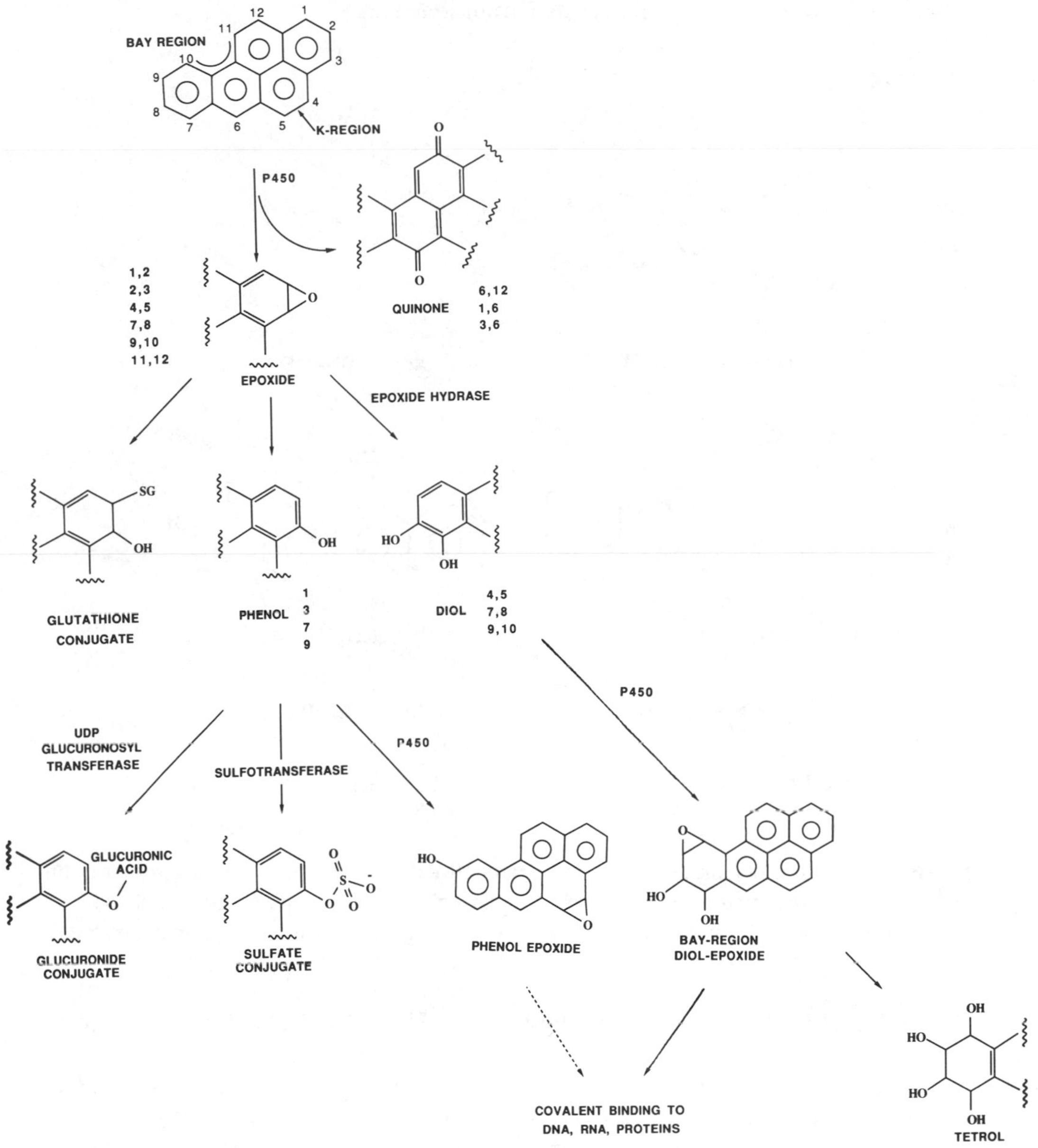

FIGURE 1. Major pathways of metabolic activation and conjugation of BaP.

depicted for BaP (Figure 1).[2] Methyl-substituted hydrocarbons, such as 7,12-dimethyl-benz(a)anthracene (DMBA), are metabolized by the same pathways, but are also subject to oxidation of the methyl groups; thus, analysis of the metabolic pathways of methyl-substituted PAH tends to be very difficult because of the complexity of the metabolite profiles.[1] The reactive DNA-binding metabolites of methylated hydrocarbons identified to date have been ''bay-region'' diolepoxides,[6] such as those shown for DMBA in Figure 2.

Most studies of PAH metabolism in aquatic organisms have been carried out with subcellular fractions. In this chapter, the metabolic activation of PAH by subcellular fractions of marine and freshwater fish will be compared with terrestrial organisms, mainly rodents. The data available on PAH metabolism in cell cultures prepared from fish will then be

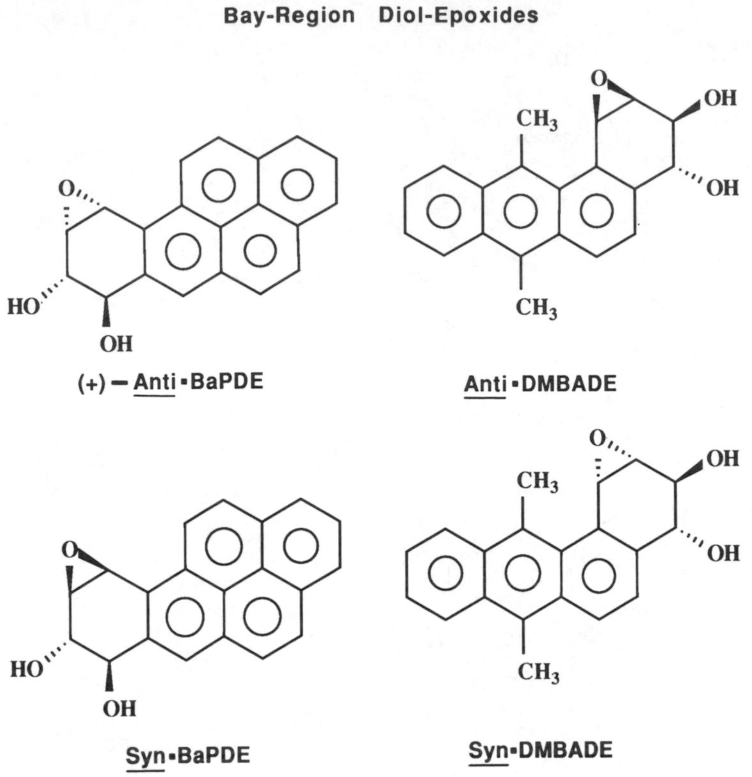

Bay-Region Diol-Epoxides

FIGURE 2.   Structures of bay-region diol-epoxides of BaP (BaPDE) and DmBa (DMBADE).

compared with data obtained from mammalian cell cultures. Finally, ongoing research on the metabolism and binding of PAH to DNA in cell cultures prepared from fish will be described and compared with studies that have been carried out using rodent embryo cell cultures.

II. METABOLISM OF BaP BY HEPATIC SUBCELLULAR FRACTIONS

Cell fractions have been used extensively to study the metabolic pathways of PAH in terrestrial and aquatic organisms; however, because the organization and integrity of the cell are destroyed during the preparation of cell fractions, data obtained in these *in vitro* assays do not always accurately represent the metabolic pathways *in vivo*. Nevertheless, because of the ease of obtaining cell fractions and because of the high concentrations of enzymatic activities present, subcellular fractions have been used to characterize enzyme systems (e.g., MFO, EH, GST) by determining (1) differences in proportion of metabolites (e.g., K-region vs. non-K-region metabolism) formed by different organisms (regiochemical control), (2) the stereochemical pathways of epoxidation and subsequent hydration to form diols and tetrols (stereochemical control), and (3) the activity and nature of these enzymes in cell fractions from different tissues of diverse organisms. Such studies have delineated important differences in the pathways of the metabolism of several PAH and in the types and the activities of enzymes present in various species (see Chapter 5 for discussion on enzymes involved in metabolism of PAH).

Most of the studies on PAH metabolism *in vitro* have been accomplished using BaP which has been shown to be a potent carcinogen in laboratory experiments with rodents (target sites: skin, lung, and mammary glands)[5] and a fish species, rainbow trout (target site: liver).[11] Hence, in this chapter we will compare primarily the *in vitro* metabolism of BaP by terrestrial and aquatic organisms in order to elucidate important differences and similarities in metabolic activation of PAH by these organisms.

A. S-9 Fraction

The postmitochondrial fraction (or S-9 fraction) contains the majority of the MFO, EH, and GST activities present in fractionated tissue. Early reports of studies where the *Salmonella typhimurium* mutagenicity testing assay has been used in combination with rat liver S-9 fraction, showed that mutagens were formed from PAH by the cytochrome P-450-dependent MFO.[10] Subsequent studies using BaP showed that certain metabolites were more mutagenic than others and that the 7β,8α-dihydroxy-9α,10α-epoxy-7,8,9,10-tetrahydroBaP (*anti*-BaPDE) was highly mutagenic and carcinogenic.[9] Because only certain metabolites of PAH were shown to be cytotoxic, mutagenic, or carcinogenic, information on the type and proportion of metabolites formed by different species becomes important in our attempts to correlate these differences with susceptibility or resistance of various organisms to PAH-induced carcinogenesis.

Analysis of metabolites formed by S-9 fractions of terrestrial and aquatic animals has been limited[12-16] (Table 1). Our earlier study with radiolabeled BaP and hepatic S-9 fraction from 3-methylcholanthrene (3-MC)-treated Sprague-Dawley (S.D.) rats showed that the metabolites separated on reversed-phase high-pressure liquid chromatography (HPLC) were the BaP-9,10-diol, BaP-4,5-diol, BaP-7,8-diol, BaP quinones, BaP-4,5-oxide, and 1-, 3-, 7-, and 9-hydroxyBaP.[16] The proportion of BaP-4,5-diol was substantially lower than either BaP-7,8-diol or BaP-9,10-diol; however, if the proportion of BaP-4,5-oxide is taken into account, the percentage of K-region metabolites formed was comparable to that of BaP-7,8-diol. Thus, the metabolite profiles obtained with S-9 fractions were qualitatively similar to those obtained with the microsomal system (Table 2). It should be noted, however, that significant amounts of radioactivity remained in the aqueous phase after organic-solvent extraction. This radioactivity was probably due to BaP metabolites being conjugated by detoxication enzymes present in the S-9 fraction.

In addition to diols, our results showing that rat liver enzymes converted BaP into 1- and 7-hydroxyBaP, along with the two frequently reported derivatives, 3- and 9-hydroxyBaP, are in agreement with the results of Selkirk et al.,[32] showing that both 1- and 7-hydroxyBaP are formed by rat liver microsomes (Table 1). In most studies, due to poor resolution of phenols, 1- and 7-hydroxyBaP often elute with 3- or 9-hydroxyBaP. Profiles of BaP metabolites obtained with hepatic S-9 fractions of 3-MC-induced S.D. rats[14] were qualitatively similar to those reported for a polychlorinated biphenyl (PCB)-induced S.D. rats,[15] except that in the latter study[15] neither 1- and 7-hydroxyBaP nor BaP-4,5-oxide were resolved and the quantitation was based on UV absorbance of metabolites.

The BaP metabolite profiles obtained with rat liver preparations[33] are qualitatively similar to those obtained with freshwater and marine fish species, such as mullet, goldfish, brown bullhead, black bullhead, starry flounder, and coho salmon (Table 1). However, except for mullet, the percentage of the K-region metabolites (BaP-4,5-diol, BaP-4,5-oxide) formed by fish liver enzymes is very low (<5%) or undetectable. High proportions of BaP 4,5-diol obtained with hepatic S-9 fractions of Aroclor 1254®-treated mullet cannot be easily explained on the basis that Aroclor 1254® contains phenobarbital (PB)-type inducers, which cause an increase in K-region metabolism of PAH in mammals,[34-36] because fish species are generally not responsive to PB-type inducers.[37,38] In a recent Ames test using rainbow trout hepatic microsomes and BaP, no increase in revertants per plate over controls were observed for PB-induced microsomes. In contrast, both 3-MC and PCB (Kaneclor 500) induction was associated with increased amounts of mutagenic intermediates formed by trout hepatic mi-

Table 1
BaP METABOLITES FORMED BY HEPATIC S-9 FRACTIONS OF RAT AND FISH

Species	Pretreatment[a]	nmol BaP per mg protein	MFO[b]	% Total metabolites								Ref.
				9,10-diol	4,5-diol	7,8-diol	Quinones[c]	9-hydroxy	1-hydroxy	3-hydroxy		
Terrestrial S.D. rat	3-MC	17—50	930	34	5.4	19	10	2.0	—[d]	29		12
	3-MC	1.0	—[e]	18	3.0	10	16	13	2.5	14		13,16
	Aroclor 1254® (PCB)	50	777	21	14	34	20	n.d.[f]	—	9.8		14
Freshwater fish (water temperature)												
Brown bullhead (25°C)	Control	17—50	5.0	54	n.d.	tr	tr	n.d.	—	tr		12
	3-MC	17—50	107	36	n.d.	31	11	n.d.	—	22		12
Black bullhead (25°C)	3-MC	17—50	64	30	3.6	27	3.8	n.d.	—	35		12
Goldfish (25°C)	Control	17—50	n.d.	n.d.	n.d.	n.d.	n.d.	n.d.	—	n.d.		12
	3-MC	17—50	49	4.7	2.0	11	40	22	—	20		
Coho salmon (13°C)	3-MC	1.0	—	33	4.8	9.0	8.5	2.2	4.8	11		13, 16
Rainbow trout (16°C)	Control	1.0	—	25	n.r.[g]	12	12	—	25	][h]		15
	PCB	1.0	—	23	n.r.	14	7.5	—	30	]		
Rainbow trout (7°C)	Control	1.0	—	20	n.r.	18	14	—	26	]		15
Marine fish (water temperature)												
Starry flounder (13°C)	3-MC	1.0	—	47	2.7	15	3.5	1.8	1.3	9.9		13
Mullet (18—22°C)	Aroclor 1254® (PBC)	50	160	6.4	19	33	29	4.9	n.d.	7.4		14

[a] 3-MC, 3-methylcholanthrene; PCB, Aroclor 1254, polychlorinated biphenyls.

[b] Pico mole BaP metabolized per milligram S-9 protein per minute.

[c] Quinones; BaP 1,6-, 3,6-, and 6,12-quinones.

[d] Not resolved from 3-hydroxy BaP.

[e] Substrate-to-protein ratio (SPR) too low for accurate measurement of MFO.

[f] n.d. — not detected.

[g] n.r. — not reported.

[h] Percents of BaP phenols were determined by silica-gel TLC.

Table 2
BaP METABOLITES FORMED BY HEPATIC MICROSOMES OF RODENTS AND FISH

Species	Treatment[a]	nmol BaP per mg protein	MFO[b]	% total metabolites								Ref.
				9,10-diol	4,5-diol	7,8-diol	Quinones	9-hydroxy	1-hydroxy[c]	3-hydroxy		
Terrestrial												
S.D. rat	Control	680	480	3.3	7.7	2.1	24	6.0	—	57		17
	3-MC		2140	19	6.2	9.3	15	6.6	—	45		
	Control	270	3350	7.8	2.1	3.9	27	7.8	—	52		18
	3-MC		8010	12	5.5	8.0	21	13	—	41		
	PB		5420	9.2	9.4	3.9	28	2.8	—	47		
	Control	15	210	19	6.1	9.5	20	1.4	—	45		19
	3-MC		2160	20	4.3	16	13	3.9	—	44		
Long-Evans rat	Control	350	1160	8.3	6.3	4.2	34	13	—	35		20
	3-MC		4970	18	6.5	14	19	5.8	—	37		
	PB		1390	6.8	23	3.0	28	8.3	—	30		
	Control	950	525	14	9	10	36	6	—	27		21
	3-MC		3650	22	10	13	26	2	—	28		
	PB		1060	10	15	5	38	2	—	30		
Wistar rat	Control	80—800	n.r.[d]	8.9	9.5	4.3	32	6.2	—	27		22
	3-MC		n.r.	16	11	11	27	15	—	15		
Fischer 344 rat	Control	32—160	560	6.5	2.9	2.9	34	3.4	—	37		23
	3-MC		2330	13	4.8	7.3	8.5	5.2	—	21		
	PB		1030	3.9	17	2.2	18	0.96	—	29		
S.D. rat	Control	32—160	590	11	11	4.7	18	2.6	—	54		23
	3-MC		2260	20	6.0	7.9	9.6	2.6	—	24		
	PB		950	4.9	15	1.7	17	0.56	—	32		
C57BL/6J mouse	Control	950	1030	0.2	2.2	4.1	46	6.0	—	41		24
	3-MC		5070	1.2	2.6	9.3	32	15	—	40		
DBA/2J mouse	Control	950	1050	0.2	2.9	3.4	45	4.1	—	44		24
	3-MC		870	6.6	1.7	5.1	47	8.0	—	31		
S.D. rat	3-MC	1000	n.r.	15	6.2	11	12	8.0	—	47		25
NIH Swiss mouse	3-MC	1000	n.r	9.0	1.2	10	9.3	15	—	60		25
Syrian hamster	3-MC	1000	n.r.	5.4	29	4.6	3.7	1.7	—	56		25
	Control	270	4380	2.5	16	2.7	30	5.0	—	45		18
	3-MC		3320	3.0	21	2.7	33	4.2	—	37		
S.D. rat	Control	210	600	11	11	4.7	14	8.3	8.3	19		26

Aquatic											
English sole	Control	210	190	17	1.1[e]	26	7	2.3	21	26	26
	BaP	210	550	17	1.3[e]	26	6.7	1.0	18	26	
Starry flounder	Control	210	180	24	3.3[e]	22	7	5	19	20	26
	BaP	210	630	26	4.1[e]	27	96	1.8	9.8	22	
Skate	Control	100	230	9.9	3.8	15	3.8	19	—	49	27
	DBA		8080	11	6.3	14	28	18	—	34	
Winter flounder	Control	55	n.r.	21	n.d.[f]	21	17	6	—	35	28
Killifish	Control	55	n.r	19	n.d.	20	27	5	—	29	28
Scup	Control	55	1190	29	n.d.	28	15	3	—	26	28
	Control	86—120	250[g]	22	n.d.	33	27	3	—	14	
	3-MC	80—800	960[g]	6		12	16	11	—	55	29
Southern flounder	Control		15	26	4.3	27	12	12	—	32	22
	3-MC	1000	840	28	3.0	34	13	7.8	—	13	
Channel catfish	Control		n.r.	2.0	3.8	7.5	13	15	—	14	30
	3-MC		n.r.	3.8	6.0	12	16	19	—	14	
Brown trout	Control	15	800	22	1.0	28	14	2.2	—	33	19
Rainbow trout	Control	210	55	17.5	n.d.	27	9.8	n.d.	17.5	28	31
	BaP	210	340	21	1.2[e]	33	6.5	n.d.	12.0	24.2	
Speckled sanddab	Control	30	43	21	n.d.	26	n.d.	n.d.	—	53	31
	3-MC	30	100	18	n.d.	28	6	n.d.	—	48	
California killifish	Control	30	120	19	n.d.	20	8	n.d.	—	52	31
	3-MC	30	400	24	n.d.	28	9	4	—	35	

a   3-MC, 3-methylcholanthrene; PB, phenobarbital; DBA, dibenz(a,c)anthracene.

b   Picomole BaP metabolized per microsomal protein per minute; activities were determined by radiometric method.

c   HLPC system did not resolve 1-hydroxy BaP from 9- and 3-hydroxy BaP. Data from HPLC methodology indicate that 1-hydroxy BaP cochromatographs with 3-hydroxy BaP; 7-hydroxy BaP was not detected in References 26 and 30.

d   n.r., not reported.

e   Does not cochromatograph with 4,5-diol in normal-phase HPLC.

f   n.d., not detected.

g   MFO activities were determined fluorometrically.

crosomes. Thus, the net effect of Aroclor 1254®, which contains many chlorobiphenyl isomers and congeners that exert the inductive effects of both PAH and PB, on fish should be similar to that of PAH, resulting in increased production of the non-K-region metabolites of BaP, mediated by cytochromes P-448. Although a number of studies have reported induction of hepatic BaP hydroxylase activity in fish treated with PCB,[12,14,15,37-39] very little information on its effect on the formation of BaP metabolites is available. One study[15] shows that significant amounts of non-K-region metabolites were produced with liver enzymes of both untreated and PCB-induced fish, whereas only trace amounts of the K-region metabolites were formed (Table 3).

Egaas and Varanasi[15] reported that fish held at 16°C responded to the inducer Aroclor 1254® within 24 h, showing a three- to fourfold increase in the rate of BaP metabolism catalyzed by liver enzymes; however, fish held at 7°C and treated with PCB showed no induction for several days after PCB treatment (Table 3), although at 1 week there was an induction in the ability of a few fish to metabolize BaP. It is likely that, when rainbow trout were held at the lower temperature, the time required to accumulate minimum concentrations of Aroclor 1254® in liver for induction was longer, thereby yielding a delayed response to the PCB treatment at 7°C. James and Bend[43] reported that the rate of absorption of 3-MC from muscle after an i.m. injection was considerably greater in fish held at 28°C than in those held at 22°C. Varanasi et al.[44] have shown that the lowering of water temperature significantly reduces that rate of uptake of ingested aromatic hydrocarbons in liver of flatfish. Further, the results show that the lowering of water temperature from 16 to 7°C significantly increased BaP metabolism by liver enzymes. The greater activation of BaP by the liver extracts prepared from rainbow trout maintained at the lower temperature may be due to a higher proportion of polyunsaturated fatty acids (PUFA) in microsomal membranes. Hazel[40] reported a higher proportion of PUFA in the phospholipids associated with membranes in liver of cold-acclimated trout compared to that in warm-acclimated fish. Willis[41] reported a linear relationship between the rate of BaP oxidation and the percentage of PUFA in the endoplasmic reticulum of the rat. As with MFO activity, EH activity can also be influenced by both the amount and the type of lipid present in the endoplasmic reticulum.[42] Preliminary results[145] show a significantly larger hepatic EH activity in untreated trout at 7°C than at 16°C. Moreover, the ratios of phenols to diols formed by liver enzymes of fish at both 7 and 16°C were similar.[15] Thus, environmental temperature did not appear to alter the balance between the EH and MFO systems in trout. Nevertheless, these findings, along with the results on the effect of temperature on basal hydrocarbon metabolism, serve to make an important point: the thermal history of fish must be considered when xenobiotic metabolism is studied and when hepatic MFO activity is used as an indicator of environmental pollution. Further, delayed response of cold-acclimated fish to chemical inducers may help explain the absence of induction reported in fish held at low water temperatures in short-term laboratory experiments. It appears that while a number of factors, such as age and sex of the test organisms,[45] structure of inducer,[34,35,44] and the route of administration of inducers may influence PAH metabolism by both aquatic and terrestrial species, environmental temperature may have the most profound effect on the ability of aquatic organisms to respond to xenobiotic chemicals.

Because of the limited data on BaP metabolism by S-9 fractions of either terrestrial or aquatic species, a detailed discussion of differences in metabolism by various species is inappropriate here. Moreover, because of the presence of conjugating enzymes in S-9 fraction, the metabolite profiles obtained do not necessarily represent the distribution of the cytochrome P-450 enzymes that are responsible for the formation of primary metabolites. For example, arene oxides and epoxides are substrates for conjugation with GSH, whereas diols and phenols can be conjugated by sulfation or glucuronidation.[1] Because of the high GSH levels in hepatocytes, conjugation with GSH may represent a major detoxication

Table 3
EFFECTS OF PCB AND ENVIRONMENTAL TEMPERATURE ON *IN VITRO* FORMATION OF BaP METABOLITES BY LIVER OF TROUT (SALMO GAIRDNERI)[a]

Fish	No. of fish	Temp. (°C)	Range of binding value	% of total radioactivity			% of metabolites in ethyl acetate		
				Unmetabolized BaP	Ethyl acetate-soluble metabolites	Aqueous phase	Phenols	7,8-diol	9,10-diol
Untreated	13	7	(0.08—0.51)	32 ± 22[b,c,d]	46 ± 16[d]	21 ± 7[d]	28 ± 8	16 ± 4	22 ± 4
PBC-treated, not induced[e]	13	7	(0.07-0.63)	56 ± 22	36 ± 16	15 ± 5	28 ± 6	6 ± 3	20 ± 3
Untreated	13	16	(0.03-0.11)	79 ± 10	12 ± 8	9 ± 3	25 ± 6	21 ± 8	18 ± 5
PCB-treated, induced[f]	3	7	(2.3-5.1)	19 ± 15	55 ± 5	26 ± 12	28 ± 2	17 ± 6	25 ± 3
Untreated PCB-treated, not induced[f]	3	16	(0.04-0.18)	70 ± 26	12 ± 11	13 ± 6	17 ± 15	12 ± 10	16 ± 13
PCB-treated, induced[f]	13	16	(0.30-5.2)	9 ± 11	44 ± 18[g]	47 ± 24[g]	19 ± 13	11 ± 5	21 ± 10

a   Adapted from Egaas and Varanasi.[14]

b   Statistical comparisons were made using Pearson's correlation analysis on binding values (picomoles BaP per milligram DNA per milligram protein) and relative proportions of BaP and its metabolites in each sample.

c   Mean ± S.D.

d   Significantly ($p < 0.05$) different from the corresponding value of 16°C.

e   Liver extracts giving binding values less than three times the corresponding control value.

f   Liver extracts giving binding values greater than three times the corresponding control value.

g   Significantly ($p < 0.05$) different from the corresponding values for uninduced fish at 16°C.

pathway using S-9 fractions. Such conjugation may alter the metabolite profile by removing the arene oxides, such as BaP 7,8-oxide, before hydration to form diols or rearrangement to phenols. Although S-9 fractions have been used as an activating system for mutagenicity testing, the types of unconjugated metabolites present do not represent the true distribution of metabolites formed in the absence of conjugation systems. Thus, pathways of metabolite formation can be masked by removal of intermediates or sequestering precursors of metabolites by conjugation (e.g., thiol compounds).

B. Microsomal Fraction

Studies of the metabolism of PAH by hepatic microsomal enzymes have provided the greatest source of information regarding the stereochemical and regiochemical selectivity of metabolism of these xenobiotics in various species. Studies by several groups[1,2,6,8,9] have detailed the stereochemical course of BaP metabolism by rat liver enzymes, with emphasis on pathways leading to diol and diol-epoxide formation.[46] For example, BaP is shown to be stereoselectively metabolized to form predominantly the $(-)$-BaP-7,8-diol. This diol is then converted by MFO to the $(+)$-*anti*-BaPDE, which has been shown in mutagenicity assays, using hamster V-79 cells, to possess the highest mutagenic activity of the diol-epoxide isomers.[9] Moreover, $(+)$-*anti*-BaPDE is believed to be a major ultimate carcinogenic metabolite of BaP.[2,8,9]

Holder et al.[17] optimized parameters for BaP metabolism using hepatic microsomes from either control, 3-MC-, or PB-induced rat and determined the distribution of metabolites. They found that induction by 3-MC causes a marked increase in both the proportion and rate of formation of benzo-ring diols (i.e., BaP-9,10-diol and BaP-7,8-diol) relative to increases in other metabolites. In contrast, induction of hepatic enzymes by PB was accompanied by an increase in the proportion and rate of formation of BaP-4,5-diol, and a decrease in the proportion of benzo-ring diols. Because of an increase in BaP turnover, however, the rates of formation of both BaP-9,10-diol and BaP-7,8-diol were comparable to control values. Such data indicate that pretreatment with chemical inducers may alter the metabolic pathways which are relevant to carcinogenesis. Haake et al.[36] in their study of BaP metabolism by hepatic microsomes from rats treated with individual polychlorinated and polybrominated biphenyls or with Aroclor 1254® and Firemaster BP$_6$®, have shown that inductive properties are remarkably dependent on the structures of the inducers and can result in high regioselectivity in the formation of metabolites. These authors confirmed previous findings[35] showing that pretreatment of rodents with PB or certain halogenated biphenyls (e.g., *ortho* substitution on two or more sites) causes a marked increase in the formation of K-region metabolites of BaP; however, halogenated biphenyls which are substituted only in both *para* and two or more *meta* positions resemble 3-MC in their induction pattern. Mixed-type inducers, namely, Aroclor 1254®, or PCB congeners such as 2,3,3′,4,4′pentachlorobiphenyl, not only increased the formation of both Bay- and K-region diols as expected, but also substantially enhanced the formation of the 6,12-quinone, suggesting a preferential increase in pathways leading to radical cation formation.[36]

In addition to the structure of the inducer, the species or strain of test animal also significantly influences both the level and type of induction.[47,48] For example, Holder et al.[24] showed that 3-MC induction of the CB (genetically responsive) mouse increased the proportion of BaP-7,8-diol two- to threefold, but decreased the proportion of BaP-9,10-diol over a tenfold substrate concentration range. In contrast, hepatic microsomes from the DB (genetically unresponsive) mouse showed an opposite trend at high substrate concentration, where BaP-9,10-diol was increased 30-fold by 3-MC induction, and BaP-7,8-diol increased only 2-fold. Moreover, liver microsomes from the DB mouse formed a lower proportion of BaP-7,8-diol at high substrate concentration than those from the CB mouse, regardless of pretreatment. At high substrate-to-protein ratio (SPR), calculated in terms of nanomole of substrate per milligram of microsomal protein, the majority of the radioactivity was extractable by organic solvents. At low SPR, however, a greater amount of secondary oxidation occurred which was accompanied by a larger proportion of the radioactivity remaining in the aqueous phase, presumably bound to protein. Dent et al.[23] have shown that a higher proportion of BaP-diols was produced by hepatic microsomes from S.D. rats than Fischer-344 rats (Table 2). Moreover, a higher proportion of metabolites resulting from the formation of BaP-9,10-epoxide (e.g., BaP-9,10-diol and 9-hydroxyBaP) was produced by liver microsomes of S.D. rats than from Fischer rats. Additionally, the results showed that hepatic microsomes from PB-induced S.D., Long-Evans, and Fischer rats produced higher proportions of BaP-4,5-diol compared to those produced by untreated animals.[18,20,23]

In contrast to mice and rats, uninduced or 3-MC-induced hepatic microsomes of Syrian hamsters form a preponderance of BaP-4,5-diol, over a wide range of SPR (Table 2). Selkirk et al.,[25] using a SPR value of 1000, compared the metabolism of BaP by hepatic microsomes from 3-MC-induced mice, rats, and hamsters. These authors reported that hamsters formed less 9-hydroxyBaP, BaP quinones, BaP-7,8-diol, and BaP-9,10-diol than either mice or rats. No significant species-specific difference was observed in the proportion of 3-hydroxyBaP formed by microsomes. However, the proportion of BaP-4,5-diol was 5 and 24 times higher than that formed by hepatic microsomes of rats and mice, respectively. Prough et al.[18] obtained similar metabolite profiles with both uninduced and 3-MC-induced Syrian hamsters, using a SPR of 270, except for a significantly higher proportion of BaP quinones than was reported by Selkirk et al.[25] Although this difference in the proportion of BaP quinones between the two studies was not discussed by Prough et al.,[18] differences in SPR and in experimental procedures may be responsbile for this discrepancy. Hepatic microsomes from untreated or 3-MC-induced hamsters produced a four to eight times higher proportion of BaP 4,5-diol than those from rats. Hepatic MFO activities for uninduced hamsters were slightly higher than the values for the rats, whereas MFO activities for the rats were approximately two times greater than that for hamsters when both animals were induced with 3-MC (Table 2). Hence, for both uninduced and 3-MC-induced animals, the total amount of BaP-4,5-diol formed by hamsters was greater than that formed by rats.

It is apparent that a number of factors, such as SPR, species or strain of test animals, and the structure of inducers, affect the metabolism of PAH. Although all rodent species produce qualitatively similar metabolite profiles and, in most cases, the rate of BaP metabolism increases after exposure of these organisms to either 3-MC or PB, indicating induction of cytochromes P-450, the quantitative differences observed indicate that different rodent species have different PAH-metabolizing capabilities. These differences in hepatic cytochrome P-450 enzymes in various species and strains may play a significant role in determining the differences in the susceptibilities of extrahepatic tissues (e.g., lung, skin) to PAH-induced carcinogenesis, since different concentrations of carcinogenic metabolites would be available for circulation to, and deposition in, extrahepatic tissues.

Studies of the metabolism of BaP by hepatic microsomes from various fish species have shown that over a 1000-fold SPR range, no qualitative differences exist between metabolites formed by various fish species (Table 2). The major metabolites formed included BaP-7,8-diol, BaP-9,10-diol, 1- and 3-hydroxyBaP, and BaP quinones (Table 2). Other minor metabolites include 9-hydroxyBaP and BaP-4,5-diol; however, the identity of the latter metabolite has not been confirmed. For example, the BaP-7,8-diol, 1-hydroxyBaP, and 3-hydroxyBaP, formed when coho salmon liver microsomes were incubated with BaP, were identified by comparison of their excitation and emission spectra with the corresponding spectra of authentic standards.[49] Structural identification of these three metabolites and BaP-9,10-diol and 9-hydroxyBaP, formed *in vivo* in English sole, was achieved by separation of these five primary metabolites from hydrolyzed bile and subsequent comparison with authentic standards using fluorescence and mass spectroscopy.[50] Our efforts to obtain sufficient quantity of 7-hydroxyBaP formed from BaP either *in vivo* or *in vitro* by fish species such as English sole or starry flounder, have not been successful. At present no other studies have reported a clear identification of this metabolite in fish. Hence, it appears that in contrast to rat liver enzymes,[33] fish liver enzymes do not generally produce significant amounts of 7-hydroxyBaP.[51] Moreover, 9-hydroxyBaP appears to be a minor metabolite in a number of fish systems (Table 2), whereas 1- and 3-hydroxyBaP were the two major phenols identified in those experiments in which good resolution of phenols was achieved by reversed-phase HPLC.

The data given in Table 2 indicates the similarity of BaP metabolites formed by liver enzymes of fish and rodent species. However, quantitative differences between the amounts of metabolites formed in *in vitro* assays at optimum conditions for fish and rodents, indicate that differences do exist in the specific content and activity of cytochrome P-450 enzymes.[29,52] Although total cytochrome P-450 content is generally greater in hepatic microsomes of rats than in hepatic microsomes of fish, the content and activity of certain individual enzymes may be comparable between species.

The data in Table 2 show that the BaP metabolites produced by dibenz(*a,c*)anthracene(DBA)-induced skate are qualitatively and proportionally similar to those found with 3-MC-induced S.D. rats.[27] The metabolites formed include a significant proportion of a metabolite having the same retention time as the standard BaP-4,5-diol on reversed-phase HPLC. Moreover, it has been reported that hepatic microsomes of the channel catfish also produce significant amounts of BaP-4,5-diol after induction with 3-MC.[31] However, the data in Table 2 reveal that the majority of the fish species studied show a marked lack of K-region metabolism. The lack of K-region metabolism is consistent with the fact that most fish species have been found to be refractive to MFO induction by PB.[38,53] Hence, the K-region metabolism reported for skate and catfish may either suggest that unlike other fish species, skate and catfish contain a cytochrome P-450 enzyme that metabolizes PAH at the K-region, or it may be that the radioactivity cochromatographing with the BaP-4,5-diol standard in the reversed-phase HPLC of metabolites formed by these species, actually contains other unidentified BaP metabolite(s). Studies from our laboratory have shown that when BaP metabolites formed by fish liver enzymes are analyzed by reversed-phase HPLC, detectable amounts of BaP-derived radioactivity cochromatographs with the BaP-4,5-diol standard.[50,54] However, when the diol fraction was chromatographed on normal-phase HPLC, no radioactivity chromatographing with BaP-4,5-diol was observed[50,54] (Figure 3). Hence, using normal-phase HPLC, we have shown that BaP-4,5-diol was not formed from BaP to an appreciable extent either *in vivo* or *in vitro* by English sole, starry flounder, and the Mount Shasta strain of rainbow trout.[26,54,55] For *in vitro* studies, this finding was consistent over a 200-fold range of SPR using liver enzymes from various teleost fish species.

The proportion of BaP quinones produced by hepatic microsomes of fish species is generally less than that produced by rodents (Table 2). The major route of quinone formation is believed to be initiated by oxidation at the C-6 position of BaP to form the 6-oxy BaP free radical.[56] Using electron spin resonance (ESR) spectrometry, the 6-oxy BaP free radical has been detected in the metabolism of BaP by rat liver enzymes; however, using English sole liver microsomes, no free-radical signal was observed in the oxidation of BaP.[146] Because the pathway of quinone formation from the 6-oxy BaP free radical is considered nonenzymatic, the differences in the proportion of quinones is directly related to the differences in the ability of the organisms to form the 6-oxy BaP free radical.

Both the experimental conditions and the analytical procedures used for separation and identification of metabolites affect the metabolite profiles obtained. Therefore, comparison of xenobiotic metabolism by different species is most valid when experiments are conducted in the same laboratory. The following studies are discussed in more detail, because in each case BaP metabolism was studied in both fish and rodent species under comparable experimental conditions.

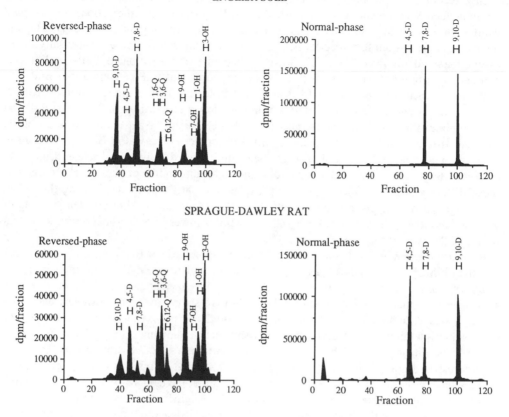

FIGURE 3.   Reversed-phase and normal-phase HPLC of BaP metabolities formed by English sole and Sprague-Dawley rat hepatic enzymes. Tritiated BaP (80 µ*M*) was incubated in a standard MFO assay and the metabolites were separated by reversed-phase HPLC. The diol region from a reversed-phase HPLC run was collected and rechromatographed using normal-phase HPLC.

The tendency of fish species to produce higher proportions of benzo-ring diols than rodent liver was demonstrated in a recent study with a feral species, southern flounder, sampled from a relatively uncontaminated estuary in Florida.[22] The authors reported that both untreated and 3-MC-treated flounder having a wide range of hepatic MFO activities (14 to 840 pmol BaP per minute·milligram of protein) showed a very similar pattern of BaP metabolism with a marked preference for benzo-ring oxidation which accounted for 60 to 70% of the metabolites formed (Table 2). In contrast, hepatic microsomes from untreated and 3-MC-treated Wistar rats produced only 20 and 40% of the metabolites through benzo-ring oxidation (Table 2). Furthermore, when the fish were treated with 3-MC, an increase in overall metabolism was accompanied by an increase in the proportion of BaP-7,8-diol and a decrease in 9- and 3-hydroxyBaP.[57] As with southern flounder, increases in BaP metabolism and in the formation of BaP-7,8-diol and 3-hydroxyBaP were observed for Wistar rats treated with 3-MC, except that with rats the proportion of BaP-9,10-diol, BaP-4,5-diol, and 9-hydroxyBaP was also increased significantly. The opposite effects of 3-MC treatment on 9-hydroxyBaP formation between rat and flounder may be due to differences in the enantiomeric composition of the BaP-9,10-oxide formed with rat and fish microsomes. It has been shown that the (+)-BaP-7,8-oxide has a higher affinity for rat microsomal EH than the (−)-BaP-7,8-oxide.[58] Thus, differences in the ratio of BaP-9,10-diol to 9-hydroxyBaP between fish and rat may be due to differences in the enantiomeric purity of the BaP-9,10-oxide formed, and also to the differences in the substrate affinities for rat and fish microsomal EH.

In a recent study, Varanasi and colleagues[26] compared microsomal metabolism of BaP by uninduced S.D. rat with two feral fish species, English sole and starry flounder, sampled from relatively uncontaminated areas in Puget Sound, Washington, (Figure 4). Hepatic microsomes from both fish species, held at 12°C, metabolized BaP to a substantially lower extent than did rat liver (Table 2); however, the rate of formation of BaP-7,8-diol was comparable among all three species. Thus, the proportion of BaP converted into BaP-7,8-diol was three- to fourfold higher for fish liver than for rat liver (Table 2). Moreover, both fish species produced virtually no BaP-4,5-diol and lower proportions of 9-hydroxyBaP and BaP quinones than rats. These differences in BaP metabolism by hepatic microsomes of the two fish species and rat are consonant with the differences observed in BaP metabolism *in vivo*, showing that BaP-7,8-diol was the major diol formed in liver and released into bile of both fish species, whereas BaP-4,5-diol is the major diol in bile of rat.[26,50,59,60] Moreover, these differences in BaP metabolism help explain, in part, the results showing that when these organisms were given equivalent doses of BaP (i.p.), the level of covalent binding of BaP metabolites to hepatic DNA was 40 to 80 times lower in S.D. rats compared to that found in the two fish species.[26]

In addition when both fish species were given the same dose of BaP p.o. or i.p., chemical modification of hepatic DNA was two- to fourfold higher in English sole than in starry flounder.[26] Further, the proportion of BaP-7,8-diol in enzymatically hydrolyzed bile was significantly higher in BaP-exposed English sole than starry flounder. However, these differences observed in BaP metabolism *in vivo* could not be explained by comparison of microsomal metabolism of BaP by the two fish species, showing no significant difference in either the rate of BaP metabolism or in the proportion of BaP-7,8-diol formed (Table 2). An interesting finding was that while hepatic MFO and EH activities for untreated English sole and starry flounder held in aquaria were comparable, the hepatic GST activity for starry flounder was almost three times higher than that for English sole.[61,147] Studies with mammalian systems show that both cellular GSH level and GST activity are inversely related to DNA binding of BaP metabolites, although there appears to be a better negative correlation between GST activity and DNA modification.[62] Hence, higher hepatic GST activity in flounder may cause more effective conjugation of BaP-7,8-oxide with GSH resulting in less BaP-7,8-diol available for conjugation with glucuronic acid and subsequent release into bile, as well as less BaPDE formed. In addition, BaPDE could also be more effectively conjugated with GSH in flounder liver than in sole liver, thereby further reducing available BaPDE for binding to hepatic DNA in flounder. Thus, it appears that the differences in detoxication rather than activation of BaP into reactive intermediates (e.g., epoxides) result in higher chemical modification of hepatic DNA in English sole, if it is assumed that rate of excision-repair of modified DNA is comparable for pleuronectids. These studies emphasize the need to evaluate detoxication pathways, along with activation, of carcinogenic compounds to allow proper interpretation of information regarding differences in tumorigenic susceptibilities of various species and tissues emerging from laboratory studies or from epizootiological surveys.

III. METABOLISM OF BaP BY EXTRAHEPATIC CELL FRACTIONS

In recent years, tumorigenicity studies conducted with BaP and DMBA demonstrated that these PAH cause liver cancer in rainbow trout and two viviparous fish species, *Poeciliopsis lucida* and *P. monarcha,* respectively.[11,63] Of particular interest is the study showing that trout treated with BaP either p.o. or i.p. had a 25 to 50% incidence of liver cancer, since this PAH is present at significantly high levels in bottom sediments of industrialized waterways where a high prevalence of liver neoplasia is reported in both freshwater and marine fish species, such as brown bullheads and English sole.[64,65] PAH, such as BaP or DMBA,

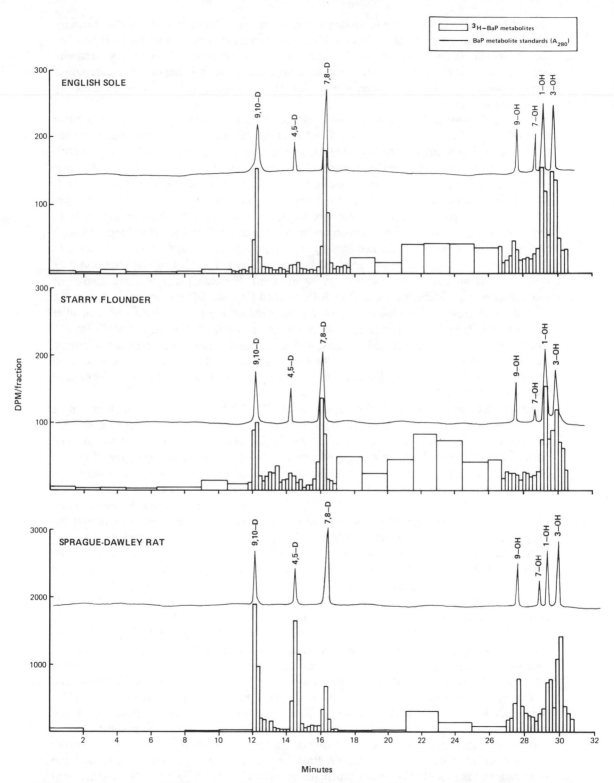

FIGURE 4.   Reversed-phase HPLC analyses of BaP metabolites formed by hepatic microsomes of English sole, starry flounder, and Sprague-Dawley rat. [³H]BaP was incubated at a concentration of 80 μM with hepatic enzymes and co-factors, then extracted with methanol:chloroform. The open bars represent the radioactivity in each fraction, whereas the trace represents UV monitoring at 280 nm of the internal BaP metabolite standards. (From Varanasi, U., Nishimoto, M., Reichert, W. L., and Eberhart, B.-T. L., *Cancer Res.*, 46, 3817, 1986. With permission.)

are not hepatocarcinogens in adult rodents, but are potent pulmonary and epidermal carcinogens in certain strains of mice and rats. These differences in target tissues for PAH-induced carcinogenesis between rodent and fish species are of considerable interest. As metabolic activation of PAH is a prerequisite for its carcinogenic action, it is important to ascertain if target tissues, such as rodent lung, rodent skin, and fish liver, metabolize BaP differently from rodent liver.

The microsomal MFO activity in 3-MC-induced A/HeJ mouse lung, which is susceptible to PAH-induced carcinogenesis, was 1.4 and 2.7 times higher than the corresponding values for C57BL/6J and DBA/2J mice (Table 4), which are both resistant to pulmonary tumorigenesis. However, after 3-MC induction, the proportions of metabolites formed by lung microsomes between different strains of mice were identical. In all cases, treatment with 3-MC resulted in decreased proportions of quinones and increased proportions of 3-hydroxyBaP. In addition, the proportion of BaP-7,8-diol formed was the same between strains. In comparisons between tissues, liver microsomes have greater MFO activity than lung, although the proportion of BaP diols formed by lung microsomes was at least two times greater than that formed by liver. This can be attributed to the greater EH to MFO activity ratio in lung than in liver, although other factors may be involved. For example, our recent study[26] of the comparative metabolism of BaP in S.D. rat and English sole shows that although the sole liver microsomes produced a greater proportion of benzo-ring diols and a smaller proportion of 9-hydroxyBaP than rat liver microsomes (Table 2), the ratio of MFO activity to EH activity is the same for both species, indicating that this ratio alone does not determine the proportion of metabolites formed. Other factors, such as the enantiomeric composition of the arene oxides formed by liver enzymes,[58] may play a significant role in the metabolism of PAH such as BaP.

Epidermal microsomes from the mouse strain C56BL/6N, which has been shown to be susceptible to PAH-induced epidermal tumors, produced the same pattern of BaP metabolites as the DBA/2N mouse and the neonatal S.D. rat, both of which are resistant to PAH-induced skin cancer (Table 4). Furthermore, the proportion of BaP-7,8-diol is the same for all species after 3-MC or PB induction. Minor differences between species include a decrease in the proportions of BaP-4,5-diol and BaP quinones in C57BL/6N mouse after 3-MC or PB induction, whereas in DBA/2N mouse and S.D. rat, there was an increase in these metabolites. An interesting fact observed following the treatment of PB to rodents is that the proportion of BaP-4,5-diol is not significantly increased when microsomes from extrahepatic tissues are used, although a marked increase is observed with liver microsomes. A significant increase in the rates of BaP metabolism, calculated based on per milligram of microsomal protein, was observed for both hepatic and extrahepatic microsomes after PB induction.

Comparisons of metabolism by S.D. rat and hamster lung microsomes show that MFO activities for the control animals were comparable; however, after 3-MC induction, a 5-fold increase in pulmonary MFO activity was observed for rat compared to only a 2.7-fold increase for hamster (Table 4). The proportion of products derived from BaP-9,10-oxide that were formed by 3-MC-induced rat lung microsomes was two times that of hamster, whereas the proportion of BaP quinones was twofold higher with hamsters than with rats. In a recent study, Hall and Grover compared the metabolism of BaP by microsomes of skin, lung, and liver of male Parkes mouse.[145] Both lung and skin are target tissues for PAH-induced carcinogenesis; however, neither the proportion of diols formed nor the rates of formation of BaP diols correlated with tissue susceptibility to PAH-induced cancer. Although stereochemical evidence showed that the BaP-7,8-diol enantiomer was the R,R isomer, the precursor to the *anti*-BaPDE, other biochemical differences (e.g., rate of DNA repair) must be considered in conjunction with the pathways of PAH metabolism observed *in vitro*.

Virtually no information is available on individual BaP metabolites formed *in vitro* by extrahepatic tissues of fish, except for studies with microsomes from kidney, heart, and gill of scup (Table 4).[69] All tissues contained significantly lower levels of microsomal MFO activity than scup liver; however, the proportions of metabolites formed by all microsomes were similar to that found with liver (Table 2). This is in contrast to terrestrial species, such as S.D. rat, C57BL/6J mouse, DBA/2J mouse, and Syrian hamster, where the ratios of proportion of metabolites formed are different between liver and extrahepatic microsomes.

In summary, the differences in the pattern of BaP metabolism by extrahepatic tissues of rodents (Table 3) are not consistent when sensitive vs. resistant strains are compared. However, comparison of data in Tables 2 and 3 shows that rodent lung and skin as well as fish liver tissues metabolize BaP at a considerably lower rate than does rodent liver, and that, in contrast to rodent liver, pretreatment with PB does not result in increased formation of BaP-4,5-diol with microsomes of rodent lung, of epidermis, or of fish liver.[27,38] It is obvious that an examination of the metabolism and activation of a xenobiotic *in vitro* is not sufficient to enable an understanding of the differences in susceptibility of various species, strains, and tissues to carcinogenesis. Studies on the activation and detoxication of xenobiotics *in vivo*, together with information on the transport of xenobiotics and their reactive intermediates from liver to extrahepatic sites, and their persistence or removal from these sites are necessary in understanding the etiology of PAH-induced tumorigenesis.

IV. METABOLISM OF OTHER PAH

As discussed earlier, much of the research on the metabolism of PAH *in vivo* by mammals has been limited to compounds known to be carcinogenic. However, studies with fish species have included several PAH, such as naphthalene, methylnaphthalene, and phenanthrene, because these compounds are major constituents of petroleum. Concerns with regard to toxicity of these PAH and their metabolites in aquatic organisms after an oil spill have prompted a number of studies to be aimed at evaluating the metabolism of lower molecular weight PAH in fish and invertebrates. However, most of these studies were confined to *in vivo* metabolism and, thus, very little comparative information is available on *in vitro* metabolism of these PAH.

The metabolism of 2-methylnaphthalene (2-MN) was studied using trout and S.D. rat liver microsomes.[70,71] The major metabolites formed by both species included the 2-MN-3,4-diol, 2-MN-5,6-diol, 2-MN-7,8-diol, and the 2-hydroxymethyl derivative (2-OH-MN). The rates of formation of all the diols were approximately ten times faster in rat than in trout. After induction with β-naphthoflavone, a 3-MC-type inducer, the rates of formation of all the diols in both rat and trout were significantly greater than those for untreated animals. However, with rainbow trout hepatic microsomes, the formation of the 2-MN-5,6- and 7,8-diols was preferentially increased (approximately two to threefold) compared to the 2-MN-3,4-diol. When trout and rat were treated with PB, no significant increases were observed in rates of formation of all three diols in trout and of the 5,6- and 7,8-diols in rat; however, a significant induction of the 2-MN-3,4-diol was observed with rat liver microsomes. Thus, it appears that formation of the 3,4-diol of 2-MN is mediated by the cytochrome P-450 responsible for the formation of BaP-4,5-diol. The difference in induction by PB of the metabolism of 2-MN by trout and rat is reflective of that seen using BaP as the substrate.

In addition to diols, rat liver microsomes also produce significant amounts of 2-OH-MN which were not produced by trout microsomes.[71] Both PB and β-naphthoflavone induction increased the rate of formation of 2-OH-MN significantly in rat liver microsomes. When hepatic cytochromes P-448 and P-450 were isolated from β-naphthoflavone and PB-treated rat and used as the source for metabolism studies, the 2-OH-MN was formed at a much greater rate when cytochrome P-450 was used in place of cytochrome P-448.

Krahn et al.[72] studied metabolism of 2,6-dimethylnaphthalene (DMN) using liver microsomes from freshwater rainbow trout to show that two major metabolites, the 2-hydroxy-

Table 4
BaP METABOLITES FORMED BY MICROSOMES OF EXTRAHEPATIC TISSUES OF RODENTS AND FISH

Species[a]	Treatment[b]	nmol BaP per mg protein	MFO[c]	% total metabolites						Ref.
				9,10-diol	4,5-diol	7,8-diol	Quinones[d]	9-hydroxy	3-hydroxy	
Terrestrial										
C57BL/6J mouse lung (−)	Control	50	9	10	10	12	24	4	33	21
	3-MC	50	191	10	10	14	15	7	39	
DBA/2J mouse lung (−)	Control	50	5	6	4	4	36	4	18	21
	3-MC	50	96	10	9	14	11	6	39	
A/HeJ mouse lung (+)	Control	50	5	10	8	10	24	6	30	21
	3-MC	50	260	10	10	14	16	9	38	
S.D. rat lung (+)	Control	16—80	23	4.7	7.8	6.5	31	16	35	18
	3-MC	16—80	113	15	8.0	9.7	15	15	38	
	PB	16—80	29	3.8	6.2	6.6	29	17	37	
Syrian hamster lung (+)	Control	16—80	22	4.1	6.9	8.7	44	9.8	27	18
	3-MC	16—80	59	9.0	5.9	9.3	37	5.1	34	
Swiss-webster mouse lung (+)	Control	80	160	9.4	6.1	7.3	16	4.5	31	66
Syrian hamster trachea (+)	BaP	12.5—250	86	5.3	2.6	7.2	50	8.6	18	67
C57BL/6N mouse epidermis (+)	Control	53	46[e]	2.5	8.5	8.5	25	29	15	68
	3-MC	53	193[e]	2.1	6.3	8.8	14	58	18	
	PCB	53	821[e]	1.7	2.1	8.6	9.2	56	21	
DBA/2N mouse epidermis (+)	Control	53	133[e]	3.0	4.1	7.5	11	50	20	68
	3-MC	53	214[e]	2.4	6.3	8.7	15	45	17	
	PCB	53	268[e]	1.9	5.3	6.1	20	42	18	
S.D. rat epidermis (−)	Control	53	68[e]	3.6	7.1	8.6	12	42	24	68
	3-MC	53	768[e]	3.5	4.5	7.7	15	43	24	
	PCB	53	1086[e]	4.5	4.4	8.7	7.7	46	24	

Aquatic

Scup kidney	Control	130	228	31	n.d.	22	6	6	35
Scup gill	Control	80	32	28	n.d.	25	9	1	34	69
Scup ventricle	Control	10—30	11[e]	31	n.d.	31	19	[	18	]
Scup atrium	Control	10—30	18[e]	39	n.d.	27	17	[	16	]

[a] (+) and (−) denotes susceptibility of tissue to BaP-induced carcinogenesis.

[b] 3-MC, 3-methylcholanthrene; PB, phenobarbital.

[c] Picomole BaP metabolized per milligram microsomal protein per minute; activities were determined by radiometric methods.

[d] Quinones; BaP 1,6-, 3,6-, and 6,12-quinones.

[e] Activities were determined fluorometrically.

methyl DMN and the 3-hydroxyDMN, were formed. The 2-OH methyl derivative of DMN was formed at twice the rate of the 3-hydroxyDMN. Interestingly, no diols were formed at the 3,4-position. This is in contrast to the studies with rat hepatic microsomes, where the DMN-3,4-diol comprised 8.7 to 26.6% of the total metabolites formed, depending on the type of pretreatment the rats received.[73] In addition, the 3-hydroxyDMN was also formed in substantial amounts (8.4 to 22.1%). Induction of hepatic EH by PB or Prudhoe Bay crude oil could explain the relationship between 3-hydroxyDMN and DMN-3,4-diol. However, the effect of EH on the ratio of phenol to diol formation does not occur with 3-MC induction. Formation of a high proportion of 3-hydroxyDMN was accompanied by formation of a low proportion of DMN-3,4-diol and a high level of EH activity.[73] Although no explanation for this anomaly was given, several hypotheses are possible: (1) other metabolites may chromatograph with the 3-hydroxyDMN standard. A prime suspect would be the 4-hydroxyDMN. Resonance stabilization of the carbonium ion intermediate formed in the rearrangement of the DMN-3,4-oxide to the phenol would favor the formation of the 4-hydroxyDMN over the 3-hydroxyDMN; (2) the level of activity of EH and the proportion of diol formed are not always directly related. Studies have shown that the ($+$) and ($-$) BaP-7,8-oxide have different affinities for microsomal EH and have different rates of hydrolysis.[58] DMN-3,4-oxide formed by rat hepatic microsomes induced by 3-MC may represent a mixture of isomers. If this is the case, then one isomer may be preferentially hydrolyzed by EH, whereas the second isomer may rearrange to a phenolic product; (3) epoxidation of 2,6-DMN by 3-MC-induced rat hepatic microsomes may occur across the 2,3-bond, although this pathway would be unfavorable due to steric hindrance. However, if the DMN-2,3-oxide is formed, then rearomatization of the arene oxide would favor the formation of the 3-hydroxyDMN.

The data obtained using naphthalene derivatives show that fish metabolize these compounds at a lower rate than rodents, a finding similar to that observed for BaP and indicating that similar enzymes are responsible for the oxidation of these different PAH. Moreover, certain differences indicating regioselectivity in the formation of metabolites of naphthalenic hydrocarbons have been observed between rat and salmonids, as well as between different strains of salmonids. Whether these differences are significant in determining the toxicity of naphthalenic hydrocarbons in fish remains to be seen.

With the exception of the three PAH discussed in this chapter, no information is available on the metabolic activation of other PAH in both aquatic and terrestrial organisms. For example, DMBA and 3-MC, which are studied extensively in rodent systems, are synthetic PAH and not considered environmental pollutants. Important petroleum constituents, such as phenanthrene, are studied in aquatic organisms, but only in *in vivo* studies. PAH such as chrysene, fluoranthrene, and benzofluoranthrenes, detected in the sediment from urban areas, are worthy candidates for future studies. Recent interest in PAH metabolism in aquatic organisms should generate a broader base of information in the next few years.

V. DNA BINDING AND ADDUCT FORMATION

Current theories on the initiation of cancer by PAH suggest that one important step is the covalent interaction of the metabolites of these carcinogens with cellular DNA.[74] Researchers have used DNA binding as an indicator of the carcinogenic potential of compounds using subcellular fractions prepared from various tissues of different animals. However, there is a lack of correlation between the extent of binding of carcinogens to DNA *in vitro* and oncogenesis *in vivo*.[75] Nevertheless, determination of the extent of modification and types of DNA adducts formed *in vitro* has defined the pathways of PAH activation in various species.

A. S-9 Fraction
Studies on the binding of BaP intermediates to DNA mediated by S-9 fractions have been

limited. Early studies from our laboratory[13,16] have shown that hepatic S-9 fraction from untreated S.D. rats metabolized BaP to yield a binding value (picomole BaP equivalents per milligram DNA) of 0.30; treatment with 3-MC resulted in an 11-fold increase in the binding (Table 5). Hepatic S-9 fractions from two marine fishes, starry flounder and English sole, and a freshwater fish, coho salmon, held at 13°C, catalyzed binding of BaP intermediates to salmon sperm DNA to yield values of 0.75, 0.30, and 0.10, respectively. The binding values for both pleuronectid fishes were equal or greater than the value reported for uninduced rat hepatic microsomes. Thus, it is apparent that although overall metabolism of BaP is lower with fish liver preparations than with rat, the higher proportions of BaP-7,8-diol formed are reflected in the comparable or greater DNA binding values that are obtained with fish liver enzymes. Another important observation of this early study was that the induction of all three fish species, held at 13°C with either 3-MC, BaP, or Prudhoe Bay crude oil, resulted in substantially higher *in vitro* binding values than the value obtained for the 3-MC-treated rat, suggesting that, on induction, levels of reactive intermediates of BaP formed by fish liver preparations were significantly higher than those formed by rat liver enzymes (Table 5).

The data in Table 5 show that when starry flounder, coho salmon, and English sole were held at 8°C and sampled 24 h after an i.p. injection of 3-MC, binding values were obtained of 2.7, 1.5, and 0.8, respectively.[16] These values are considerably lower than the values obtained for fish induced at 13°C, demonstrating a strong influence of water temperature on the inducibility or on the ability of fish liver enzymes to metabolize PAH. As discussed earlier, Egaas and Varanasi[15] conducted a detailed study with rainbow trout and demonstrated the effect of environmental temperature on *in vitro* metabolism of BaP and DNA binding. Hepatic S-9 fractions from untreated trout, held at 16 and 7°C, yielded DNA binding values of 0.35 and 1.85, respectively (Table 5). When trout held at 16°C were treated with Aroclor 1254® and livers were excised at 24 to 120 h after a single injection of the inducer, significant increases in the binding values occurred. Whereas no increase was observed in binding values for hepatic S-9 fractions prepared from fish held at 7°C induced with PCB these results of *in vitro* metabolism and activation serve primarily as an indication that a number of variables, such as chemical inducers and habitat temperature, markedly influence the ability of fish liver enzyme to activate a carcinogen. These factors should be taken into account when assessing the formation of toxic metabolites that interact with critical cellular constituents in feral fish residing in polluted waterways.

B. Microsomal Fraction

Binding of PAH to DNA mediated by microsomal enzymes was first shown by Grover and Sims[83] and Gelboin,[84] using rat liver as the source of enzymes. Pretreatment of rodent species with 3-MC substantially increased the binding of both BaP and DMBA to calf thymus DNA compared to the values obtained with uninduced control microsomes (Table 5). Alexandrov et al.[77] have shown that 3-MC pretreatment of CB-hooded and Wistar rats increased the binding of BaP to DNA by 5.4- and 6.3-fold, respectively. Moreover, recent results from our laboratory[148] show that hepatic microsomes from S.D. rats, pretreated with BaP, yielded a tenfold greater binding of ^{3}H-BaP to DNA than did control rat liver microsomes. Ahokas et al.[19] reported a 35-fold increase in BaP-DNA binding using 3-MC-induced S.D. rat liver microsomes (Table 5). Differences in doses of 3-MC for induction and a greater extent of secondary oxidation at the lower SPR values used in some experiments may contribute to the differential increases in DNA binding observed in these studies. As with 3-MC, PB also increases the extent of DNA binding mediated by hepatic microsomes from rodents (Table 5). This increase in DNA binding paralleled the increase in MFO activity after PB induction.

Marked species-specific differences have been noted in the ability of fish liver microsomes

Table 5
COVALENT BINDING OF BaP METABOLITES TO DNA CATALYZED BY HEPATIC ENZYMES OF RAT AND FISH

Species	Treatment[a]	Dose of inducers (mg/kg)	Time of incubation (min)	SPR[b]	pmol BaP-equiv. per mg DNA	Ref.
S-9 Fraction						
Terrestrial						
S.D. rat	Control	—	15	1	0.30	3
	3-MC	10	15	1	3.45	
Aquatic[c]						
English sole (13°C)[d]	Control	—	15	1	0.30	13
	PBCO	10	15	1	5.25	
	3-MC[e]				0.08	
Starry flounder (13°C)	Control	—	15	1	0.75	13
	3-MC	10	15	1	8.10	
	BaP	10	15	1	8.50	
	3-MC[e]	10	15	1	2.65	
Coho salmon (13°C)	Control	—	15	1	0.10	13
	3-MC	10	15	1	4.85	
	BaP	10	15	1	5.30	
	3-MC[e]	10	15	1	1.50	
Rainbow trout (16°C)	Control	—	15	1	0.35	15
	PCB	10	15	1	9.5	
Rainbow trout (7°C)	Control	—	15	1	1.85	15
	PCB	10	15	1	1.40	
Microsomes						
Terrestrial						
CB rat	3-MC	20	10	320	8.0	76
CB rat	Control	—	30	320	9.4	77
	3-MC	40—53	30	320	54	

	Treatment	Doseᵃ	SPRᵇ	—	—	—
Wistar rat	Control	—	30	320	12	76
	3-MC	40	30	320	76	
	PB	240	30	320	20	80
	3-MC	60	30	10.5	98	
			30	10.5	79	
Sprague-Dawley rat	Control	—	30	15	0.22	19
	3-MC	75	30	15	7.8	81
	Control	—	90	5.5	37	82
	3-MC	60	10	400	50	144
	Control	—	30	5	12	
	BaP	2	30	5	53	
Aquatic						
Brown trout (4—6°C)ᵈ	Control	—	30	15	0.68	19
Roach (4—6°C)ᵈ	Control	—	30	15	0.02	19
English sole (14°C)ᵈ	Control	—	30	5	3.7	144

a   3-MC, 3-methylcholanthrene; PBCO, Prudhoe Bay crude oil; BaP, benzo(a)pyrene; PCB, polychlorinated biphenyls (Aroclor 1254®).

b   SPR, substrate-to-protein ratio (nanomole BaP per milligram protein).

c   English sole and starry flounder are marine water species, whereas salmonids and roach were held in fresh water.

d   Temperature at which fish were prior to removal of their livers.

e   Fish held at 8°C.

to catalyze the binding of BaP intermediates to DNA (Table 5). Ahokas et al.[19] showed that microsomes from untreated brown trout, held at 4 to 6°C, yield a threefold greater binding of BaP to DNA than do uninduced rat microsomes. In contrast, liver microsomes from another fish species, the roach held at low water temperature, show almost no catalytic ability in terms of MFO activity or DNA binding. Liver microsomes from control English sole, held at 14°C, yield binding values comparable to values obtained for S.D. rat at a SPR of 5.[148] These findings, given in Table 5, show that rodent hepatic microsomes catalyze the binding of BaP to DNA at a greater rate and to a greater extent than fish liver microsomes. However, studies with S-9 fractions show that liver preparations from most fish species yield higher binding values than rodents (Table 5). The presence of conjugating enzymes in S-9 fractions may be responsible for the contradictory results that have been obtained with microsomes and S-9 fractions. It appears from studies using susceptible and resistant animals that gross binding levels do not necessarily correlate susceptibility to PAH-induced oncogenesis, but that the types of DNA adducts formed and persistence of these adducts in tissues may be important factors in initiation of carcinogenesis.[85,86]

C. DNA Adducts

After the development of a chromatographic procedure for the separation of hydrocarbon-modified deoxyribonucleosides from unmodified deoxyribonucleosides,[87] Sims et al.[88] demonstrated that the major metabolite of BaP that was bound to DNA in embryo cell cultures was BaPDE (Figure 2). Subsequent studies with rodent microsomes have shown that the major intermediates of BaP that are bound to DNA are *anti*-BaPDE and 9-hydroxyBaP-4,5-oxide (Figure 5).[79,80,89] Studies with 3-MC-induced mice and rats have shown that hepatic microsomes produce at least twice as much phenol-oxide adducts as BaPDE adducts, regardless of the SPR values. It should also be noted that control rat liver microsomes produce a significant amount of BaP-4,5-oxide adducts, which is similar to the profile obtained with control or 3-MC-induced rat lung microsomes.

Legraverend et al.[90] showed that using 3-MC-induced mouse liver enzymes, three distinct BaP-DNA adducts were formed. At low substrate concentration, the major adducts were formed from the interaction of DNA with BaPDE and BaP-phenol epoxide(s). These were formed at approximately the same rate. As the SPR increased, the proportion of phenol-epoxide adduct(s) increased relative to the BaPDE adduct. In addition, the proportion of BaP-4,5-oxide-DNA adduct also increased. Similar results were obtained using DMBA as the substrate.[91] At low DMBA concentration, binding to DNA occurred mainly via the diol epoxide pathway. However, at higher SPR, binding occurred mainly through the K-region metabolite, the DMBA-5,6-oxide.[91] This increase in the proportion of K-region adducts relative to the diol-epoxide adducts of these two PAHs arises from the decrease in the rate of secondary oxidation of the diol precursors due to the high substrate-to-enzyme ratio.

Meehan et al.[92] showed that using female S.D. rat liver microsomes and BaP, only a single DNA adduct was formed. Using reversed-phase HPLC, the adduct was identified as arising from *anti*-BaPDE. This result is in contrast to the studies which showed that rat liver enzymes catalyzed the formation of two BaP-DNA adducts.[80,89] Meehan et al.[92] suggested that the magnesium chloride concentration used in incubation mixtures may have affected the qualitative nature of the adducts formed. However, later studies[93] using BaP as the substrate showed that the magnesium chloride concentration affects only the quantitative and not the qualitative aspect of DNA adduct formation. Because the SPR used in these two studies were the same, the differences in the types of DNA adducts reported is perplexing.

Several studies have reported on the types of BaP-DNA adducts formed by fish liver enzymes. For example, Ahokas et al.[19] reported that brown trout liver microsomes catalyzed the binding of BaP to DNA and that enzymatic hydrolysis of DNA and subsequent analysis by Sephadex LH-20® column chromatography showed two BaP-DNA adducts formed in

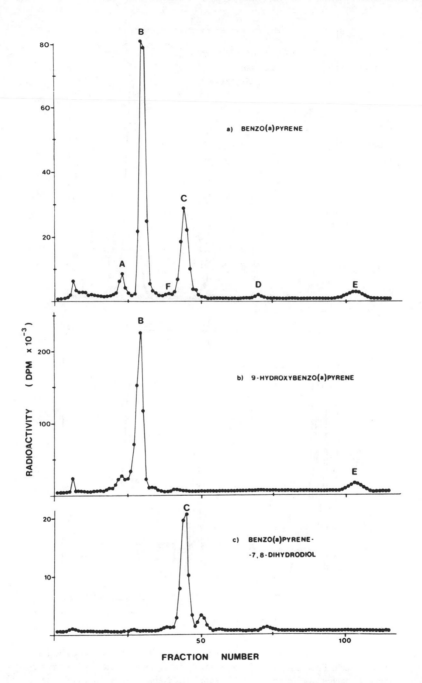

FIGURE 5.   Reversed-phase HPLC analyses of BaP-deoxyribonucleoside adducts formed by rat hepatic microsomes from (A) BaP, (B) 9-hydroxyBaP, and (C) BaP-7,8-dihydro-diol. (Adapted from Ashurst, S. W. and Cohen, G. M., *Chem.-Biol. Interact.*, 29, 117, 1980.)

equal amounts (Figure 6). These were tentatively identified as the BaPDE and 9-hydroxyBaP-4,5-oxide adducts. However, since BaP-4,5-diol was not detected in analyses of BaP metabolites formed by trout liver microsomes, the identity of the phenol-epoxide adduct is in question.

Nishimoto and Varanasi[94] analyzed BaP-DNA adduct formation by liver microsomes from

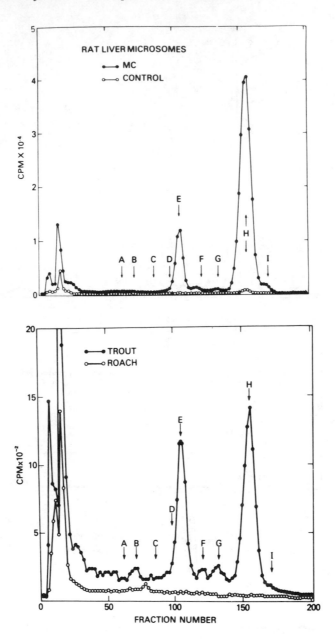

FIGURE 6.   BaP-deoxyribonucleoside adducts formed by hepatic microsomes of (A) control and 3-MC-induced rat, and (B) trout and roach. Peak E represents a phenoloxide-DNA adduct, whereas Peak H represents the *anti-BPDE-dGuo* adduct. (Adapted from Ahokas, J. T., Saarni, H., Nebert, D. W. and Pelkonen, O., *Chem.-Biol. Interact.*, 25, 103, 1979.)

English sole, a species having a high incidence of liver neoplasms when sampled from polluted waterways. English sole used in this experiment were caught from a relatively pristine area of Puget Sound. Sole hepatic microsomes metabolized BaP at a low SPR to products that bind to deproteinized salmon sperm DNA. Analysis by both reverse-phase HPLC and by immobilized boronate column chromatography showed a single major peak and that $60 \pm 5\%$ of the total radioactivity applied onto the HPLC column chromatographed with the indentical retention time of the *anti*-BaPDE-dGuo adduct (Figure 7). This diol-epoxide isomer has been implicated as the major ultimate carcinogenic metabolite of BaP in mouse epidermis.

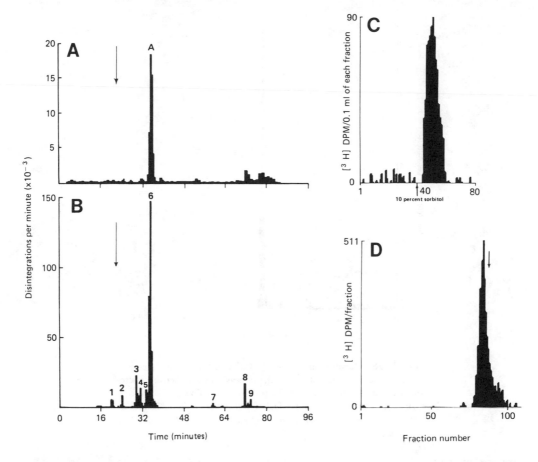

FIGURE 7. Typical reversed-phase HPLC and immobilized boronate column chromatography profiles of BaP-deoxyribonucleoside adducts formed by English sole hepatic microsomes. (A) C_{18} HPLC analysis of BaP-deoxyribonucleoside adducts formed in the presence of sole liver microsomes; (B) C_{18} HPLC analysis of adduct standards formed by the reaction of (+) -anti-BaPDE with DNA; (C) immobilized boronate column chromatography analysis of adducts formed by sole hepatic microsomes; the first 40 fractions were eluted with 1 M morpholine, followed by 1 M morpholine:10% sorbitol; (D) C_8 HPLC analysis of the morpholine-10% sorbitol fraction from immobilized boronate column chromatography. The arrows in (A) and (B) represent the position of elution of the trans-2-tetrol, the major hydrolysis product of anti-BaPDE, whereas in (C) and (D), the arrows represent the position of elution of the [^{14}C]-(+)-anti-BPDE-dGuo chromatographs. (From Nishimoto, M. and Varanasi, U., Polynuclear Aromatic Hydrocarbons: 9th Int. Symp. Chemistry, Characterization and Carcinogenesis, Cooke, M. W. and Dennis, A. J., Eds., Battelle Press, Columbus, Ohio, 801, 1982. With permission.)

Additional evidence for stereoselective metabolism of BaP by English sole liver was obtained by analyzing the tetrols formed from BaP, racemic BaP-7,8-diol, and BaP-7,8-diol isolated from bile of BaP-exposed sole (designated as biosynthetic BaP-7,8-diol). Of the four possible bay region tetrols that can be formed by metabolism of BaP-7,8-diol, only trans-2-tetrol, the major hydrolysis product of anti-BaPDE, was detected when either BaP or biosynthetic BaP-7,8-diol was the substrate. When racemic BaP-7,8-diol was used as a substrate, the hydrolysis products of both syn- and anti-BaPDE were formed;[8] however, a greater proportion of the hydrolysis products of anti-BaPDE than of syn-BaPDE were detected. Thakker et al.[8] have reported similar distributions of tetrols derived from racemic BaP-7,8-diol metabolized by liver microsomes of 3-MC treated rats. Moreover, in good agreement with the results obtained with 3-MC-induced rat liver microsomes,[8] a substantially greater proportion of anti-BaPDE-DNA adduct was formed compared to the syn-BaPDE-DNA adduct when racemic ^{3}H-BaP-7,8-diol was the substrate for sole liver microsomes

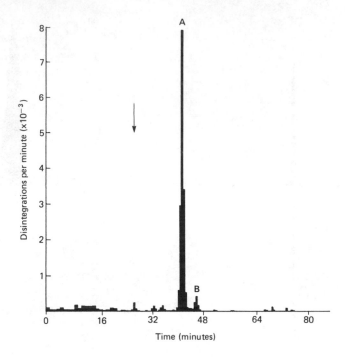

FIGURE 8.   Reversed-phase HPLC profile of BaP 7,8-diol-deoxyri-
bonucleoside adducts formed in the presence of sole hepatic microsomes. The arrow represents the position of elution of the *trans*-2-
tetrol, the major hydrolysis product of *anti*-BaPDE. (From Nishimoto,
M. and Varanasi, U., *Biochem. Pharmacol.*, 34, 263, 1985. With
permission.)

(Figure 8). This demonstrates that *syn*-BaPDE-adducts are formed in substantially smaller
amounts than the *anti*-BaPDE-DNA adducts even when their required precursors are present
in equal amounts.

The differences in the profiles of the tetrols formed from the biosynthetic BaP-7,8-diol
when compared to the racemic BaP-7,8-diol indicate that the BaP-7,8-diol isolated from the
bile of fish was predominantly a single stereoisomer, namely the (−)-BaP-7,8-diol; however,
the exact stereochemistry of most of the BaP metabolites formed by fish liver remains to
be delineated.

Because of the apparent lack of K-region metabolism in sole, BaP-4,5-oxide and 9-
hydroxyBaP-4,5-oxide adducts would not be expected to be formed *in vitro*. Moreover,
because of the apparent absence of K-region metabolism as well as the lack of induction of
MFO by PB in other fish species, these types of adducts would be expected to form only
a minor percentage of the total BaP metabolites that would be bound to DNA *in vitro* and
in vivo in experiments carried out with other fish species.

In a recent study, BaP metabolite profiles and DNA adduct profiles were obtained from
hepatic microsomes of English sole from a relatively unpolluted site and two contaminated
areas in Puget Sound, WA. The single, major BaP-DNA adduct formed by fish from the
reference and contaminated sites was identified by chromatography with standards and by
mild acid hydrolysis to be the (+)-*anti*-BaPDE-dGuo adduct. Although MFO activities of
hepatic microsomes of English sole from the reference site were significantly lower than
those from polluted sites, no significant changes in the metabolism of BaP were observed
with respect to induction by xenobiotic chemicals such as PAH and PCB present in the
contaminated sites.

Comparison of the *in vitro* metabolism of BaP by English sole and starry flounder shows no significant differences in the types of metabolites or DNA adducts formed between species; the major DNA adduct formed *in vitro* by either species was the (+)-anti-BaPDE-dGuo. Furthermore, epizootological evidence has shown that English sole sampled from areas having high concetrations of PAH and other aromatic compounds have a high prevalence of liver neoplasms, whereas starry founder appear to be relatively less susceptible to liver cancer in contaminated sites.[64,95] It is apparent, therefore, that differences in other parameters, such as levels or types of conjugating enzymes, rates of DNA repair, and cell cycling in liver of fish from contaminated sites, need to be studied using isolated hepatocytes.

VI. METABOLISM OF PAH IN CELL CULTURES

A. Cell Cultures

Cell cultures derived from aquatic species have been used to only a limited extent for studies of the metabolic activation of PAH. The first cell line established from cold-blooded vertebrates was the rainbow trout gonadal (RTG-2) cell line derived from pooled gonadal tissues of yearling rainbow trout by Wolf and Quimby in 1962.[96] Since that time, a number of other cell lines derived from fish have been described and made available to other investigators through the repository maintained by the *American Type Culture Collection*.[97] These include cell lines derived from a number of organs from a variety of species. Some of those which have been used in studies of hydrocarbon metabolism include: RTG-2 (gonadal tissue, rainbow trout, *Salmo gairdneri*, CCL 55), BF-2 (caudal trunk, bluegill fry, *Lepomis macrochirus*; CCL 91), and FHM (epithelial cells, fathead minnow, *Pimephales promelas*; CCL 42). Two of these were used in the initial comparative study of BaP metabolism in cell cultures derived from various species by Diamond and Clark in 1970.[98] In addition to those cell lines mentioned above, a number of other fish cell lines are available from the *American Type Culture Collection* and several laboratories are developing additional lines such as the nine salmonid cell lines described by Lannan et al.[99] Wolf and Mann have summarized the cell lines available from fishes.[100]

Early passage or primary cultures of cells derived from fish can also be used for studies of carcinogen metabolism. Methods have been described for the preparation of primary cultures from trout[101] and from marine fishes.[102] Wolf and Quimby provide detailed methods for preparing and growing these cultures in the *Tissue Culture Association Manual*.[103,104] Martin et al.[105] have recently reported an embryo-primary cell culture system for sheepshead minnow, *Cyrinodon variegatus* in which a single embryo is placed in a well of a 24-well culture plate, and the primary culture that results from attachment and outgrowth of cells can be maintained for up to 30 days. In this system, the major portion of the embryo remains intact and it is possible to observe simultaneously the effects of a carcinogen on both the intact embryo and primary cultures from the same organism.[105]

The liver is of special interest in studies of hydrocarbon metabolism (and other carcinogen metabolism), for it is the major organ of xenobiotic metabolism in fish, as well as mammals; however, fish liver is a target tissue of PAH-induced carcinogenesis, whereas livers of mammals are relatively resistant. Numerous investigators have used freshly isolated liver cells from fish for biochemical studies; most were carried out within the first day of culture. Bailey et al.[106] examined the metabolism and DNA binding of aflatoxin B_1 in isolated hepatocytes from rainbow trout. Klaunig and coworkers[107,108] modified a two-step hepatic portal perfusion method developed for rodents[109] for the isolation of hepatocytes from rainbow trout and channel catfish, *Ictalurus punctatus* and characterized the cells obtained. Although they failed to attach firmly to a plastic tissue culture substrate, trout hepatocytes had survival rates of greater than 50% after 8 days in culture. A number of carcinogens including BaP were shown to induce a dose-related increase in unscheduled DNA synthesis in hepatocytes from trout and catfish.[107] The activity of cytochrome P-450 decreased rapidly in hepatocyte cultures from both trout and channel catfish with a five- to sixfold reduction in enzyme activity during the first day of culture. However, even after 4 days, the trout hepatocytes were able to activate sufficient aflatoxin B_1 to give measurable levels of unscheduled DNA synthesis.

Similar culture techniques are available for a great number of cells from many organs and species including rodents and humans. There are also numerous cell lines from these species maintained in the *American Type Culture Collection* repository and other repositories.[97] The culture techniques are too numerous to adequately review and the reader is referred to reference works such as those by Pollack[110] and the *Tissue Culture Association Manual* for details of culture techniques and to the *American Type Culture Collection Catalogue*[97] for a list of currently available cell lines.

B. Metabolism of PAH

Many cell cultures have been tested for metabolism and metabolic activation of PAH. Rodent fibroblasts were shown to metabolize BaP[111] and DMBA[112] and to activate DMBA to metabolites bound to DNA.[113] The use of mammalian cell culture systems for studies of hydrocarbon metabolism was reviewed by Diamond and Baird.[114] More recent developments in hydrocarbon metabolism are reviewed by Dipple et al.[2] and Harvey.[6] Hydrocarbon metabolism and DNA interactions have now been examined in a large number of cells derived from many tissues of both humans and rodents.[2,6]

The initial finding that fish cells in culture were able to metabolize BaP was reported by Diamond and Clark in 1970.[98] In this study, they compared the ability of cells obtained from birds, rodents, reptiles, amphibia, and fish to convert BaP to water-soluble metabolites. This was assayed by treating the cells with [³H]BaP and then extracting culture medium samples with chloroform/methanol. Unmetabolized BaP and some nonpolar metabolites were extracted into the chloroform phase, while the more polar metabolites were retained in the aqueous phase. The amount of radioactivity recovered in each phase was then measured by liquid scintillation counting. Based upon the amount of BaP metabolized to water-soluble derivatives per 10^7 cells in 24 h, cells were classed as having low (<0.25 μg), intermediate (0.25 to 0.5 μg), or high (> 0.5 μg) metabolism. The RTG cells (rainbow trout gonad) had high metabolizing activity and the BF cells (bluegill fry) had intermediate activity. The amount of BaP metabolized in the RTG cells was comparable to that in the highest metabolizing rodent cell, primary hamster embryo cells. In general, cells with high metabolic activity were susceptible to the induction of cytotoxicity by BaP and the RTG cells demonstrated a dose-dependent decrease in cell number from 93% of control at 0.1 μg BaP per milliliter to 30% of control at 5.0 μg BaP per milliliter.[98] Clark and Diamond[115] also demonstrated that the rate of BaP metabolism was dependent upon the temperature of incubation in both RTG cells and rodent embryo cells. In RTG cells the maximum rate of metabolism was observed at 26 to 28°C and in rodent embryo cells it was between 38 to 42°C. Thus, in both types of cells the temperature that supported maximum rate of BaP metabolism was about 4 to 8°C above the temperature that supported maximum cell growth, but in both rodent embryo cells and RTG cells metabolism of BaP was nearly maximum at the optimum temperature for cell growth.[115]

Analyses of the hydrocarbon metabolites formed in cells in culture after exposure to radioisotopically labeled hydrocarbon are carried out by extraction of organic solvent-soluble metabolites from the culture medium with ethyl acetate/acetone or chloroform/methanol. The organic solvent-soluble metabolites are then analyzed by chromatography, originally by TLC and recently mainly by HPLC. Thornton et al.[116] analyzed the metabolites formed from BaP in cells derived from rainbow trout (RTG-2), bluegill fry (BF-2) and fathead minnow (FHM) by this procedure. Both the RTG-2 cells and the BF-2 cells converted 63%

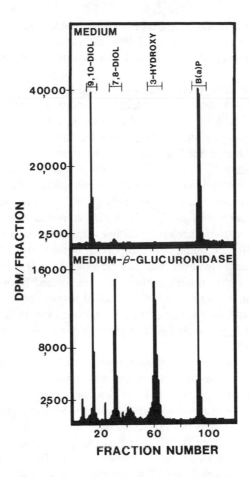

FIGURE 9. HPLC profiles of the chloroform-extractable BaP metabolites in the medium in which bluegill cells (BF-2) were exposed to [³H]BaP for 48 h. Top panel, medium without β-glucuronidase treatment; bottom panel, medium with β-glucuronidase treatment. Samples were prepared and chromatographed on a C18 reversed-phase column with a linear gradient of 55 to 95% methanol:water. Elution positions of BaP metabolite markers are shown at the top.

of the BaP (0.5 nmol BaP/per milliliter culture medium) to water-soluble metabolites within 24 h; the FHM cells metabolized a much smaller proportion of the BaP during this period. Analysis of the ethyl acetate-extractable metabolites by reversed-phase HPLC demonstrated that the major organic solvent-soluble metabolite formed by all three cells was BaP-9,10-diol.[116] Treatment of the water-soluble metabolites with β-glucuronidase to cleave the glucuronic acid conjugates present converted 67, 42, and 19% of those in the BF-2, FHM, and RTG-2 cell culture media, respectively, into ethyl acetate-extractable metabolites. Analyses of these by HPLC demonstrated that all three cell lines contained glucuronide conjugates of BaP-7,8-diol and of 3-hydroxyBaP (FHM) or 9-hydroxyBaP (RTG-2 and BF-2).[116] BaP-quinone glucuronides were also detected in media samples from all three cell lines. No significant amount of BaP-4,5-diol was detected either free or as a glucuronide conjugate in any of the cell lines. Thus, these cell lines formed both the 9,10-diol and 7,8-diol and either the 3- or 9-phenol, as well as small amount of quinones. The HPLC profiles of the BaP metabolites present in the media of BF-2 cell cultures after 24 h of exposure to BaP are shown in Figure 9.[149] The only major metabolite present in the culture medium is BaP-9,10-diol (Figure 9, top). After treatment of this medium sample with β-glucuronidase the presence of two additional metabolite peaks, BaP-7,8-diol and 3-hydroxy BaP, is clearly demonstrated in the HPLC profile (Figure 9, bottom). These studies have demonstrated that the peak originally designated as 9-hydroxyBaP in the studies of Thornton et al.[116] is actually 3-hydroxyBaP, and confirm that BaP-7,8-diol glucuronide is a major metabolite of BaP in BF-2 cell cultures.

Early-passage hamster embryo cell cultures exposed to BaP at the same concentration as the fish cell lines described above under identical conditions metabolized more than 80% of the BaP within 24 h; 90% of the metabolites formed were water soluble.[117] The major ethyl-acetate-extractable metabolite recovered from the media was BaP-9,10-diol; only very small amounts of BaP-7,8-diol and 9-hydroxy BaP were present in the media of cells treated with this low concentration of BaP.[117,118] β-glucuronidase treatment converted more than 40% of the metabolites to ethyl acetate-extractable metabolites, mainly 9-hydroxyBaP and 3-hydroxyBaP at a ratio of about 2:1, respectively.[117,118] Thus, hamster embryo cell cultures exposed to BaP at the same low dose for 24 h also converted most of the BaP to water-soluble metabolites. As with the three fish cell lines, the only major organic solvent-soluble metabolite was BaP-9,10-diol, and a large proportion of the water-soluble metabolites were glucuronic acid conjugates. However, the hamster embryo cells differed from the fish cell lines in that most of the glucuronide conjugates were formed from hydrocarbon phenols, whereas the fish cell lines also formed a substantial proportion of BaP metabolites from BaP-7,8-diol. Studies with two mouse cell lines and one rat cell line exposed to the same low dose of BaP for 24 h also indicated that glucuronide conjugates were formed, mainly with hydrocarbon phenols.[118]

Analysis of the water-soluble metabolites formed from BaP in cells in culture has usually been done by enzymatic hydrolysis of the glucuronide or sulfate conjugates or by separation of the various classes of conjugates by column chromatography.[119,120] Recently, an ion-pair HPLC system resolved the glucuronide, sulfate, and GSH conjugates of BaP metabolites formed from BaP in cell cultures from rodents, fish, and humans.[121] Analysis of the water-soluble metabolites formed from BaP (0.5 μg/ml medium) in BF-2 cells exposed for 24 h demonstrated the presence of two major metabolite peaks. The first peak (17% of the total BaP metabolites) contained mainly BaP-7,8-diol glucuronide, and the second peak (47% of the total metabolites) contained mainly 3-hydroxyBaP-glucuronide (32%) and small amounts of glucuronides of 9-hydroxyBaP (2%) and BaP-quinones (5%).[121] No distinct peaks of either BaP sulfate or GSH conjugates were observed in the ion-pair HPLC analyses of the medium from these cells. Thus, cell fractions as well as cell lines derived from fish species produce similar profiles of BaP metabolites with BaP-7,8-diol and 3-hydroxyBaP as major metabolites and 9-hydroxyBaP and quinones as minor metabolites.

The same procedure was used to analyze the water-soluble metabolites formed from BaP in third-passage cell cultures prepared from Syrian hamster, Wistar rat, and Sencar mouse embryos and the human hepatoma cell line Hep-G2.[121] After 24 h of exposure to 0.5 μg BaP/ml medium, all cultures had metabolized at least 64% of the BaP and at least 84% of the metabolites were water soluble.[121] All three types of rodent embryo cell cultures formed large proportions of BaP-phenol-glucuronides (42 to 52% of total metabolites), but only the mouse embryo cells formed a significant amount of BaP-7,8-diol glucuronides (4%).[121] None of these cell cultures formed detectable amounts of BaP-sulfate conjugates, and only the mouse embryo cell culture medium contained small amounts of material that eluted as distinct peaks in the area where GSH conjugates of BaP elute. In contrast, the Hep-G2 cell media contained no detectable glucuronide conjugates of BaP: the major conjugate peak contained mainly the sulfate of 3-hydroxyBaP (37% of total metabolites). Small amounts of 9-hydroxyBaP, BaP-quinone, and BaP-diol sulfate conjugates were also observed, as were several peaks of metabolites that eluted in similar regions as GSH conjugate standards.[121]

Numerous studies have been carried out to elucidate the pathways of metabolism of BaP to water-soluble derivatives in cell cultures from humans and other mammalian species, and it is not possible to review all of them in this chapter.[2,6,119,122] It is often impossible to compare results directly because of the use of different analytical techniques, concentrations of BaP, times of exposure, and other factors which can alter the quantitation of BaP metabolites. In general, rodent embryo cell cultures tend to form large proportions of BaP-phenol glucuronide conjugates, whereas cells from humans often form higher proportions of sulfate and/or GSH conjugates. The most striking difference in the profile of BaP metabolites formed in the three fish cell lines for which data are available[116,121] and rodent and human cells[2,6,119,121,122] is the formation of high proportions of metabolites from the 7,8-diol. In rodent cells in culture, rodent tissues *in vivo,* and human cells in culture the major reactive metabolite involved in DNA binding and the induction of biological effects is the 7,8-diol-9,10-epoxide.[6] Thus, formation of a glucuronide conjugate of the 7,8-diol could provide a mechanism for prevention of the formation of this ultimate carcinogenic metabolite of BaP in fish. To determine if this pathway of metabolism also occurs in primary cell cultures from fish, hepatocyte cultures were prepared from brown bullhead, *Ictalurus nebulosus* and exposed to 0.14 μg BaP/ml medium for 24 h.[123] The cultures metabolized 63% of the BaP; 88% of the metabolites were water soluble. Ion-pair HPLC demonstrated that comparable amounts of diol and phenol glucuronide conjugates were formed, mainly BaP-7,8-diol and 3-hydroxyBaP glucuronides. The 7,8-diol glucuronide represented 20% of the total metabolites.[123] Thus, formation of a glucuronide conjugate of BaP-7,8-diol appears to be a major pathway of metabolism of BaP in cell cultures derived from a number of species of fish. The potential significance of this will be discussed in the section on BaP-DNA interactions.

VII. EFFECTS OF PAH ON CELLS IN CULTURE

Hydrocarbons have been found to induce a number of effects on mammalian cells in culture. These include the induction of cytotoxicity, DNA damage, DNA repair, chromosomal alterations and sister chromatid exchange, mutation, and *in vitro* transformation.[1,6] As described in the section on metabolism, Diamond and Clark[98] demonstrated that BaP induced cytotoxicity with a dose-dependent relationship in RTG cells, a cell line with high hydrocarbon-metabolizing capacity. Kocan et al.[124] examined the toxicity of a number of mutagens including the hydrocarbons BaP and 3-MC and the hydrocarbon derivative 3-hydroxyBaP in cell lines from bluegill fry (BF-2), rainbow trout (RTG-2), and steelhead trout embryo (STE). Both PAH and 3-hydroxyBaP were toxic to the RTG-2 and BF-2 cells, but much less toxic to STE cells. Thus, fish cell lines which metabolize hydrocarbons appear to be susceptible to the induction of toxicity by these chemicals. This is similar to the effects of hydrocarbons on mammalian cells which metabolize them.[114] Kocan et al.[125] also demonstrated that the sequestration and release of BaP in vertebrate cell cultures was dependent upon the serum concentration, but the processes involved were similar in low passage human skin fibroblasts and the bluegill cell line BF-2.

The ability of carcinogenic PAH to induce mutations in mammalian cells in which they are metabolically activated or in mammalian cell-mediated mutation assays is well established.[6,126] Examples of various types of assays are described in *Cellular Systems for Toxicity Testing.*[127] The authors were unable to find any reports of the use of fish cells as activator cells in cell-mediated mutation assays, a type of assay which would allow direct comparison of their ability to activate hydrocarbons to mutagens with that of mammalian cells from various species and tissues. However, Kocan et al.[128] have been able to demonstrate that BaP is mutagenic to BF-2 cells in culture. Exposure of BF-2 cultures to BaP at concentrations from 0.3 to 50 μg/ml medium resulted in the induction of 3 to 6 ouabain-resistant mutants per 10^6 survivors. This represented a two- to sevenfold increase over the spontaneous mutation rate, but was much lower than the 210 mutants per 10^6 survivors induced by *N*-methyl-*N*′-

nitro-*N*- nitrosoguanidine (MNNG). Thus, BaP is mutagenic to BF-2 cells in culture, but it appears to be a relatively weak mutagen in this system.

Hydrocarbons have also been shown to induce damage to chromosomes in mammalian cell culture systems.[1] The assays used are reviewed in References 129 and 130. Kocan et al.[131] examined the induction of anaphase aberrations in chromosomes in RTG-2 cell cultures exposed to a number of mutagens. They found that BaP induced a significant increase in the frequency of anaphase aberrations in these cells. Kocan et al.[132] have also used this assay to demonstrate that sediment extracts from various contaminated sites in Puget Sound, Washington are able to induce anaphase aberrations in RTG-2 cells as well as cytotoxicity in RTG-2 and BF-2 cells. Thus, hydrocarbons are able to induce chromosomal damage in cell cultures from aquatic organisms that are known to be able to metabolically activate these compounds. The relationship of metabolism pathways of PAH to the induction of chromosomal damage, has not yet been examined in cell cultures derived from fish or other aquatic organisms.

Another measure of damage to cellular DNA caused by carcinogen treatment is the induction of DNA repair. One of the most common tests for this type of damage in mammalian cell culture systems is the unscheduled DNA synthesis (UDS) assay. This assay, which measures the incorporation of tritiated thymidine into DNA of nonreplicating cells in response to carcinogen treatment, has recently been reviewed.[133,134] A number of PAH have been shown to induce UDS in rat hepatocytes.[133] Walton et al.[135] examined the ability of four carcinogenic chemicals and UV light to induce UDS in human fibroblasts, a Chinese hamster ovary-derived cell line (CHO), RTG cells, FHM cells, and cell lines derived from chum salmon heart and rainbow trout ovary. The greatest response was observed in the human fibroblasts, the CHO cells had a lower response, and all the fish cell lines responded the least (all four responded about equally). They concluded that the magnitude of DNA repair synthesis in fish cell lines was much lower than that in rodent and human cells.[135] Klaunig[107] examined the effect of several carcinogens including BaP, on the induction of UDS in primary hepatocyte cultures from rainbow trout, and channel catfish. He found that BaP induced a dose-dependent increase in UDS in hepatocyte cultures from both species.[107] A similar BaP dose-dependent increase in UDS was reported in oyster toadfish, *Opsanus tau* hepatocytes by Kelly and Maddock.[136] Thus, hydrocarbons are capable of causing DNA damage in fish cells in culture as demonstrated by this DNA repair assay.

Although only a limited number of studies have been carried out with cell cultures from aquatic organisms, it is evident that many of the assays used to measure the effects of hydrocarbons on mammalian cell cultures are also applicable to studies of the effects of hydrocarbons on cells from aquatic organisms. Based upon the very limited amount of data available about the effects of hydrocarbons on fish cells in culture and the limited number of cell cultures that have been tested, it is not possible at this time to draw conclusions about the similarities and differences between cell cultures from aquatic and terrestrial species.

VIII. BINDING OF PAH TO DNA IN CELLS IN CULTURE

The interaction of carcinogenic PAH with DNA is believed to be one of the initial steps in the process of tumor induction by these agents.[6] Hinton et al.[30] demonstrated that BaP became bound to DNA in primary hepatocyte cultures from channel catfish, but the identity of the interacting products was not established. Analysis of these DNA interaction products or adducts can provide information about the relationship of specific adducts to tumor induction and also provides a method to determine the reactive metabolites formed from hydrocarbons in cells. Although the role of specific hydrocarbon-DNA adducts in the induction of carcinogenesis has not yet been established, analysis of the formation and persistence of PAH-DNA adducts in cells from different species allows comparison of the different pathways of PAH activation, PAH-DNA binding, and PAH-DNA adduct repair and may help to explain the differences in the susceptibility of various species to PAH-induced tumorigenesis.

The major DNA-binding metabolite formed in cell cultures from most terrestrial organisms is (+)-*anti*-BaPDE (Figure 2).[2,122] The formation and persistence of the product formed by reaction of the 10-position of (+)-*anti*-BaPDE with the 2-amino of deoxyguanosine in DNA has been found to correlate with the induction of mutation.[137,138] However, smaller amounts of other DNA interaction products are also formed and several have been shown to have high mutagenic potency.[139] To permit the analysis of the BaP-DNA adducts formed in cells, a technique involving separation of the adducts formed by the *syn*- and *anti*-isomers of BaPDE (or DMBADE[140]) on a boronate column followed by analysis of the individual adducts by HPLC was developed in our laboratory.[141] Application of this technique to the analysis of the BaP-DNA adducts formed in Sencar mouse embryo cells exposed to 0.5 μg [³H]BaP/ml for 24 h[141] is illustrated in Figure 10, I. The DNA was isolated from the cell cultures and enzymatically degraded to deoxyribonucleosides. The BaP-deoxyribonucleoside adducts were isolated on a short Sephadex LH20® column and an aliquot analyzed by reversed-phase HPLC (Figure 10, IA).[141] The remainder of the adducts were placed on an immobilized boronate column and those formed from *syn*-BaPDE eluted in a morpholine buffer, and were analyzed by HPLC (Figure 10, IB). The *anti*-BaPDE adducts were then eluted with a morpholine/sorbitol buffer and analyzed by HPLC (Figure 10, IC). The *syn*-BaPDE adducts represented 18% of the total adducts; the *anti*-BaPDE adducts represented 82%.[141] The (+)-*anti*-BaPDE-dGuo adduct was the only major *anti*-BaPDE adduct present, and the *syn*-BaPDE-dGuo adduct accounted for most of the *syn*-BaPDE adducts.[141] Thus, the major DNA binding metabolite in mouse embryo cell cultures was (+)-*anti*-BaPDE, a result compatible with other results in mammalian cell cultures.[122]

Other rodent embryo cell cultures formed a more complex mixture of BaP-DNA adducts. Analysis of the BaP-DNA adducts formed in Syrian hamster embryo cells in culture exposed to the same dose of BaP for the same time as the Sencar mouse embryo cells[141] is shown in Figure 10, II. The morpholine buffer fraction, representing 39% of total adducts, contained two major adduct peaks, the larger of which eluted in the same position as a *syn*-BaPDE-dGuo marker.[141] The morpholine/sorbitol buffer representing 60% of the adducts contained two major peaks; the larger one eluted with the [¹⁴C]-(+)-*anti*-BaPDE-dGuo marker. Thus, the hamster embryo cell DNA contained two major BaP-DNA adducts and substantial amounts of several minor adducts.[141,142]

The BaP-DNA adduct profile obtained from treatment of Wistar rat embryo cells in culture for 24 h at 0.5 μg [³H]BaP/ml demonstrates the formation of a number of different BaP-DNA adducts (Figure 10, III A).[141] HPLC analysis alone (Figure 10, III A) gave only limited resolution of the adducts indicated as b and c. By use of the immobilized boronate chromatography-HPLC procedure, Pruess-Schwartz et al.[141] were able to completely resolve and quantitate these stereochemically different BaPDE-DNA adducts.[141,142] Six major BaP-deoxyribonucleoside adducts were detected and identified. Three (b, d, e) eluted in the morpholine buffer (Figure 10, III B) and represented 50% of the total adducts; two (a, c) eluted in the morpholine/sorbitol buffer and (Figure 10, III C) represented 49% of the total adducts. Pruess-Schwartz and Baird[143] have confirmed the identity of some of these adducts by the use of cochromatography and acid hydrolysis techniques. Adduct c cochromatographed with a marker of (+)-*anti*-BaPDE-dGuo, adduct d with a marker of *syn*-BaPDE-dGuo, and adduct b cochromatographed with a *syn*-BaPDE-dGuo or a *syn*-BaPDE-dCyt. Adducts a and e did not coelute with known BaP-deoxyribonucleoside marker standards. Studies of adduct a by acid hydrolysis demonstrated that it was formed by a different mechanism of activation than a simple bay-region diol-epoxide.[143]

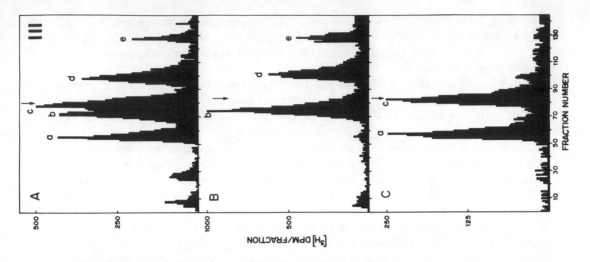

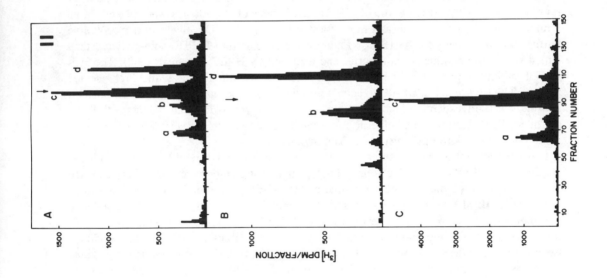

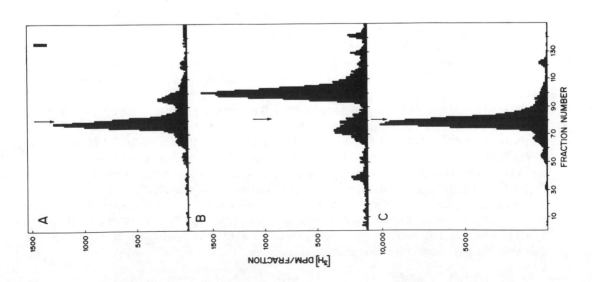

FIGURE 10. (I), Syrian hamster embryo cells (II), Wistar rat embryo cells, and (III) Sencar mouse embryo cells. HPLC profiles of BaP:deoxyribonucleoside adducts present after exposure to 0.5 μg [^{3}H]BaP (5.68 Ci/mmol) per ml medium for 24 h. The DNA was isolated and degraded, and the BaP:deoxyribonucleosides were separated by chromatography on Sephadex® LH-20. An aliquot (A), and the remainder was subjected to boronate chromatography. The adducts in the 1 M morpholine fractions and 1 M morpholine:10% sorbitol fractions were each concentrated and analyzed by reversed-phase HPLC (B, 1 M morpholine fractions; C, 1 M morpholine: 10% sorbitol fractions). Arrow shows elution position of [^{14}C]-(+)-anti-BaPDE-dGuo adduct marker. (From Pruess-Schwartz, D., Sebti, S. M., Gilham, P. T., and Baird, W. M., Cancer Res., 44, 4108, 1984. With permission.)

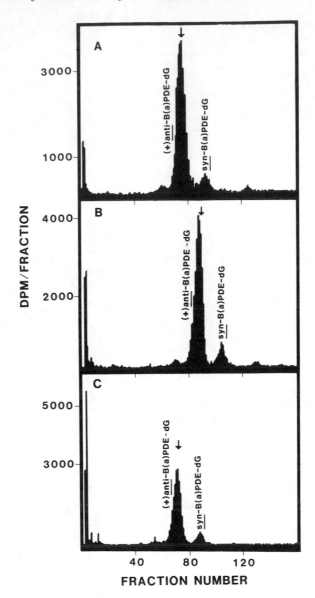

FIGURE 11.   HPLC profiles of [³H]BaP-DNA adducts isolated
from BF-2 cell cultures exposed to 0.5 μg [³H]BaP (5 Ci/mmol)
per ml medium for (A) 24 (B) 48, and (C) 120 h. Arrow shows
the elution position of a [¹⁴C]-(+)-*anti*-BaPDE-dGuo adduct marker
added to each sample.

To examine the pathways of metabolic activation of BaP to reactive derivatives in cells
from fish, we have investigated the binding of [³H]BaP to the DNA of fish cells in culture.
The BF-2 cell line was exposed to [³H]BaP at a concentration of 0.5 μg/ml (specific activity. 5 Ci/mmol) for 24, 48, and 120 h, and the BaP-DNA adducts were analyzed
(Figure 11). The BaP-DNA adduct profiles contained two major adduct peaks: the larger
peak cochromatographed with a [¹⁴C]-*anti*-BaPDE-dGuo adduct marker. The second peak
eluted in the same position as a *syn*-BaPDE-dGuo marker. Further characterization of these
adducts is currently being carried out by immobilized boronate chromatography and HPLC.
The BF-2 cell BaP-DNA adduct profile (Figure 11) is qualitatively similar to the Sencar

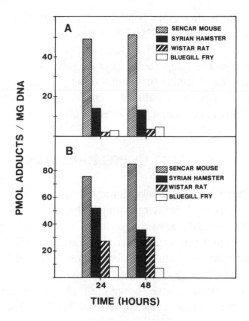

FIGURE 12.   Amount of (+)-*anti*-BaPDE-dGuo adduct per mg of DNA present (A) and the total amount of BaP-DNA adducts formed per mg DNA (B) in embryo cell cultures from several species of rodents and the BF-2 cell line after 24 and 48 h of exposure of cells to [³H]BaP.

mouse embryo cells (Figure 10, IA). Sebti et al.[142] have found that in Sencar mouse embryo cells in culture the (+)-*anti*-BaPDE-dGuo adduct represented 82 to 86% of the adducts after 24 h of BaP exposure, and the *syn*-BaPDE-dGuo adduct represented less than 13% of the adducts. In the BF-2 cells in culture, the (+)-*anti*-BaPDE-dGuo adduct represents 60 to 70% of the adducts after 24 h of exposure and *syn*-BaPDE-dGuo 8 to 9%. Although the Sencar mouse embryo cells and the BF-2 cells have similar relative proportions of (+)-*anti*-BaPDE-dGuo adduct, the amount of this adduct present per milligram DNA was much larger in the Sencar mouse embryo cells than in the BF-2 cells after both 24 and 48 h of exposure (Figure 12A). Both the Sencar mouse and Syrian hamster embryo cells have higher levels of (+)-*anti*-BaPDE-dGuo than the BF-2 cells, but the Wistar rat embryo cells had a similar level to the BF-2 cells.

The level of binding of BaP to the DNA of rodent and fish cells in culture varies between species (Figure 12B). All three rodent embryo cell cultures had greater levels of binding of BaP to DNA than the BF-2 cells. This was not due to lack of oxidation of the BaP by the BF-2 cells, for BF-2 cells in culture were able to metabolize up to 90% of the BaP by 48 h.[116] However, the extent of conjugation of BaP metabolites to water-soluble conjugates was greater than 70% and the glucuronide conjugate of a proximate carcinogenic metabolite of BaP, BaP-7,8-diol, was one of the major water-soluble metabolites in BF-2 cells. The high extent of glucuronide conjugation of BaP-7,8-diol could account for the low level of BaP-DNA binding in BF-2 as compared with embryo cells. This indicates that glucuronidation of BaP-7,8-diol in fish is highly efficient; however, because fish liver appears susceptible to tumor induction by PAH, other factors such as adduct persistence may be very important in tumor-susceptible tissues.

To determine if the temperature of incubation of BF-2 cell culture affects the level of DNA binding of [³H]BaP, cells were incubated at 22 and 35°C for 24 and 48 h at a concentration of 0.5 µg [³H]BaP/ml medium. The level of BaP-DNA binding was threefold higher at 24 h and twofold higher at 48 h in cells treated at 35°C than in those treated at 22°C. The increase in total binding at 35°C was due mainly to an increase of two- and threefold in the amount of (+)-*anti*-BaPDE-dGuo adduct per milligram DNA at 24 and 48 h of exposure, respectively. The temperature-induced increase in metabolism appears to be stereoselective based upon the selective formation of the (+)-*anti*-BaPDE-DNA adduct. The exact mechanism for the temperature-induced increase in BaP metabolism and BaP-DNA binding has not been determined. The DNA interaction products of BaP in other fish cell lines are presently being investigated.

IX. CONCLUDING REMARKS

Analyses of PAH metabolism in both subcellular fractions and cell cultures from a number of species of aquatic organisms, have demonstrated the ability of many aquatic organisms to metabolize PAH. These studies have demonstrated that similar metabolites are formed from BaP in both microsomal preparations and cell cultures from aquatic and terrestrial organisms, but the proportions of particular metabolites formed differ greatly between species.

Studies with subcellular fractions have shown the regio- and stereochemical properties of xenobiotic metabolizing enzymes in terrestrial and fish species. Species-specific differences such as relative content and activity of individual enzymes may have significant implications in the activation of PAH into mutagenic and carcinogenic metabolites. The absence of PB-inducible isozymes of cytochrome P450 in fish shifts the balance of metabolism to benzo-ring metabolites, which have been shown to possess the higher carcinogenic potential of the metabolites of BaP. Moreover, differences in the types of PAH-DNA adducts formed by hepatic microsomes may indicate that differences exist in the pathways of activation of PAH *in vivo*. Although other factors, such as detoxication potential, will influence the activation of PAH *in vivo*, data obtained *in vitro* provide a foundation for the identification of the pathways for PAH metabolism and DNA adduct formation.

Cell culture studies to date have had only a limited role in providing our present understanding of PAH metabolism and the induction of biological effects by PAH in aquatic organisms. However, the limited number of studies which have been carried out demonstrate that cell cultures from fish will be a valuable resource for such studies. Some fish cell cultures have high hydrocarbon-oxidizing activity and form similar BaP oxidation products to those observed in mammalian cell cultures. In addition, cell cultures derived from a number of species of fish form high proportions of conjugated hydrocarbon metabolites. The formation of high levels of the glucuronic acid conjugate of BaP-7,8-diol (a proximate carcinogenic metabolite of BaP in mammals) in cell cultures from several species of fish, and the low level of binding of BaP to DNA in these cultures, may indicate that some fish tissues are able to detoxify this metabolite through glucuronide conjugation. Moreover, the species-specific differences in hepatic GST activities noted in bottom fishes may explain, in part, differential detoxication of PAH, such as BaP, *in vivo*. Studies of the biological effects of PAH on fish cell cultures have demonstrated the induction of cytotoxicity, DNA damage, chromosomal damage, and mutations. Thus, studies which combine the analysis of PAH metabolism, DNA binding, and the induction of specific biological effects, should provide a valuable method for establishing the role of specific metabolic pathways in the activation and detoxification of PAH in aquatic organisms.

The data reviewed in this chapter clearly indicate the value of both subcellular fraction studies and cell culture studies for determining how PAH are metabolized in aquatic species and how they induce biological effects in these species. At present, studies have been carried out with only a limited number of species sampled from relatively pristine areas and only a few hydrocarbons. In view of the recent interest in the effects of pollution, in general, and of PAH, in particular on aquatic organisms, it is evident that both subcellular fraction and cell culture studies with aquatic organisms will increase. The present results will be extended to other hydrocarbons, especially the methyl-substituted hydrocarbons, as well as to other xenobiotic chemicals. Moreover, in some cases where there are unique pairs of susceptible and resistant organisms available. The comparison of hydrocarbon activation in microsomal fractions and cell cultures from these organisms should provide valuable insight into the mechanisms of PAH-induced toxicity. Comparison of results obtained with aquatic organisms with those obtained in terrestrial species will also help to establish the role of specific PAH metabolism pathways in both the activation and detoxication of these carcinogens. Hopefully, within a few years a data base will be obtained in both subcellular fractions and cell cultures from aquatic organisms which will allow such comparisons with the data already available for many mammalian species, including humans. Thus, studies of PAH metabolism, PAH-DNA binding, and the induction of biological effects by PAH in *in vitro* systems prepared from aquatic organisms from urban and nonurban areas, will aid in determining how these carcinogens affect aquatic organisms and also in understanding how they induce cancer in terrestrial organisms, including humans.

ACKNOWLEDGMENTS

The authors are grateful to Drs. Tracy Collier and William L. Reichert (National Marine Fisheries Service/National Oceanic and Atmospheric Administration [NOAA]) for critical review of this chapter. The authors thank Marilyn Hines for typing the manuscript. This work was supported in part by Public Health Service grants CA40228 and CA28825 from the National Cancer Institute, Department of Health and Social Services.

REFERENCES

1. **Gelboin, H. V. and Ts's, P. O. P., Eds.,** *Polycyclic Hydrocarbons and Cancer,* Vol. 1 to 3, Academic Press, New York, 1978 to 1981.
2. **Dipple, A., Moschel, R. C., and Bigger, C. A. H.,** Polynuclear aromatic carcinogens, in *Chemical Carcinogens,* Vol. 1, 2nd ed. Searle, C. E., Ed., American Chemical Society, Washington, D.C., 1984, 41.
3. **Falk, H. L., Kotin, P., Lee, S. S., and Nathan, A.,** Intermediary metabolism of benzo(*a*)pyrene in the rat, *J. Natl. Cancer Inst.,* 28, 699, 1963.
4. **Harper, K. H.,** The intermediary metabolism of 3:4-benzpyrene; the biosynthesis and identification of the X_1 and X_2 metabolites, *Br. J. Cancer,* 12, 645, 1958.
5. **Largenbach, R., Nesnow, S., and Rice, J. M., Eds.,** *Organ and Species Specificity in Chemical Carcinogenesis,* Plenum Press, New York, 1983.
6. **Harvey, R. G., Ed.,** *Polycyclic Hydrocarbons and Carcinogenesis,* ACS Symp. Ser. 283, American Chemical Society, Washington, D.C., 1985.
7. **Watabe, T., Hakamata, Y., Hiratsuka, A., and Ogura, K.,** A 7-hydroxymethyl sulphate ester as an active metabolite of the carcinogen, 7-hydroxymethylbenz(*a*)anthracene, *Carcinogenesis (London),* 7, 207, 1986.
8. **Thakker, D. R., Yagi, H., Levin, W., Wood, A. W., Conney, A. H., and Jerina, D. M.,** Polycyclic aromatic hydrocarbons: metabolic activation to ultimate carcinogens, in *Bioactivation of Foreign Compounds,* Anders, M. W., Ed., Academic Press, New York, 1985, 177.
9. **Conney, A. H.,** Induction of microsomal enzymes by foreign chemicals and carcinogenesis by polycyclic aromatic hydrocarbons: G. H. A. Clowes Memorial Lecture, *Cancer Res.,* 42, 4875, 1982.
10. **Ames, B. N., Durston, W. E., Yamasaki, E., and Lee, F. D.,** Carcinogens are mutagens: a simple test system combining liver homogenates for activation and bacteria for detection, *Proc. Natl. Acad. Sci. U.S.A.,* 70, 2281, 1973.

11. **Hendricks, J. D., Meyers, T. R., Shelton, D. W., Casteel, J. L., and Bailey, G. S.,** Hepatocarcinogenicity of benzo(*a*)pyrene to rainbow trout by dietary exposure and intraperitoneal injection, *J. Natl. Cancer Inst.,* 74, 839, 1985.

12. **Swain, L. and Melius, P.,** Characterization of benzo(*a*)pyrene metabolites formed by 3-methylcholanthrene-induced goldfish, black bullhead and brown bullhead, *Comp. Biochem. Physiol.,* 79C, 151, 1984.

13. **Varanasi, U., Gmur, D. J., and Krahn, M. M.,** Metabolism and subsequent binding of benzo(*a*)pyrene to DNA in pleuronectid and salmonid fish, in *Polynuclear Aromatic Hydrocarbons: 4th Int. Symp. on Analysis, Chemistry and Biology,* Bjorseth, A. and Dennis, A. J., Eds., Battelle Press, Columbus, Ohio, 1979, 455.

14. **Tan, B., Kilgore, M. V., Elain, D. L., Jr., Melius, P., and Schoor, W. P.,** Metabolities of benzo(*a*)pyrene in Arocolor 1254-treated mullet, in *Proc. 4th Annu. Symp. Aquatic Toicology,* Branson, D. R. and Dickson, K. L., Eds., ASTM, Philadelphia, Pennsylvania, 1979, 239.

15. **Egaas, E. and Varanasi, U.,** Effects of polychlorinated biphenyls and environmental temperature on *in vitro* formation of benzo(*a*)pyrene metabolites by liver of trout *(Salmo gairdneri),* Biochem. Pharmacol., 31, 561, 1982.

16. **Varanasi, U. and Gmur, D. J.,** Metabolic activation and covalent binding of benzo(*a*)pyrene to deoxyribonucleic acid catalyzed by liver enzymes of marine fish, *Biochem. Pharmacol.,* 29, 753, 1980.

17. **Holder, G., Yagi, H., Dansette, P., Jerina, D. M., Levin, W., Lu, A. Y. H., and Conney, A. H.,** Effects of inducers and epoxide hydrase on the metabolism of benzo(*a*)pyrene by liver microsomes and a reconstituted system: analysis by high pressure liquid chromatography, *Proc. Natl. Acad. Sci. U.S.A.,* 71, 4356, 1974.

18. **Prough, R. A., Patrizi, V. W., Okita, R. T., Masters, B. S. S., and Jakobsson, S. W.,** Characteristics of benzo(*a*)pyrene metabolism by kidney, liver, and lung microsomal fractions from rodents and humans, *Cancer Res.,* 39, 1199, 1979.

19. **Ahokas, J. T., Saarni, H., Nebert, D. W., and Pelkonen, O.,** The *in vitro* metabolism and covalent binding of benzo(*a*)pyrene to DNA catalyzed by trout liver microsomes, *Chem.-Biol. Interact.,* 25, 103, 1979.

20. **Pezzuto, J. M., Yang, C. S., Yang, S. K., McCourt, D. W., and Gelboin, H. V.,** Metabolism of benzo(*a*)pyrene and (−)-trans-7,8-dihydroxy-7,8-dihydrobenzo(*a*)pyrene by rat liver nuclei and microsomes, *Cancer Res.,* 38, 1241, 1978.

21. **Levin, W., Lu, A. Y. H., Ryan, D., Wood, A. W., Kapitulnik, J., West, S., Huang, M.-T., Conney, A. H., Thakker, D. R., Holder, G., Yagi, H., and Jerina, D. M.,** Properties of the liver microsomal monooxygenase system and epoxide hydratase: factors influencing the metabolism and mutagenicity of benzo(*a*)pyrene, in *Origins of Human Cancer,* Hiatt, H. H., Watson, J. D., and Winsten, J. A., Eds., Cold Spring Harbor Laboratory, Cold Spring Harbor, New York, 1977, 659.

22. **Little, P. J., James, M. O., Pritchard, J. B., and Bend, J. R.,** Benzo(*a*)pyrene metabolism in hepatic microsomes from feral and 3-methylcholanthrene-treated southern flounder, *Paralichthys lethostigma, J. Environ. Pathol. Toxicol. Oncol.,* 5, 309, 1984.

23. **Dent, J. G., Graichen, M. E., Schnell, S., and Lasber, J.,** Constitutive and induced hepatic microsomal cytochrome P-450 monooxygenase activities in male Fischer-344 and CD rats. A comparative study, *Toxicol. Appl. Pharmacol.,* 52, 45, 1980.

24. **Holder, G. M., Yagi, H., Jerina, D. M., Levin, W., Lu, A. Y. H., and Conney, A. H.,** Metabolism of benzo(*a*)pyrene. Effect of substrate concentration and 3-methylcholanthrene pretreatment on hepatic metabolism by microsomes from rats and mice, *Arch. Biochem. Biophys.,* 170, 557, 1975.

25. **Selkirk, J. K., Croy, R. G., Wiebel, F. J., and Gelboin, H. V.,** Differences in benzo(*a*)pyrene metabolism between rodent liver microsomes and embryonic cells, *Cancer Res.,* 36, 4476, 1976.

26. **Varanasi, U., Nishimoto, M., Reichert, W. L., and Eberhart, B.-T. L.,** Comparative metabolism of benzo(*a*)pyrene and covalent binding to hepatic DNA in English sole, starry flounder and rat, *Cancer Res.,* 46, 3817, 1986.

27. **Bend, J. R., Ball, L. M., Elmamlouk, T. H., James, M. O., and Philpot, R. M.,** Microsomal mixed-function oxidation in untreated and polycyclic aromatic hydrocarbon-treated marine fish, in *Pesticide and Xenobiotic Metabolism in Aquatic Organisms,* Khan, M. A. Q., Lech, J. J., and Menn, J. J., Eds., Washington, D.C., 1979, 297.

28. **Stegeman, J. J. and Woodin, B. R,** Patterns of benzo(*a*)pyrene metabolism in liver of the marine fish *Stenotomus versicolor, Fed. Proc.,* 39, 1752, 1980.

29. **Stegeman, J. J., Klotz, A. V., Woodin, B. R., and Pajor, A. M.,** Induction of hepatic cytochrome P-450 in fish and the indication of environmental induction in scup *(Stenotomus chrysops), Aquat. Toxicol.,* 1, 197, 1981.

30. **Hinton, D. E., Klaunig, J. E., Jack, R. M., Lipsky, M. M., and Trump, B. F.,** *In vitro* evaluation of the channel catfish *Ictalurus punctatus* (Rafinesque) as a test species in chemical carcinogenesis studies, in *Proc. 4th Annu. Sym. Aquatic Toxicology,* Branson, D. R. and Dickson, K. L., Eds., ASTM, Philadelphia, Pennsylvania, 1979, 226.

31. **von Hofe, E. and Puffer, H. W.,** *In vitro* metabolism and *in vivo* binding of benzo(*a*)pyrene in the California killifish (*Fundulus parvipinnis*) and speckled sanddab *(Citharicthys stigmaeous), Arch. Environ. Contam. Toxicol.,* 15, 251, 1986.

32. **Selkirk, J. K., Croy, R. G., and Gelboin, H. V.,** High-pressure liquid chromatographic separation of 10 benzo(*a*)pyrene phenols and the identification of 1-phenol and 7-phenol as new metabolities, *Cancer Res.,* 36, 922, 1976.

33. **Schmeltz, I., Tosk, J., and Williams, G. M.,** Comparison of the metabolic profiles of benzo(*a*)pyrene obtained from primary cell cultures and subcellular fractions derived from normal and methylcholanthrene-induced rat liver, *Cancer Lett.,* 5, 81, 1978.

34. **Alvares, A. P., Bickers, D. R., and Kappas, A.,** Polycholorinated biphenyls: a new type of inducer of cytochrome P-448 in the liver, *Proc. Natl. Acad. Sci. U.S.A.,* 70, 1321, 1973.

35. **Parkinson, A., Robertson, L., Safe, L., and Safe, S.,** Polychlorinated biphenyls as inducers of hepatic microsomal enzymes: structure-activity rules, *Chem.-Biol. Interact.,* 30, 271, 1980.

36. **Haake, J. M., Merrill, J. C., and Safe, S.,** The *in vitro* metabolism of benzo(*a*)pyrene by polychlorinated and polybrominated biphenyl induced rat hepatic microsomal monoxygenases, *Can. J. Physiol. Pharmacol.,* 63, 1096, 1985.

37. **Elcombe, C. R., Franklin, R. B., and Lech, J. J.,** Induction of hepatic microsomal enzymes in rainbow trout, in *Pesticide and Xenobiotic Metabolism in Aquatic Organisms,* Khan, M. A. Q., Lech, J. J., and Menn, J. J., Eds., Washington, D. C., 1979, 319.

38. **Buhler, D. R. and Rasmusson, M. E.,** The oxidation of drugs by fishes, *Comp. Biochem. Physiol.,* 25, 223, 1968.

39. **Tan, B. and Melius, P.,** Benzo(*a*)pyrene metabolism in hepatic S-9 fractions of Aroclor 1254-treated mullet (*Mugil cephalus*), in *Polynuclear Aromatic Hydrocarbons: 6th Int. Symp. on Physical and Biological Chemistry,* Cooke, M., Dennis, A. J., and Fisher, G. L., Eds., Battelle Press, Columbus, Ohio, 1982, 801.

40. **Hazel, J. R.,** Influence of thermal acclimation on membrane lipid composition of rainbow trout liver, *Am. Physiol. Soc.,* 236, R91, 1979.

41. **Willis, E. D.,** The role of the polyunsaturated fatty acid composition of the endoplasmic reticulum in the regulation of the rate of oxidative drug and carcinogen metabolism, in *Microsomes, Drug Oxidations and Chemical Carcinogenesis,* Coon, M. J., Conney, A. H., Estabrook, R. W., Gelboin, H. V., Gillette, J. R., and O'Brien, P. J., Eds., Academic Press, New York, 1980, 545.

42. **Lu, A. Y. H. and Levin, W.,** The resolution and reconstitution of the liver microsomal hydroxylation system, *Biochem. Biophys. Acta,* 344, 205, 1974.

43. **James, M. O. and Bend, J. R.,** Polycyclic aromatic hydrocarbon induction of cytochrome P-450-dependent mixed-function oxidases in marine fish, *Toxicol. Appl. Pharmacol.,* 54, 117, 1980.

44. **Varanasi, U., Gmur, D. J., and Reichert, W. L.,** Effect of environmental temperature on naphthalene metabolism by juvenile starry flounder *(Platichthys stellatus), Arch. Environ. Contam. Toxicol.,* 10, 203, 1981.

45. **Thomas, P. E., Reik, L. M., Ryan, D. E., and Levin, W.,** Regulation of the forms of cytochrome P-450 and epoxide hydrolase in rat liver microsomes: effect of age, sex, and induction, *J. Biol. Chem.,* 256, 1044, 1981.

46. **Thakker, D. R., Levin, W., Yagi, H., Conney, A. H., and Jerina, D. M.,** Regio- and stereoselectivity of hepatic cytochrome P-450 toward polycyclic aromatic hydrocarbon substrates, in *Advances in Experimental Medicine and Biology: Biological Reactive Intermediates IIA,* Parke, D. V., Kocsis, J. J., Jollow, D. J., Gibson, C. G., and Witmer, C. M., Eds., Plenum Press, New York, 1982, 529.

47. **Rasmussen, R. E. and Wang, I. Y.,** Dependence of specific metabolism of benzo(*a*)pyrene on the inducer of hydroxylase activity, *Cancer Res.,* 34, 2290, 1974.

48. **Selkirk, J. K., Croy, R. G., Roller, P. P., and Gelboin, H. V.,** High-pressure liquid chromatographic analysis of benzo(*a*)pyrene metabolism and covalent binding and the mechanisms of action of 7,8-benzoflavone and 1,2-epoxy-3,3,3-trichloropropane, *Cancer Res.,* 34, 3474, 1974.

49. **Krahn, M. M., Schnell, J. V., Uyeda, M. Y., and MacLeod, W. D., Jr.,** Determination of mixtures of benzo(a)pyrene, 2,6-dimethylnaphathalene and their metabolites by high-performance liquid chromatography with fluorescence detection, *Anal. Biochem.,* 113, 27, 1981.

50. **Gmur, D. J. and Varanasi, U.,** Characterization of benzo(*a*)pyrene metabolites isolated from muscle, liver, and bile of a juvenile flatfish, *Carcinogenesis (London),* 3, 1397, 1982.

51. **Tjessum, K. and Stegeman, J. J.,** Improvement of reverse-phase high pressure liquid chromatographic resolution of benzo(*a*)pyrene metabolites using organic amine: application to metabolites produced by fish, *Anal. Biochem.,* 99, 129, 1979.

52. **Steward, A. R., Dannan, G. A., Buzelian, P. S., and Guengerich, F. P.,** Changes in the concentration of seven forms of cytochrome P-450 in primary cultures of adult rat hepatocytes, *Mol. Pharmacol.,* 27, 125, 1985.

53. **Stegeman, J. J.,** Polynuclear aromatic hydrocarbons and their metabolism in the marine environment, in *Polycyclic Hydrocarbons and Cancer,* Vol. 3, Gelboin, H. V. and Ts'o, P. O. P., Eds., Academic Press, New York, 1981, 1.

54. **Varanasi, U., Stein, J. E., Nishimoto, M., and Hom, T.,** Benzo(*a*)pyrene metabolites in liver, muscle, gonads and bile of adult English sole (*Parophrys vetulus*), in *Polynuclear Aromatic Hydrocarbons: 7th Symp. on Formation, Metabolism and Measurement,* Cooke, M. W. and Dennis, A. J., Eds., Battelle Press, Columbus, Ohio, 1982, 1221.

55. **Varanasi, U., Reichert, W. L., Dempcy, R. O., Bailey, G. S., and Hendricks, J. D.,** *In vivo* and *in vitro* metabolism of benzo(*a*)pyrene and covalent binding to hepatic DNA in Mount Shasta strain of rainbow trout and Sprague-Dawley rat, *Proc. Am. Assoc. Cancer Res.,* 27, 108, 1986.

56. **Leskos, S., Caspary, W., Lorentzen, R., and Ts'o, P. O. P.,** Enzymatic formation of 6-oxo-benzo(*a*)pyrene radical in rat liver homogenates from carcinogenic benzo(*a*)pyrene, *Biochemistry (U.S.A.),* 14, 3978, 1975.

57. **James, M. O. and Little, P. J.,** Perturbation of benzo(*a*)pyrene (BaP) metabolism in rat and fish hepatic microsomes by monooxygenase (MO) inhibitors which stimulate epoxide hydrolase (EH), *Fed. Proc.,* 40, 697, 1981.

58. **Levin, W., Buening, M. K., Wood, A. W., Chang, R. L., Kedzierski B., Thakker, D. R., Boyd, D. R., Gadaginamath, G. S., Armstrong, R. N., Yagi, H., Karle, J. M., Slaga, T. J., Jerina, D. M., and Conney, A. H.,** An enantiomeric interaction in the metabolism and tumorigenicity of (+)- and (−) - benzo(*a*)pyrene 7,8-oxide, *J. Biol. Chem.,* 255, 9067, 1980.

59. **Chipman, J. K., Frost, G. S., Hirom, P. C., and Millburn, P.,** Biliary excretion, systemic availability and reactivity of metabolites following intraportal infusion of [³H]benzo(*a*)pyrene in the rat, *Carcinogenesis (London),* 2, 741, 1981.

60. **Boroujerdi, M., Kung, H. C., Wilson, A. G. E., and Anderson, M. W.,** Metabolism in DNA binding of benzo(*a*)pyrene *in vivo* in rat, *Cancer Res.,* 41, 951, 1981.

61. **Collier, T. K., Stein, J. E., Wallace, R. J., and Varanasi, U.,** Xenobiotic metabolizing enzymes in spawning English sole (*Parophrys vetulus*) exposed to organic solvent extracts of marine sediments from contaminated and reference areas, *Comp. Biochem. Physiol.,* 84C, 291, 1986.

62. **Hesse, S., and Jernstrom, B.,** Role of glutathione s-transferases: detoxification of reactive metabolities of benzo(*a*)pyrene-7,8-dihydrodiol by conjugation with glutathione, in *Biochemical Basis of Chemical Carcinogenesis,* Greim, H., Jung, R., Framer, M., Marquardt, H., and Desch, F., Eds., Raven Press, New York 1984/1985.

63. **Schultz, M. E. and Schultz, R. J.,** Induction of hepatic tumors with 7,12-dimethylbenz(*a*)anthracene in two species of viviparous fishes (genus: *Poecilpiosis*), *Envion. Res.,* 27, 337, 1982.

64. **Malins, D. C., McCain, B. B., Brown, D. W., Chan, S.-L., Myers, M. S., Landahl, J. T., Prohaska, P. G., Friedman, A. J., Rhodes, L. D., Burrows, D. G., Gronlund, W. D., and Hodgins, H. O.,** Chemical pollutants in sediments and diseases of bottom-dwelling fish in Puget Sound, Washington, *Environ. Sci. Technol.,* 18, 705, 1984.

65. **Black, J. J.,** Field and laboratory studies of environmental carcinogenesis in Niagara river fish, *J. Great Lake Res.,* 9, 326, 1983.

66. **Sydor, W., Jr., Lewis, K. F., and Yang, C. S.,** Effects of butylated hydroxyanisole on the metabolism of benzo(*a*)pyrene by mouse lung microsomes, *Cancer Res.,* 44, 134, 1984.

67. **Mass, M. J. and Kaufman, D. G.,** (³H)benzo(*a*)pyrene metabolism in tracheal epithelial microsomes and tracheal organ cultures, *Cancer Res.,* 38, 3861, 1978.

68. **Bickers, D. R., Mukhtar, H., and Yang, S. K.,** Cutaneous metabolism of benzo(*a*)pyrene: comparative studies in C57BL/6N and DBA/2N mice and neonatal Sprague-Dawley rats, *Chem.-Biol. Interact.,* 43, 263, 1983.

69. **Stegeman, J. J., Woodin, B. R., Klotz, A. V., Wolke, R. E., and Orme-Johnson, N. R.,** Cytochrome P-450 and monooxygenase activity in cardiac microsomes from the fish *Stenotomus chrysops, Mol. Pharmacol.,* 21, 517, 1982.

70. **Breger, R. K., Franklin, R. B., and Lech, J. J.,** Metabolism of 2-methylnaphthalene to isomeric dihydrodiols by hepatic microsomes of rat and rainbow trout, *Drug Metab. Dispos.,* 9, 88, 1981.

71. **Melancon, M. J., Williams, D. E., Buhler, D. R., and Lech, J. J.,** Metabolism of 2-methylnaphthalene by rat and rainbow trout hepatic microsomes and purified cytochromes P-450, *Drug Metab. Dispos.,* 13, 542, 1985.

72. **Krahn, M. M., Collier, T. K., and Malins, D. C.,** Aromatic hydrocarbon metabolites in fish: automatic extraction and high-performance liquid chromatographic separation into conjugate and non-conjugate fractions, *J. Chromatog.,* 236, 441, 1982.

73. **Shamsuddin, Z. A. and Rahimtula, A. D.,** Metabolism of 2,6-dimethylnaphthalene by rat liver microsomes and effect of its administration of glutathione depletion *in vivo, Drug Metab. Dispos.,* 14, 724, 1986.

74. **Farber, E. and Sarma, D. S. R.,** Chemical carcinogenesis: the liver as a model, *Pathol. Immunopathol. Res.,* 5, 1, 1986.

75. **Buty, S. G., Thompson, S., and Slaga, T. J.,** The role of epidermal arylhydrocarbon hydroxylase in the covalent binding of polycyclic hydrocarbon to DNA and its relationship to tumor initiation, *Biochem. Biophys. Res. Commun.,* 70, 1102, 1976.

76. **King, H. W. S., Thompson, M. H., and Brookes, P.,** The benzo(a)pyrene deoxyribonucleoside products isolated from DNA after metabolism of benzo(a)pyrene by rat liver microsomes in the presence of DNA, *Cancer Res.,* 34, 1263, 1975.

77. **Alexandrov, K., Brookes, P., King, H. W. S., Osborne, M. R., and Thompson, M. H.,** Comparison of the metabolism of benzo(a)pyrene and binding to DNA caused by rat liver nuclei and microsomes, *Chem. Biol. Interact.,* 12, 269, 1976.

78. **Alexandrov, K. and Frayssinet, C.,** Microsome-dependent binding of benzo(a)pyrene and aflatoxin B_1 to DNA and benzo(a)pyrene binding to aflatoxin-conjugated DNA, *Cancer Res.,* 34, 3289, 1974.

79. **King, H. W. S., Osborne, M. R., Beland, F. A., Harvey, R. G., and Brookes, P.,** ($\pm$)-7α,8β-Dihydroxy-9β,10β-epoxy-7,8,9,10-tetrahydrobenzo(a)pyrene is an intermediate in the metabolism and binding to DNA of benzo(a)pyrene, *Proc. Natl. Acad. Sci. U.S.A.,* 73, 2679, 1976.

80. **Ashurst, S. W. and Cohen, G. M.,** A benzo(a)pyrene-7,8-dihydrodiol-9,10-epoxide is the major metabolite involved in the binding of benzo(a)pyrene to DNA in isolated viable rat hepatocytes, *Chem. Biol. Interact.,* 29, 117, 1980.

81. **Jaggi, W., Lutz, W. K., and Schlatter, C.,** Comparative studies on the covalent binding of the carcinogen benzo(a)pyrene to DNA in various model systems, *Experimentia,* 35, 631, 1979.

82. **Jernstrom, B., Vadi, H., and Orrenius, S.,** Formation in isolated rat liver microsomes and nuclei of benzo(a)pyrene metabolites that bind to DNA, *Cancer Res.,* 36, 4107, 1976.

83. **Grover, P. L. and Sims, P.,** Enzyme-catalyzed reactions of polycyclic hydrocarbons with deoxyribonucleic acid and protein *in vitro, Biochem. J.,* 110, 159, 1968.

84. **Gelboin, H. V.,** A microsome-dependent binding of benzo(a)pyrene to DNA, *Cancer Res.,* 29, 1272, 1969.

85. **Eastman, A., Sweetenham, J., and Bresnick, E.,** Comparison of *in vivo* and *in vitro* binding of polycyclic hydrocarbons to DNA, *Chem. Biol Interact.,* 23, 345, 1978.

86. **Lutz, W. K.,** *In vivo* covalent binding of organic chemicals to DNA as a quantitative indicator in the process of chemical carcinogenesis, *Mutat. Res.,* 65, 289, 1979.

87. **Baird, W. M. and Brookes, P.,** Isolation of the hydrocarbon-deoxyribonucleoside products from the DNA of mouse embryo cells treated in culture with 7-methylbenz(a)anthracene-^{3}H, *Cancer Res.,* 33, 2378, 1973.

88. **Sims, P., Grover, P. L., Swaisland, A., Pal, K., and Hewer, A.,** Metabolic activation of benzo(a)pyrene proceeds by a diol-epoxide, *Nature (London),* 252, 326, 1974.

89. **Thompson, M. H., King, H. W. S., Osborne, M. R., and Brookes, P.,** Rat liver microsome mediated binding of benzo(a)pyrene metabolites to DNA, *Int. J. Cancer,* 17, 270, 1976.

90. **Legraverend, C., Nebert, D. W., Boobis, A. R., and Pelkonen, O.,** DNA binding of benzo(a)pyrene metabolities. Effects of substrate and microsomal protein concentration *in vitro*, dietary contaminants, and tissue differences, *Pharmacology,* 20, 137, 1980.

91. **Bigger, C. A. H., Tomaszewski, J. E., and Dipple, A.,** Variation in route of microsomal activation of 7,12-dimethylbenz(a)anthracene with substrate concentration, *Carcinogenesis (London),* 1, 15, 1980.

92. **Meehan, T., Straub, K., and Calvin, M.,** Benzo(a)pyrene diol epoxide covalently binds to deoxyguanosine and deoxyadenosine in DNA, *Nature (London),* 269, 725, 1977.

93. **Ashurst, S. W. and Cohen, G. M.,** Magnesium ions affect the quantitative but not the qualitative microsome mediated binding of benzo(a)pyrene to DNA, *Chem.-Biol. Interact.,* 28, 279, 1979.

94. **Nishimoto, M. and Varanasi, U.,** Benzo(a)pyrene metabolism and DNA adduct formation mediated by English sole liver enzymes, *Biochem. Pharmacol.,* 34, 263, 1985.

95. **McCain, B. B., Myers, M. S., Varanasi, U., Brown, D. W., Rhodes, L. D., Gronlund, W. D., Elliot, D. G., Palseson, W. A., Hodgins, H. O., and Malins, D. C.,** Pathology of two species of flatfish from urban estuaries in Puget Sound, Federal Interagency Energy/Environmental Research and Development Report, EPA-6001 7-82-001, Environmental Protection Agency, 1982.

96. **Wolf, K. and Quimby, C.,** Established eurythermic line of fish cells *in vitro, Science (Washington, D.C.),* 135, 1065, 1962.

97. **Hay, R., Macy, M., Corman-Weinblatt, A., Chen, T. R., and McClintock, P., Eds.,** American Type Culture Collection, Catalogue of Cell Lines & Hybridomas, 5th ed., 1985.

98. **Diamond, L. and Clark, H. F.,** Comparative studies on the interaction of benzo(a)pyrene with cells derived from poikilothermic and homeothermic vertebrates. I. Metabolism of benzo(a)pyrene, *J. Natl. Cancer Inst.,* 45, 1005, 1970.

99. **Lannan, C. N., Winton, J. R., and Fryer, J. L.,** Fish cell lines: establishment and characterization of nine cell lines from salmonids, *In Vitro,* 20, 671, 1984.

100. **Wolf, K. and Mann, J. A.,** Poikilothermic vertebrate cell lines and viruses; a current listing for fishes, *In Vitro,* 16, 168, 1980.

101. **Wolf, K., Quimby, C., and Pyle, E. A.,** Preparation of monolayer cell cultures from tissues of some lower vertebrates, *Science (Washington, D.C.),* 132, 1890, 1960.
102. **Clem, L. W., Moewus, L., and Sigel, M. M.,** Studies with cells from marine fish in tissue culture, *Proc. Soc.,* 108, 762, 1961.
103. **Wolf, K. and Quimby, M. C.,** Primary monolayer culture of fish cells initiated from minced tissues, *Tissue Cult. Assoc. Man.,* 2, 445, 1976.
104. **Wolf, K. and Quimby, M. C.,** Procedures for subculturing fish cells and propagating fish cell lines, *Tissue Cult. Assoc. Man.,* 2, 471, 1976.
105. **Martin, B. J., Ellender, R. D., Hillebert, S. A., and Guess, M. M.,** Primary cell cultures from the teleost, *Cyprinodon variegatus:* culture establishment and application in carcinogen exposure studies, *Natl. Cancer Inst. Monogr.,* 65, 175, 1984.
106. **Bailey, G., Tayler, M., and Loveland, P.,** Dietary modification of alfatoxin B_1 carcinogenesis: mechanism studies using isolated hepatocytes from rainbow trout, *Natl. Cancer Inst. Monogr.,* 65, 379, 1984.
107. **Klaunig, J. E.,** Establishment of fish hepatocyte cultures for use in *in vitro* carcinogenicity studies, *Natl. Cancer Inst. Monogr.,* 65, 163, 1984.
108. **Klaunig, J., Ruch, R. J., and Goldblatt, P. J.,** Trout hepatocyte culture: isolation and primary culture, *In Vitro Cell. Dev. Biol.,* 2, 221, 1985.
109. **Williams, G. M., Bermudez, E., and Scaramuzzino, D.,** Rat hepatocyte primary cell cultures. Improved dissociation and attachment technologies and the enhancement of survival by culture medium, *In Vitro,* 13, 809, 1978.
110. **Pollack, R., Ed.,** *Readings in Mammalian Cell Culture,* 2nd ed., Cold Spring Harbor Laboratory, Cold Spring Harbor, New York, 1981.
111. **Andrianov, L. N., Belitsky, G. A., and Ivanova, O. J.,** Metabolic degradation of 3,4-benzopyrene in the cultures of normal and neoplastic fibroblasts, *Br. J. Cancer,* 21, 566, 1967.
112. **Diamond, L., Sardet, C., and Rothblat, G. H.,** The metabolism of 7,12-dimethylbenz(a)anthracene in cell cultures, *Int. J. Cancer,* 3, 838, 1968.
113. **Diamond, L. Defendi, V., and Brookes, P.,** The interaction of 7,12-dimethylbenz(a)anthracene with cells sensitive and resistant to toxicity induced by this carcinogen, *Cancer Res.,* 27, 890, 1967.
114. **Diamond, L. and Baird, W. M.,** Chemical carcinogenesis *in vitro,* in *Growth, Nutrition, and Metabolism of Cells in Culture,* Vol. 3, Rothblat, G. H. and Cristofalo, V. J., Eds., Academic Press, New York, 1977, 421.
115. **Clark, H. F. and Diamond, L.,** Comparative studies on the interaction of benzpyrene with cells derived from poikilothermic and homeothermic vertebrates. II. Effect of temperature on benzypyrene metabolism and cell multiplication, *J. Cell. Physiol.,* 77, 385, 1971.
116. **Thornton, S. C., Diamond, L., and Baird, W. M.,** Metabolism of benzo(a)pyrene by fish cells in culture, *J. Toxicol. Environ. Health,* 10, 157, 1982.
117. **Baird, W. M., Chern, C.-J., and Diamond, L.,** Formation of benzo(a)pyrene-glucuronic acid conjugates in hamster embryo cell cultures, *Cancer Res.,* 37, 3190, 1977.
118. **Baird, W. M., Chemerys, R., Erickson, A. A., Chern, C.-J., and Diamond, L.,** Differences in pathways of polycyclic aromatic hydrocarbon metabolism as detected by analysis of the conjugates formed, in *Polynuclear Aromatic Hydrocarbons, 3rd Int. Symp. on Chemistry and Biology — Carcinogenesis and Mutagenesis,* Jones, P. W. and Leber, P., Eds., Ann Arbor Science Publishers, Ann Arbor, Michigan, 1979, 507.
119. **Autrup, H.,** Separation of water soluble metabolites of benzo(a)pyrene formed by cultured human colon, *Biochem. Pharmacol.,* 28, 1727, 1979.
120. **Merrick, B. A. and Selkirk, J. K.,** HPLC of benzo(a)pyrene glucuronide, sulfate, and gluthathione conjugates and water-soluble metabolites from hamster embryo cells, *Carcinogenesis (London),* 6, 1303, 1985.
121. **Plakunov, I., Smolarek, T. A., and Baird, W. M.,** Separation by ion-pair high-performance liquid chromatography of the glucuronide, sulfate and gluthathione conjugates formed from benzo(a)pyrene in cell cultures from rodents, fish, and humans, *Carcinogenesis (London),* 8, 59, 1987.
122. **Cooper, C. S., Grover, P. L., and Sims, P.,** The metabolism and activation of benzo(a)pyrene, in *Progress in Drug Metabolism,* Vol. 7, Bridges, J. W. and Chasseaud, L. F., Eds., John Wiley & Sons, New York, 1983.
123. **Baird, W. M., Smolarek, T. A., Plakunov, I., Hevezi, K., Kelley, J., and Klaunig, J.,** Formation of benzo(a)pyrene-7,8-dihydrodiol glucuronide is a major pathway of metabolism of benzo(a)pyrene in cell cultures from bluegill fry and brown bullhead, *Aquat. Toxicol.,* 11, 398, 1988.
124. **Kocan, R. M., Landolt, M. L., and Sabo, K. M.,** *In vitro* toxicity of eight mutagens/carcinogens for three fish cell lines, *Bull. Environ. Contam. Toxicol.,* 23, 269, 1979.
125. **Kocan, R. M., Chi, E. Y., Eriksen, N., Benditt, E. P., and Landolt, M. L.,** Sequestration and release of polycyclic aromatic hydrocarbons by vertebrate cells *in vitro, Environ. Mutagen.,* 5, 643, 1983.

126. **Huberman, E. and Sachs, L.,** Cell-mediated mutagenesis of mammalian cells with chemical carcinogens, *Int. J. Cancer,* 13, 326, 1974.

127. **Williams, G. M., Dunkel, V. C., and Ray, V. A.,** Cellular systems for toxicity testing, *Ann. N.Y. Acad. Sci.,* 407, 1983.

128. **Kocan, R. M., Landolt, M. L., Bond, J., and Benditt, E. P.,** *In vitro* effect of some mutagens/carcinogens on cultured fish cells, *Arch. Environ. Contam. Toxicol.,* 10, 663, 1981.

129. **Evans, H. J.,** Cytogenetic methods for detecting effects of chemical mutagens, in *Cellular Systems for Toxicity Testing, Ann. N.Y. Acad. Sci.,* Vol. 407, Williams, G. M., Dunkel, V. C. and Ray, V. A., Eds., 1983, 131.

130. **Preston, R. J., Au, W., Bender, M. A., Brewen, J. G., Carrano, A. V., Heddle, J. A., McFee, A. F., Wolff, S., and Wassom, J. S.,** Mammalian *in vivo* and *in vitro* cytogenetic assays: a report of the U.S. EPAs Gene-Tox Program, *Mutat. Res.,* 87, 143, 1981.

131. **Kocan, R. M., Landolt, M. L., and Sabo, K. M.,** Anaphase aberrations: a measure of genotoxicity in mutagen-treated fish cells, *Environ. Mutagen,* 4, 181, 1982.

132. **Kocan, R. M., Sabo, K. M., and Landolt, M. L.,** Cytotoxicity/genotoxicity: the application of cell culture techniques to the measurement of marine sediment pollution, *Aquat. Toxicol.,* 6, 165, 1985.

133. **McQueen, C. A. and Williams, G. M.,** The use of cells from rat, mouse, hamster and rabbit in the hepatocyte primary culture/DNA-repair test, in *Cellular Systems for Toxicity Testing, Vol. 407,* Ann. N.Y. Acad. Sci., 407, 119, 1983.

134. **Mitchell, A. D., Casciano, D. A., Meltz, M. C., Robinson, D. E., San, R. H. C., Williams, G. M., and VanHalle, E. S.,** Unscheduled DNA synthesis test: a report for the U.S. Environmental Protection Agency Gene-Tox Program, *Mutat. Res.,* in press.

135. **Walton, D. G., Acton, A. B., and Hans, F. S.,** DNA repair synthesis in cultured mammalian and fish cells following exposure to chemical mutagens, *Mutat. Res.,* 124, 153, 1983.

136. **Kelly, J. J. and Maddock, M. B.,** *In vitro* induction of unscheduled DNA synthesis by genotoxic carcinogens in the hepatocytes of the oyster toadfish *(Opsanus tau), Arch. Environ. Contam. Toxicol.,* 14, 555, 1985.

137. **Yang, L. L., Maher, V. M., and McCormick, J. J.,** Relationship between excision repair and the cytotoxic and mutagenic effect of the *anti* 7,8-diol-9,10-epoxide of benzo(*a*)pyrene in human cells, *Mutat. Res.,* 94, 435, 1982.

138. **Yang, L. L., Maher, V. M., and McCormick, J. J.,** Error-free excision of the cytotoxic, mutagenic N^2-deoxyguanosine DNA adduct formed in human fibroblasts by ($\pm$)-7-β,8α-digydroxy-9α,10α-epoxy-7,8,9,10-tetrahydrobenzo(*a*)pyrene, *Proc. Matl. Acad. Sci. U.S.A.,* 77, 5933, 1980.

139. **Stevens, C. W., Bouck, N., Burgess, J. A., and Fahl, W. E.,** Benzo(*a*)pyrene diol-epoxide: different mutagenic efficiency in human and bacterial cells, *Mutat. Res.,* 152, 5, 1985.

140. **Sawicki, J. T., Moschel, R. C., and Dipple, A.,** Involvement of both *syn-* and *anti* dihydrodiol-epoxides in the binding of 7,12-dimethylbenz(*a*)anthracene to DNA in mouse embryo cell cultures, *Cancer Res.,* 43, 3212, 1983.

141. **Pruess-Schwartz, D., Sebti, S. M., Gilham, P. T., and Baird, W. M.,** Analysis of benzo(*a*)pyrene:DNA adducts formed in cells in cultured by immobilized boronate chromatography, *Cancer Res.,* 44, 4104, 1984.

142. **Sebti, S. M., Pruess-Schwartz, D., and Baird, W. M.,** Species- and length of exposure-dependent differences of benzo(*a*)pyrene:DNA adducts formed in embryo cell cultures from mice, rats, and hamsters, *Cancer Res.,* 45, 1594, 1985.

143. **Pruess-Schwartz, D. and Baird, W. M.,** Benzo(*a*)pyrene:DNA adduct formation in early-passage Wistar rat embryo cell cultures: evidence for multiple pathways of activation of benzo(*a*)pyrene, *Cancer Res.,* 46, 545, 1986.

144. **Nishimoto, M. and Varanasi, U.,** unpublished data.

145. **Egaas, E. and Varanasi, U.,** unpublished data.

146. **Roubal, W. T.,** personal communication.

147. **Collier, M., and Varanasi, U.,** unpublished data.

148. **Nishimoto, M. et al.,** unpublished results.

149. **Smolarek, T. A., Hevezi, K., and Baird, W. M.,** unpublished results.

Chapter 7

FACTORS INFLUENCING EXPERIMENTAL CARCINOGENESIS IN LABORATORY FISH MODELS

George S. Bailey, Douglas E. Goeger, and Jerry D. Hendricks

TABLE OF CONTENTS

I. INTRODUCTION

The occurrence over two decades ago of tumor epizootics in hatchery populations of rainbow trout led not only to the ultimate identification of aflatoxins as a new class of human carcinogens,[1-3] but also to the development of the rainbow trout as an alternate model for the study of chemical carcinogenesis.[4-6] A number of additional fish species, primary aquarium fish, have also been shown to be responsive to a few selected carcinogenic agents (for a recent review see Reference 7).

Results from these studies indicate that fish models offer a number of advantageous features to supplement the use of traditional rodent models for the study of carcinogenesis. These include (1) unique aquatic habitat; (2) large numbers of animals at low cost; (3) convenient access to all life stages; (4) ability to conduct vertebrate tumor studies on submicrogram quantities of rare suspect compounds; (5) unique genetic systems, including cloned lines and thousands of offspring per mating; (6) tolerance to a wide range of habitat conditions and pollutants; and (7) nonmammalian comparative status.

The availability of aquatic vertebrates known to be responsive to carcinogens is especially relevant for understanding the environmental health significance of genotoxic aquatic pollutants. Recent discoveries of tumor epizootics in feral fish populations[8-10] have raised concern that such epizootics may be chemical in origin, possibly from industrial discharge or deposits. However, despite strong indications from associative epizootiological studies,[11,12] there is as yet no direct evidence that neoplasia in feral fish populations derives from a single, specific carcinogenic pollutant. Unlike the case for human neoplasia, such evidence may be obtainable by direct experimentation with the subjects of interest, fish. However, full exploitation of these systems will depend upon production of laboratory models and on an understanding of associated cellular and molecular mechanisms of carcinogenesis relevant to these species.

Unfortunately, fish models suitable to address many of these problems are presently either unavailable or not sufficiently developed. For example, though liver tumor incidences among populations of flatfish in the Puget Sound area[9] of Washington, have been shown to correlate strongly with levels of various hydrocarbon pollutants, the precise etiological agent(s) have not been identified. Ultimately, demonstration of cause and effect would require that the suspect etiological agent(s) be shown to induce tumors in the appropriate host organ under exposure protocols and doses which mimic environmental conditions. The minimum conditions for development of a laboratory flatfish model to address this question are that culture through the model's full life-cycle in the laboratory be avaialble, that the nutritional requirements for the species be known, and that genetic stocks be established and characterized. For further development it would be necessary to investigate in the laboratory the effects of variables which may significantly impact feral population response to carcinogens. These include the influence of diet and nutritional status on population response, the sensitivities of each life-stage to relevant modes of carcinogen exposure, the impact of potentially important environmental variables such as water temperature during exposure and/or growth phases, and the impact of noncarcinogenic-modulating compounds on carcinogen metabolism, detoxication, distribution, DNA damage and repair, and tumor promotion and progression. There are at present no marine species for which more than a modest amount of such information currently exists.

Among freshwater fishes, the rainbow trout is the most thoroughly characterized model for carcinogenesis,[13-16] but in terms of the above needs, the current state of knowledge can only be regarded as incomplete. Especially lacking are more extensive studies of polycyclic aromatic hydrocarbon (PAH) carcinogenicity in this (and other) fish species.

The purpose of this article is to review the limited number of PAH tumor studies in various fish models, including trout, and especially to focus on those variables which appear from the available data to have significant impact on the response of laboratory populations to PAH and related carcinogens. It will be essential that such variables be more thoroughly investigated before the associations between feral fish tumor incidences and the presence of PAH and other genotoxins in aquatic environments can be understood.

II. PAH CARCINOGENICITY STUDIES IN FISH MODELS

A. Aquarium and Marine Species

The majority of laboratory tumor studies in aquarium fish have examined the carcinogenicities of nitrosamines and other alkylating agents.[17-21] Although PAH as a class are ubiquitous and frequently carcinogenic compounds (40% of the 39 PAH tested in rodent models by 1972 were weakly to strongly positive[22]), only a few few of these compounds have been tested for carcinogenicity in fish models. Exposure of *Poeciliopsis* to aqueous solutions of 7,12-dimethyl benzanthracene (DMBA) induced hepatic tumors in 22 of 46 fish surviving after 7 to 8 months, and 2 fish showed lymphosarcoma[24] (Table 1). Sham-exposed controls had no neoplasms. A single s.c. injection of this carcinogen was also shown to elevate the incidence of melanotic neoplasms in the skin of croaker, *Nibea mitsukurii* (Table 1), compared to fish receiving vehicle plus the antibacterial nifurpirinol, or nifurpirinol only (incidences were 73, 23, and 13%, respectively).[12] In contrast to these results, 3-methylcholanthrene (3-MC) and DMBA failed to induce tumors in guppies, *Poecilia reticulata* and zebra fish, *Lebistes reticulatus* after 32 to 80 weeks by any route of exposure.[23] However, painting of 3-MC or benzo(a)pyrene (BaP) on the skin twice weekly for 3 to 6 months was shown to induce epitheliomas in *Gasterosteus aculeatus* and *Rhodeus amarus*, though not in *Cyprinus carpio*.[25] These laboratory studies demonstrate that PAH are carcinogenic to several fish species, inducing sarcomas, carcinomas, and epitheliomas.

B. Rainbow Trout

In rainbow trout, as in other fish models, very few studies of PAH carcinogenicity have been conducted (see Table 1). DMBA carcinogenicity and oil refinery extract cocarcinogenicity were recently investigated in rainbow trout by embryo microinjection.[26,27] Tumor responses in this study were very low. Of 28 individuals receiving DMBA (500 ng per embryo), a single eosinophilic focus and a single hepatic carcinoma were observed at 12 months. Preincubation of the DMBA with rat liver S-9 prior to injection produced two foci and one carcinoma among 32 fish after 12 months. A much stronger response to BaP was demonstrated in a recent separate study,[28] where exposure of Shasta strain rainbow trout to BaP either by feeding or by repeated i.p. injection induced substantial incidences of hepatocellular carcinoma at 18 months (25 and 45%, respectively). One treated individual also showed a single hepatic fibrosarcoma and swim bladder papilloma, while one vehicle-injected control had a single hepatocellular carcinoma. Black and co-workers,[29] using embryo microinjection, recently confirmed the carcinogenicity of a BaP in rainbow trout.

Since PAH are demonstrated by these studies to be carcinogenic to fish species in the laboratory, they may be potential carcinogens for feral fish populations. BaP and DMBA are not especially potent carcinogens in fish species when compared, for example, to aflatoxin B_1 (AFB$_1$). However, trout were responsive to BaP alone and did not require any promotional stimulus for expression of hepatocarcinogenesis. By contrast, PAH are hepatocarcinogens in rodents only following promotion or embryonic exposure.[30,31] Reasons for this somewhat enhanced response of trout to PAH hepatocarcinogenesis compared to rodents, may lie in the relatively high capacity of microsomes from uninduced trout for procarcinogen activation, coupled with relatively poor ability for excision repair.[32-34]

Table 1
PAH-INDUCED NEOPLASIA IN LABORATORY FISH MODELS

Carcinogen	Species	Exposure	Time (months)	Tumor type[a]	Incidence[b]	Ref.
BaP	Rainbow trout	Dietary, 1000 ppm	18	HC	22/111 (20%)	28
	Rainbow trout	i.p. injection, 12 per month, 1 mg	18	HC	13/28 (46%)	28
	Rainbow trout	i.p., 12 per month, 1 mg	18	SPA	1/28 (4%)	28
	Rainbow trout	i.p., 12 per month, 1 mg	18	Liver FS	1/28 (4%)	28
	Rainbow trout	Embryo injection, 10 µg	9	HC	2/23 (8.7%)	29
DMBA	Rainbow trout	Embryo injection,[c] 0.5 µg	12	Liver EN	2/32 (6%)	26
	Croaker	s.c. injection,[d] 1—2 mg	7	CP	11/15 (73%)	12
	Poeciliopsis	Water exposure, 5 ppm; 20, 10, 5.5 h	7	HC	7/21 (33%)	24

[a]   Tumor types: HC, hepatocellular carcinoma; SPA, stomach papillary adenoma; FS, fibrosarcoma; EN, eosinophilic nodule; CP, chromatophoroma.

[b]   Unless otherwise noted, controls showed no neoplasms.

[c]   Solution of DMBA was "activated" by incubation with rat liver S9 prior to injection.

[d]   Controls receiving vehicle plus treatment with nifupirinol, or nifurpirinol alone, showed 9/40 (23%) and 32/246 (13%) chromatophoromas.

Table 2
HEPATOCELLULAR CARCINOMA INCIDENCES IN RAINBOW TROUT EXPOSED AT DIFFERENT DEVELOPMENTAL AGES TO AFB_1

Developmental stage	Age at exposure (d)	Immersion exposure conditions	Tumor incidence after 12 months
Embryonic	1	0.5 ppm AFB_1, 1 h	0/59 (0%)
	3		2/53 (4%)
	5		11/58 (19%)
	7		11/55 (20%)
	9		11/62 (18%)
	11		11/60 (18%)
	13		10/60 (17%)
	15		19/60 (32%)
	17		29/63 (46%)
	19		29/59 (49%)
	21		33/55 (60%)
	Control		0/60 (0%)
Fry	42	0.01 ppm AFB_1, 30 min	23/73 (32%)
	42		26/75 (35%)
	98		10/66 (15%)
	98		14/64 (22%)

Data are from Wales et al.[6] for embryos, unpublished for fry.

III. VARIABLES INFLUENCING THE RESPONSE OF FISH TO CARCINOGENS IN THE LABORATORY

Controlled laboratory studies indicate that the response of fish to particular carcinogens can be strongly influenced by a number of variables such as diet, growth rate, temperature, and age at carcinogen exposure. While few of these studies have involved PAH per se, it is probable that the response of feral populations to these compounds would be profoundly influenced by many of the same factors which modulate the response to other carcinogens in laboratory fish models. The following section reviews some of the variables indicated as important in the response of laboratory fish models to chemical carcinogens.

A. Developmental Age at Exposure

Sinnhuber and co-workers[6] were among the first to show that exposure of trout embryos to static solutions of carcinogens could result in development of tumors after several months of growth and expansion of initiated cell populations. Metcalfe and Sonstegard,[26] using microinjection, demonstrated that exposure of embryos to as little as 13 ng of AFB_1 or 500 ng of DMBA per embryo could lead to tumor development in a few individuals (3 of 28 and 1 of 28 fish, respectively) after 12 months. Such results suggest that all life stages in fish populations may be susceptible to carcinogenesis from environmental genotoxins. Whether embryonic exposure among feral populations is more significant than exposure at later life-stages is at present unclear.

We have recently initiated studies to systematically examine the relationship between age of exposure and tumor induction in rainbow trout. Table 2 shows that for fry exposed to static solutions of AFB_1, the carcinogenic response decreases with increasing age at exposure. These results may be taken to indicate that juvenile fish resemble rodents in being less susceptible to tumor initiation with increasing age. The situation is somewhat reversed in embryos. As shown in Table 2, the response of trout embryos to AFB_1 carcinogenesis *increases* with stage of development, with a large increase in sensitivity occurring at the time of liver organogenesis, at day 14.

The biochemical mechanisms underlying these age-related changes in carcinogen sensitivity in trout have yet to be established. Before the effects of age at exposure on tumor response are fully understood, studies are needed on the relationship between carcinogen dose applied (e.g., carcinogen concentration in static exposure solution), dosage (dose taken up per unit body weight), DNA damage *in the relevant target tissue nuclear DNA,* and ultimate tumor incidences among fish populations exposed at different ages. Such studies are important in focusing our attention on the most vulnerable life-stages in feral populations. For example, study of the stomach contents, liver mixed-function oxidase (MFO) levels, and bile contents of flatfish captured in the wild may provide important information for assessing xenobiotic environmental load, but could be less important for explaining tumor incidence than assessment of exposure of pelagic embryos to PAH in the surface film, where high concentrations can occur.

B. Route of Carcinogen Exposure

Routes of exposure potentially important to feral populations include involuntary uptake (maternal or post-partum uptake by embroys, gill uptake, direct contact of skin with substrate and water) as well as through the drinking and feeding behavior of hatched fish. Exposure route is important because it may affect the tissue specificity of a given carcinogen and may be a determinant in the maximum dosage to which a particular life-stage is likely to be exposed. For example, exposure by microinjection[26,29] was useful to demonstrate that BaP and DMBA can be carcinogenic to trout embryos, but clearly does not serve to indicate whether carcinogenic dosages of PAH could ever be reached through embryo contact with water, surface films, or sediments containing PAH in the wild.

The studies of Hendricks et al.[28] on BaP carcinogenicity indicated liver as the primary target by both dietary and i.p. injection exposure (Table 3). The latter route, while not directly relevant to environmental exposure, also produced fibrosarcoma and stomach papilloma in one individual, along with liver neoplasms, indicating some effect of exposure route in fish for this compound. Much more striking effects have been observed for *N*-nitroso compounds. As seen in Table 3, exposure of rainbow trout to the well-known mutagen and carcinogen *N*-methyl,*N'*-nitro,*N*-nitrosoguanidine (MNNG) produces a spectrum of target tissue responses which depend on route of exposure. Whereas dietary exposure results solely in stomach papillary adenomas, embryo exposure induces hepatocellular carcinomas, nephroblastomas, and stomach papillary adenomas. By comparison, immersion of fingerlings in an aqueous solution of MNNG dramatically shifts the primary target organ to kidney, with lower incidences in stomach and liver. This route also uncovers additional target organs, namely, swim bladder and gill tissue. The incidences of neoplasms in liver and kidney appear to be especially dependent on route of exposure by this compound. The tissue specificity of *N*-nitroso-diethylamine (DENA) is similarly highly dependent on exposure route; embryo and dietary exposure produce only hepatomas, whereas fry immersion also induces papillary adenomas in stomach and swim bladder (Table 3). By contrast to these carcinogens, exposure of trout to AFB_1 by dietary, embryo, fry, or i.p. routes in our laboratory historically has revealed only liver neoplasms, primarily hepatocellular carcinoma. Presumably, this is because liver is the predominant site for metabolic activation of AFB_1 in trout regardless of route of administration. It is clear from these studies that route of exposure is a major determinant of target organ specificity for some carcinogens in laboratory models and, hence, may be an important determinant of target organ tumor incidences among feral populations.

C. Genetic Variation

Genetic variation is a well-documented determinant of PAH tumorigenicity in mice, but

Table 3
ORGANOSPECIFICITY FOR BaP, DENA, AND MNNG CARCINOGENICITY IN TROUT, WITH VARYING ROUTES OF EXPOSURE

Exposure route	Tumor incidence in various organs				
	Liver	Stomach	Kidney	Swim bladder	Gill
BaP[a]					
Dietary	28/111 (25%)	—	—	—	—
i.p. injection	13/28 (46%)	—	—	1/28 (4%)	—
DENA[b]					
Embryo immersion	68/79 (86%)	—	—	—	—
Fry immersion	41/42 (98%)	23/42 (55%)	—	10/42 (24%)	—
Fry dietary	44/80 (55%)	—	—	—	—
MNNG[c]					
Embryo immersion	80/118 (68%)	47/118 (40%)	22/191 (12%)	—	—
Fry immersion	8/117 (7%)	47/117 (40%)	59/115 (38%)	6/117 (5%)	1/117 (1%)
Fry dietary	—	110/110 (100%)	—	—	—

[a] Data are from Hendricks et al.[28] and are after 18 months of exposure.
[b] Unpublished data. Exposure conditions were 2700 ppm, 24 h; 100 ppm, 3 weeks; and 1100 ppm, 12 months, respectively.
[c] Unpublished data. Exposure conditions were 100 ppm, 30 min; 50 ppm, 30 min; and 500 ppm, 18 months, respectively. For the dietary exposure, tumor incidence was determined at 18 rather than 12 months as for all other studies.

this parameter has received little study in fish. Schultz and Schultz[24,35] examined the response of two species of *Poeciliopsis (monacha* and *lucida)* and four strains of *lucida* to DMBA. Although the number of treated and surviving individuals was too few to be definitive, no striking differences were observed in response. A more extensive study involving nine inbred, wild, and intercrossed strains of *P. lucida* by these authors failed to reveal any substantive differences in response to DENA.[24]

We have initiated studies in rainbow trout on the genetics of response to AFB$_1$ among outcrossed and inbred strains. As seen in Table 4, embryo treatment of several strains of trout with identical levels of AFB$_1$ resulted in differing tumor responses when assessed 12 months after exposure. Among the various strains tested, including inbred lines and wild steelhead strains, tumor response ranged from 11 to 64%. By comparison, variation among genera appears to be more striking, with rainbow trout and coho salmon (*Oncorhynchus kistuch*) differing in sensitivity to dietary AFB$_1$ by perhaps two orders of magnitude.[32]

These studies indicate that for salmonid fish, intraspecific genetic variation in response to AFB$_1$ may be much less significant than differences at the species or genus level. Aquarium fish studies with DENA indicated even less genetic variability in tumor responsiveness. Without question these data are limited by the number of strains and carcinogens tested. Although tumor studies were not carried out, Pedersen et al.[36] did report modest to large differences among six strains of rainbow trout in metabolism of selected xenobiotics. For example, the conversion of BaP to the 3-OH metabolite by hepatic enzymes *in vitro* differed

Table 4

SENSITIVITY OF SEVERAL STRAINS OF RAINBOW TROUT TO AFB$_1$ CARCINOGENICITY BY EMBRYO EXPOSURE[a]

Strain	Hepatocarcinoma incidence at 12 months	
Experiment 1 (1984)		
FGS 14	23/75	(30%)
FGS 16	10/90	(11%)
FGS 23	15/94	(16%)
FGS 24	14/73	(19%)
FGS 34	14/84	(17%)
INB 30	22/70	(31%)
INB 55	21/87	(24%)
Shasta	77/191	(40%)
Experiment 2 (1983)		
Steelhead	9/80	(11%)
Albino	14/59	(24%)
Shasta	62/97	(64%)

[a]   0.5 ppm, 30 min.

by greater than 50-fold among strains. The inducibility of this activity by 3-MC also varied about fivefold among these strains. The genetic basis of these differences was not clearly established by these studies, however, especially since the strains tested were not reared under identical laboratory conditions, differed in size and age, and were acclimated for only 1 month prior to study. Additional long-term, controlled laboratory tumor studies are necessary before the possible impact of genetic variation among feral fishes in response to PAH can be estimated.

D. Nutritional Status, Growth Rate, and Carcinogen Response

The overall nutritional status, rate of growth, and rate of proliferation of target tissue may well have a profound effect on the response of a particular population following carcinogen exposure. Examples of such effects are seen in the trout studies shown in Table 5. Among fish exposed to BaP or AFB$_1$ there is an effect of overall growth rate on the probability of tumor development at the time of termination. To facilitate comparison among experiments, fish were divided into four groups with respect to the mean body weight of the population at termination: those whose body weights were greater than one standard deviation (SD) above the mean, within one SD above the mean, within one SD below the mean, and greater than one SD below the mean. In virtually all experiments historically, including seven of the eight experiments shown here, a clear trend toward higher tumor incidence is seen in those fish with greater growth over the course of the experiment. In some cases the difference between smallest and largest groups within an experiment can be very striking (e.g., 0 or 4 vs. 39%).

It is important to consider if the size effect could be due to variation in dosage at time of carcinogen exposure. Thus, by dietary exposure, one might postulate that larger fish have higher tumor incidences at termination, because they demonstrate more aggressive feeding behavior and thus would receive a greater total dose of dietary carcinogen. However, this has not been directly determined. More importantly, it is not known if the actual *dosage*

Table 5

EFFECT OF GROWTH RATE ON RESPONSE OF TROUT TO BaP AND AFB$_1$

Carcinogen	Exposure route	Promotional stimulus	Tumor incidence vs. body weight[a]			
			< Mean −1 SD	Mean −1 SD	Mean +1 SD	<Mean +1 SD
BaP	Dietary	—	5%	17%	33%	33%
	(1000 ppm, 18 mo)		(1/20)	(4/24)	(17/52)	(5/15)
AFB$_1$	Dietary	—	3.6%	13%	25%	39%
	(20 ppb, 4 wks)		(1/28)	(8/64)	(10/64)	(12/31)
	Dietary	—	0%	16%	50%	39%
	(10 ppb, 4 wks)		(0/24)	(10/61)	(23/46)	(12/31)
	Embryo	—	53%	56%	68%	78%
	(5 ppm, 30 min)		(9/17)	(15/27)	(23/34)	(14/18)
	Embryo	BNF	31%	68%	86%	92%
			(5/12)	(21/31)	(19/22)	(12/16)
	Embryo	PB	69%	67%	64%	53%
			(11/16)	(22/33)	(18/28)	(8/15)
	Embryo	DDT	87%	92%	88%	100%
			(13/15)	(22/24)	(30/34)	(11/11)
	Embryo	I3C	50%	76%	79%	92%
			(6/12)	(28/37)	(26/33)	(12/13)

[a] Tumor incidences 12 months after AFB$_1$ initiation, for groups in four weight categories relative to the mean weight of the total group population.

received (dose/body weight) varies substantially among fish of various sizes, or if the smaller class at initiation remains the smaller class at termination. In this context it is significant that the size effect is also apparent when carcinogen exposure is involuntary and probably more uniform (embryo immersion), though variation among size groups is less pronounced than with voluntary (dietary) exposure. The size effect is also observed for different types of initiators, including BaP and AFB$_1$ (Table 5) and MNNG (data not shown), and when promoters of carcinogenesis in trout such as dichlorodiphenyltrichloroethane (DDT) are included in the diet following initiation (Table 5).

The influence of dietary protein, a major nutritional component, on carcinogenesis in trout has previously been described.[37] Increasing protein content in the formulated diet was shown to have a direct stimulating effect on the final tumor incidence in a carcinogen-treated population of fish. In these studies, groups of trout initiated with 20 ppb AFB$_1$ and subsequently fed isocaloric diets containing 40, 50, 60, or 70% dietary protein had final tumor incidences of 33, 48, 68, and 90%, respectively (unpublished results). Similar results have been obtained substituting MNNG for AFB$_1$ as initiator, and for two different sources of protein, namely purified casein or fish protein concentrate. The effect of dietary protein appears to be primarily promotional in nature, remains even after correction for any differences in growth rates among groups fed varying protein levels, and is independent of mode of carcinogen exposure (data not shown).

A clear need exists to expand these initial observations to include other carcinogens, including PAH, and other dietary factors besides protein. However, the limited results available strongly indicate that variation in nutritional status and growth rates among various populations of a given species would be an important determinant in response to environmental genotoxins. This may serve as a confounding factor not only in comparisons of tumor incidences among feral populations, but in reproducibility of results among investigators working with other fish laboratory models whose nutritional requirements are not established or reliably controlled.

E. Dietary Micronutrients and Non-Nutrient Modifiers of Carcinogenesis

Ample evidence is available from studies in rodent models that the process of carcinogenesis can be profoundly modified by a number of factors present in the diet which may not themselves be carcinogenic. These compounds include the vitamins A, C, and E, antioxidants, flavonoids, indoles, the trace element selenium, chlorinated hydrocarbons including pesticides and polychlorinated biphenyls (PCB), barbiturates, phenols, and hormones (for recent reviews see References 38 and 39).

The rainbow trout model has been used extensively to examine tumor promotion and inhibition by several such dietary modulators. Three carcinogens have been used in our trout studies, namely, AFB_1, MNNG, and DENA. We have not conducted tumor modulation studies with PAH as initiating carcinogen in our laboratory. The effects of modulators on initiation processes were studied by dietary administration prior to and during carcinogen exposure. Promotional effects were examined by initiating with carcinogen first (usually as embryos), followed by a period of modulator treatment in the diet of hatched fry or fingerlings.

The results in Table 6 show that dietary pretreatment of trout with PCB, the flavonoid β-napthoflavone (BNF), or indole-3-carbinol (I3C) leads to a reduced carcinogenic response in trout exposed to AFB_1, compared to controls receiving AFB_1 alone. This anticarcinogenic effect is dose-responsive to inhibitor concentration (Table 6), and reproducible. Not all rodent tumor modulators in rodents were effective in trout. For example, the antioxidants butylated hydroxyanisole (BHA) and ethoxyquin (EQ), the barbiturate phenobarbital (PB), and soybean protease inhibitor, have proven inhibitory in certain rodent carcinogen protocols, but were not effective in trout (data not shown).

In the trout model the effect of so-called "inhibitors" was determined to be highly dependent on timing of modulator and carcinogen exposure. Post-initiation exposure of trout to BNF, DDT, 17-β-estradiol, and I3C, provided promotional enhancement of tumor response, rather than inhibition (unpublished results). This effect was seen using either AFB_1 or MNNG as initiator. Similarly, post-initiation enhancement occurred for these "inhibitors" whether initiation was at the embryonic or fry feeding stage. Thus, some environmental modulators may act as either inhibitors or promoters, depending on timing of exposure relative to genotoxin insult.

The effectiveness of a modulator when given prior to initiation depends on the carcinogen used for initiation. For example, treatment with Aroclor 1254® prior to and along with AFB_1 inhibited hepatoma induction in trout (Table 6), but had no effect on induction by the direct mutagen MNNG (not shown). Simultaneous treatment of trout with Aroclor 1254® and DENA actually enhanced the hepatocarcinogenic response.[40] These results are readily understood in terms of the differing requirements of AFB_1, MNNG, and DENA for metabolic activation, detoxication, and DNA adduct formation. Aroclor 1254® and BNF are well-known inducers of cytochrome P-448 and associated MFO activities in trout liver.[37,41-43] This induction is accompanied by altered AFB_1 pharmacokinetics *in vivo*, enhanced production of detoxication products in liver, enhanced bile elimination of glucuronide conjugates, and reduced level of AFB_1-DNA adduct formation in isolated hepatocytes and *in vivo* in liver.[44,45] MNNG is a direct-acting mutagen not requiring metabolic activation, and its carcinogenicity would not be expected to be influenced by inhibitors which operate through induction. By contrast, DENA activation appears to procede via a P-448-dependent[46] α-hydroxylation reaction; stimulation of this activity in trout may well account for the enhancing effects of PCB on DENA carcinogenicity. Mechanisms by which modulators enhance or promote carcinogenesis in trout are currently not well understood.

At present, no studies examining the effects of tumor modulators on PAH carcinogenicity in fish have been published. However, a large number of studies have demonstrated that PCB and other modulators can influence the metabolism of genotoxic and nongenotoxic

Table 6
EFFECTS OF SELECTED MODULATORS ON DIETARY AFB$_1$-INDUCED HEPATOCELLULAR CARCINOMA RESPONSE IN RAINBOW TROUT

| Experiment | Period of dietary exposure | | | Tumor incidence |
	Pre-initiation	Initiation	Post-initiation	
1	Control	Control	Control	0/118 (0%)
	Control	AFB$_1$[a]	Control	3/118 (45%)
	BNF (50 ppm)	BNF + AFB$_1$	BNF	23/117 (20%)
	BNF (500 ppm)	BNF + AFB$_1$	BNF	9/120 (8%)
	I3C (2000 ppm)	I3C + AFB$_1$	I3C	6/118 (5%)
2	Control	Control	Control	0/99 (0%)
	Control	AFB$_1$[b]	Control	9/99 (9%)
	BNF (50 ppm)	BNF + AFB$_1$	Control	7/101 (7%)
	BNF (500 ppm)	BNF + AFB$_1$	Control	1/100 (1%)
	I3C (2000 ppm)	I3C + AFB$_1$	Control	3/99 (3%)
	Control	AFB$_1$	BNF (50 ppm)	18/101 (18%)
	Control	AFB$_1$	BNF (500 ppm)	30/100 (30%)
	Control	AFB$_1$	I3C (2000 ppm)	58/100 (58%)
3	Control	Control	Control	0/56 (0%)
	Control	AFB$_1$[c]	Control	37/79 (47%)
	PCB (100 ppm)	PCB + AFB$_1$	Control	20/77 (26%)

[a] Pre-initiation was for 6 weeks, initiation was for 10 d at 20 ppb, AFB, post-initiation was for 6 weeks. Data are from Reference 37.
[b] Pre-initiation was for 8 weeks, initiation was for 4 weeks at 20 ppb AFB, post-initiation was for 12 weeks. Data are unpublished.
[c] Pre-initiation was for 12 weeks, initiation was for 2 weeks at 20 ppb AFB$_1$.

PAH in fish. Egaas and Varanasi[47] showed that PCB induction in trout produced microsomes with enhanced rates of BaP metabolism and DNA adduct formation *in vitro*. Statham et al.[48] showed that induction of hepatic microsomal P-450 and BaP hydroxylase activities by 2,3-benzanthracene (BaA) resulted in increased metabolism and biliary excretion of 2-methyl-naphthalene *in vivo*. BNF pretreatment produced similar effects on the metabolism, disposition, and bile elimination of naphthalene, 2-methylnaphthalene, and 1,2,4-trichlorobenzene in trout.[49,50]

The potential effects of MFO inducers on the response of fishes to PAH and other carcinogens may be significant for all life-stages, including embryonic exposure. Gurney et al.[51] studied the distribution of a PCB isomer during sexual maturation in trout, and concluded that a portion of the parental burden was transferred to the gametes. This transfer of inducing agents may have significant effects on the response of embryos to subsequent environmental carcinogen challenge. Indeed, treatment of *Fundulus heteroclitus* embryos with Arclor 1254® or Number 2 fuel oil has been shown to alter their ability to metabolize BaP.[52]

There is a clear need that the tumor and mechanism studies on modulation of AFB$_1$ carcinogenesis in trout be extended to studies of PAH carcinogenesis in this and other fish species. Present data indicate that variation in exposure to tumor modulators may be a significant variable when comparing tumor incidences among feral fish populations.

F. Temperature Effects on the Carcinogenic Process in Fish

Seasonal fluctuation in temperature is an environmental parameter potentially significant in the carcinogenic response in poikilotherms. The influence of water temperature on the initiation and post-initiation stages of carcinogenesis has been examined for DENA tumor induction in the aquarium fish, medaka, *Oryzias latipes*.[53] Fish were initiated by exposure to DENA solutions at high (22 to 25°C) or low (6 to 8°) temperature for 8 weeks, and held at high or low temperature for an additional 12 weeks for tumor development. Fish initiated and held at high temperature had the greatest extent of tumor development, whereas those initiated and held at low temperature failed to develop tumors. Fish initiated at low and shifted to high temperature showed only a few tumors, whereas those treated in reverse had tumor development approaching that of the high-high group. Thus, a high temperature during initiation, possibly leading to greater DENA metabolism and DNA damage, appears to have been more important than high temperature during the cell proliferation phase for eventual tumor development. Khudoley[54] also demonstrated a significant effect of water temperature on carcinogenesis in the zebra fish, *Danio rerio* and the guppy, *Poecilia reticulata*. *N*-nitrosodimethylamine (DMNA), DENA, and nitrosomorpholine (NM) were used as initiators, and fish were initiated and held at 17, 22, or 27°C, without any temperature shifts. For all carcinogens, tumor incidences at 17°C were about half those at 27°C, and the latency period was about 5 weeks longer. Hendricks and co-workers[14] reported a single experiment examining temperature effects on AFB_1 carcinogenesis in trout embryos. Embryos were initiated by exposure to a 0.5-ppm solution of AFB_1 at 12, 15, or 18°C for 15 min. All groups were then returned to 12°C water for 12 months. Even this brief temperature shift, without time for adaptation to occur, resulted in an apparent lowering of tumor latency period with increasing temperature, and a slight elevation of final tumor incidence at termination.

Biochemical and physiological mechanisms underlying these temperature effects on carcinogenesis in fish were not reported. One possibility is that temperature-mediated alterations in xenobiotic uptake, metabolism, detoxication, or excretion processes lead to different levels of initial DNA damage or persistence. Tissue retention of xenobiotics, including BaP, can be higher at cold than at warm temperatures for marine organisms.[55,56] Niimi and Palazzo[57] demonstrated that half-lives for elimination of hexachlorobenzene and Mirex® in rainbow trout decreased with increasing temperature of acclimation. Excretion of phenolphthalein as the glucuronide in bile *in vivo*,[58] and of 7-ethoxycoumarin metabolites from perfused liver *in vitro*,[59] was shown to increase with acclimation temperature. Since the glucuronation reaction itself exhibited ideal temperature compensation,[58] these changes may be more dependent on temperature modulation of transport processes than on changes in phase I or phase II metabolism *in vivo*.

Little work appears to have been conducted on the effects of temperature on DNA adduct formation in fish. Microsomes isolated from trout held at 7°C were shown to catalyze BaP-DNA adduct formation *in vitro* approximately fivefold more than microsomes from trout held at 16°C.[47] However, both reactions were actually carried out at the "temperature optimum" of 29°C, and also lacked the total cellular compartmentalization which can be crucial for assessing competing reactions in branched metabolic pathways. The significance of these results for carcinogen metabolism and initiation *in vivo* is thus unclear. We have examined the short-term effects of temperature shift on AFB_1-DNA adduct formation in isolated trout hepatocytes. As seen in Table 7, the overall rate of AFB_1 metabolism to produce DNA adducts was about twofold greater when hepatocytes were incubated at 22°C, compared to 12°C. Significantly, however, no change was seen in the relative rates of production of the various phase I and phase II metabolites of AFB_1. In this study, all hepatocytes were taken from fish acclimated at one temperature (12°C), so the effects would not reflect long-term *in vivo* temperature adaptations. Studies on long-term temperature adaptation using isolated trout hepatocytes are in progress.[62]

Table 7
EFFECT OF INCUBATION TEMPERATURE ON METABOLISM OF
AFB$_1$ BY FRESHLY ISOLATED TROUT HEPATOCYTES

Incubation temperature (°C)	Polar metabolites	AFM$_1$	AFB$_1$	AFL	DNA adducts[a]
Percent of Total Aflatoxin Recovered[b]					
12	7.4 (1.2)	1.2 (0.2)	76 (3)	10 (1)	0.59 (0.07)
22	12.2 (0.06)	2.6 (0.3)	61 (2)	18 (2)	1.13 (0.15)
Percent of Metabolite in Total Pool[c]					
12	39 (7)	6 (1)		55 (8)	
22	37 (3)	8 (1)		56 (6)	

[a] Mean (SEM) micromoles AFB$_1$ bound per gram DNA; N = 5 for each treatment.

[b] Mean (SEM) of percent total radiolabel recovered from HPLC analysis of 1-h incubation of [^{3}H]-AFB$_1$ with hepatocytes. Viability 95% or better using dye exclusion; N = 5 for each treatment. All means at 12°C were significantly different ($p < 0.05$, polar metabolites; $p < 0.01$, all others) from those at 22°C, using analysis of variance of a random block design.

[c] Mean (SEM) of percent each metabolite contributes to total metabolite pool (i.e., AFB$_1$ excluded).

Seasonal and/or sex-dependent differences in drug-metabolizing capacity have also been reported for trout.[60,61] For example, hepatic monooxygenase activities in trout were shown to exhibit ideal temperature compensation during water cooling in autumn, in contrast to UDP glucuronyltransferase which fluctuated cyclically with seasonal changes. It should be stressed that these effects were apparent only if enzyme assays were conducted at the relevant environmental temperature. Although further studies are needed, the results to date indicate that temperature variation is likely to be an important environmental determinant for tumor response to genotoxins in feral fish populations, and that the biochemical and physiological mechanisms underlying these effects are accessible to experimental discovery.

IV. SUMMARY AND CONCLUSIONS

The likely impact of an environmental carcinogen on any fish population will depend on the amount of exposure received by each susceptible life-stage, the ability of each stage to absorb and metabolize the carcinogen and to repair any ensuing damage, and the genetic consequences for tumor induction per unit of genomic damage in target organs at each developmental age. The studies reviewed here demonstrate that in laboratory fish models, several additional factors may have significant impact on the population response to a given chemical carcinogen. These include, but may not be limited to, developmental age at exposure, route of exposure, genetic variation, nutritional status and growth rate, nutrient and non-nutrient modulators of carcinogenicity, and water temperature during initiation and/or growth. When considered individually, the effect of each parameter on experimental carcinogenicity seems not to be great, generally less than one order of magnitude. However, it may be a rare circumstance that only one such variable will be operative in comparison of tumor incidences among feral fish populations. More frequently, some combination of several such variables may operate, along with carcinogen exposure, to determine the tumor incidence in populations under comparison. Since most of these variables function independently, in some instances their environmentally random occurrence may combine to profoundly alter the tumor incidence among populations under comparison, and confound any attempt to ascribe tumor incidence solely to variation in exposure to a suspect environmental carcinogen.

The major impediment to identification of specific etiological agents in feral fish neoplasia may well be the difficulties imposed by the existence of highly complicated mixtures of genotoxic compounds. This will be superimposed on an intermittent and variable exposure of host species to compounds which can modulate fish response to environmental carcinogens. By analogy to human epidemiological studies, careful recognition and control for each known confounding variable will be essential for causative tumorigenic agents in the aquatic environment to be identified confidently.

ACKNOWLEDGMENTS

The unpublished studies reported here were partially supported by U.S. Public Health Service research grants ES00092, ES00541, ES00210, ES03850, CA34732, and CA398398. We are grateful for the expert assistance of John Casteel, Patricia Loveland, Ted Will, Toshiko Morita, and Janet Wilcox in some of these studies. This is Technical Paper No. 7985, Oregon Agricultural Experiment Station, Oregon State University.

REFERENCES

1. **Rucker, R. R., Yasutake, W. T., and Wolf, H.,** Trout hepatoma — a preliminary report, *Prog. Fish Cult.,* 23, 3, 1961.
2. **Linsell, C. A. and Peers, F. B.,** Field studies on human cancer, in *Origins of Human Cancer. Book A: Incidence of Cancer in Humans,* Hiatt, H. H., Watson, J. D., and Winsten, J. A., Eds., Cold Spring Harbor Laboratory, Cold Spring Harbor, New York, 1977, 549.
3. **Wogan, G. N.,** The induction of liver cancer by chemicals, in *Liver Cell Cancer,* Cameron, H. M., Linsell, C. A., and Warwick, G. P., Eds., Elsevier, New York, 1976, 121.
4. **Sinnhuber, R. O. et al.,** Trout bioassay of myocotoxins, in *Mycotoxins in Human and Animal Health,* Rodricks, J. V., Hesseltine, C. W., and Mehlman, M. A., Eds., Pathotox Publishers, Park Forest South, Illinois, 1977, 731.
5. **Sinnhuber, R. O. et al.,** Neoplasms in rainbow trout, a sensitive animal model for environmental carcinogenesis, *Ann. N.Y. Acad. Sci.,* 389, 1977.
6. **Wales, J. H. et al.,** Aflatoxin B_1 induction of hepatocellular carcinoma in the embryos of rainbow trout *(Salmo gairdneri), J. Natl. Cancer Inst.,* 60, 1133, 1978.
7. **Hoover, K. L.,** Ed., Use of small fish species in carcinogenicity testing, *J. Natl. Cancer Inst. Monogr.,* 65, 1984.
8. **Couch, J. A. and Harshbarger, J. C.,** Effects of carcinogenic agents on aquatic animals: an environmental overview, *Environ. Carcinogenesis Rev.,* 3, 63, 1985.
9. **Malins, D. C., McCain, B. B., and Brown, D. W.,** Chemical pollutants in sediments and diseases of bottom-dwelling fish in Puget Sound, Washington, *Environ. Sci. Technol.,* 18, 705, 1984.
10. **Murchelano, R. A. and Wolke, R. E.,** Epizootic carcinoma in the winter flounder, *Pseudopleuronectes americanus, Science,* 228, 587, 1985.
11. **Malins, D. C. et al.,** Toxic chemicals in marine sediment and biota from Mukilteo, Washington: relationships with hepatic neoplasms and other hepatic lesions in English Sole *(Parophrys vetulus), J. Natl. Cancer Inst.,* 74, 487, 1985.
12. **Kimura, I. et al.,** Correlation of epizootiological observations with experimental data: chemical induction of chromatophoromas in the Croaker, *Nibea mitsukurii, J. Natl. Cancer Inst. Monogr.,* 65, 139, 1984.
13. **Hendricks, J. D.,** The use of rainbow trout *(Salmo gairdneri)* in carcinogen bioassay, with special emphasis on embryonic exposure, in *Phyletic Approaches to Cancer,* Dawe, C. J. et al., Eds., Japan Scientific Society Press, Tokyo, 1981, 227.
14. **Hendricks, J. D. et al.,** Rainbow trout embryos: advantages and limitations for carcinogenesis research, *J. Natl. Cancer Inst. Monogr.,* 65, 129, 1984.

15. **Bailey, G. S. et al.**, Dietary modification of aflatoxin B_1 carcinogenesis: mechanism studies with isolated hepatocytes from rainbow trout, *J. Natl. Cancer Inst. Monogr.*, 65, 379, 1984.

16. **Hendricks, J. D., Meyers, T. R., and Shelton, D. W.**, Histological progression of hepatic neoplasia in rainbow trout *(Salmo gairdneri)*, *J. Natl. Cancer Inst. Monogr.*, 65, 321, 1984.

17. **Sato, S. et al.**, Hepatic tumors in the guppy *(Lebistes reticulatus)* induced by aflatoxin B_1, dimethylnitrosamine, and 2-acetylaminofluorene, *J. Natl. Cancer Inst.*, 50, 765, 1973.

18. **Park, E.-H. and Kim, D. S.**, Hepatocarcinogenicity of dimethylnitrosamine to the self-fertilizing hermaphroditic fish *Rivulus marmoratus* (Teleostomi: Cyprinodontidae), *J. Natl. Cancer Inst.*, 73, 871, 1984.

19. **Ishikawa, T., Shimamine, T., and Takayama, S.**, Histologic and electron microscopy observations on diethylnitrosamine-induced hepatomas in small aquarium fish *(Oryzias latipes)*, *J. Natl. Cancer Inst.*, 55, 909, 1975.

20. **Schwab, M. et al.**, Genetics of susceptibility in the platyfish/swordtail tumor system to develop fibrosarcoma and rhabdomyosarcoma following treatment with *N*-methyl-*N*-nitrosourea (MNU), *Z. Krebsforsch.*, 91, 301, 1978.

21. **Hawkins, W. E. et al.**, Development of aquarium fish models for environmental carcinogenesis: tumor induction in seven species, *J. Appl. Toxicol.*, 5, 261, 1985.

22. **Neff, J. M.**, *Polycyclic Aromatic Hydrocarbons in the Aquatic Environment*, Applied Sciences Publishers, London, 1979, 216.

23. **Pliss, G. B. and Khudoley, V. V.**, Tumor induction by carcinogenic agents in aquarium fish, *J. Natl. Cancer Inst.*, 55, 129, 1975.

24. **Schultz, R. J. and Schultz, M. E.**, Characteristics of a fish colony of *Poeciliopsis* and its use in carcinogenicity studies with 7,12-dimethylbenz[a]anthracene and diethylnitrosamine, *J. Natl. Cancer Inst. Monogr.*, 65, 5, 1984.

25. **Ermer, R.**, Versuche mit cancerogenen Mitteln bei kurzlebigen Fischarten, *Zool. Anz.*, 184, 175, 1970.

26. **Metcalfe, C. D. and Sonstegard, R. A.**, Microinjection of carcinogens into rainbow trout embryos: an in vivo carcinogenesis assay, *J. Natl. Cancer Inst.*, 73, 1125, 1984.

27. **Metcalfe, C. D. and Sonstegard, R. A.**, Oil refinery effluents: evidence of co-carcinogenic activity in the trout embryo microinjection assay, *J. Natl. Cancer Inst.*, in press.

28. **Hendricks, J. D. et al.**, Hepatocarcinogenicity of benzo[a]pyrene to rainbow trout by dietary exposure and intraperitoneal injection, *J. Natl. Cancer Inst.*, 74, 839, 1985.

29. **Black, J. J., Maccubbin, A. E., and Schiffert, M.**, A reliable, efficient, microinjection apparatus and methodology for the in vivo exposure of rainbow trout and salmon embryos to chemical carcinogens, *J. Natl. Cancer Inst.*, 75, 1123, 1985.

30. **Marquardt, H., Sternberg, S. S., and Phillips, F. S.**, 7,12-Dimethylbenz(a)anthracene and hepatic neoplasia in regenerating rat liver, *Chem. Biol. Interact.*, 2, 401, 1970.

31. **Klein, M.**, Susceptibility of strain BGAF/J hybrid infant mice to tumorigenesis with 1,2-benzanthracene, deoxycholic acid, and 3-methylcholanthrene, *Cancer Res.*, 23, 1701, 1963.

32. **Bailey, G. S. et al.**, The sensitivity of rainbow trout and other fish to carcinogens, *Drug Metab. Rev.*, 15, 725, 1984.

33. **Coulombe, R. A. et al.**, Comparative mutagenicity of aflatoxins using a *Salmonella*/trout hepatic enzyme activation system, *Carcinogenesis*, 3, 1261, 1982.

34. **Woodhead, A. D., Setlow, R. B., and Grist, E.**, DNA repair and longevity in three species of cold-blooded vertebrates, *Exp. Gerontol.*, 15, 301, 1980.

35. **Schultz, M. E. and Schultz, R. J.**, Induction of hepatic tumors with 7,12-dimethybenz[a]anthracene in two species of viviparous fishes *(Genus Poeciliopsis)*, *Environ. Res.*, 27, 337, 1982.

36. **Pedersen, M. G. et al.**, Hepatic biotransformation of environmental xenobiotics in six strains of rainbow trout *(Salmo gairdneri)*, *J. Fish Res. Board Can.*, 33, 666, 1976.

37. **Bailey, G. S. et al.**, Mechanisms of dietary modification of aflatoxin B_1 carcinogenesis, in *Genetic Toxicology*, Fleck, R. A. and Hollaender, A., Eds., Plenum Press, New York, 1982, 149.

38. **Wattenberg, L. W.**, Chemoprevention of cancer, *Cancer Res.*, 45, 1, 1985.

39. **Shubik, P.**, Progression and promotion, *J. Natl. Cancer Inst.*, 73, 1005, 1984.

40. **Shelton, D. W., Hendricks, J. D., and Bailey, G. S.**, The hepatocarcinogenicity of diethylnitrosamine to rainbow trout and its enhancement by Aroclors 1242® and 1254®, *Toxicol. Lett.*, 22, 27, 1984.

41. **Lech, J. J. and Bend, J. R.**, Relationship between biotransformation and the toxicity and fate of xenobiotic chemicals in fish, *Environ. Health Perspect.*, 34, 115, 1980.

42. **Addison, R. F. et al.**, Induction of hepatic mixed function oxidase activity in trout *(Salvelinus fontinalis)* by Aroclor 1254® and some aromatic hydrocarbon PCB replacements, *Toxicol. Appl. Pharmacol.*, 63, 166, 1982.

43. **Williams, D. E. and Buhler, D. R.**, Purified form of cytochrome P-450 from rainbow trout with high activity toward conversion of aflatoxin B_1 to aflatoxin B_1-2,3-epoxide, *Cancer Res.*, 43, 4752, 1983.

44. **Shelton, D. W. et al.**, Mechanisms of anticarcinogenesis: the distribution and metabolism of aflatoxin B_1 in rainbow trout fed Aroclor 1254®, *Carcinogenesis*, 7, 1065, 1986.

45. **Nixon, J. E. et al.**, Inhibition of aflatoxin B$_1$ carcinogenesis in rainbow trout by flavone and indole compounds, *Carcinogenesis*, 5, 615, 1984.

46. **Fiala, E. S., Reddy, B. S., and Weisburger, J. H.**, Naturally occurring anticarcinogenic substances in foodstuffs, *Annu. Rev. Nutr.*, 5, 295, 1985.

47. **Egaas, E. and Varanasi, U.**, Effects of polychlorinated biphenyls and environmental temperature on in vitro formation of benzo(a)pyrene metabolites by liver of trout *(Salmo gairdneri)*, *Biochem. Pharmacol.*, 31, 561, 1982.

48. **Statham, C. N. et al.**, Effect of polycyclic aromatic hydrocarbons on hepatic microsomal enzymes and disposition of methylnaphthalene in rainbow trout in vivo, *Xenobiotica*, 8, 65, 1978.

49. **Melancon, M. J., Jr. and Lech, J. J.**, Uptake, biotransformation, disposition, and elimination of 2-methylnaphthalene and naphthalene in several fish species, in *Aquatic Toxicology*, ASTM STP 667, Marking, L. L. and Kimerle, R. A., Eds., American Society for Testing and Materials, 1979, 5.

50. **Melancon, M. J. and Lech, J. J.**, Uptake, metabolism, and elimination of ^{14}C-labeled 1,2,4-trichlorobenzene in rainbow trout and carp, *J. Toxicol. Environ. Health*, 6, 645, 1980.

51. **Gurney, P. D. et al.**, Effects of egg and sperm maturation and spawning on the distribution and elimination of a polychlorinated biphenyl in rainbow trout *(Salmo gairdneri)*, *Toxicol. Appl. Pharmacol.*, 47, 261, 1979.

52. **Binder, R. L. and Stegeman, J. J.**, Induction of aryl hydrocarbon hydroxylase activity in embryos of an estuarine fish, *Biochem. Pharmacol.*, 29, 949, 1980.

53. **Kyono-Hamaguchi, Y.**, Effect of temperature and partial hepatectomy on the induction of liver tumors in *Oryzias latipes*, *J. Natl. Cancer Inst. Monogr.*, 65, 337, 1984.

54. **Khudoley, V. V.**, Use of aquarium fish, *Danio rerio* and *Poecilia reticulata*, as test species for evaluation of nitrosamine carcinogenicity, *J. Natl. Cancer Inst. Monogr.*, 65, 65, 1984.

55. **Varanasi, U., Gmur, D. J., and Reichert, W. L.**, Effect of environmental temperature on naphthalene metabolism by juvenile starry flounder *(Platichthys stellatus)*, *Arch. Envion. Contam. Toxicol.*, 10, 203, 1981.

56. **Little, P. J. et al.**, Temperature-dependent disposition of [^{14}C]benzo(a)pyrene in the spiny lobster, *Panulirus argus*, *Toxicol. Appl. Pharmacol.*, 77, 325, 1985.

57. **Niimi, A. J. and Palazzo, V.**, Temperature effect on the elimination of pentachlorophenol, hexachlorobenzene and Mirex® by rainbow trout *(Salmo gairdneri)*, *Water Res.*, 19, 205, 1985.

58. **Curtis, L. R.**, Glucuronidation and biliary excretion of phenolphthalein in temperature-acclimated steelhead trout *(Salmo gairdneri)*, *Comp. Biochem. Physiol.*, 76C, 107, 1983.

59. **Andersson, T., Forlin, L., and Hansson, T.**, Biotransformation of 7-ethoxycoumarin in isolated perfused rainbow trout liver, *Drug Metab. Dispos.*, 11, 494, 1983.

60. **Koivusaari, U., Harri, M., and Hanninen, O.**, Seasonal variation of hepatic biotransformation in female and male rainbow trout *(Salmo gairdneri)*, *Comp. Biochem. Physiol.*, 70C, 149, 1981.

61. **Stegeman, J. J. and Chevion, M.**, Sex differences in cytochrome P450 and mixed-function oxidase activity in gonadally mature trout, *Biochem. Pharmacol.*, 29, 553, 1980.

62. **Curtis, L.**, personal communication.

Chapter 8

PAH, METABOLITES, AND NEOPLASIA IN FERAL FISH POPULATIONS*

Paul C. Baumann

TABLE OF CONTENTS

---

* This paper is the official writing of an employee of the U.S. Government. As such, it resides in the public domain and is not subject to copyright.

I. CHEMICALLY INDUCED CARCINOGENESIS IN FERAL FISH

A. Field Studies A Historical Perspective

Reports of tumors in individual feral fish and descriptions of these lesions appeared in the literature at least as early as 1793,[1] and became common in the late 1800s and early 1900s.[2,3] The earliest report on a population of a particular species with an elevated frequency of neoplasia may have been a publication by Keysselitz[4] on the prevalance of epidermal papillomas on the lips of barbels, *Barbus fluviatialis* from the Mosel River, West Germany. The author stated that the cells of the tumor contained intracellular inclusion bodies, suggesting a viral etiology.

The first description of an epizootic of neoplasia probably caused by chemical carcinogens was that of Lucke' and Schlumberger[5] concerning epitheliomas on the lip and mouth of brown bullheads, *Ictalurus nebulosus*, which was published in 1941. Although the authors did not mention chemicals or pollutants, they stated that no cytoplasmic or nuclear inclusions were found. Also, the fish were obtained from the Delaware and Schuykill Rivers near Philadelphia and appeared to be abundant in some river sections and absent in others.[5] Such a distribution in a heavily industrialized area would be consistent with a chemical etiology. The neoplasms included small outward-growing tumors with little or no invasion of adjacent tissue, as well as tumors that were definitely invasive. All of these were diagnosed as epidermoid carcinomas by the authors, who believed that the smaller tumors were early stages of neoplasms which later became malignant.[2]

The first suggestion that chemical carcinogens caused neoplasia in feral fish came in 1963.[6] The following year, Dawe et al.[7] published their diagnoses of cholangiocellular neoplasms in the livers of 3 of 12 white suckers (*Castostomus commersoni*) and a minimal-deviation hepatoma in a brown bullhead, all collected from Deep Creek Lake, Maryland, and proposed that these lesions might have been induced by carcinogenic contaminants. Among possible causes considered by the authors were a protozoan, carcinogenic hydrocarbons from motorboats, chlorinated hydrocarbons such as dichlorodiphenyltrichloroethane (DDT), and rotenone. An important contribution was the authors' suggestion that bottom-feeding fish may be useful indicators of environmental carcinogens. This assertion has since been reinforced in a wide range of studies (Table 1).

Brown et al.[8] conducted the first large-scale study linking increased incidence of neoplasia in a fish community with environmental pollutants. They collected more than 18 species of fish from the Fox River, Illinois, a heavily industrialized system, and from Lake of the Woods, Ontario, a relatively pristine reference location. Of 4639 fish sampled from the Canadian reference watershed, slightly over 1% had tumors compared with 4.4% of the 2121 fish collected from the Fox River. Differences in tumor frequency was greatest in brown bullheads in which the incidence of neoplasia in the Fox River (12.2%) was more than sixfold greater than that in Lake of the Woods.[9] The incidence of stomach, liver, and skin tumors was especially high in the Fox River fish; hepatic neoplasms were the most prevalent lesion in the brown bullhead. Among the organic chemicals listed from the Fox River were the two polycyclic aromatic hydrocarbons (PAH) naphthalene (NPH) and benz(a)anthracene (BaA) present at levels of 100 and 10 ppb in the water (averaged over a 12-month period). Unfortunately, no chemical analyses were performed on the sediment or on fish food organisms.

Since the mid-1970s a number of investigators have reported epizootics of neoplasia in feral fish of both marine and freshwater species, for which anthropogenic carcinogens were a potential casual factor. Often a single population has displayed neoplastic growths in several different cell types; however, papillomas, epithelial cancers, and liver cancers (both biliary and hepatic) seem to occur most commonly.

A survey of white suckers in Lake Ontario by Sonstegard[10] showed that the frequency of

Table 1
LOCATIONS HAVING EPIZOOTICS OF NEOPLASIA IN FISH AND ASSOCIATED CONTAMINANTS OR CARCINOGENS

Location	Species	Types of tumors	Types of contaminants	Ref.
Fox River, IL	Brown bullhead	Hepatomas	Aromatic and aliphetic hydrocarbons	8
Lake Ontario	White sucker	Papillomas	PAH	10
Hudson River, NY	Atlantic tomcod	Hepatomas	PCB	12
Torch Lake, MI	Sauger and walleye	Hepatomas	Zanthenes, pine tars/ oils, and asbesti-forms	18
Boston Harbor, MA	Winter flounder	Hepatocarcinoma and cholangiocarcinoma	Unspecified hepato-toxin	20
Tuskegie, AL	Black bullhead	Papillomas	Chlorinated organics	21, 22
Puget Sound, WA, Duwamish River, Mukilteo, Eagle Harbor, etc.	English sole Rockm sole Staghorn sculpin	Hepatocellular and biliary neoplasms including carcinomas and mesenchymal neoplasms	PAH, metals, and nitrogen heterocycles	32, 33 35,36
Eastern Lake Erie, Buffalo River, and Niagara River	Freshwater drum White sucker Brown bullhead	Neurolemmoma, papilloma, cholangioma, hepatoma	PAH, aromatic amines	15, 49
Black River, OH	Brown bullhead	Hepatocellular, biliary, and squamous neoplasms including carcinomas	PAH	62

Note: PAH, polycyclic aromatic hydrocarbons; PCB, polychlorinated biphenyls.

papillomas was highest in the industrialized Oakville-Burlington region (35 to 50%) and decreased substantially in either direction along the shoreline. Although a viral etiology was postulated for the tumors, contaminants in the sediment were mentioned as a possible contributing factor, and PAH were listed as potent carcinogens known to occur in Great Lakes sediments. Sonstegard[10] also observed that in resting and feeding suckers the upper lip was more often in contact with bottom sediments than the lower lip. Of the tumor-bearing suckers from Oakville, neoplasms were much more frequent on the upper lip (61.4%) than on the lower lip (38.6%). Subsequent studies by Cairns[11] have determined that 12% of white sucker and 1% of brown bullhead from Hamilton Harbor have hepatocarcinoma or cholangiocarcinoma.

Incidental observations of Atlantic tomcod, *Microgadus tomcod* during studies undertaken to determine the effects of power plants on several Hudson River fish species revealed the presence of liver tumors. In January and February of 1978, Smith et al.[12] sampled 269 fish, divided the livers into four categories by gross appearance, and then subsampled livers from each category for histological examination. Hepatocellular carcinoma was diagnosed in 25% of the livers. The investigators wrote that elevated concentrations of polychlorinated biphenyls (11 to 98 ppm) in the tomcod livers suggested a causal relationship between PCB and the hepatomas. A variety of other contaminants, including PAH were present in the Hudson River and New York Bight.

The first reports of liver neoplasia in English sole, *Parophrys vetulus* came from Puget Sound, Washington during the late 1970s.[13,14] Case histories of these studies and the subsequent research they generated are presented in more detail later. Two other research programs are also highlighted later: one on fish populations in eastern Lake Erie and the

Niagara River, where elevated tumor frequencies were first published in 1980;[15] and the other on research related to elevated neoplasia in brown bullheads from the Black River, Ohio, where field studies began in 1980.[16]

Most epizootic outbreaks of neoplasia seem to occur in fish that feed on benthic invertebrates that live in bottom sediments (e.g., suckers, bullheads, tomcods, various flounders, and soles). Tomljanovich,[17] however, reported an outbreak in saugers, *Stizostedion canadense* that are piscivores from Torch Lake, Michigan. These findings led to a survey of Torch Lake conducted in 1979 when Black et al.[18] examined 20 saugers, almost all of which showed grossly evident signs of neoplasia. Abnormal livers of eight saugers and three walleyes were examined histologically and were found to contain hepatocellular carcinomas. In 1980, 3 of 3 saugers and 3 of 11 walleyes sampled also had hepatocellular carcinomas. Fish of both species also had dermal ossifying fibromas and perivisceral masses resembling mesotheliomas. Torch Lake had been extensively used as a repository for copper mining wastes. Possible causative factors mentioned by the authors for the cancers included organic chemical formulations (xanthates) and pine tars or oils used in the flotation process that separated native copper from associated minerals, and fine fibrous particulates (asbestiforms) that occur in the mineral wastes.[19]

Winter flounder, *Pseudopleuronectes americanus*, in Boston Harbor, Massachusetts[20] also had epizootic neoplasia. Only liver lesions were detected; 10% of the 200 fish sampled had either preneoplastic or neoplastic lesions. Cholangiocarcinomas or hepatocarcinomas were found in 7.5% of the fish. In winter flounders from Connecticut (New Haven Harbor) and Rhode Island (upper Narrangansett Bay) the incidence of hepatic neoplasia was 3.4%. In contrast, no such lesions were found in 93 winter flounders collected from unpolluted sites along the south shore of central and eastern Long Island, New York, Casco Bay, Maine, and Georges Bank (off Massachusetts). The authors stated that the high incidence of liver neoplasia in Boston Harbor fish was consistent with the action of a hepatotoxin, but they provided no data on residues or contaminants.

An unusually high incidence of oral papillomas (75%) in a population of black bullheads (*Ictalurus melas*) living in the final oxidation pond of the Tuskegie, Alabama wastewater treatment facility was first reported by Grizzle et al.[21] In later investigations, Tan et al.[22] determined that cytochrome P-450 and aryl hydrocarbon hydroxylase (AHH) activity was elevated compared to bullhead from a nearby pollution-free pond. They suggested that chemical agents in the sewage pond might be responsible for induction of the monooxygenase system. Further research by Grizzle et al.[23] showed that the incidence of oral papillomas had declined from the 73% noted in 1979 to 1980 to 23% in March 1983. This decline followed a reduction, in November 1979, of residual chlorine in the effluent that flowed out of the chlorine contact chamber and entered the final oxidation pond, from 1.3 to 3.1 mg/l to 0.25 to 1.2 mg/l. No evidence was found for a viral etiology of the papilloma. However, water samples from both the inlet and the outlet of the oxidation pond were mutagenic in the Ames test.[23] The authors suggested that the oral papillomas were chemically induced, and that the chemicals responsible were related to the chlorination process.

B. Laboratory Studies: An Overview

Although many reports that describe the occurrence of tumors in feral fish suggest chemically induced neoplasms, a definite cause-and-effect relationship has been difficult to establish. Nevertheless, laboratory studies have repeatedly demonstrated that (1) many fish have the enzymatic capability to biotransform procarcinogens to proximate carcinogens,[24] and (2) known mammalian carcinogens have been shown to produce tumors in hatchery-reared fish under laboratory conditions. The carcinogenic effect of many compounds, including aflatoxin B_1, and acetylaminofluorine, nitrosamines, and PAH, has been demonstrated in many species of fish.[24] Although tumor types seen in laboratory studies were not

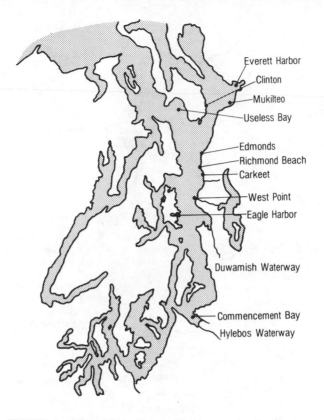

FIGURE 1. Map of Puget Sound, Washington, showing locations sampled during studies of neoplasia in fish. (Adapted from Malins, D.C., et al.[71])

always the same as those seen in comparable mammalian studies, there is little doubt that compounds that are carcinogenic to mammals can be carcinogenic to fish. Except for one study with brown bullhead,[25] however, tumors have not been chemically induced in feral fish under laboratory conditions. Details of the biotransformation pathways for several carcinogens are described later. It is now widely accepted that many xenobiotics, including carcinogens, are biotransformed via pathways that are similar to those seen in mammals.[26-31] One large gap in the understanding of chemical carcinogenesis in fish is the lack of knowledge about the dose-effect and temporal aspects of carcinogen exposures in the wild.

II. CASE HISTORIES: FIELD AND LABORATORY STUDIES OF PAH AND NEOPLASIA

The hypothesis that anthropogenic PAH cause epizootics of neoplasia in feral fish rests primarily on research at three locations: Puget Sound, Washington; eastern Lake Erie, including portions of the Buffalo and Niagara Rivers; and the Black River, Ohio (Figures 1 to 3). At each of these sites, high frequencies of neoplasms have been associated with elevated levels of PAH (Table 2). A variety of studies have been completed or are in progress at these sites to probe this association between PAH and tumors. I will present case histories for research in each of these areas, and then summarize the information.

A. Puget Sound, Washington
Liver abnormalities in English sole from Puget Sound were first reported in the mid-

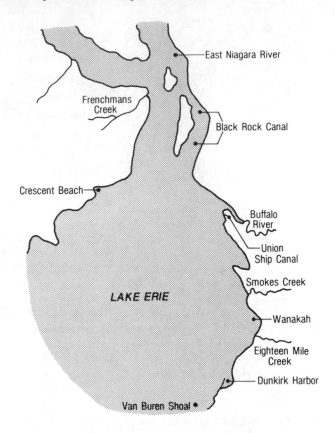

FIGURE 2.   Areas sampled in studies of neoplasia in feral fish on eastern Lake Erie and the Niagara River. (Adapted from Black, J. J.[49])

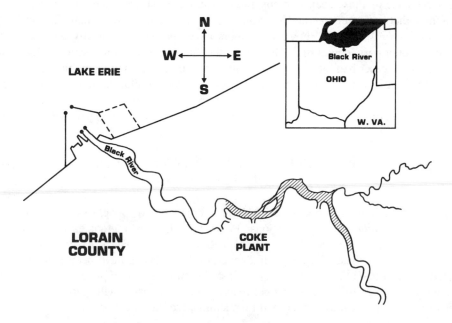

FIGURE 3.   Map of the Black River, Ohio. Shaded area indicates location **sampled** for fish.

Table 2

CONCENTRATIONS OF SELECTED PAH IN SEDIMENT (µg/g DRY WT.) FROM AREAS WITH EPIZOOTICS OF NEOPLASIA IN FISH (E) AND REFERENCE LOCATIONS (R).

Sites	Phenanthrene	Fluoranthene	Pyrene	Chrysene	BaA[a]	BbF[b]	BkF[c]	BaP[d]	Ref.
Puget Sound									
Duwamish Waterway (E) (n = 5)	1.2	1.8	1.7	NA	1.3	NA	NA	0.66	71, 72
Hylebos Waterway (E) (n = 4)	2.0	2.0	2.2	2.0	1.2	—	3.0[e]	—	0.57	71, 72
Mukilteo Waterway (E) (n = 2)	2.4	3.35	2.8	1.4	0.83	—	1.3[e]	—	0.36	35, 71
Eagle Harbor (E) (N = 3)	13	NA	27	NA	8.0	NA	NA	3.1	71
Port Madison (R)	0.04	NA	0.07	NA	0.05	NA	NA	0.02	71
President Point (R)	0.15	0.22	0.09	0.14	0.07	—	0.1[e]	—	0.04	35, 71
Great Lakes/fresh water									
Smokes Creek (E)[e]	0.93	7.6	2.0	18	1.5	1.9	0.73	1.6	49
Union Ship Canal (E)[f]	7.5	33	24	14	7.1	11	3.4	6.4	49
Buffalo River (E)	23	28	38	9.5	7.5	6.5	3.4	6.8	49
Black Rock Canal (E)[f]	3.4	9.9	11	2.7	3.2	3.8	2.4	3.4	49
Black River (E)[g]	390	220	140	51	51	—	75[e]	—	3	16
Lake Ontario (R) 0—05 cm	NA	0.28	0.056	0.22	NA	NA	NA	0.337[h]	73
Lake Ontario (R) 10—15 cm	NA	0.058	0.029	0.088	NA	NA	NA	ND	73
Buckeye Lake (R)	0.04	0.11	0.072	0.028	0.021	—	0.36[e]	—	0.014	62

Note: If multiple samples were taken, the concentration represents the arithmetic mean. NA, data not available; ND, compound not detected.

[a] BaA, benz(a)arthracene.
[b] BbF, benzo(b)fluoranthene.
[c] BkF, benzo(k)fluoranthene.
[d] BaP, benzo(a)pyrene.
[e] Combined concentrations of all berzofluoranthenes.
[f] Converted from ng/g wet wt. assuming 55% moisture content of sediment.
[g] Point source.
[h] Combined concentrations of all benzpyrenes.

Table 3

SPEARMAN'S RANK CORRELATION COEFFICIENTS (r_s) FOR ALL
SIGNIFICANT ($p < 0.05$) CORRELATIONS FOUND BETWEEN
PREVALENCES OF HEPATIC LESIONS IN ENGLISH SOLE AND
CHEMICAL CONCENTRATIONS IN BOTTOM SEDIMENT

Lesion type	Chemical group[a]	r_s	Significance level
Neoplasms	AH	0.48	0.048
Megalocytic hepatosis	AH	0.54	0.016
Steatosis/hemosiderosis	AH	0.49	0.032
One or more hepatic lesions	AH	0.58	0.016
One or more hepatic lesions	Metals	0.54	0.016

Note: Significance levels were adjusted for the number of statistical tests performed.

[a]   Chemical groups selected by factor analysis included: aromatic hydrocarbons (AH), metals, selected metals plus PCB and chlorinated compounds.

Adapted from Reference 33.

1970s.[13,14] Histopathologic examination of the livers of 62 English sole sampled from the Duwamish River Estuary showed that 20 (32%) had hepatomas. The authors suspected environmental contaminants as the etiologic agent, but initially focused on the elevated PCB residues in English sole reported by the U.S. Environmental Protection Agency.[32]

A 4-year (1979 to 1982) multidisciplinary study examined the relation between pollutants and neoplasms in Puget Sound. Both sediments and bottom-dwelling fish (English sole; rock sole, *Lepidopsetta bilineata*; and Pacific staghorn sculpin, *Leptocottus armatus*) were collected from 43 stations, and then combined into 19 geographic subareas for data analysis.[33] Fish and sediment were analyzed for PAH, organochlorines, and a number of elements. Pathological examination of individuals from all three fish species included histopathology of grossly visible lesions and of normal-looking tissue samples from major organs. A battery of statistical tests, including the G test for heterogeneity, cluster analysis, factor analysis, and Spearman's rank correlation procedure, were used to distinguish differences in frequencies of neoplasia, group chemicals in the sediment, and to compare prevalances of disease types with concentrations of xenobiotics.

Lesions detected included a variety associated with parasites or microoorganisms, as well as areas of cellular alteration and neoplasms (largely hepatic). The hepatic lesions included hepatocellular neoplasms, biliary neoplasms, and mixed carcinomas having both cholangiocellular and hepatocellular components. The frequency of hepatic neoplasms was greatest in English sole from the Duwamish Waterway (16%, n = 136) and Everett Harbor (12%, n = 66) (Figure 1). Frequencies in English sole from other sampling areas ranged from 0 to 5.5%. Neoplasms in rock sole were highest in Everett Harbor (4.8% n = 43); staghorn sculpin with hepatic neoplasms were found only in Commencement Bay, where the highest frequency was in Hylebos Waterway (17%, n = 116) (Figure 1).

Statistically significant correlations were found between concentrations of PAH in sediment and both total hepatic lesions and neoplasms in English sole (Table 3). Total hepatic lesions in English sole were also significantly correlated with metals. The other three significant correlations seen were between PAH concentrations and other types of liver lesions in English sole (megalocytic hepatosis, steatosis/hemosiderosis) and Pacific staghorn sculpin (hyperplasia/FCA). No significant correlations were found between neoplasms or hepatic lesions and chlorinated hydrocarbons for any of the species;[33] however, the most recent statistical analyses show that there may be a negative relationship between PCB levels in sediment and liver lesions, indicating a possible inhibitory effect of PCB on PAH-induced liver neoplasia.[34]

Recently, two more contaminated sites with a high incidence of neoplasia have been investigated in Puget Sound: Mukilteo[35] and Eagle Harbor (Figure 1).[36] In English sole near Mukilteo, frequencies were 7.5% for hepatic neoplasms and 16.7% for areas of cellular alteration. Incidence of these same lesions was 27 and 44%, respectively, in Eagle Harbor sole. English sole taken from a lightly polluted reference location showed no evidence of neoplasia. Sediments from both polluted sites contained high concentrations of aromatic hydrocarbons, and the sediments from Eagle Harbor (which were contaminated with creosote) also contained a variety of azaarenes such as carbazole and acridine.

Sediment from contaminated areas such as the Duwamish River basin and Eagle Harbor contained 4- to 5-ring PAH, including benzo(a)pyrene (BaP) and benz(a)arthracene (BaA) (Table 2). In laboratory studies, the ability of English sole to accumulate BaP from sediment was demonstrated by Stein et al.,[37] who exposed the fish to environmentally realistic levels of radiolabeled BaP and PCB (both separately and together) associated with sediment. The BaP and its metabolites reached equilibrium levels in tissue after the first day, indicating continued absorption and metabolism. Tissue concentrations of BaP-derived radioactivity were increased when fish were simultaneously exposed to both organics. Later laboratory studies conducted by Varanasi and colleagues[38] showed that English sole placed on sediment from the Duwamish River to which radiolabeled BaP was added, readily took up PAH from the sediment, as measured by BaP-derived radioactivity in livers and BaP fluorescence in bile. The fact that BaP radioactivity was present almost entirely as metabolites indicated that sediment-associated BaP and presumably other PAH were not only available to English sole, but could be metabolized by them.

Furthermore, English sole force-fed radiolabeled BaP were able to biotransform it into a variety of metabolites including BaP 7,8-dihydrodiol,[39,40] a penultimate carcinogen of BaP. Moreover, BaP intermediates formed by English sole covalently bound to hepatic DNA.[39-41] When the bile of English sole and starry flounder, *Platichthys setllatus* that had been fed BaP were compared, English sole (which had a higher incidence of hepatic neoplasia in Puget Sound) contained a higher proportion of BaP 7,8-diol conjugates in bile and a higher proportion of BaP intermediates bound to liver DNA than did starry flounder.[42] Further studies indicated that BaP 7,8-dihydrodiol and its conjuugates were more prevalent than BaP 9,10-dihydrodiol-particularly in the bile and liver of juvenile English sole where the ratio was 2:1.[37,43] Liver microsomes of juvenile sole were also shown to catalyze the formation of BaP-DNA adduct, tentatively identified by the authors as *anti*-BPDE-deoxyguanosine, casually linked to BaP-induced carcinogenesis in rodent skin.[44]

Further combined field and laboratory work linked PAH metabolites in feral sole with high tumor frequencies. In a series of studies by Krahn et al.,[45,46] adult English sole were collected from ten variably polluted- and four reference locations in Puget Sound that were selected on the basis of previous studies. Liver lesions were identified by using standard histophatological techniques, high-pressure liquid chromatography (HPLC)-fluorescence was used to estimate relative concentrations of PAH metabolites, and gas-liquid chromatography-mass spectrometry (GLC-MS) was used to identify specific metabolites of selected PAH. In the first study, bile of fish from two polluted sites displayed naphthalene-, phenanthrene-, and BaP-like fluorescence at levels 9, 14, and 19 times, respectively, the levels in bile of fish from two reference areas.[45] Also, at the only site where fish with hepatic neoplasia were captured (the Duwamish River), the concentrations of xenobiotics with BaP-like fluorescence was significantly higher in fish with liver lesions than in those without liver lesions. Individual metabolites of fluorene, phenanthrene, anthracene, biphenyl, and dimethylnaphthalene were identified. The authors believed that TMS derivatives of 4- and 5-ring metabolites were probably present, but could not be identified, due partly to their low concentrations and the low intensities of their molecular ions.

In a second study the sampling was extended to ten additional sites.[48] Mean concentrations of aromatic metabolites in bile measured at BaP wavelengths varied more than 100-fold in English sole from the different locations (Table 4, Figure 1). The locations where sole had the highest metabolite concentrations in BaP equivalents—Eagle Harbor, the Duwamish Waterway, and Clinton—were also the locations where the frequency of neoplasia was highest in this species. Frequencies of idiopathic lesions among sites were compared with bile metabolite levels by Spearman's rank correlation; the most significant correlations of bile metabolite levels occurred for neoplasms ($r_s = 0.85$; $p < 0.002$) and for megalocytic hepatosis ($r_s = 0.89$; $p < 0.001$). However, concentrations of metabolites in the bile of English sole were highly variable within locations and were not correlated with concentrations of aromatic hydrocarbons in the sediment. The authors suggested that the lack of correlation was caused by patchy distribution of PAH in the sediment and by movement and migration of fish.

This work seemed to substantiate the results of earlier laboratory studies that demonstrated the bioconcentration of xenobiotics in trout bile.[47] In particular, the metabolites of naphthalene and methylnaphthalene were present at high concentrations in the bile of exposed trout, and the authors suggested using analysis of metabolites in bile as a technique to monitor petroleum pollution. Although measurements of metabolite concentrations in bile provide a more direct index of exposure to PAH than do PAH concentrations in sediment, more extensive research needs to be done on rates of uptake, metabolism, and depuration to understand how migration might affect metabolite distribution and concentrations. Furthermore, the correlation between metabolite concentration in bile and frequency of neoplasia might be partly fortuitous, considering the rapid metabolism of BaP,[37,46] the latent period common to tumor formation, and the mobility and seasonal dietary changes of fish. However, if the fish remain essentially localized over their lifetime, or if their movements have a consistent seasonal cycle, PAH metabolites found in the bile might well indicate previous exposure, and therefore serve as a good indicator of neoplasia. A useful experiment might be to take juvenile fish from populations with high and low levels of PAH metabolities in the bile, raise them to adults in a clean environment, and then compare neoplasia frequencies. This might help to determine whether metabolite levels in bile are predictive of later tumor development.

Stomach contents of feral English sole captured in both Eagle Harbor and near Mukilteo contained high concentrations of PAH which reached microgram-per-gram concentrations (dry weight) for some individual compounds such as phenanthrene, fluoranthene, pyrene, BaA and chrysene.[35,36] Not surprisingly, levels of PAH metabolites were also high in the bile of sole from the same locations. These results emphasize the importance of dietary uptake of PAH.

In associated experiments, Varanasi and co-workers[38] explored the uptake of PAH by a diverse group of benthic invertebrates exposed to contaminated sediment from the Duwamish River, to which radiolabeled BaP had been added. Amphipods, *Rhepoxynius abronius* and *Eohaustorius washingtonianus*, clams, *Macoma nasuta*, and shrimp, *Pandalus platyceros* were sampled after 1- and 4 weeks of exposure. Shrimp tissue contained no detectable levels of parent PAH. Among the other organisms, tissue concentrations of PAH differed significantly, with *E. washingtonianus* > *R. abronius* > *M. nasuta*. Extensive metabolism of BaP was shown to have occurred in shrimp. Both amphipods metabolized BaP to a lesser extent, and no BaP metabolites were detectable in clam tissue. Thus, tissue concentrations of parent PAH were, in general, negatively correlated with the extent of metabolism of BaP. The fact that this relationship did not hold among the clams and amphipods indicated to the authors that other variables such as feeding strategies and excretion rates also affected accumulation of PAH. Nevertheless, the results indicated that invertebrates, many of which are known to food organisms for bottom-feeding fish, were able to accumulate high con-

Table 4

CONCENTRATION (ng/g DRY WT.) OF SELECTED AROMATIC HYDROCARBONS (AH) IN SEDIMENT AND CONCENTRATIONS (ng/g WET WT.) OF AH METABOLITES (MEASURED AT BaP WAVELENGTHS) AND SELECTED IDIOPATHIC LIVER LESIONS (%) IN CORRESPONDING POPULATIONS OF ENGLISH SOLE FROM PUGET SOUND, WASHINGTON

Site	Sediment			English Sole			
	Fluoranthene[a]	Pyrene[a]	BaP[a]	No.	Metabolites[b] $\bar{x}$ ±SD	Neo[c]	FCA[c]
Eagle Harbor	59,000	32,000	2,300	22	2,100 ± 1,500[d]	18.2	18.2
Duwamish Waterway	440	470	73	58	1,400 ± 2,200	20.7	32.8
Clinton	16	19	13	16	1,300 ± 1,700	12.5	25.0
Inner Everett Harbor	1,000	590	<14	20	520 ± 410	5.0	20.0
Outer Everett Harbor	210	250	<5	17	270 ± 220	5.9	17.6
Richmond Beach	37	63	24	21	270 ± 700	0	28.6
West Point	6,100	8,300	2,200	20	240 ± 160	0	5.0
Carkeek	18	28	22	18	110 ± 120	0	5.6
President Point	220	93	41	20	100 ± 89	0	0
Edmonds	84	110	34	21	91 ± 82	0	4.8
Useless Bay	19	17	5	16	67 ± 45	0	6.2

[a] Aromatic hydrocarbons which have metabolites that fluoresce at the BaP wavelength pair.

[b] Fluorescence response at the BaP wavelength pair, 380/430 nm.

[c] Neo, neoplasms; FCA, foci of cellular alterations.

[d] Malins, D.C., personal communication, 1986.

Adapted From Reference 46.

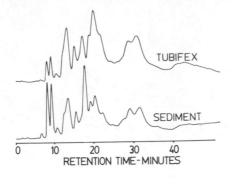

FIGURE 4.   HPLC chromatograms of PAH-containing fractions from tubifex worm and sediment samples taken from the Buffalo River. (From Black, J. J., Holmes, M., and Paigen, B., *Environ. Sci. Res.*, 16, 559, 1980. With permission.)

centrations of PAH and might, therefore, be an important source of both PAH and metabolites for English sole.

B. Eastern Lake Erie and Niagara River

In a preliminary study of the Buffalo River and the nearby Black Rock Ship Canal (Figure 2), Black et al.[15] found a high incidence of neoplasia among several species of benthic fishes. Among the relatively few fish collected, 20 to 25% of freshwater drum, *Aplodinotus grunniens* showed a dermal lesion described by the authors as a neurolemmoma. They also observed papillomas on white suckers and epidermal tumors on two of nine brown bullheads. One of these epidermal lesions was malignant and had invaded the dorsum of the upper jaw through the palate.

In the same study, HPLC and GC/MS were used to document the presence of over 40 PAH in Buffalo River sediment, including the carcinogen BaA. The HPLC chromatograms of PAH-containing fractions from tubificid worms, a common fish food organism, were very similar to chromatograms of sediment (Figure 4). Sediments from nine stations were extracted with an organic solvent and the resulting residues then concentrated into DMSO.[48] When these sediment extracts were tested for mutagenic activity in the Ames assay, a strong source of mutagenic material was indicated below an industrial discharge. HPLC characterization of these sediments indicated much higher levels of UV-absorbing materials consistent with retention times for PAH.

Field sampling was expanded in 1980 and 1981 to nine locations in eastern Lake Erie and the upper Niagara River (Figure 2).[49] Eleven neoplastic and/or preneoplastic lesions were found among six species of fish and a hybrid common carp × goldfish, *Carassius auratus*. Two were epizootic, dermal neoplasms described as chromatophoromas or neurolemmomas in freshwater drum, and oral papillomas in white sucker. Tumor frequencies were significantly higher ($p \times 0.05$) in freshwater drum collected from five polluted stations (East Niagara River, Black Rock Canal, Buffalo River, Smokes Creek, and Wanakah [n = 305]), where the incidence was 8.8% higher than in drum from two reference locations, Crescent Beach and Van Buren Shoal [n = 891)], where the incidence was 2.2%. Tumor frequencies in drum were higher at individual sites such as Wanakah (16.7%, n = 30) and E. Niagara River (11.6%, n = 103). The frequency of neoplasms in drum increased with increasing length (and, presumably, age). On the other hand, the frequency of oral papillomas in white suckers was less variable among locations, ranging from 5.3 to 9.4% (n = 50).

Additional sampling of the Buffalo River in 1983 resulted in the collection of 28 brown

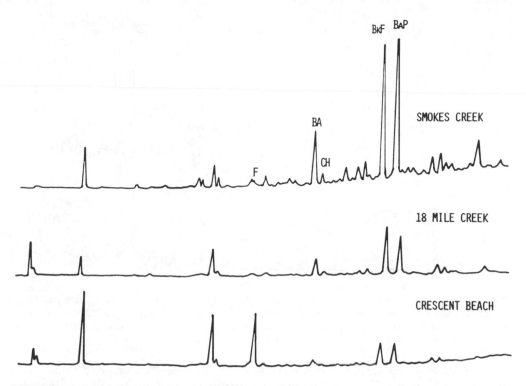

FIGURE 5. HPLC chromatograms comparing levels of PAH in stomach contents of fish from three Lake Erie tributaries. F, fluoranthene; BaA, benzanthracene; CH, chrysene; BkF, benzo(k)fluoranthene; BaP, benzo(a)pyrene. (From Maccubbin, A. E., Black, P., Trzeciak, L., and Black, J. J., *Environ. Contam. Toxicol.*, 34, 876, 1985. With permission.)

bullheads.[50] Gross examination indicated a 20% incidence of liver tumors, and some external tumors, of which 35% were oral. Histopathology identified both cholangiocellular and hepatocellular neoplasms, many with eosinophilic characteristics.

Black[49] also determined the PAH levels in sediments from a number of eastern Lake Erie and upper Niagara River locations. As measured in quantitative HPLC studies, PAH were found in greater concentrations along the more heavily industrialized U.S. shoreline than along the Canadian shore. Some individual PAH (such as fluoranthene) were measured in tens of parts per million levels (wet weight) in the Union Ship Canal and the Buffalo River. BaA and BaP were recorded at parts per million concentrations at the Union Ship Canal and Buffalo River and in the Black Rock Canal.

To determine whether PAH in sediment was available to bottom-feeding fish, Maccubbin et al.[51] examined stomach contents of white suckers from a location with high PAH levels in the sediment (Smokes Creek) and two control locations (Crescent Beach and 18 Mile Creek) (Figure 2). Analysis by HPLC revealed that stomach contents from Smokes Creek fish included a more complex group of PAH and had the higher concentrations of BaP, BaA and benzo(k)fluoranthene (BkF) (Figure 5). Also, the chromatographic PAH profiles of the stomach contents of Smokes Creek fish were similar to those of Smokes Creek sediment (Figure 6). The authors concluded that white suckers were receiving at least a portion of their PAH body burden by the ingestion of benthic organisms and/or sediment contaminated with PAH. They further suggested that stomach contents of bottom-feeding fish might be useful in monitoring PAH levels in areas where sediment samples are difficult to obtain.

To demonstrate the possibility of a chemical etiology for the tumors seen in wild fish populations, Black[25,52] applied sediment extract from the Buffalo River to brown bullhead

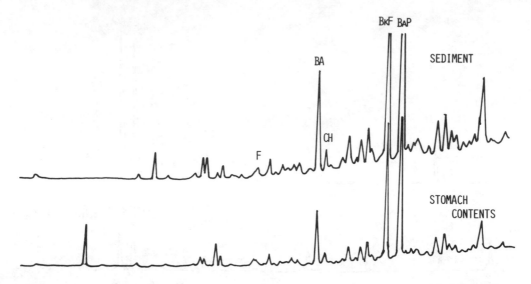

FIGURE 6.   HPLC chromatograms comparing levels of PAH in sediment (1 × conc.) and stomach contents (20 × conc.) samples from Smokes Creek. F, fluoranthene; BaA, benzanthracene; CH, chrysene; BkF, benzo(k)fluoranthene; BaP, benzo(a)pyrene. (From Maccubbin, A. E., Black, P., Trzeciak, L., and Black, J. J., *Environ. Contam. Toxicol.*, 34, 876, 1985. With permission.)

by skin painting. Extractable residue was obtained by Soxhlet extraction of dried sediment for 12 h with hexane:acetone, 1:1 volume. The extract was a 5% solution of the extractable residue. This solution contained a mixture of PAH that was qualitatively similar, but two to five times more concentrated than that in dry sediment; however, there was some enrichment of compounds with higher molecular weights.[25,53] BaP for example, was present at 4.4 μg/g in dry sediment and at 15.4 μg/ml in the extract. The extract was applied once a week to an area bounded by the upper lip and the occiput. Control fish were painted only with solvent.

During the first 12 months of treatment, only mild changes were observed in bullheads exposed to the extract.[25] After 12 to 14 months, however, the skin of several bullheads became irregularly thickened, and some sections showed hyperplasia, an increase in cellular basophilia, loss of dermal pegs, and decreased numbers of alarm substance cells. Of the 22 surviving fish (from an original group of 69) 5 were grossly hyperplastic after 14 to 18 months of exposure, and 2 of these had developed multiple papillomas over the entire painted surface area. By the 24th month, when the experiment was ended, papillomas had developed in 8 of the 22 surviving fish. Fish painted only with solvent remained essentially normal throughout the experiment. The authors noted that, although the experiment was insufficient to enable identification of the compounds responsible for initiating the hyperplastic changes seen, various PAH are recognized skin carcinogens of humans and laboratory animals.[54]

In a small-scale test, bullheads from a noncontaminated location were exposed to the river sediment extract mixed in their diet at a concentration of 5 mg/g.[52] Six bullheads were dosed for 12 months and five for 4 months. One of the fish sacrificed after 4 months had a basophilic hepatocellular focus and a small cholangioma. Two of the bullheads exposed for 12 months had eosinophilic and basophilic nodules similar to those in rodents described as type A hepatocellular neoplasms.[55]

To characterize further the carcinogenicity potential of the sediment extract,[56] Black et al. performed two experiments with 2- to 3-month-old Swiss IcR$_{Ha}$ female mice. Fifty mice were used per treatment, except for the 200-μg/ml BaP treatment (n = 25). Solutions were applied to the shaved backs of the mice. In one experiment the animals were treated five times a week with different carcinogen solutions in 0.1 ml of acetone. Results were reported as frequency of papillomas after 30 weeks, although an undetermined percentage of tumors progressed to squamous cell carcinomas including some with metastases to lung and other organs.[56,57] Controls treated with acetone did not develop tumors. A 2% solution of Buffalo River sediment extract was more carcinogenic than a 66-μg/ml BaP solution (30 and 12% tumor incidence, respectively). A solution combining 66 μg/ml BaP and 2% extract was more carcinogenic (86% tumor incidence) than expected if the effects of the carcinogens were additive. This same solution was also more carcinogenic than 200 μ/ml BaP (64% tumor incidence).

In a second experiment, the Buffalo River extract was assessed for initiation and promotion activity, again with 50 Swiss mice for each treatment.[56] Sediment extract was used both as a promoter with dimethylbenz(a)anthracene (DMBA), and as an initiator with tetradecanoyl phorbol acetate (TPA) as the promoter. A single dose of 0.25 ml of the extract solution was given 3 weeks before treatment with the promoting solution. The promoting solution was administered five times per week in 0.1 ml acetone; the experiment was terminated after 30 weeks. No tumors resulted in controls treated with DMBA when acetone was a promoter, nor in those treated with acetone when TPA was a promoter. A low tumor incidence (8%) was obtained by using a 2% sediment extract solution as the initiating dose and a 1.5-μg/ml TPA solution. When 0.05% DMBA in acetone was used as the initiating agent, the tumor incidence was 40% when 1.5 μg TPA/ml of acetone was used as the promoter, but a significantly higher incidence 52%, when the 2% sediment extract solution was used as the promoter. The tumor frequency in mice in this experiment was significantly greater than that obtained in the first experiment when the sediment extract was applied as a complete carcinogen. The authors interpreted this difference as resulting from the presence of a cocarcinogenic effect-promoting activity in the river sediment extract.[56]

C. Black River Near Lake Erie

A survey of organics in the sediment of the Black River, Ohio (Figure 3), performed by the U.S. Environmental Protection Agency in 1974, revealed a wide range of PAH.[58] Several of these were found at concentrations exceeding 10 ppm—including dibenzanthracene and BaP which are mammalian carcinogens. Subsequently, Baumann et al.[16] began a study of the brown bullhead population in the river in 1980. Bullheads were measured, sexed, and examined for liver and external tumors, and spines were removed for aging. Sediment samples were collected in the immediate vicinity of the outfall from a USX coking plant and in an area approximately 1 km upstream. Subsamples (five fish each) of smaller (younger) and larger (older) bullheads were composited for chemical analysis. Analysis of all samples was by GC-MS.

A total of 26 PAH were identified in the sediment in concentrations of a few to hundreds of parts per million. BaA and BaP were present at 51 and 43 μg/g (dry weight), respectively.[18] These levels were greater by three or four orders of magnitude than those in sediment taken from an upstream location or in sediment from Buckeye Lake, the reference site. The profile of PAH found in Black River sediment matched that of PAH detected by Lao et al.[59] in effluents from coke and coal tar production facilities. In sediment samples composited from the river section starting at the coking plant outfall through 0.5 km downstream, average PAH and nitrogen heterocycle concentrations were about one tenth of those in the sediments previously described that were collected at the point source.[16,60] This difference is not surprising, because concentrations of phenanthrene (PHN), BaA and BaP have been found to decline exponentially downstream from a point source.[61] Nevertheless, extracts taken from these sediments with these lower concentrations proved to be both mutagenic and carcinogenic.[56,60]

Black et al.[56] prepared an organic solvent-soluble extract from Black River sediment,

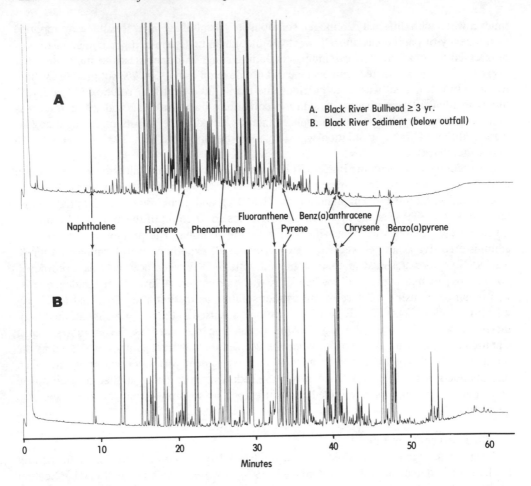

FIGURE 7.   Gas chromatograph of PAH compounds in sediment and brown bullhead (whole fish) from the Black River, Ohio.

according to the procedure previously described for Buffalo River sediment. Black River sediment extract contained BaA at 5.7 ng/1.0 × 10⁻⁴ ml, benzo(b)fluoranthene (BbF) at 4.9 ng/1.0 × 10⁻⁴ ml, and BaP at 4.5 ng/1.0 × 10⁻⁴ ml; fluoranthene had the highest concentration of the PAH (31.8 ng/1.0 × 10⁻⁴ ml). When the shaved backs of Swiss mice were painted with a 2% solution of the extract in 0.1 ml acetone five times a week, as part of the Black et al. experiment described previoulsy,[56] papillomas were produced in 2 of 25 mice after only 9 weeks[19] and in 80% of the mice tested after 30 weeks.[56] This extract was more carcinogenic than the 200-mg/ml solution of BaP used as a positive control (64% tumor frequency at 30 weeks).

The chromatographic profiles of PAH for Black River bullheads (whole fish) were similar to that for sediment (Figure 7); acenaphthylene, phenanthrene, fluoranthene, and pyrene were present in part-per-million concentrations (wet weight).[26] Heavier PAH were present at lower concentrations than in sediments; BaP averaged about 12 ppb in fish 3 years old or older. Parent PAH compounds were low to undectable in livers[35,36] and muscle tissue[36] of English sole even from fish with high PAH levels in stomach contents. The higher concentrations of parent compounds in brown bullheads may reflect both metabolic differences between the species and higher sediment PAH concentrations near the point source in the Black River. The change in relative PAH concentrations (brown bullhead vs. sediment) between lighter and heavier compounds (Figure 7) argues against the whole fish analysis simply reflecting sediment in the gut. Rather, the longer chain PAH may be metabolized more rapidly.

Total DDT was present in Black River fish at 130 ppb and other organochlorines were present at tens of parts-per-billion levels. Total PCB reached a level of 1.3 ppm, but dioxins and dibenzofurons were present only at parts-per-trillion levels.[62] Similarly, arsenic was below the detection limit of 10 ppt,[62] and cadmium and mercury did not exceed averages for 112 stations nationwide recorded in the National Pesticide Monitoring Program.[63]

Liver tumors in brown bullheads from the Black River were diagnosed by Dr. John Harshbarger as cholangiomas and cholangiocarcinomas.[62] Mitotic figures occurred throughout the tumors and invasion of surrounding normal tissue was common.[64] Lip and dermal tumors included both papillomas and carcinomas.[62] Incidence of grossly observable liver tumors in 2-year-old bullheads (2%, n = 263) was significantly lower ($p < 0.01$) than in older bullheads (33%, n = 94).[16] None of the brown bullheads collected from Buckeye Lake, Ohio, a reference location, had visually observable hepatomas (n = 329).

During 2 years of additional research on tumor incidence in brown bullheads[62] field collections focused on fish 3 years old or older in 1981 (n = 175) and 1982 (n = 233). Over the 3-year period, the incidence of grossly observable liver tumors ranged from 12 to 18% in 3-year-olds and from 28 to 44% in 4-year-olds. Liver tumors observed visually were diagnosed as cholangiocellular neoplasms. The increase in tumor incidence between 3- and 4-year-olds was significant for all 3 years ($p < 0.05$). External tumors were also common; frequency in 4-year-olds was as high as 32% for lip tumors and 18% for other skin tumors (both in 1980).

A subsample of 125 bullheads 250 mm long or longer was collected in 1982 regardless of the presence or absence of grossly visible liver tumors, and was processed for histopathology and aging. On average, each liver was sectioned at five levels for diagnosis. Liver cancer occurred in over 38% of the fish, and preneoplastic or neoplastic liver alterations occurred in 84% of the fish.[65] Hepatocellular and cholangiocellular lesions occurred both as advanced neoplasms and preneoplastic alterations. Cholangiocellular neoplasms were more frequent as carcinomas; their incidence in 3- and 4-year-old fish (19 and 33%, respectively) closely matched the frequencies of grossly observable liver tumors documented for those age groups.[65] Electron microscope examiantion of cholangiocellular carcinomas from several fish revealed no evidence of virus.[62]

Brown bullheads from the Black River and Buckeye Lake transported to the Columbia National Fishery Research Laboratory and acclimated to well water for 7 days before processing were found to have measurable hepatic microsomal BaP hydroxylase activity.[66] This activity (expressed as nanomoles of 3-OH BaP equivalents produced per minute per milligram of microsomal protein) was highly variable among individual fish, ranging from 0.2 to 23 in Black River fish and from 0.2 to 30 in Buckeye Lake fish. Black River fish were not significantly different from Buckeye Lake fish in BaP hydroxylase activity per gram of tissue; however, the hepatosomatic index was significantly greater ($p < 0.001$) for Black River bullheads. A greater relative liver weight would allow more enzyme activity per fish if activity remained equivalent on a per gram of tissue basis.[66]

In an additional study in progress, hepatic AHH activity is being compared between newly captured brown bullheads from the Black River and from a reference location (Old Woman Creek, which is also a Lake Erie tributary). A crude homogenate of liver tissue was incubated in a field with radiolabeled BaP.[67] Preliminary results confirmed that BaP was metabolized, and that AHH activity was greater (on a per gram protein basis) in Black River fish than in the control. In a related study, bile from brown bullheads captured in the Black River is being analyzed by HPLC and fluorescence detection for the presence of BaP metabolites. Preliminary findings indicate that BaP metabolites are present in the bile of these fish.[68]

III. SUMMARY AND RESEARCH NEEDS

The three case histories highlighted here have a number of similarities. Frequency of liver neoplasia, including both cholangicellular and hepatocellular lesions, was high in bottom-dwelling fishes at each location. Bottom sediments in the areas that these fish populations inhabited contained elevated levels of PAH, including known mammalian carcinogens such as BaP. The PAH concentrations were typically several orders of magnitude greater than those recorded from less polluted areas. Skin painting experiments with PAH extracted from sediments produced papillomas in brown bullheads (Buffalo River sediment) and in Swiss mice (Buffalo and Black River sediments).

Furthermore, elevated levels of PAH were found in the stomach contents of English sole and white suckers from populations with high frequencies of neoplasia. Elevated PAH levels have been demonstrated in fish tissue from all three research areas. Experiments with English sole and brown bullhead[69] have demonstrated the ability of both species to metabolize BaP to carcinogenic derivatives, and have documented the subsequent formation of DNA adducts. Bile of feral fish from all locations has been found to contain PAH metabolites. A recent significant experiment by Hendricks et al.[70] demonstrated that hepatic carcinoma can be induced in rainbow trout, *Salmo gairdneri* by exposure to dietary BaP. Although this research taken as a whole does not demonstrate cause and effect, the field studies detailed above and the results of laboratory investigations can best be interpreted by assuming the pathology to be the result of a carcinogen taken by fish from their environment by way of the diet, or by water and sediment intake or contact, and activated by hepatic enzymes.

Many different aspects require additional research. The actual progression of neoplasia in fish is still poorly understood. Nomenclature is often based on similarities to better-studied mammalian systems, and experimental work is needed to confirm or refute the implied assumptions. The usefulness of such indicators as enzyme induction and metabolite levels in bile will depend on research defining how rapidly these indices change after exposure and after withdrawal from exposure of various fish species. More work is needed on the route of exposure of xenobiotics for feral fish and on how this route might influence the process of metabolic transformation. Attention must be given to contaminants in the food chain and to bioaccumulation or metabolic transformations in food organisms that occur before ingestion by fish. Laboratory studies defining dose-response curves for various carcinogens and for various species, should include both single compounds and mixtures. Understanding carcinogenesis of PAH in fish also requires additional studies on adduct formation and the mechanisms of initiation.

ACKNOWLEDGMENTS

I thank Usha Varanasi and John J. Black for providing unpublished data and for manuscript review. John J. Black also provided figure copies. John J. Lech and John C. Harshbarger reviewed the manuscript and supplied valuable information for the text.

REFERENCES

1. **Bell, W.,** Description of a species of *Chaetodon, Philos. Trans.,* p. 7, 9, 1793; cited by **Schlumberger, H. G. and Lucke, B.,** 1948.
2. **Schlumberger, H. C. and Lucke', B.,** Tumors of fishes, amphibian and reptiles, *Cancer Res.,* 8, 657, 1948.

3. **Wellings, S. R.,** Neoplasia and primitive vertebrate phylogeny: echinoderms, prevertebrates, and fishes — a review, *Natl. Cancer Inst. Monogr.,* 31, 59, 1969.

4. **Keysselitz, G.,** Uber ein epithelioma der Barben, *Arch. Protistenkd.,* 11, 326, 1908; cited by Wellings, 1969.

5. **Lucke', B. and Schlumberger, H. G.,** Transplantable epitheliomas of the lip and mouth of catfish, *J. Exp. Med.,* 74, 397, 1941.

6. **Hueper, W. E.,** Environmental carcinogenesis in man and animals, *Ann. N.Y. Acad. Sci.,* 108, 963, 1963.

7. **Dawe, C. J., Stanton, M. F., and Schwartz, F. J.,** Hepatic neoplasms in native bottom-feeding fish of Deep Creek Lake, Maryland, *Cancer Res.,* 24, 1194, 1964.

8. **Brown, E. R., Hazdra, J. J., Keith, L., Greenspan, I., Kwapinski, J. B. G., and Beamer, P.,** Frequency of fish tumors found in a polluted watershed as compared to nonpolluted Canadian waters, *Cancer Res.,* 33, 189, 1973.

9. **Brown, E. R., Keith, L., Hazdra, L., and Arndt, T.,** Tumors in fish caught in polluted waters: possible explanations, in *Comparative Leukemia Research,* Ito, Y. and Dutcher, R. M., Eds., University of Tokyo Press, Tokyo, 1975, 47.

10. **Stonstegard, R. A.,** Environmental carcinogenesis studies in fishes of the Great Lakes of North America, *Ann. N.Y. Acad. Sci.,* p. 261, 1977.

11. **Cairns, V.,** Annual Review, Great Lakes Fisheries Res. Branch, Department of Fisheries and Oceans, Burlington, Ontario, Canada, 1984.

12. **Smith, C. E., Peck, T. H., Klauda, R. J., and McLaren, J. B.,** Hepatomas in Atlantic tomcod *Microgadus tomcod* (Walbaum) collected in the Hudson River estuary in New York, *J. Fish Dis.,* 2, 313, 1979.

13. **McCain, B. B., Pierce, K. V., Wellings, S. R., et al.,** Hepatomas in marine fish from an urban estuary, *Bull. Environ. Contam. Toxicol.,* 18, 1, 1977.

14. **Pierce, K. V., McCain, B. B., and Wellings, S. R.,** Pathology of hepatomas and other liver abnormalities in English sole (*Parophrys vetulus)* from the Duwamish River estuary, Seattle, Washington, *J. Natl. Cancer Inst.,* 60(6), 1445, 1978.

15. **Black, J. J., Holmes, M., and Paigen, B.,** Fish tumor pathology and aromatic hydrocarbon pollution in a Great Lakes estuary, *Environ. Sci. Res.,* 16, 559, 1980.

16. **Baumann, P. C., Smith, W. D., and Ribick, M.,** Hepatic tumor levels and polynuclear aromatic hydrocarbon levels in two populations of brown bullheads *(Ictalurus nebulosus),* in *Polynuclear Aromatic Hydrocarbons: Physical and Biological Chemistry,* Cooke, M., Dennis, A. J., and Fisher, G. L., Eds., Battelle Press, Columbus, Ohio, 1982, 93.

17. **Tomljanovich, D. A.,** Growth Phenomena and Abnormalities of the Sauger *Stizostedion canadense* (Smith) of the Keweenaw Waterway, M. S. thesis, Michigan Technical University, Houghton, 1974.

18. **Black, J. J., Evans, E. D., Harshbarger, J. C., and Zeigel, R. F.,** Epizootic neoplasms in fishes from a lake polluted by copper mining wastes, *J. Natl. Cancer Inst.,* 69(4), 915, 1982.

19. **Black, J. J.,** Fish tumors as known field effects of contaminants, in Symp. Proc. World Conf. on Large Lakes, Mackinac Island, Michigan, May 1986, in press.

20. **Murchelano, R. A. and Wolke, R. E.,** Epizootic carcinoma in the winter flounder, *Pseudopleuronectes americanus, Science,* 228, 587, 1985.

21. **Grizzle, J. M., Schwedler, T. E., and Scott, A. L.,** Papillomas of black bullheads, *Ictalurua nebulosus (Rafinesque),* living in a chlorinated sewage pond, *J. Fish Dis.,* 4, 345, 1981.

22. **Tan, B., Melius, P., and Grizzle, J.,** Hepatic enzymes and tumor histopathology of black bullheads with papillomas, in *Polynuclear Aromatic Hydrocarbons: Chemical Analysis and Biological Fate,* Cooke, M. and Dennis, A. J., Eds., Battelle Press, Columbus, Ohio, 1981, 377.

23. **Grizzle, J. M., Melius, P., and Strength, D. R.,** Papillomas on fish exposed to chlorinated waste water effluent, *J. Natl. Cancer Inst.,* 73(5), 1133, 1984.

24. **Lech, J. J. and Vodicnik, M. J.,** Biotransformation of chemicals by fish: an overview, in Use of Small Fish Species in Carcinogenecity Testing, Greenwald, P., Ed., proc. of a symp. held at Lister Hill Center, Bethesda, Maryland, December 8 to 10, 1981, 355.

25. **Black, J. J.,** Epidermal hyperplasia and neoplasia in brown bullheads *(Ictalurus nebulosus)* in response to repeated applications of a PAH containing extract of polluted river sediment, in *Polynuclear Aromatic Hydrocarbons: Formation, Metabolism, and Measurement,* Cooke, M. W. and Dennis, A. J., Eds., Battelle Press, Columbus, Ohio, 1983, 99.

26. **James, M. O., Fouts, J. R., and Bend, J. R.,** Xenobiotic metabolizing enzymes in marine fish, in *Pesticides in the Aquatic Environment,* Plenum Press, New York, 1977, 171.

27. **Lech, J. J. and Bend, J. R.,** The relationship between biotransformation and the toxicity and fate of xenobiotic chemicals in fish, *Environ. Health Perspect.,* 34, 115, 1980.

28. **Stegman, J. J.,** Polynuclear aromatic hydrocarbons and their metabolism in the marine environment, in *Polycyclic Hydrocarbons and Cancer,* Gelboin, H. V. and Ts'o, P. O., Eds., Academic Press, New York, 1981, 1.

29. **Ahokas, J. T., Saarni, H., and Nebert, D. W.,** The in vitro metabolism and covalent binding of benzo(a)pyrene to DNA catalyzed by trout liver microsomes, *Chem. Biol. Interact.,* 25, 103, 1979.

30. **Williams, D. E. and Buhler, D. R.,** Benzo(a)pyrene-hydroxylase catalyzed by purified isozomes of cytochrome P-450 from B-naphthoflavone-fed rainbow trout, *Biochem. Pharmacol.,* 33, 3743, 1984.

31. **Gmur, D. J. and Varanasi, U.,** Characterizations of benzo(a)pyrene metabolites isolated from muscle, liver, and bile of a juvenile flatfish, *Carcinogenesis,* 3(12), 1397, 1982.

32. **Pattie, B. H.,** Estuarine monitoring program (July 1, 1974 to June 30, 1975), project completion report, State of Washington Department of Fisheries, Marine Fish Program, Seattle, 1975.

33. **Malins, D. C., McCain, B. B., Brown, D. W., Chan, S., Myers, M. S., Landahl, J. T., Prohaska, P. G., Friedman, A. J., Rhodes, L. D., Burrows, D. G., Gronlund, W. D., and Hodgins, H. O.,** Chemical pollutants in sediments and diseases of bottom-dwelling fish in Puget Sound, Washington, *Environ. Sci. Technol.,* 18(9), 705, 1984.

34. **Malins, D. C., McCain, B. B., Myers, M. S., Brown, D. W., Krahn, M. M., Roubla, W. T., Schiewe, M. H., Landahl, J. T., and Chan, S.-L.,** Field and laboratory studies of the etiology of liver neoplasia in marine fish from Puget Sound, *Environ. Health Perspect.,* 71, 5, 1987.

35. **Malins, D. C., Krahn, M. M., Brown, D. W., Rhodes, L. D., Myers, M. S., McCain, B. B., and Chan, S.,** Toxic chemicals in marine sediment and biota from Mukilteo, Washington: relationships with hepatic neoplasms and other hepatic lesions in English sole *(Parophrys vetulus), J. Natl. Cancer Inst.,* 74(2), 487, 1985.

36. **Malins, D. C., Krahn, M. M., Myers, M. S., Rhodes, L. D., Brown, D. W., Krone, C. A., McCain, B. B., and Chan, S.,** Toxic chemicals in sediments and biota from a creosote-polluted harbor: relationships with hepatic neoplasms and other hepatic lesions in English sole *(Parophrys vetulus), Carcinogenesis,* 6(10), 1463, 1985.

37. **Stein, J. E., Hom, T., and Varanasi, U.,** Simultaneous exposure of English sole *(Parophrys vetulus)* to sediment-associated xenobiotics I. Uptake and disposition of ^{14}C-polychlorinated biphenyls and ^{3}H-benzo(a)pyrene, *Mar. Environ. Res.,* 13, 97, 1984.

38. **Varanasi, U., Reichert, W. L., Stein, J. E., Brown, D. W., and Sanborn, H. R.,** Bioavailability and biotransformation of aromatic hydrocarbons in benthic organisms exposed to sediment from an urban estuary, *Environ. Sci. Technol.,* 19(9), 836, 1985.

39. **Varanasi, U. and Gmur, D. J.,** Metabolism activation and covalent binding of benzo(a)pyrene to deoxyribonucleic acid catalyzed by liver enzymes of marine fish, *Biochem. Pharmacol.,* 29, 753, 1980.

40. **Varanasi, U., Gmur, D. J., and Krahn, M. M.,** Metabolism and subsequent binding of benzo(a)pyrene to DNA in pleuronectid and salmonid fish, in *Polynuclear Aromatic Hydrocarbons: Chemistry and Biological Effects,* Bjorseth, A. and Dennis, A. J., Eds., Battelle Press, Columbus, Ohio, 1980, 455.

41. **Varanasi, U., Stein, J. E., and Hom, T.,** Covalent binding of benzo(a)pyrene to DNA in fish liver, *Biochem. Biophys. Res. Commun.,* 103, 780, 1981.

42. **Varanasi, U., Nishimoto, M., Reichert, W. L., and Le Eberhart, B.-T.,** Comparative metabolism of benzo(a)pyrene and covalent binding to hepatic DNA in English sole, starry flounder and rat, *Cancer Res.,* 46, 3817, 1985.

43. **Varanasi, U., Stein, J. E., Nishimoto, M., and Hom, T.,** Benzo(a)pyrene metabolites in liver, muscle, gonads, and bile of adult English sole *(Parophrys vetulus),* in *Proc. 7th Int. Symp. on Formation, Metabolism, and Measurement of Polynuclear Aromatic Hydrocarbons,* Cooke, M. W. and Dennis, A. J., Eds., Battelle Press, Columbus, Ohio, 1982, 1221.

44. **Nishimoto, M. and Varanasi, U.,** Benzo(a)pyrene metabolism and DNA adduct formation mediated by English sole liver enzymes, *Biochem. Pharmacol.,* 34, 203, 1985.

45. **Krahn, M. M., Myers, M. S., Burrows, D. G., and Malins, D. C.,** Determination of metabolites of xenobiotics in the bile of fish from polluted waterways, *Xenobiotica,* 14, 633, 1984.

46. **Krahn, M. M., Rhodes, L. D., Myers, M. S., Moore, L. K., MacLeod, W. D., and Malins, D. C.,** Associations between metabolites of aromatic compounds in bile and the occurrence of hepatic lesions in English sole *(Parophrys vetulus)* from Puget Sound, Washington, *Environ. Contam. Toxicol.,* 15, 61, 1986.

47. **Statham, C. N., Melancon, M. J., and Lech, J. J.,** Bioconcentration of xenobiotics in trout bile: a proposed monitoring aid for some waterbourne chemicals, *Science,* 193, 680, 1976.

48. **Ames, B. N., McCann, J. H., and Yamasaki, E.,** Methods for detecting carcinogens and mutagens with the Salmonella/mammalian microsome mutagenicity test, *Mutat. Res.,* 31, 347, 1975.

49. **Black, J. J.,** Field and laboratory studies of environmental carcinogenesis in Niagara River fish, *J. Great Lakes Res.,* 9(2), 326, 1983.

50. **Baumann, P. C., Black, J. J., and Harshbarger, J. C.,** unpublished data, 1983.

51. **Maccubbin, A. E., Black, P., Trzeciak, L., and Black, J. J.,** Evidence for polynuclear aromatic hydrocarbons in the diet of bottom-feeding fish, *Environ. Contam. Toxicol.,* 34, 876, 1985.

52. **Black, J. J.,** Environmental implications of neoplasia in Great Lakes fish, *Mar. Environ. Res.,* 14, 529, 1984.

53. **Black, J. J.,** Aquatic animal neoplasia as an indicator for carcinogenic hazards to man, *Hazard Assessment Chem. Curr. Dev.,* 3, 181, 1984.

54. Occupational Safety and Health Administration, OSHA issues tentative carcinogen list (1978), *Chem. Eng. News,* 56, 20, 1978.

55. **Squire, R. A. and Levitt, M. H.,** Report of a workshop on classification of specific hepatocellular lesions of rats, *Cancer Res.,* 35, 3214, 1975.

56. **Black, J. J., Fox, H., Black, P., and Bock, F.,** Carcinogenic effects of river sediment extracts in fish and mice, in *Water Chlorination Chemistry: Environmental Impact and Health Effects,* Jolley, R. L., Bull, R. J., Davis, W. P., Katz, S., Roberts, M. H., Jr., and Jacobs, V. A., Eds., Lewis Publishers, Chelsea, Michigan, 1984, 415.

57. **Black, J. J.,** personal communication, 1986.

58. **Brass, J. H., Elbert, W. C., Feige, M. A., Glick, E. M., and Lington, A. W.,** United States Steel Lorain Ohio Works, Black River Survey: Analysis for Hexane Organic Extractables and Polynuclear Aromatic Hydrocarbons, U.S. Environmental Protection Agency, Cincinnati, Ohio, 1974.

59. **Lao, R. C., Thomas, R. S., and Monkman, J. L.,** Computerized gas chromatographic-mass spectrometer analysis of polycyclic aromatic hydrocarbons in environmental samples, *J. Chromatogr.,* 112, 681, 1975.

60. **Baumann, P. C.,** unpublished data, 1985.

61. **Black, J. J.,** Movement and identification of a creosote-derived PAH complex below a river pollution point source, *Arch. Environ. Contam. Toxicol.,* 11, 161, 1982.

62. **Baumann, P. C., Smith, W. D., and Parland, W. K.,** Tumor frequencies and contaminant concentrations in brown bullheads from an industralized river and a recreational lake, *Transcr. Am. Fish. Soc.,* 116, 79, 1987.

63. **Lowe, T. P., May, T. W., Brumbaugh, W. G., and Kane, W. D.,** National contaminant biomonitoring program: concentrations of seven elements in freshwater fish, 1978—1981, *Arch. Environ. Contam. Toxicol.,* 14, 363, 1985.

64. **Harshbarger, J. C.,** unpublished data, 1985.

65. **Baumann, P. C. and Harshbarger, J. C.,** Frequencies of liver neoplasia in a feral fish population and associated carcinogens, *Mar. Environ. Res.,* 17, 324, 1985.

66. **Fabacher, D. L. and Baumann, P. C.,** Enlarged livers and hepatic microsomal mixed-function oxidase components in tumor-bearing brown bullheads from a chemically contaminated river, *Environ. Toxicol. Chem.,* 4, 703, 1985.

67. **Jodon, T. A., Baumann, P. C., and Landrum, P.,** unpublished data, 1985.

68. **Johnston, E. P. and Baumann, P. C.,** unpublished data, 1985.

69. **Varanasi, U. and Black, J. J.,** unpublished data, 1986.

70. **Hendricks, J. D., Myers, T. R., Shelton, D W., Casteel, J. L., and Bailey, G. S.,** The hepatocarcinogenesis of benzo(a)pyrene to rainbow trout by dietary exposure and intraperitoneal injection, *J. Natl. Cancer Inst.,* 74(4), 839, 1985.

71. **Malins, D. C., McCain, B. B., Brown, D. W., Varanasi, U., Krahn, M. M., Myers, M. S., and Chan, S.,** Sediment-associated contaminants and liver diseases in bottom-dwelling fish, in *Proc. Intl. Workshop On In-Situ Sediment Contaminants,* Aberystwyth, Wales, 1984.

72. **Malins, D. C., McCain, B. B., Brown, D. W., Sparks, A. K., Hodgins, H. O., and Chan, S.,** Chemical contaminants and abnormalities in fish and invertebrates from Puget Sound, Technical Memorandum OMPA-19, National Oceanic and Atmospheric Administration, Boulder, Colorado, 1982.

73. **Hallet, D. J. and Brecher, R. W.,** Aquatic ecosystem objective for polynuclear aromatic hydrocarbons in the Great Lakes Ecosystem, International Joint Commission, Windsor, Ontario, 1982, 39.

74. **Malins, D. C.,** personal communication, 1986.

Chapter 9

HYDROCARBONS IN MARINE MOLLUSKS: BIOLOGICAL EFFECTS AND ECOLOGICAL CONSEQUENCES

M. N. Moore, D. R. Livingstone, and J. Widdows

TABLE OF CONTENTS

I. INTRODUCTION

Marine mollusks are being used increasingly as sentinel organisms for the detection of environmental deterioration. This involves both measurement of accumulated contaminants and assessment of sublethal biological effects.[1-3]

Polycyclic aromatic hydrocarbons (PAH) represent a major class of environmental contaminant originating from both petrogenic and pyrogenic sources.[4] Marine organisms, including mollusks, accumulate PAH in their body tissues and concentrate them to a marked degree over seawater levels.[5,6] The quantity, form, biological fate, and effects of toxic environmental contaminants such as PAH are determined by interactions between a number of processes, including uptake (absorption), distribution, accumulation (storage), biotransformation (metabolism), and elimination (excretion and depuration). Our understanding of many of these processes is limited in marine organisms. Many PAH are capable of being transformed to carcinogenic or cocarcinogenic derivatives in mammals by the cytochrome P-450 monooxygenase- and epoxide hydratase systems.[7] These systems are also present in most living organisms, although our knowledge of the metabolic fates of PAH in mollusks is extremely limited.[8] In addition, the accumulation of high concentrations of PAH in molluskan tissues may represent a hazard for human consumers as well as for the organism itself.[5,6]

Much of our present understanding of toxic effects of hydrocarbons (including PAH) and other contaminants in marine mollusks is as yet very simplistic in comparison to mammalian studies. Clearly, there is a very long way to go before we can claim to comprehend the biological consequences of the often very large number of chemical contaminants that may occur in a given environmental situation.[9] This is further exacerbated by the occurrence of other environmental factors that either may interact with the contaminants to influence bioavailability, such as processes in sediments and binding to biotic and abiotic particulate matter, or modify the functioning of the organism, such as temperature and salinity changes.[10-12]

This chapter is not intended to be a comprehensive documentation of biological effects induced by PAH, but rather considers one particular integrated approach to the areas of bioaccumulation, biotransformation, and biological effects and their ecological consequences. In addition, it must be emphasized that while the chapter focuses on PAH, many of the environmental examples described relate to petroleum-derived xenobiotics, which include not only PAH, but also aliphatic hydrocarbons, as well as heterocyclic compounds. The possibility of contributory effects due to these other components cannot, therefore, be ruled out. This situation also presents some terminological problems, hence, PAH is used where it is most appropriate, while the terms hydrocarbons and petroleum hydrocarbons are used where particular circumstances necessitate this lack of specificity. The prime objective of this work has been to identify and investigate the processes involved in the responses to hydrocarbons at different levels of biological organization (i.e., molecular, subcellular, cellular, tissue/organ, individual, and population), whether these can be conceptually linked in order to gain an understanding of how the animal responds to hydrocarbons and to attempt to predict what the possible ecological consequences might be. An additional objective has been the development of indices of harmful effects at the various hierarchical levels of biological organization; this aspect has been extensively reviewed elsewhere.[3,8,13,14]

II. MOLECULAR AND BIOCHEMICAL EFFECTS

The aim of this section is to review the current state of knowledge of the enzymes involved in PAH metabolism, their responses to PAH, and the molecular consequences or effects of such responses. Emphasis is placed on the common mussel, *Mytilus edulis* and other mytilids, for which there is the most information, but recent data for other bivalve- and gastropod mollusks are also presented. For a more general review of molluskan organic xenobiotic metabolism the reader is referred to elsewhere in the book and to a recent review.[8] The potential of the detoxication enzymes for use as specific indicators of biological impact by hydrocarbon pollution[15] is briefly discussed, but a more complete consideration can be found elsewhere.[8,16,17] The responses of biochemical systems not involved in the metabolism of PAH, such as the enzymes of intermediary metabolism, the adenosine phosphates, and free amino acids, are not considered, and the reader is referred to recent publications.[16-18]

A. Enzymes of Detoxication and Activation

The metabolism of xenobiotics has been classified into Phase I (biotransformation) and Phase II (conjugation) reactions. The first phase involves oxidation by various monooxygenase reactions including epoxidation, hydroxylation, and dealkylation and is catalyzed by the cytochrome P-450 monooxygenase- or mixed-function oxidase (MFO) enzyme system. The terminal component of the system, cytochrome P-450, exists in multiple forms. The resulting products may then be converted to dihydrodiols and/or conjugated with glutathione or glucuronic acid by, respectively, epoxide hydratases (E.C.4.2.1.64), glutathione S-transferases (E.C.2.5.1.12 and E.C.4.4.1.7), and UDP-glucuronyl transferases (E.C.2.4.1.17); further monooxygenation of some products, conjugation with other chemicals such as sulfate, and metabolism of conjugates (e.g., conversion of glutathione conjugates to mercapturic acids) are also possible. Paradoxically, during the course of the metabolic transformations, reactive electrophilic intermediates may be formed which are more toxic, mutagenic, or carinogenic than the parent compounds, and the MFO system and associated enzymes must therefore be viewed as part of a detoxication/toxication system, the value of which rests on the balance of the enzymes present and the chemistry of the metabolites produced. An important feature of the system in that the activities of the enzymes and the concentration of, for example, specific isoenzymes of cytochrome P-450 may be increased by exposure of the animal to the xenobiotic.

The MFO system has been studied most in mollusks and four types of oxidative activities (epoxidation, hydroxylation, O-dealkylation, N-dealkylation) have been detected *in vitro* in eight species of marine bivalve, three gastropod species, and one cephalopod species.[8,19] Studies on *M. edulis*[19-21] and *M. galloprovincialis*[22] indicate that the molluskan cytochrome P-450 monooxygenase system is similar in a number of respects to that typically described for mammals and other organisms, viz., cytochromes P-450 and b_5 are present; flavoprotein P-450 and b_5 reductases are indicated to be present from the NADH-ferricyanide reductase, NADH- and NADPH-cytochrome c reductase and NADPH-neotetrazolium reductase activities; a positive correlation is observed between tissue-distributed microsomal NADPH-cytochrome c reductase and benzo(a)pyrene hydroxylase (BaPH) activities; BaPH activity is inhibited by known inhibitors of cytochrome P-450 (carbon monoxide, SKF-525A, α-naphthoflavone) and is oxygen dependent; and some of the metabolites of benzo(a)pyrene (BaP) produced *in vitro* are consistent with cytochrome P-450 monooxygenation. Cytochromes P-450 and b_5 have also been detected in five other species of bivalves and gastropods, and the cytochrome concentrations and enzyme activities are similar in all species examined (Table 1).

Tissue and subcellular characterization studies of *M. edulis* demonstrated that, in common with other phyla, the cytochrome P-450 monooxygenase system is primarily membrane-bound in the endoplasmic reticulum and is localized predominantly in one tissue-type, the digestive gland or hepatopancreas,[19,20] which in bivalves is a major site of hydrocarbon (predominantly PAH) uptake. The tissue differences are particularly marked when considered in terms of the microsomal yield of the MFO components or activities,[20] e.g., for BaPH of female mussels (in pmol min^{-1} g^{-1} wet wt; mean $\pm$ SEM, n = 4) digestive gland, 273.1

Table 1
CONCENTRATIONS OF CYTOCHROME P-450 AND b₅ AND SPECIFIC ACTIVITIES OF BENZO(a)PYRENE HYDROXYLASE AND NADPH-CYTOCHROME c REDUCTASE IN DIGESTIVE GLAND MICROSOMES OF SEVERAL SPECIES OF BIVALVES AND GASTROPODS

Species	P-450 (pmol mg^{-1} protein)	b$_5$ (pmol mg^{-1} protein)	NADPH-cytochrome c reductase (nmol min^{-1} mg^{-1} protein)	BaPH (pmol min^{-1} mg^{-1} protein)	Ref.
Mytilus edulis	47—134	26—76	10.3—15.4	3—25	8, 25
Mytilus galloprovincialis	101	37	8.1	35	19
	47	205	11.8	24	22
Mytilus californianus	3—88	75—400	—	—	23
Macrocalista maculata	—	40^a	—	—	24
Arca zebra	79	81	4.0	—	19
	106	76	8.3	—	19
Cardium edule	20—85	80—94	5.3—10.6	6—10	26, 52
Littorina littorea	8—54	26—70	7.5—1.2	—	25—27

Note: The data are for animals collected from the field from both clean and polluted sites. Mean values are quoted and the ranges represent different populations or sampling times.

^{a} Viscera.

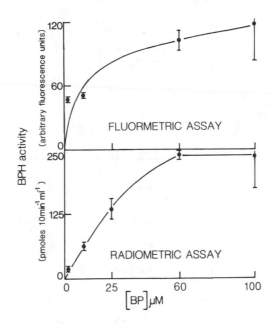

FIGURE 1.   Dependence of *M. edulis* digestive gland microsomal benzo(a)pyrene hydroxylase (BaPH) activity on benzo[a]pyrene concentration [BaP]. Different samples used for fluormetric and radiometric studies. Means are ± range (duplicate cuvettes). 50 m*M* *tris*-HCl pH 7.6, 0.2 m*M* NADPH, 25°C.[198]

± 18.8; gills, 7.3 ± 3.2; mantle, 19.7 ± 6.7; remaining tissues, 4.5 ± 2.3. In *M. edulis,* cytochrome P-450 also occurs in gills,[19] and reductase and BaPH activities, but not cyto-chromes P-450 and b₅, have been detected in blood cells.[19,20] The MFO system in *M. edulis* and *M. galloprovincialis* varies with sex and season. Microsomal BaPH, NADPH-cyto-chrome *c* reductase, and NADH-cytochrome *c* reductase activities were higher in female digestive gland than male in *M. edulis,* at least at one time of the year, and subcellular fractionation studies indicated the same for cytochrome P-450.[20] BaPH activity changes seasonally, with maximum values occurring in summer and autumn followed by a decline in late winter or early spring, possibly associated with the approach of spawning.[8,19,28] Less marked seasonal changes are also indicated for NADPH-cytochrome *c* reductase activity and cytochrome P-450.[27,28]

Information on other detoxication enzymes is considerably more limited. An alternative route for the monooxygenation of aromatic amines and other compounds has been indicated by the detection of FAD-containing monooxygenase (E.C.1.14.13.8, dimethylaniline mon-ooxygenase, *N*-oxide forming) activity in digestive gland microsomes of *M. edulis*[29] and *M. galloprovincialis.*[30] Microsomal epoxide hydratase and cytosolic glutathione *S*-transferase activities have been detected in digestive gland or whole tissues of *M. edulis, M. gallo-provincialis,* and *Mya arenaria.*[28,31] Cytosolic UDP-glucuronyl transferase activity is present in *M. galloprovincialis,*[30] in contrast to the terrestrial slug, *Arion ater,* for which it is suggested that this enzyme has been replaced by the analogous UDP-glucosyltransferase, giving rise to glucosides rather than glucuronides as the xenobiotic conjugate.[32]

B. BaPH Activity and *In Vitro* Benzo(a)pyrene Metabolism

Although a number of MFO activities have been measured in mollusks, BaPH has been the most studied. In digestive gland microsomes of *M. edulis* the activity is very low for a short period around spawning, but for most of the year varies between about 10 and 20 pmol total metabolites min⁻¹ mg⁻¹ protein, reaching a maximum of about 30 in autumn.[8] During these latter times it is possible to study the properties of the BaPH activity, and some of these are presented in Figures 1 to 4.

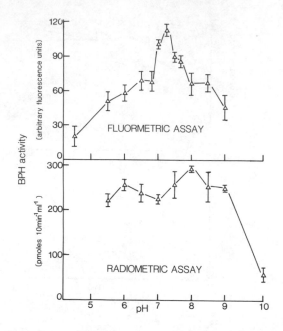

FIGURE 2.   Dependence of *M. edulis* digestive gland microsomal benzo(a)pyrene hydroxylase (BaPH) activity on pH. Different samples used for fluormetric and radiometric studies. Means are ± range (duplicate cuvettes). Overlapping 50 m*M* buffers used and buffer interactions removed by normalization relative to triethanolamine-HCl; Na-citrate-citric acid (pH 4.5 to 5.5), Na cacodylate-HCl (5.5 to 6.5), triethanolamine-HCl (6.5 to 7.5), *tris*-HCl (7.5 to 9.5), 0.2 m*M* NADPH, 25°C.[198]

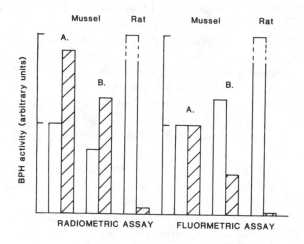

FIGURE 3.   Dependence of *M. edulis* digestive gland and rat liver microsomal BaPH activity on presence of 0.2 mM NADPH. (□): with NADPH, (▨): without NADPH; (A) freshly prepared microsomes; B, previously frozen microsomes. Differences between assay significant at $p = 0.05$ (n = 4), one-way analysis of variance. (Data from Livingstone, D. R., *Mar. Pollut. Bull.*, 16, 158, 1985. With permission.)

Digestive gland microsomes were prepared and BaPH assays carried out as described by Livingstone and Farrar[20] and Livingstone et al.[25] BaPH activity was measured using both the radiometric assay of Van Cantfort et al.[33] which detects total metabolites, and the fluorometric assay of Dehnen et al.[34] which primarily detects the 3- and 9-phenols.[35] Both reaction rates are linear for about 10 min[17] and both activities showed Michaelis-Menten kinetics with respect to BaP with values for apparent K_m BaP of 41 µ*M* (radiometric) and

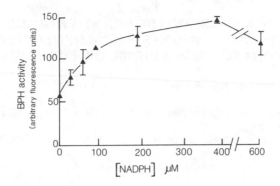

FIGURE 4.   Dependence of benzo(a)pyrene hydroxylase (BaPH) activity (fluorometric assay) of previously frozen microsomes of *M. edulis* digestive gland on NADPH. Means are ± range (duplicate
cuvettes). Microsomes were prepared and frozen overnight before use. 50 m*M* *tris*-HCl pH 7.6, 60 μm
BaP, 25°C.[198]

9 μ*M* (fluorometric) (Figure 1; kinetic analysis as described in Livingstone and Clarke[36]).
The fluorometric apparent K_m is of the same order as those described for the crab, *Carcinus
maenas*[37] and the copepod, *Calanus helgolandicus*,[38] but higher than that for the barnacle,
Balanus eburneus.[39] The fluorometric BaPH activity varied with pH with an optimum at
pH 7.25; in contrast, the radiometric BaPH activity changed little between pH 6 and 9
(Figure 2). Sharp pH profiles for BaPH activity with optima between 6.7 and 7.8 for both
fluorometric and radiometric assays are seen for a number of different organisms, viz., the
fish, *Esox lucius*,[40] the insect larva, *Spodoptera eridania*,[41] and the crabs, *Callinectes sapidus*[42]
and *C. maenas*.[37] An NADPH-independent BaPH activity occurs in *M. edulis* digestive
gland microsomes, which contrasts with the absolute requirement for NADPH of the microsomal BaPH activity of vertebrates such as rat (Figure 3) and of other invertebrates such
as crustacea;[42] the low requirement of bivalve BaPH activity for NADPH has been observed
by other workers.[19,28,43] Again, a difference was seen between the fluorometric and radiometric assays with respect to NADPH dependence: whereas NADPH (at the concentration
used) inhibited the radiometric activity, no inhibition of the fluorometric assay occurred,
and in previously frozen microsomes an NADPH dependence could be demonstrated (Figures
3 and 4). The NADPH-independent BaPH activity (fluorometric assay, no NADPH) of *M.
edulis* digestive gland microsomes was inhibited by the cytochrome P-450 inhibitor SKF-
525A,[199] viz., percentage inhibition of BaPH activity (means and range, n = 2) 19 ± 4%
(0.1 m*M* SKF-525A) and 75 ± 3% (1 m*M*).

The NADPH-independent BaPH activity could be due to an endogenous source of reducing
power. However, the different pH profiles, apparent K_m BaP values, and effects of NADPH
on the fluorometric and radiometric assays possibly indicate that more than one process is
involved in the metabolism of BaP *in vitro* by digestive gland microsomes of *M. edulis*.
The same conclusion was reached by Stegeman[19] based on the properties of BaPH activity
and the *in vitro* metabolites produced from BaP. Identified metabolites included the 9-,
10-, 4,5-, and 7,8-dihydrodiols and the 9-hydroxy and other phenols which are consistent
with cytochrome P-450 monooxygenaton and epoxide hydratase function. However, 65%
of the total metabolites produced were quinones, including the unusual 6,12-quinone, which
contrasts with 20 to 30% for various fish species[44] and less than 5% for *B. eburneus*.[39]
Similar BaP metabolite profiles with high levels of quinones have been observed for digestive
gland homogenates of *Crassostrea virginica*.[45] The predominance of quinones and the variable dependence of BaPH activity on NADPH and the variable inhibition by carbon monoxide has led Stegeman[19] to speculate the involvement of peroxidative catalyst(s) in addition
to cytochrome P-450 monooxygenation in BaP metabolism. Other mechanisms also exist

Table 2
RESPONSES OF THE DIGESTIVE GLAND
MICROSOMAL MFO SYSTEM OF *MYTILUS EDULIS*
EXPOSED TO ~30 PPB DIESEL OIL FOR 4 MONTHS[a]

Microsomal parameter	Control	~30 ppb
NADH-ferricyanide reductase (nmol min^{-1} mg^{-1} protein)	741 ± 33	764 ± 25
NADH-cytochrome c reductase (nmol min^{-1} mg^{-1} protein)	95.4 ± 8	109.4 ± 5.9
NADPH-cytochrome c reductase (nmol min^{-1} mg^{-1} protein)	10.3 ± 1.2	13.5 ± 0.9[b]
P-450 (pmol mg^{-1} protein)	46.6 ± 14.3	93.0 ± 14.3[c]
P-450 λ_{max} nm	449 ± 1	445 ± 1[b]
b$_5$ (pmol mg^{-1} protein)	25.8 ± 2.3	40.2 ± 5.8[c]
BaPH (arbitary fluorescence units, unit time^{-1} mg^{-1} protein)	54.1 ± 21.7	56.4 ± 3.8
NADPH-neotetrazolium reductase (relative units of absorbance)	11.1 ± 1.0	15.3 ± 0.6[b]

[a]   Data from Livingstone, D. R., et al., *Mar. Environ. Res.*, 17, 101, 1985. With permission.
[b] $p < 0.05$.
[c] $p < 0.1$.

that could result in the *in vitro* metabolism of BaP e.g., lipid peroxidation,[47] and further studies are required to resolve the situation.

C. Responses of the Detoxication/Activation Enzymes to Hydrocarbon Exposure

With the exception of the responses of the MFO-associated NADPH-neotetrazolium reductase activity to xenobiotics,[13,48] relatively few molluskan exposure studies have been carried out and some can only be regarded as pilot experiments. Increases in cytochromes P-450 and b$_5$ in digestive gland microsomes of *M. galloprovincialis* have been indicated in response to laboratory exposures to paraffin, anthracene, perylene, and 3-methylcholanthrene, and in field exposures to petroleum hydrocarbons.[23] Slight elevations of aryl hydrocarbon hydroxylase activity have been reported in *C. virginica* digestive gland in response to BaP, 3-methylcholanthrene (3-MC) and PCB[45,46] and in the digestive gland of mussels and other bivalves exposed to polychlorinated biphenyls (PCB), polybrominated biphenyls, and petroleum hydrocarbons.[49] A responsiveness of the MFO system to phenobarbital has been indicated in *M. edulis* by sister chromatid exchange studies.[50]

More recently, studies have been carried out in which the responses of the components (cytochromes, flavoprotein reductases) of the MFO system and MFO (BaPH) activity have been examined together.[25,26] Mussels (*M. edulis*) and periwinkles (*Littorina littorea*) were experimentally exposed, for short- and long-term exposure periods, to diesel oil under conditions closely resembling the field (outdoor flow-through tanks, natural seawater and food levels, wave and local tidal stimulation). PAH were taken up into the tissues,[25] and with long-term exposure, increases were seen in cytochrome P-450 and b$_5$ concentrations and NADPH-cytochrome c and NADPH-neotetrazolium reductase activities, but not in BaPH activity (Tables 2 and 3). The increases in NADPH-dependent reductase activities in *L. littorea* were consistently higher at the higher oil dosage level[25,26] (Table 3). The NADPH-cytochrome c and NADPH-neotetrazolium reductase activities are thought to measure some aspect, though not exactly the same aspect, of the *in situ* catalytic activity of NADPH-cytochrome P-450 reductase,[51] and considering the data from the exposure experiments a

Table 3
RESPONSES OF THE DIGESTIVE GLAND MICROSOMAL MFO SYSTEM OF *LITTORINA LITTOREA* EXPOSED TO ~30 AND ~130 PPB DIESEL OIL FOR 16 MONTHS[a]

Microsomal parameter	Control	~30 ppb	~130 ppb
NADH-ferricyanide reductase	1272 ± 99	1222 ± 88	1247 ± 296
NADH-cytochrome c reductase	117 ± 15	129 ± 14	128 ± 27
NADPH-cytochrome c reductase	7.5 ± 0.4	11.3 ± 0.7[b]	21.8 ± 1.9[c]
NADPH-neotetrazolium reductase	7.0 ± 1.0	20.4 ± 1.3[c]	27.6 ± 2.2[c]
b_5	26.4 ± 2.2	28.1 ± 2.7	48.0 ± 7.8[c]
	46.2 ± 8.9	71.3 ± 15.7	50.6 ± 5.8

Note: Units of microsomal parameters are as for Table 2. Values are means ± S.E.M. (n = 5 or 4).

[a] Data from Livingstone, D. R., et al., *Mar. Environ. Res.*, 17, 101, 1985. With permission.
[b] $p < 0.01$.
[c] $p < 0.05$.

positive correlation of 0.922 was observed between the two enzyme activities ($p < 0.001$).[25] Varying responses (increases) of cytochrome P-450 and the NADPH-dependent reductases, but not BaPH activity, were seen with short-term exposures of up to 8 d.[25,26] Sex and seasonal interactions were evident for *M. edulis*, with little or no response when the mussels were ripe with gametes, but not for *L. littorea*.[26] Long-term recovery over a period of weeks to months resulted in all the components and activities returning to control levels, and the decreases paralleled the long-term depuration of PAH.[27] Similar responses have also been observed in two other mollusk species. Cytochrome P-450 and b_5 contents and NADPH-cytochrome c reductase activity, but not BaPH activity, were elevated in digestive gland microsomes of the cockle, *Cardium edule* exposed to petroleum hydrocarbons in the field.[26,52] NADPH-cytochrome c reductase activity, but not BaPH activity, was increased in digestive gland microsomes of the oyster drill, *Thais haemostoma* exposed to water-soluble fraction of southern Louisiana crude oil.[53]

The wavelength maximum of the cytochrome P-450 peak in *M. edulis* was 2 to 3 nm lower in diesel oil exposed than in control or recovered mussels[25] (Table 2), suggesting a conformational change in the protein or the synthesis of a new form of P-450. Such wavelength changes in response to 3-MC-type inducers, and the existence of P-450 isoenzymes, are well characterized in mammals[54] and certain fish;[44] different molecular forms of P-450 also exist in crustacea[55,56] and are indicated in polychaete worms.[55] A second major carbon monoxide-binding reduced molecule peak is evident in the P-450 assay of molluskan digestive gland microsomes, the wavelength maximum of which has been recorded as 420 nm in *M. galloprovincialis*[22,23] and 416 to 418[20,25] and 424 nm[19] in *M. edulis*. The size of this peak varies from about 5% to more than 100% of the P-450 peak.[19,20] Assuming it is cytochrome P-420, the denatured or altered form of P-450, this gives total cytochrome P-450 contents of about 150 to 300 pmol mg^{-1} protein (References 19 and 23 and calculated from Livingstone et al.[25] using an extinction coefficient of 111 mM^{-1} cm^{-1}) compared with about 100 pmol mg^{-1} or less for "native" P-450 (Table 1). Parallel increases and decreases in the low wavelength and 450-nm peaks, with exposure and recovery from diesel oil, are seen in both mussels and periwinkles,[25] suggesting that it is, in fact, denatured P-450. Conversely, the low wavelength peak has a wide tissue and subcellular distribution in *M. edulis*,[20] suggesting that a part of the digestive gland microsomal 416 to 424 peak might be due to some other hemoprotein such as another cytochrome or a peroxidase.

From the studies carried out to date, the general trend of response of the digestive gland of bivalve- and gastropod mollusks to petroleum hydrocarbon exposure (predominantly PAH) appears to be an increase in cytochromes P-450 and b_5 and cytochrome P-450 reductase activity, but little or no change in BaPH activity. The responses are therefore similar to other phylogenetic groups with respect to the cytochromes, but differ with respect to BaPH and possibly NADPH-cytochrome c reductase activity.[44] NADPH-cytochrome c reductase activity is unchanged in response to 3-MC-type exposure in certain species of mammal, reptile, and amphibia;[59] increases,[44] does not change,[59] or decreases[60] in fish species; and is reported to be unchanged, or have very low activity, in crustacea and polychaetes.[55,61] The lack of change in BaPH activity could have several origins. It could be peculiar to mollusks and simply reflect the different specificity of their MFO system: certainly, the NADPH dependence of the BaPH activity indicates fundamental differences from higher organisms (see above). It could be due to complications arising from possible seasonal changs in BaPH activity,[8] but this would seem unlikely given the consistent responses to hydrocarbons (diesel oil) of, for example, *L. littorea* over the year.[26] Finally, the apparent presence of cytochrome P-420 must be noted, indicating that there may have been a loss of cytochrome P-450, and therefore of some catalytic functons, in the digestive gland microsomal preparations. The apparent conversion of cytochrome P-450 to P-420 may be due to an intrinsic instability of the former, but equally could be the result of the action of contaminating lysosomal hydrolases.[20] These enzymes are present in high activity in the digestive gland[20] and in *M. edulis* are unaffected by the generally used trypsin-type inhibitor PMSF.[199]

In additon to the digestive gland, attention has also been focused on the blood cells of *M. edulis* and *L. littorea*. Increases in NADPH-neotetrazolium reductase activity have been seen in the former in response to exposure to, or injecton of, anthracene, phenobarbital, 2,3-dimethylnaphthalene (2,3-MeN), phenanthrene (PHN), and water-soluble fraction of North Sea crude oil,[21,48] and in the latter in response to PHN and petroleum hydrocarbons in the field.[62] Increases in blood cell glucose-6-phosphate dehydrogenase-specific activity of *M. edulis* have been observed in response to 2,3-MeN and water-soluble fraction of North Sea crude oil, but not in response to temperature.[17,21] The specificity of thcse responses and the presence of BaPH (see above) and aldrin epoxidase[21] activities in blood cells of *M. edulis* are clearly indicative of a role in organic xenobiotic metabolism, but the chemical nature of this awaits further research.

A responsiveness of phase II conjugase enzymes has been demonstrated only in the terrestrial *A. ater*, treatment with phenobarbital increasing digestive gland microsomal UDP-glucosyltransferase activity.[32]

D. *In Vivo* Metabolism and Molecular Effects of Hydrocarbon Exposure

A consideration of *in vivo* metabolism raises the questions of the cellular functions of the *in vitro* identified enzymes, the pathways involved in the metabolism of the compounds, the rates of flux through the pathways, and the nature, fate, and effects of the metabolites produced. As is obvious from the following discussion, very little detailed information is available on any of these aspects.

The evidence for the existence of a cytochrome P-450 monooxygenase system in mollusks is substantial, from enzyme, metabolite, and exposure studies (see before). The microsomal yields, cytochrome concentrations, and enzyme activities are comparable to those of other marine invertebrate phyla and are those to be expected in relation to the general level of molluskan enzyme activity.[20] However, the quantitative and qualitative nature of the role(s) of the system *in vivo* remains to be established. A function in the metabolism of endogenous compounds, such as fatty acids and steroids,[54] is suggested in *M. edulis* and other mytilids by the sex- and seasonal variation in the MFO system,[8,28] the demonstration of the conversion of arachidonic acid to prostaglandins,[63,64] and the stimulation of BaP metabolism in digestive gland microsomes by arachidonic acid;[19] sterols have also been identified in the tissues of many mollusks.[65] *In vivo* evidence for the metabolism of exogenous compounds such as hydrocarbons and the involvement of the cytochrome P-450 monooxygenase system is elusive. Most studies have failed to detect metabolites,[66,67] or concluded that the low levels formed could have been bacterial in origin.[68,69] In contrast, phase I oxidative metabolism has been indicated *in vivo* in *M.californianus* (aldrin conversion to dieldrin),[24] the whelk, *Buccinum undatum* (biphenyl metabolism),[70] and *C. virginica* (tributyl tin oxide metabolism).[71]

The difficulty in detecting oxidative metabolites *in vivo* and the observation that bivalves accumulate hydrocarbons to high concentrations, the bioconcentration process apparently being explainable by a simple lipid/water equilibrium model,[72] appear to indicate that hydrocarbon metabolism *in vivo* is negligible or absent. However, the accumulation of hydrocarbons does not, per se, indicate the absence of metabolism, only that there is an imbalance between the rates of uptake and metabolism. Whereas uptake appears to be a passive process and, therefore, presumably similar in different organisms with similar lipid concentrations,[72] metabolism is enzymic and intrinsic to the particular organism. Uptake rates are dependent on the hydrocarbon concentration of exposure and for bivalves at environmentally realistic chronic hydrocarbon concentrations (0.2 to 20 ppb[16,72]) the rates appear to be of the same order as their BaPH activities, and for oil-spill concentrations (20 to 500 ppb[16]) are considerably greater, viz., for *M. edulis*, rates of uptake in pmol min^{-1} g^{-1} wet wt. (1) aliphatics: 0.6 to 12 (1 to 20 ppb) and 12 to 113 (20 to 200 ppb) (calculated from Stegeman and Teal[73] using an average molecular weight of 234); (2) aromatics (3- to 4-ring): 0.6 to 4 (0.4 to 4 ppb) (calculated from McLeese and Burridge[74]); compared with whole animal BaPH activities in pmol min^{-1} g^{-1} wet weight of 10 to 15 (seasonal minimum = 1 and maximum = 29) (calculated from Livingstone[8] and Livingstone and Farrar[20]). Given that *in vivo* fluxes for most reactions and pathways are less than maximally measured *in vitro* enzyme activities,[75] and that fish MFO activities are at least an order of magnitude greater than those of bivalves,[44] the much greater long-term accumulations of PAH in bivalves than fish are readily understandable.

Such considerations cannot account for the absence of metabolites, however, particularly when compared to crustacea where *in vivo* metabolism of PAH is readily detectable.[67,76] Various explanations are possible. The *in vivo* fluxes may, indeed, be much lower than the *in vitro* BaPH activities, either because the latter are artifactually high due to, for example, *in vitro* lipid peroxidation, or because factors operate *in vivo* to inhibit MFO function. The MFO activities of crustacea, although of the same order as mollusks, may be sufficiently higher to make detection of metabolites easier. Comparative data are limited, but indicate that on the average, BaPH activities are higher in crustacea.[20,37,42] This indication is supported by general enzyme data, the specific activities of various enzymes of intermediary metabolism also being consistently higher in crustacea than bivalves, e.g., means ± S.E.M., number of species used for calculation in brackets, crustacean value first; arginine kinase (E.C. 2.7.3.3), 351 ± 51(4) and 63.4 ± 10.8(30),[77] glutamic-oxaloacetic transaminase (E.C. 2.6.1.1), 18.9 ± 3.7(4) and 6.5 ± 0.8(30),[77] glutamic-pyruvic transaminase (E.C. 2.6.1.2), 20.1 ± 8.8(3) and 3.9 ± 0.8(30),[77] phosphofructokinase (E.C. 2.7.1.11), 9.7 ± 1.2(10) and 4.1 ± 1.2(11),[78] and phosphorylase (E.C. 2.4.1.1), 9.0 ± 1.7(8) and 2.4 ± 0.9(9).[78] The nature of the metabolites produced and their fate in terms of further metabolism or macromolecular binding may be sufficiently different in mollusks to render detection more difficult. In this respect, the observation that chemicals extracted from *M. edulis* tissues by chloroform/methanol are not activated to mutagens by *M. edulis* microsomes, but those extracted by the different and presumably more rigorous nitric acid treatment are,[79] is of interest. Finally, there is the possibility that metabolites are more readily lost from mollusks than crustacea into the surrounding water.

Information on the nature, fate, and effects of the metabolites produced, and the mechanisms involved, is similarly limited. That particular PAH or their metabolites can directly or indirectly affect the genetic material is indicated by a number of observations at different levels of genetic and cellular organization. Neoplastic diseases are found in bivalves and there is some correlation with PAH exposure or body burden,[80,81] although the etiology of these lesions is controversial. Epithelial tumors were induced in the gastropod, *Ampullarius australis* by exposure to 20-MC[82] Embryo abnormalities occur in the periwinkle, *L. saxatilis* in association with deteriorating environmental water quality including the presence of hydrocarbons.[83,84] Chromosomal aberrations, in the form of increased aneuploidy, were found in *M. edulis* with high body burdens of aromatic hydrocarbons.[85]

Mutagenic chemicals have been detected in the tissues of *M. edulis* and other mollusks from the field using a variety of microbial assay systems.[86] Significant mutagenic activity was found in mussels from sites associated with pollution, whereas mussels from other sites had little or no mutagenic activity.[86] Mutagenic activity was lower in tissues extracted with chloroform/methanol than in those extracted with concentrated nitric acid, and whereas incubation with rat liver microsome increased the mutagenic activity in both types of tissue extract, incubation with mussel microsomes only increased it in the latter,[79] suggesting different substrate specificities or metabolic products for the two enzyme systems. The level of mutagens in *M. edulis* tissues and the ability of the microsomes to activate premutagens varied seasonally, being higher in summer than winter,[79,86] and therefore showed a degree of correlation with seasonal BaPH activities (see above). Highest mutagenic activity was found in the digestive gland.[79] The chemical nature of the mutagenic activity was not identified, but was indicated not to be dependent on heavy metals or nitrosamines.[86] The digestive glands of *M. edulis, M. mercenaria,* and *C. virginica* produce the proximate carcinogen BaP 7,8-dihydriol in *in vitro* incubations,[19,46] but show only minimal mutagenic activation of BaP and 3-MC in the Ames bacteria (*Salmonella typhimurium* strain TA98) mutagenicity assay.[29,30,46,87] In contrast, marked mutagenic activation is seen with aromatic amines such as 2-aminoanthracene, 2-aminofluorene, 2-acetylaminofluorene, aminobiphenyl, and 4-amino-*trans*-stilbene.[29,30,46,87] The aromatic amine mutagenic activating ability in *M. mercenaria* was greatest in the digestive gland and was heat labile, NADPH dependent, and not affected by carbon monoxide,[87] suggesting that the principal enzymic activity responsible is the FAD-containing monooxygenase (E.C. 1.14.13.8) and not the cytochrome P-450 monooxygenase system. This has, in fact, been proposed for *M. edulis*[29] and *M. galloprovincialis*[30] based on Ames test studies and the detection in high activity of the FAD-containing monooxygenase enzyme in the bivalves. The specific activities in the mytilids were calculated from the measured substrate-stimulated NADPH oxidation rates and were, in fact, remarkably high, viz., using N,N-dimethylaniline as substrate, 960 nmol min^{-1} mg^{-1} microsomal protein for *M. edulis* digestive gland[29] compared with 9.8 to 25 for porcine liver microsomes[88,89] and 107 to 895 for the purified porcine enzyme.[90,91] Much lower activities for N,N-dimethylaniline N-demethylase were obtained for *M. galloprovincialis* digestive gland microsomes when the rate of formaldehyde formation was measured, viz., 0.6 nmol min^{-1} mg^{-1} protein,[22] suggesting that stoichiometry of substrate-stimulated NADPH oxidation rates with N,N-dimethylaniline consumption or formaldehyde formation should be demonstrated in bivalves, as has been done for mammals.[92] However, despite this discrepancy, the data of Ade et al.[22] for *M. galloprovincialis* indicate that the mixed-function amine oxidase activity is relatively high in bivalve mollusks, viz., ratio of N,N-dimethylaniline N-demethylase to BaPH activities is 0.31 (rat liver) and 25 (mussel digestive gland) (calculated from Reference 22). The FAD-monooxygenase activity in bivalves appears not to be inducible,[29,30,87] as is the case with mammals, and therefore other mechanisms must be involved in, for example, the enhanced mutagenic activation of cyclophosphamide in *M. edulis* following exposure to phenobarbital.[50]

E. Detoxication/Activation Enzymes as Specific Indicators of Organic Xenobiotic Effect

The use of biochemical systems which respond to only one type of environmental contaminant has received attention and been recommended for environmental impact assessment studies.[15,16,93] The development of measurements of the MFO system as indicators of organic xenobiotic effect is considerably advanced in fish,[94,95] and, given the widespread use of mussels and other mollusks in environmental monitoring, would be an equally desirable facility for use with these organisms.[96] The problems of developing and applying such an index (or indices) essentially fall into two categories: the practical considerations of the detection of the biochemical differences or changes against the background variability, and the interpretation of the observed differences. Seasonality is the greatest practical problem but can be accommodated for on the basis of a knowledge of the seasonal cycle of the biochemical parameter and the current reproductive state of the bivalve population.[17] More complex is the question of the interpretation of the observed differences which requires a fundamental understanding of the detoxication enzyme system. It is obvious that while the molluskan cytochrome P-450 monooxygenase system is similar to those of other phyla, e.g., presence of cytochromes and flavoproteins, fundamental differences also exist (e.g., nature of BaPH activity, responsiveness of NADPH-cytochrome c reductase) which at present limit interpretation. However, the results from the exposure studies are sufficiently encouraging to suggest a practicable index indicating specific effect due to organic xenobiotics is a possibility in the future. The responses appear to fulfill the first criterion of an index of stress response and are indicated from general, mutagenic, and cytogenic studies to fulfill the second,[2,16] viz., respectively, the biochemical changes are a response to change in an environmental factor, and they may have a detrimental effect on some aspect of animal fitness such as reproduction or survival.

III. SUBCELLULAR AND CELLULAR EFFECTS

A. Cell Injury

Alterations in specific aspects of cellular structure-function offer a means of identifying and characterizing adaptive responses or reactions to cell injury by PAH and other environmental contaminants. It should be possible to observe structural-functional alterations in individual cell-types or groups of cells at an early stage of a response, before an integrated cellular alteration would manifest itself at the level of organ or whole-animal physiological processes.[48] Some of these cellular reactions may be generalized, whereas others are likely to be specific for PAH.[13,48]

When cells are injured they undergo a series of biochemical and structural alterations. These can be classified into two phases: a reversible phase preceding cell death at which time an irreversible phase commences.[97] Trump and Arstila have defined the "point of no return" or point of cell death as the point beyond which changes are irreversible, even if the injurious stimulus is removed and the cell is returned to a normal environment.[97] This section is largely concerned with the reversible phase associated with sublethal cell injury. Cells are able to continue their existence following many types of injury by means of adaptive physiological responses. Examples of such adaptations in mammalian cells include hypertrophy, atrophy, fatty changes, proliferation of smooth endoplasmic reticulum, increased lysosomal autophagy, aging, neoplastic transformations, and accumulation of materials such as lipofuscin.[97] A number of these changes involve functional and structural alterations in intracellular membranes, particularly those of the lysosomal system in, for example, autophagy and the endoplasmic reticulum in induction of cytochrome(s) P-450.[3,8,13,97-99]

Xenobiotic-induced cellular pathology reflects disturbances of structure-function at the molecular level of biological organization. In many instances, the earliest detectable changes or "primary events" are associated with a particular class of subcellular organelle such as the lysosome or endoplasmic reticulum.[100] Investigations in mammals have revealed that much of the damaging action of xenobiotics is produced by highly reactive derivatives. It is these activated chemical forms that are responsible for the initiation of what may be termed the primary intracellular disturbances. These may spread rapidly into a complex network of associated secondary and higher-order disturbances which become progressively more difficult for the cell to reverse or modify.

There are numerous ways in which the structure and function of organelles and cells can be disturbed by contaminants such as PAH and these have been grouped by Slater[100] into four main categories:

1.   Depletion or stimulation of metabolites or coenzymes
2.   Inhibition or stimulation of enzymes and other specific proteins
3.   Activation of a xenobiotic to a more toxic molecular species
4.   Membrane disturbances

It is primarily with categories 2 and 4 that we will be concerned in considering injury induced by PAH to the lysosomal system where membrane damage results in major changes in the structure and function of the lysosomal compartment.

B. Reaction of Lysosomes to Cell Injury

Mammalian lysosomes are noted for their responsiveness to many types of cell injury and molluskan lysosomes have also been shown to respond to a wide range of injurious agents including PAH and petroleum hydrocarbons.[101-105] The cells of many molluskan tissues are especially rich in lysosomes.[106,107] Such tissues include the digestive gland, pericardial gland, kidney, adipogranular cells in the mantle, and ovarian eggs (oocytes).[106-114] Tissues in many other invertebrates also have numerous lysosome-rich cells such as in the gastrodermis of coelenterates.[115,116]

There are a number of ways that lysosomes can react to cellular injury; these can be divided into basically three categories. Lysosomal responses can be considered as decreases or increases in: (1) lysosomal contents such as hydrolytic degradative enzymes, (2) rate of membrane fusion events with either the cell membrane or other components of the vacuolar system, and (3) lysosomal membrane permeability.[104] For a variety of reasons, however, lysosomal reactions to cell injury are not well understood due to the many forms of cell injury and wide variety of organisms studied. Furthermore, one type of lysosomal change may assume the appearance of another type.[104,105]

Lysosomal injury may be either primary or secondary and it has been difficult generally to decide which situation applies for many injurious stimuli.[104] Even in mammals, which are the most-studied systems, the extent and ultimate cause of lysosomal reactions to injury are frequently unknown.[104]

Many exogenous and endogenous chemical agents are known that influence lysosomal fusion and motility, as well as modifying membrane permeability, and these have aided considerably in achieving a better understanding of lysosomal response patterns.[105,117]

Lysosomal reactions to cellular injury have been categorized above into three groups. Reactions to particular stimuli, such as PAH and petroleum hydrocarbons, may be considered as appropriate adaptive responses leading to the reestablishment of cellular homeostasis, a conservative response in which the cell is protected from further injury (e.g., detoxication and/or excretion), or an inappropriate response leading to further cellular deterioration and ultimately cell death.[97,104,105] The remainder of this section will attempt to describe observed reactions of molluskan lysosomes, induced by PAH and petroleum hydrocarbons in terms of alterations (increases or decreases) in lysosomal contents, rate of fusion events, or membrane permeability.

Table 4
TOTAL RELATIVE ACTIVITIES OF LYSOSOMAL ACID PHOSPHATASE AND β-GLUCURONIDASE IN THE DIGESTIVE CELLS OF THE PERIWINKLE *Littorina Littorea* FOLLOWING EXPOSURE TO PAH[a]

Experimental treatment/sample site (Shetland Oil Terminal)	Total relative activities as a % of the control (experimental) or clean references (field samples) values (mean ± S.D., n = 5)	
	β-Glucuronidase	Acid phosphatase
Control	100.0 ± 1.1	—
Phenanthrene (400 μg·1⁻¹; three snails; 1⁻¹; 5 d)	114.7 ± 4.9[b]	—
Ronas Voe (uncontaminated reference)	100.0 ± 4.5	100 ± 8.8
The Kames (Tidal Pool)	167.4 ± 18.6[b]	—
Scatsta Voe	88.4 ± 44.0	114.3 ± 7.6[c]
The Kames	121.8 ± 17.0[b]	—
Tanker Jetty 4	149.1 ± 17.8[b]	124.2 ± 12.0[c]
Skaw Taing	144.3 ± 21.2[b]	—
Mavis Grind (contaminated reference)	153.2 ± 28.2 [b]	124.9 ± 13.4[c]

[a] Table adapted from References 62 and 124.
[b] $p < 0.01$, Mann-Whitney U-test comparing with either control data (experimental) or uncontaminated reference — Ronas Voe (field data).
[c] $p < 0.05$.

Many environmental xenobiotics including PAH are known to be sequestered within lysosomes under certain conditions.[48,101-103,118-123] These may, in some circumstances, be accompanied by lysosomal damage.[118-123] Exposure of marine mollusks to PAH and petroleum hydrocarbons has been demonstrated to result in increases in the activities of certain lysosomal enzymes, notably, β-glucuronidase and acid phosphatase (Table 4).[62,124] A consequence of this pattern would be to prepare the cell for the degradation of particular macromolecules, hence, making the products available for maintenance of the cell.[104,125]

Other lysosomal contents such as lipofuscin have also been observed to accumulate following exposure to PAH or petroleum hydrocarbons in several species of mollusks.[126-128] Lipofuscin or aging pigment is a product of free-radical peroxidative reactions derived from the autophagy of lipoprotein membranes.[129] In molluskan digestive cells this pigment accumulates in residual bodies or tertiary lysosomes and can be excreted by exocytosis.[107] Exposure of *L. littorea* to PHN has resulted in an elevated accumulation of lipofuscin in these cells, although this material is rapidly lost following removal of the injurious stimulus (Figure 5); this is probably indicative of enhanced autophagy.[126,127] The consequences of this accumulation are not fully understood, although the indications of elevated production of free radicals within the lysosomal environment can hardly be considered as beneficial.

The regulation of the lysosomal system is dependent on controlled fusion with other components of the vacuolar system such as phagosomes, primary lysosomes, and the plasma membrane.[104,105,117] The digestive cells of mollusks are largely concerned with heterophagy and the digestion of food material.[107] Disturbances of the fusion processes involved could have marked consequences for the nutritional status of the organism by perturbing ''normal'' intracellular digestion and the balance of autophagy to heterophagy.[48]

There are a number of indications that experimental exposure to PAH such as PHN as well as petroleum hydrocarbons in both the field and laboratory, induces profound alterations in the rate of fusion events in the lysosomal-vacuolar system of the digestive cells.[13,62,130,131] Ultrastructural studies show that the large secondary lysosomes (2 to 5 μm in diameter, approximately) in the digestive cells show marked increases in the presence of internalized

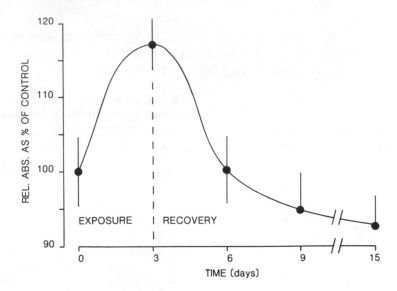

FIGURE 5.   Effect of PHN exposure and a subsequent recovery period on the
lysosomal concentration of lipofuscin in the digestive cells of *L. littorea*. Lipofuscin concentration was determined in tissue sections using scanning microden-
sitometry as Relative Absorbance (REL. ABS) of cytochemical reaction product
(Schmorl reaction). Error bars indicate ± S.E.M. (n = 10). (Adapted from Moore,
M. N., Mayernik, J. A., and Giam, C. S., *Mar. Environ. Res.,* 17, 230.)

membrane-bound vesicular components and that these secondary lysosomes become abnormally enlarged (up to 15 μm in diameter) in mussels, oysters, and periwinkles.[127,130,132]
Quantitative image analysis of these enlarged lysosomes demonstrates that both lysosomal
volume and surface area within the cells is significantly increased, while numerical density
is decreased (Figure 6).[130] This is perhaps indicative of fusion of vacuolar components to
produce these abnormally enlarged lysosomes. This type of response has also been observed
to occur when mussels are exposed to an abrupt increase in salinity and has been linked to
increased fusion of lysosomal vacuoles and autophagy.[133] These alterations are also associated
with elevated intracellular protein catabolism and formation of amino acids as measured
within the lysosomal cellular compartment.[134] The formation of these "giant" lysosomes
has also been linked with atrophy of the digestive tubule epithelium which is largely comprised of digestive cells.[130-133] This latter relationship will be discussed in more detail in a
subsequent section.

The third category of lysosomal disturbance involves membrane permeability. This property can be investigated both biochemically and cytochemically. It is, however, more realistic
to employ cytochemical procedures in molluskan digestive cells, as the large secondary
lysosomes do not readily lend themselves to the trauma of homogenization and subsequent
fractionation.[135] Cytochemical procedures for the determination of lysosomal permeability
are well established for a number of enzyme substrates in mollusks and these are all conceptually based on the Bitensky Fragility Test for lysosomal stability.[13,48,109,136,137] Lysosomal
destabilization is measured as increased permeability to certain enzyme substrates whose
products can be used to give a measurable final cytochemical reaction product.[109,135]

Destabilization of lysosomes has been demonstrated in molluskan digestive cells as a
result of injury by 2-methyl naphthalene, 2,3-dimethyl naphthalene, anthracene, phenanthrene, diesel oil emulsion, water-accommodated fraction of crude oil (North Sea — Auk),
and the petroleum hydrocarbon-contaminated field samples (Table 5).[48,62,98,99,124,130,138] Assessment of this type of injury has been confirmed as an extremely sensitive index of cellular

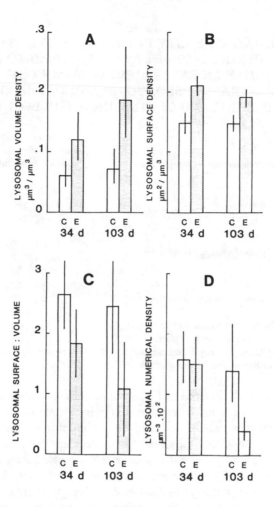

FIGURE 6. Results of stereological analyses of lysosomes in the digestive cells of mussels. Control (C) and experimental (E) mussels exposed to water-accommodated fraction (30 µg l⁻¹ total hydrocarbons) of North Sea crude oil were sampled after 34 and 103 d. (A) volume density; (B) surface density; (C) surface to volume ratio; (D) numerical density. Interval estimates are 95% confidence limits/$\sqrt{2}$. (n = 5). (Adapted from Lowe, D. M., Moore, M. N., and Clarke, K. R., *Aquat. Toxicol.*, 1, 213, 1981. With permission.)

condition and the destabilization of the lysosomal membrane apears to bear a quantitative relationship to the magnitude of the stress response, and this presumably contributes to the intensity of catabolic or degradative effects, as well as to the level of pathological change that results.[13]

The consequences of destabilization of the secondary lysosomal compartment have been investigated in several experimental studies. Subcellular fractions rich in destabilized secondary lysosomes have been shown to contain significantly increased concentrations of amino acids as compared to stable lysosomes.[134] This is indicative of enhanced intralysosomal protein catabolism. Ultrastructural investigations of cells with destabilized lysosomes indicate increased secondary lysosomal volume with evidence of increased autophagy and possible heterophagy of apoptotic cell fragments, thus providing further indications of elevated catabolic activity.[127,133]

A quantitative cytochemical approach such as the measurement of lysosomal permeability or stability, based on substrate penetrability (hydrolase latency), may also provide insight

Table 5
**LYSOSOMAL STABILITY IN THE DIGESTIVE CELLS OF
THE PERIWINKLE (*Littorina littorea*) EXPOSED TO DIESEL
OIL IN AN EXPERIMENTAL MESOCOSM
(SOLBERGSTRAND, OSLO FJORD) AND IN THE VICINITY
OF THE SHETLAND OIL TERMINAL (SULLOM VOE)[a]**

Experimental treatment/sample site	Lysosomal stability (β-glucuronidase labilization period in minutes)[b]
Control	24 (20, 25)
~30 ppb total hydrocarbons ⎫ 16 months	2.6 (2, 5)[c]
~130 ppb total hydrocarbons ⎭	2 (2, 2)[c]
Ronas Voe ⎫(uncontaminated references)	19.0 (15, 20)
Gluss Voe ⎭	24.0 (20, 25)
Scatsta Voe	5.0 (5, 5)[c]
Tanker Jetty 4	3.8 (2, 5)[c]
Mavis Grind (contaminated reference)	8.0 (5, 10)[c]

[a] Table adapted from References 25 and 62.
[b] Mean with data range in parentheses; n = 5.
[c] $p \leq 0.01$, Mann-Whitney U-test comparing with either control data (experimental)
    or data for Ronas Voe (uncontaminated reference).

into the mechanisms of PAH-induced cell injury in mollusks. Caution is required, however, in the interpretation of lysosomal damage as a primary event, when it may, in fact, be a secondary or higher order alteration. Recent investigations of lysosomal responses to specific PAH have demonstrated that the lysosomal disturbances are complex and differ markedly for PAH which are structurally dissimilar, such as the isomeric 3-ring forms anthracene and phenanthrene (Figure 7).[139] The biochemical evidence of relatively low activity for cytochrome(s) P-450 monooxygenase in molluskan digestive gland cells, when considered together with the ability of these cells to accumulate and retain very high concentrations of PAH, indicates that their loss by metabolic transformation is limited.[8] The fact that the secondary lysosomes are often lipid rich, particularly following exposure to PAH and petroleum hydrocarbons,[200] would tend to argue for a direct effect on the lysosomes by these xenobiotics, rather than a secondary effect. Aromatic hydrocarbons and PAH have been shown to penetrate synthetic phospholipid membranes and alter their physical-chemical properties, including membrane fluidity and permeability.[140,141]

The possibility also exists that reactive species could be produced by oxidation of PAH within the lysosomes themselves;[142,143] many molluskan lysosomes contain lipofuscin, which is probably being produced *in situ* by peroxidation processes. These lysosomes are noted for their accumulation of metal ions, including those of iron and copper that are known to be involved in the generation of oxygen radicals.[111,118,129] If reactive species are formed from PAH by such a process, then these could also contribute to lysosomal membrane injury. There is evidence of enhanced formation of lipofuscin-rich lysosomes and residual bodies in digestive cells following exposure of *L. littorea* to phenanthrene[126,127] (Figure 5), although the major factor here is probably increased autophagy of cytoplasmic lipoprotein membranes.[137]

Further evidence of lysosomal destabilization comes from several ultrastructural studies of the effects of PHN on secondary lysosomes in digestive cells.[127,144] These have demonstrated the presence of corrugation of the bounding membrane with possible associated blebbing activity. Increased frequency of membrane breaks has also been described, and

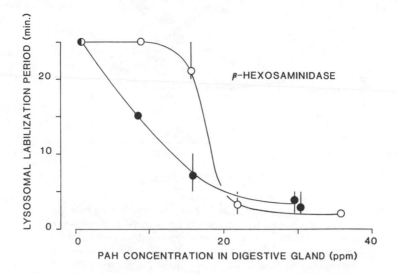

FIGURE 7.   Effects of phenanthrene (○) and anthracene (●) on lysosomal membrane stability measured as the cytochemical labilization period for latent β-*N*-acetylhexosaminidase in digestive cells of *M. edulis*. Concentrations in the digestive gland are μg·g⁻¹ wet wt; error bars indicate data range (n = 5). (Adapted from Moore, M. N. and Farrar, S. V., *Mar. Environ. Res.*, 17, 222, 1985. With permission.)

while these breaks may be artifacts of fixation and tissue processing, their relative infrequency in control lysosomes is indicative of the greater fragility of lysosomes from cells exposed to PHN. Apparent leakage of lysosomal β-glucuronidase has been demonstrated in the case of lysosomes from the digestive cells of PHN-exposed *L. littorea*; extracellular release of lysosomal β-glucuronidase was also observed in these cells.[127] Limited release of lysosomal enzymes into the cytosol and nucleoplasm has been described by Szego[145] following treatment of rat preputial gland cells with 17β-estradiol, which destabilizes the lysosomal membrane.

The consequences of such a release of lysosomal enzymes is uncertain, but is believed to lead to enhanced cell damage and possibly cell death.[104,105] Evidence from rat preputial gland cells indicates increased protein catabolism following estrogen treatment, although enzyme release in these cells in noninjurious and precedes initiation of cell division.[145] This may represent a fundamental difference in the lysosomal response to physiological agonists as opposed to xenobiotics.

In summary, the evidence of both ultrastructural, quantitative cytochemical, and morphometric approaches indicats that PAH and petroleum hydrocarbons induce profound alterations in both structure and function. These involve all three categories of lysosomal response and there are some grounds for suggesting that membrane destabilization may represent the primary injury which could lead to the other events described above. Cytochemical demonstration of lysosomal membrane destabilization has proved to be a useful investigative tool both for PAH and other injurious agents. That this procedure does, in fact, measure membrane destabilization is further supported by evidence of reversibility and restabilization by treatment with hydrocortisone, an established membrane stabilizer.[109,134]

C. Structural Alterations in Cells

Exposure of marine mollusks to single PAH and petroleum hydrocarbons results in atrophy of the epithelium of the digestive tubules.[130-132] This atrophy or epithelial "thinning" involves structural changes in the digestive cells, the major component of the epithelium. These changes have been quantified in *M. edulis* using image analysis of histological sections (Figure 8),[130] and in the clam, *M. mercenaria* using morphometry.[146] There is evidence in

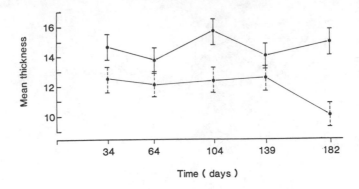

FIGURE 8.   Morphometric analysis of digestive cells showing the effects of the water-accommodated fraction (30 µg l^{-1} total hydrocarbons) of North Sea crude oil on mean epithelial thickness (arbitrary units). Interval estimates are 95% confidence limits (n = 5); (————); controls; (----) oil exposed. (Adapted from Lowe, D. M., Moore, M. N., and Clarke, K. R., *Aquat. Toxicol.*, 1, 213, 1981. With permission.)

both mussels and periwinkles that this may be a generalized response to toxic xenobiotics and other stressors such as starvation.[12,62,130,133]

Investigation of digestive cell atrophy has revealed that there is a significant increase in lysosomal volume as described in the previous section.[130,131] This increase in volume of the lysosomal compartment involved the formation of enlarged or "giant" lysosomes and this alteration is associated with membrane destabilization and increased permeability.[131] There is also evidence for increased lysosomal fusion events leading to the formation of the enlarged lysosomes.[127,133] As discussed in the preceding section, the consequences of these lysosomal disturbances would be increased autolytic and autophagic activity presumably leading to atrophy of the digestive cells.

Physiological investigation of mussels has shown that scope for growth is significantly reduced (and in some cases becomes negative) following exposure to PAH and petroleum hydrocarbons.[138,147] This situation is indicative of relatively enhanced tissue catabolism. Samples from these experiments demonstrate digestive cell atrophy and lysosomal disturbances as described above, arguing strongly for a mechanistic link from the lysosomal events through to the whole animal.[130,131,138]

Turning to considerations of effects on reproduction at the cellular level, mussels exposed to both 30 and 130 ppb diesel oil showed a reduction in the volume of storage cells in the mantle tissue, a reduction in volume of ripe gametes, and increased degeneration or atresia of oocytes (Table 6).[148] These data indicate a direct impairment of the reproductive processes and the implication is that reproductive capability would be reduced, both by degeneration of oocytes and reduction in ripe gametes, as well as by a reduction in the energy reserves available for gametogenesis as supplied by the connective tissue storage cells.[112] Ultrastructural investigations designed to explore the mechanisms of oocyte degeneration have revealed that degradative lysosomal enzymes are associated with yolk granules and with pinocytotic phenomena that occur along the basal membrane of developing oocytes.[114] Lysosomal enzymes are also associated with the degradation (atresia) and resorption of oocytes, as well as the resorption of adipogranular storage cells.[112-114] Future experiments will test whether PAH have a detrimental effect on oocyte lysosomes leading to enhanced lysosomal autolytic processes.

D. Chromosomal Damage

Understanding the biological consequences of injury to the genome by environmental chemicals is of prime importance, particularly where this damage occurs in the germinal

Table 6
REPRODUCTIVE TISSUE IN THE MUSSEL: EFFECTS OF HYDROCARBONS AND A RECOVERY PERIOD ON TISSUE VOLUMES OF THE COMPONENT CELLS[a]

Condition	Developing gametes	Ripe gametes	Adipogranular cells	Vesicular cells	Degenerating oocytes
Control	0.10 ± 0.03[b]	0.77 ± 0.27	0.11 ± 0.03	0.53 ± 0.07	0.05 ± 0.01
~30 ppb hydrocarbons	0.13 ± 0.04	0.44 ± 0.16[c]	0.05 ± 0.04[c]	0.38 ± 0.07[c]	0.35 ± 0.13[c]
~130 ppb hydrocarbons	0.13 ± 0.03	0.30 ± 0.11[c]	0.01 ± 0.01[c]	0.14 ±0.02[c]	0.31 ± 0.09[c]
Recovery 53 d	0.28 ± 0.08[c]	0.66 ± 0.18	0.11 ± 0.05	0.54 ± 0.04	0.01 ± 0.01 [c]

[a] Table adapted from Reference 148.
[b] Mean (mm^3) ± S.E.; n = 10.
[c] $p < 0.05$; one-way analysis of variance, comparing with control.

cells or gametes. Recent cytogenetic research has demonstrated damage at the chromosomal level of organization in mussel embryos obtained from adults naturally exposed to high concentrations of PAH and other xenobiotics in a polluted shipping dock.[85] This chromosomal damage comprised abnormal increases or decreases in chromosomal complement (aneuploidy). The incidence of abnormality was greater than three times the normal "spontaneous" incidence (Table 7).[85] Although these results are highly suggestive of a pollutant-related effect, a direct causal relationship was not established. Chromosomal damage has been demonstrated previously in cells following lysosomal injury and release of hydrolytic enzymes including acid DNAase.[101] Molluskan eggs and embryos are rich in lysosomes,[111,114] as stated in an earlier section, and although the evidence is still circumstantial, this might represent a possible mechanism for the chromosomal damage in embryos obtained from hydrocarbon-exposed adult mussels. Further investigation, however, would be required to establish a causal link.

IV. EFFECTS OF HYDROCARBONS ON INDIVIDUALS, POPULATIONS, AND COMMUNITIES

A. Physiological Responses

Production of matter (growth and reproduction) is a fundamental property of all living organisms and one that is necessary if a population is to persist in a given environment. The amount of production represents the difference between an individual's or a population's intake and output of matter or energy, and this will vary under different environmental conditions. The measurement of individual physiological responses, such as rates of feeding, digestion, respiration, excretion, and growth, and their integration by means of physiological energetics, can provide insight into the overall growth process and how it might be disrupted by environmental stress and pollution.

1. Feeding

A reduction in feeding rate or energy acquisition by mollusks exposed to sublethal concentrations of petroleum hydrocarbons appears to be the primary determinant of a depressed growth rate.[138,149] The effect of hydrocarbons on feeding rate has been documented for *C. virginica* as a reduction in the rate of feces and pseudofeces production;[150] for *M. edulis*[138,151,152] and *M. mercenaria*[153] as a decline in the rate of clearance of food particles from suspension; and for *Macoma balthica*,[154] *Nassarius obsoletus*,[155] and *Thais lima*[149] as a reduction in feeding activity. A depression of feeding rate is known to occur at low and environmentally realistic concentrations of 30 to 40 µg petroleum hydrocarbons per liter.[138,149,150]

Table 7
FREQUENCIES OF CHROMOSOMALLY NORMAL
AND ABNORMAL EMBRYOS OF *MYTILUS EDULIS*
ORIGINATING FROM KING'S DOCK
(CONTAMINATED) AND WHITSAND BAY (CLEAN
REFERENCE)

Site	n	Normal	Aneuploid	Haplo/polyploid
King's Dock	50	30	13	7
Whitsand Bay	50	41	4	5

Note: $p < 0.05$ χ^2 test.

Adapted from Dixon, D. R., *Mar. Biol. Lett.*, 3, 155, 1982. With permission.

The inhibition of feeding probably results from the narcotic effect of hydrocarbons, particularly aromatic hydrocarbons, which may have a direct action on cilia and muscles[156] and/or the nervous system which controls such activity.[157] Recent studies have demonstrated that aromatic hydrocarbons and some alkanes produce a narcotic effect on ionic currents and, thus, the functioning of the squid giant axon.[157,158] The narcotization effect of hydrocarbons is also reflected in some behavioral responses of mollusks, such as a reduction in byssal attachment by *M. edulis*[138,159,160] at 130 µg l^{-1} and impairment of burrowing behavior of infaunal bivalves such as *M. balthica* in response to 70 and 300 µg petroleum hydrocarbons per liter.[154,160] These effects on behavior may then make them more vulnerable to predation.

In addition to the reduction in feeding and ingestion rate there is a decline in the efficiency of food absorption by *M. edulis*[138,152,161] and *M. mercenaria*.[153] Impaired food absorption is probably the result of hydrocarbon impact on the structure and functioning of the digestive epithelial cells in the digestive gland (see Section III). Epithelial thinning or tubule dilation in response to hydrocarbons has been reported in *Mya arenaria*[162] and in *M. edulis*.[130] This response is accompanied by a loss of synchrony between digestive tubules and a decline in the number of secondary lysosomes and an increase in their volume. These cellular changes in the digestive gland are due, in part, to the increased mobilization of stored energy reserves for body maintenance during periods of stress.

2. Respiration
An enhanced rate of oxygen consumption, or energy expenditure, appears to be a common response of mollusks (e.g., *M. edulis*,[138,151] *M. arenaria*,[161,163,164] *M. balthica*,[154] *L. littorea*,[165] *T. lima*[149]) to low/moderate concentrations of petroleum hydrocarbons (i.e., ~30 to 600 µg l^{-1}), but it is not always apparent.[152] At high hydrocarbon concentrations (i.e., >1 mg l^{-1}) respiration rates may be depressed, mainly as a result of partial valve closure or a narcotization effect on ciliary activity, both of which will reduce ventilation rate and, thus, oxygen availability (*M. edulis*,[151,166] *M. californianus*,[167] *M. arenaria*[168]).

The fundamental cause of an enhanced rate of oxygen consumption in response to low hydrocarbon concentrations remains unknown, but it appears to be a direct effect of petroleum hydrocarbons on metabolism and not mediated through behavioral changes.[138,154] Hydrocarbons may increase respiration rate as a result of an uncoupling of oxidative phosphorylation[168] or an increased flux through the glycolytic pathway.[138] It is unlikely, however, that the excess oxygen consumption is due to the activity of the P-450-linked MFO system, because the oxygen requirement, based on the maximum rate of BaPH activity in *M. edulis*, accounts for only a fraction of 1% of the total oxygen consumption.[201]

3. Excretion
The rate of ammonia excretion is normally closely related to the metabolic rate of an

animal. This close relationship appears to be unaffected by exposure to low/moderate concentrations of petroleum hydrocarbons, increasing and decreasing with respiration rate and, thus, maintaining a constant O:N ratio in response to hydrocarbon contamination.[138,149] However, in response to very high concentrations (>1 mg l^{-1})[149] or after long-term (>8 months) exposure to moderate concentrations (130 μl^{-1})[159] there is an elevated rate of ammonia excretion reflecting enhanced utilization of protein reserves for maintenance and survival. However, in the context of energetics, the energy equivalents of these excretory losses are small relative to respiratory energy expenditure.

In addition to the normal excretion of nitrogenous waste products there are excretory processes concerned with the elimination of toxicants such as PAH and their metabolites from animals. The kidneys and gills have been shown to be actively involved in the elimination of nahthalene (NPH) from the body tissues of *M. edulis* and this excretory process is enhanced following chronic exposure to NPH.[169] The kinetics of hydrocarbon uptake and elimination by aquatic animals and the bioconcentration factor between water and tissues are a function of the hydrocarbon's structure and its hydrophobicity. For example, the rate of uptake and loss of mono- and di-aromatic hydrocarbons are rapid (i.e., minutes to hours) to reach a steady-state tissue concentration, whereas the rates of uptake and loss of 3- and 4-ring aromatic hydrocarbons are slow (i.e., days to weeks to attain an equilibrium between tissue and water concentrations).[169-172]

4. Growth

The ultimate effect of petroleum hydrocarbons on feeding (energy acquisition) and metabolism (energy expenditure) is to markedly reduce the energy available for growth and reproduction, often termed "scope for growth".[138,149,173] Concentrations as low as 30 μg petroleum hydrocarbons l^{-1} and within the range found in, for example, the outer Thames estuary,[174] produced a significant reduction in the scope for growth of *M. edulis*[138] and in the shell growth of *M. balthica*.[154] Similarly, juvenile *M. mercenaria* exposed to 60 μg l^{-1} had a reduced growth rate due to depressed feeding and food assimilation.[153]

The effects of petroleum hydrocarbons observed in laboratory experiments are also apparent in mesocosm experiments and in the field. For example, *M. arenaria* subjected to oil spills showed a reduction in the energy available for growth[161,164] and a decline in tissue and shell growth[175,176] which persisted for up to 5 years after the spill due to the continued presence of petroleum hydrocarbons in the sediments.

5. Recovery of Physiological Responses from Chronic Oil Exposure

Although there have been many studies of the effects of hydrocarbons on bivalves, until recently there has been relatively little information concerning the rate of recovery from oil exposure and the extent to which physiological recovery is related to the depuration of petroleum hydrocarbons from the body tissues. A recent study[152] has shown a marked reduction in the feeding rate and scope for growth of *M. edulis* exposed to two concentrations of diesel oil (30 and 130 μg l^{-1}) for 8 months. The high-oil-concentration exposed mussels had a negative scope for growth, indicating the need to utilize body reserves in order to satisfy the animal's energy requirements. This was confirmed by the observed degrowth in body tissues (i.e., utilization of body reserves) which resulted in high mortality in this group (27% over 8 months).[177] During recovery from chronic diesel oil exposure, the depuration of PAH (2- and 3-ring compounds) from the mussel's body tissues was concomitant with the recovery of physiological performance (i.e., feeding and growth), thus, demonstrating that the effect of diesel oil on the mussel's performance is related to the concentration of PAH within the body tissues and is not directly related to the hydrocarbon concentration in the water.

High-oil-concentration exposed mussels were found to recover more rapidly than the low-oil-concentration mussels, both in terms of PAH depuration and scope for growth (Figure 9).[152] This led to an "overshoot" in the feeding rate and, consequently, the scope for growth by high-oil-concentration mussels and, thus, accounted for the observed "catch-up" growth in body tissues during the two months of recovery from high-oil-concentration exposure (130 μg l^{-1}). The more rapid depuration and recovery of high-oil-concentration exposure mussels was probably the combination of physiological adaptation and/or selection in terms of individuals with enhanced rates of excretion and metabolism of PAH, and a "dilution effect" due to enhanced tissue growth (catch-up growth). The rate of recovery by low-oil-concentration exposed mussels was slower both in terms of tissue depuration and physiological performance, but both groups showed complete recovery after 55 d (i.e., their growth rates were not significantly different from control).

6. Relationship between Growth and Tissue Hydrocarbon Concentrations

Several field and laboratory studies have demonstrated a significant negative correlation between scope for growth and the concentration of specifically aromatic hydrocarbons in the tissues of mollusks (*M. arenaria*,[161] *M. edulis*,[138,179] *T. lima*[149]). Figure 10 provides a synthesis of data derived from recent mesocosm experiments (both oil exposure and recovery phases — triangles)[152] and field studies (mussel populations in the vicinity of Sullom Voe oil terminal, Shetlands — diamonds)[178] and illustrates the significant negative correlation ($r^2 = -0.95$) between scope for growth and $\log_{10}$ of the concentration of 2- and 3-ring aromatic hydrocarbons in the body tissues of *M edulis* (this simply reflects the nature of the analytical procedure rather than identifying these aromatic hydrocarbons as the sole-toxic components). Such a relationship demonstrates that petroleum hydrocarbons affect scope for growth over a wide range of tissue concentrations without an apparent threshold concentration of effect. Furthermore, it illustrates the degree of contamination in the "control" mussels in the mesocosm experiment conducted at Solbergstrand on the Oslofjord (Norway) relative to the Shetland Islands (U.K.).

It is generally very difficult to make interstudy comparison of the relationship between biological effects measurements and tissue hydrocarbon concentrations because of the lack of compatibility, particularly in the chemical data. This is primarly due to the different extraction and analytical procedures adopted by different laboratories, which are usual' semiquantitative, emphasizing either the lower or higher molecular weight fractions of aliphatic or aromatic hydrocarbons. A broad synthesis of data such as that achieved in Figure 10, where there is consistency in both biological and chemical methodology, will not be possible until further standardization of procedures for the quantification of toxic hydrocarbons of low and high molecular weight has been agreed.

7. Reproduction

Reproductive processes provide the critical interface between the individual and the population. Furthermore, the results of laboratory studies have indicated that stress, measured in terms such as scope for growth, has a significant effect on fecundity, egg size, and subsequent growth rate of thc larvae of *M. edulis*.[180] The reduction in energy available for growth and reproduction in response to diesel oil exposure (30 and 130 μg l^{-1}) has been shown to result in a significant and concentration-dependent decline in both the mass of storage tissue in *M. edulis* and the mass of gametes produced (fecundity).[148,177] *M. balthica* exposed to 30 μg petroleum hydrocarbons l^{-1} also showed gamete resorption and abnormal gamete development at 300 μg l^{-1}.[154]

Therefore, a decline in the overall energy available for production as a response to petroleum hydrocarbons results in a reduction in the energy allocated to reproduction (reproductive effort), which when combined with an increased mortality ultimately leads to a

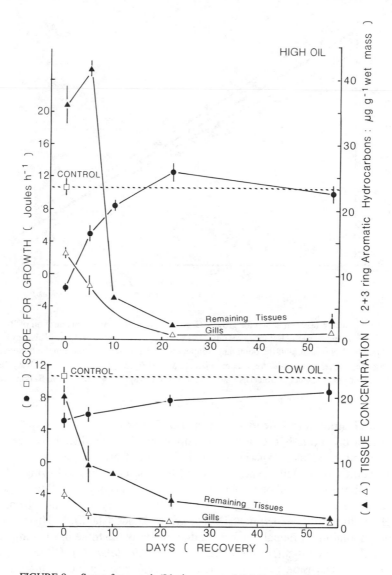

FIGURE 9.   Scope for growth (J h^{-1}; mean ± S.E.M.) of *M. edulis* (●) during recovery from 8 months of exposure to ~ 130 μg 1^{-1} (high oil) and ~ 30 μg 1^{-1} (low oil). Short-term recovery (0, 5, and 10 d; May 1983) and long-term recovery (0, 22, and 55 d; May 1984) expressed relative to control mussels (no significant differences between controls in 1983 and 1984). Depuration of 2- and 3-ringed aromatic hydrocarbon (μg g^{-1} wet mass) from gills (△) and remaining tissue (▲). Mean ± range of 2 pooled samples.[152]

reduction in the residual reproductive value[180] or the long-term prospects of producing successful offspring in future years.

B. Population Effects

The damaging effects of petroleum hydrocarbons on cellular and physiological responses, which are ultimately reflected in the growth, reproduction, and survival of the individual, suggest a major reduction in population fitness and the ability to maintain the population or colonize new habitats, as well as increased susceptibility to predation, parasitism, and disease. Such population effects have been observed in both field (*M. arenaria*)[161,175] and mesocosm studies (*M. edulis*)[152,182] in the form of reduced recruitment, increased mortality, and reduced production.

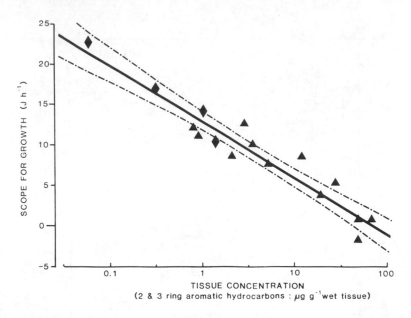

FIGURE 10.   Relationship between scope for growth (J g^{-1} dry mass h^{-1}) and concentration of 2- and 3-ringed aromatic hydrocarbons (μg^{-1} wet mass; mean ± range) in the body tissues of *M. edulis*. (This simply reflects the nature of the analytical procedure and the dominant component of the accumulated arromatic hydrocarbons rather than identifying these PAH as the sole toxic components). Mean ± 95% C.I; r = −0.95. (▲) Data from oil exposure and recovery experiments at Solbergstrand, Oslofjord, Norway.[152] (◆) Data from Sullom Voe, Shetland Islands, U.K.[178]

C. Population/Community Effects — Mesocosm Studies

The effects of petroleum hydrocarbons on populations/communities have been studied in two major benthic mesocosm experiments with the aim of bridging the gap between laboratory-based short-term toxicity experiments and field observations following oil pollution, and examining the effects of oil on trophic interactions in the ecosystem. Benthic soft-bottom communities were exposed to 190 μg l^{-1} (or 109 μg g^{-1} dry sediment) of no. 2 fuel oil for 25 weeks in the MERL experimental ecosystems (Rhode Island).[181] There was a significant decline in total macrofaunal numbers and particularly the three most common species, which included two mollusks (*Nucula annulata* and *Yoldia limatula*). Oil had a particularly marked effect on suspension feeders. In addition, the metazoan meiofauna (harpacticoids and ostracods) showed a drastic reduction in numbers, whereas the unicellular meiofauna (Foraminifera and large ciliates) increased in the oil tank probably due to decreased predation and competition. After less than 2 months without oil additions there was a strong recovery of meiofauna, but macrofauna with a longer generation time showed no signs of recovery within 2 months.

Rocky shore communities established in large concrete basins were exposed to two concentrations (30 and 130 μg l^{-1}) of diesel oil.[182] The highest concentration resulted in the total extinction of mussel and amphipod populations over a period of a year and also reduced population growth of barnacles and fucoids. The mussel population in the low-oil-concentration basin was also reduced, whereas other species showed population changes that may be related to oil pollution, but these were not severe enough for them to be distinguishable from control population variability.[182]

The results of mesocosm studies[152,182] have largely confirmed the conclusions of earlier laboratory studies,[138,154] that bivalves, including mussels, are sensitive to petroleum hydrocarbons when chronically exposed. A diesel oil concentration of 130 μg l^{-1} reduced feeding rate and scope for growth of *M. edulis* to such an extent that during long-term exposure

(>8 months) there was degrowth which ultimately resulted in mortality. However, their predators (*Asterias rubens* and *Carcinus maenas*) were both more tolerant of 130 μg l^{-1} and only declined in numbers after the decline in the mussel population. These findings are contrary to predictions made in a recent study[183] which stated that *M. edulis* was apparently not sensitive to petroleum hydrocarbons, because high concentrations (3 to 1000 mg l^{-1}) were required to kill mussels in 96-h LC$_{50}$ toxicity experiments.[184] This highlights the limitations and dangers of extrapolating from 96-h LC$_{50}$ data to the effects of chronic exposure to low hydrocarbon concentrations, especially when animals such as mussels can close their shell valves for long periods, tolerate weeks of anoxia, and survive zero food intake and undergo degrowth for >6 months. Furthermore, the authors[183] stated that the predatory starfish, *Evasterias troschelii* was more sensitive than mussels, based on 28-d LC$_{50}$ data and sublethal effects on feeding rates down to 200 μg^{-1}, and concluded that chronic oil pollution is likely to reduce predation by *Evasterias,* thus permitting *Mytilus* to expand and monopolize the lower intertidal zone. These conclusions, inferred from short-term LC$_{50}$ studies, are therefore contrary to the findings of long-term (up to 18 months) chronic oil exposures in mesocosm experiments and serve to emphasize the value of such an experimental approach.

D. Community Effects

The ecological consequences of changes in populations will be a shift in species composition and diversity within the community. The effects of petroleum hydrocarbons on community structure have been demonstrated in many field studies (e.g., effects of an oil spill,[185] petrochemical industrial waste,[186] and offshore oil platforms[187]).

The community response to oil pollution is a classical successional response to a point-source of organic pollution.[188] There is a gradient of increasing species diversity, beginning with an abiotic zone in the immediate vicinity of a point-source of heavy contamination, followed by a zone characterized by a low number of species, low abundance, and low biomass, and then a zone showing a progressive increase in the number of species and a dramatic increase in abundance and biomass of a few dominant species. Finally, with increasing distance from a discharge, the community becomes more diverse, but abundance and biomass decline.

In relation to North Sea oil activities there is significant offshore contamination of sediments close to platforms using oil-based drilling muds.[189] A review of the environmental effects of these oil-contaminated sediments around oil platforms[187] has concluded that, beyond the immediate area of physical smothering by oil-based mud cuttings, the major deleterious effects on the benthic community (e.g., low species diversity and high numbers of opportunistic species such as *Capitella* spp.) occurred within 500 m of the platforms, beyond which there was a transition zone, generally within 400 to 1000 m where "subtle" biological effects (e.g., slight changes in species diversity and community structure) could be detected. Species diversity declined progressively with a logarithmic increase in oil concentration in the sediments (Figure 11A), and there was no evidence of a threshold of effect. Where there was quantitative gas chromatography-mass spectrometry (GC-MS) data for two oil fields (Beryl and Thistle), a linear relationship between species diversity and log of NPH concentration in the sediment was also recorded (Figure 11B). It is interesting to note that where hydrocarbon concentratons (regardless of units) span several orders of magnitude above background levels, there is a parallel relationship and good agreement between biological effects such as scope for growth, measured at the individual level of organization using an indicator species, and changes in species diversity at the community level. This supports the notion that adverse effects measured at the cellular and individual levels ultimately manifest themselves at the population and community levels of organization. Cellular and physiological responses, therefore, serve to complement ecological surveys of community structure, not only by providing an early detection of effects, often before statistical significant changes occur at the community level, but also by providing some insight into the toxicological causes of changes in population and community structure. A particularly important attribute of sublethal cellular and physiological responses is that they are amenable to both laboratory and field measurement, unlike traditional toxicity testing based on LC$_{50}$ and ecological surveys based on community structure. On the one hand, they can be used to derive "dose-response" relationships in order to help identify and predict the effects of potential pollutant levels; on the other, they can be used to monitor the effects in the environment.

Oil pollution damage to temperate coastal ecosystems is considered to be reversible in about 2 to 5 years,[190,191] but more subtle effects may persist for at least a decade.[185] However, the time-scale may be very different for polar or tropical regions and for ecosystems such as coral reefs, mangroves, and salt marshes. The rate of recovery is primarily dependent on the rate of biodegradation, the rate of physical dispersion in a high- or low-energy environment, the nature of the community structure, and the generation time of important macrofauna species.

V. CONSEQUENCES FOR HUMANS

Accumulation of PAH in the tissues of marine mollusks includes a number of those that are potentially carcinogenic.[5,6,44,192] The significance of transfer of PAH to humans from this source is not well understood; however, recent reviews by Neff[5] and Stegeman[44] indicate that consumption of PAH-contaminated mollusks probably constitutes a relatively minor component of dietary PAH, in comparison to PAH in smoked foods, charcoal-broiled meats, and even many vegetables.[5,44] The production and retention of reactive or potentially carcinogenic metabolites of PAH in molluskan tissues have had little assessment;[19,45,46,87] this is mainly due to the low activities associated with the cytochrome P-450 monooxygenase system and a virtual absence of data on the *in vivo* formation of PAH metabolites.[8,44,46]

Existing data indicate that PAH in marine mollusks are unlikely to represent a major human health hazard, except possibly where animals have been exposed to very high concentrations of PAH such as those occurring following an oil spill.[5] However, there is substantial evidence for the accumulation of chemically unidentified mutagens in *M. edulis*,[86] and the identification of these mutagens and their assessment as a health hazard remain a pertinent problem.

VI. CONCLUSIONS AND PROSPECTUS

The value of using an integrated approach to the problem of assessment of sublethal effects of PAH in mollusks is clearly evident from the preceeding sections, particularly in facilitating the development of conceptual linking of responses at several levels of biological organization. Study of the responses to PAH and petroleum hydrocarbons has extended our understanding of certain aspects of physiological functions and has also highlighted the capacity for PAH to disturb biochemical and cellular processes, particularly those involving the lysosomes.

In the case of cytochrome P-450 in mollusks, our understanding of the role of this system in the "normal" physiology, such as the hydroxylation of endogenous substrates, is even less than its presumed role in processes of xenobiotic biotransformation. There is clearly considerable scope for future research into the functions of cytochrome P-450 and the regulation of these processes, from a fundamental biochemical standpoint as well as an evolutionary one. Although lysosomal function in mollusks is more clearly understood, there are still considerable gaps in our knowledge, as is also the case in mammals. Regulation of lysosomal catabolism is obviously critical in maintaining cellular homeostasis and this is

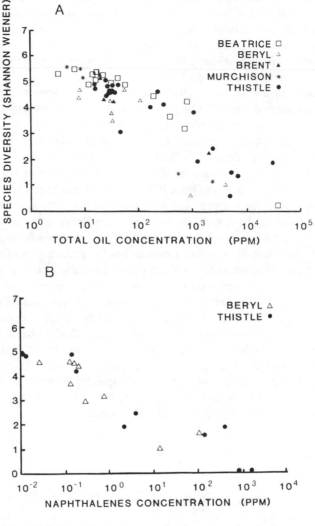

FIGURE 11. (A) Relationship between species diversity and total oil concentration in the sediments of five North Sea oilfields. (B) Relationship between species diversity and total NPH concentration in the sediments of two North Sea oil fields.

evidently perturbed by exposure to PAH. The nature of these control mechanisms, which often involve membrane-membrane interactions, is complex and unclear.[115] Lysosomally related processes involve endocytosis, phagosome-lysosome interaction, autophagy, and exocytosis.[193] In each situation, lysosomes must fuse with discrete compartments of the cytoplasm in order to carry out their function, and this appears to be dependent on the composition of the membranes involved.[193] The regulation of lysosomal function and the effect of membrane disturbance by PAH clearly represent an important area for future investigation.

There is also considerable scope for the investigation of the underlying processes involved in the uptake, loss, and bioaccumulation of PAH in tissues. A better understanding of these processes and the mechanism of PAH toxicity should provide a toxicological interpretation of the chemical data concerning complex tissue residues of PAH in mollusks.

Prediction of ecological consequences is probably one of the most difficult problems facing those involved in biological effects measurements; however, this does not invalidate the use of indices of biological effects as an ''early warning system'' for detection of environmental deterioration. The rationale for the development of such environmental safeguards is that it is first essential to understand the mechanisms of toxicity and how the animal responds to toxic insult. Once this is established by laboratory experimentation, it then becomes possible to develop effects indices based on the biochemical and cellular reactions and physiological responses. In this context, certain marine mollusks such as mussels (*M. edulis*) and periwinkles (*L. littorea*) have been found to be highly responsive indicator species in investigations of pollutant toxicology.[8,13,14]

A major difficulty in the use of marine mollusks in environmental impact assessment involves our current limited understanding of the mechanisms of toxicity of PAH in these animals; consequently, this raises problems in the interpretation of the significance or specificity of any biological responses measured, whether at the molecular, cellular, or whole-animal levels of organization. To attempt to draw parallels with mechanisms in mammalian toxicology can be misleading and may be of limited value. A further complicating factor arises from the fact that contamination from PAH seldom occurs in environmental isolation, as has already been mentioned above, and this therefore poses the question of interactive biological effects resulting from multiple xenobiotic challenge.[9,194] Indices of biological effects must also be capable of resolving such a situation, particularly those indices which are regarded as being relatively specific. Progress has been made in a number of research areas relating to these problems and details are emerging of the mechanisms and processes involved in the toxicity and detoxication of PAH and other contaminants, as well as the related subcellular, cellular, and physiological responses to PAH-induced injury.[8,13,14]

We have not attempted to be comprehensive in this chapter and, hence, due to the selective nature of the coverage, there are several areas which have either not been discussed or only mentioned briefly. These include such questions as the relationship of PAH to neoplasia in mollusks,[195-197] effects of PAH on metabolic functions other than the cytochrome P-450 system, and possible disturbances of immunological capability. Omission of these areas does not mean that we consider them to be unimportant; it is rather that only limited data are available and, hence, our understanding is very restricted.

REFERENCES

1. **Goldberg, E. D., Bowen, V. T., Farrington, J. W., Harvey, G., Martin, J. H., Parker, P. L., Risebrough, R. W., Robertson, W., Schneider, E., and Gamble, E.,** The mussel watch, *Environ. Conserv.,* 5, 101, 1978.
2. **Bayne, B. L., Livingstone, D. R., Moore, M. N., and Widdows, J.,** A cytochemical and biochemical index of stress in *Mytilus edulis* (L.), *Mar. Pollut. Bull.,* 7, 221, 1976.
3. **Bayne, B. L., Brown, D. A., Burns, K., Dixon, D. R., Ivanovici, A., Livingstone, D. R., Lowe, D. M., Moore, M. N., Stebbing, A. R. D., and Widdows, J.,** *The Effects of Stress and Polluton on Marine Animals,* Praeger Scientific, New York, 1985.
4. **Guerin, M. R.,** Energy sources of polycyclic aromatic hydrocarbons, in *Polycyclic Hydrocarbons and Cancer,* Vol. 1, Gelboin, H. V. and Ts'O, P. O. P., Eds., Academic Press, New York, 1978, 3.
5. **Neff, J. M.,** *Polycyclic Aromatic Hydrocarbons in the Aquatic Environment: Sources, Fates and Biological Effects,* Elsevier, London, 1979.
6. **Santodonato, J., Howard, P., and Basu, D.,** Health and ecological assessment of polynuclear aromatic hydrocarbons, *J. Environ. Pathol. Toxicol.,* 5, 1, 1981.
7. **Bresnick, E.,** The molecular biology of the induction of the hepatic mixed function oxidases, in *Hepatic Cytochrome P-450 Monooxygenase System,* Schenckman, J. B. and Kupfer, D., Eds., Pergamon Press, Oxford, 1982, 191.
8. **Livingstone, D. R.,** Responses of the detoxication/toxication enzyme systems of mollusks to organic pollutants and xenobiotics, *Mar. Pollut. Bull.,* 16, 158, 1985.
9. **Malins, D. C. and Collier, T. K.,** Xenobiotic interactions in aquatic organisms: effects on biological systems, *Aquat. Toxicol.,* 1, 257, 1981.

10. **Readman, J. W., Mantoura, R. F. C., and Rhead, M. M.,** The physicochemical speciation of polycyclic aromatic hydrocarbons (PAH) in aquatic systems, *Fresenius Z. Anal. Chem.,* 319, 126, 1984.

11. **Varanasi, V., Reichert, W. L., Stein, J. E., Brown, D. W., and Sanborn, H. R.,** Bioavailability and biotransformation of aromatic hydrocarbons in benthic organisms exposed to sediment from an urban estuary, *Environ. Sci. Technol.,* 19, 836, 1985.

12. **Bayne, B. L., Holland, D. L., Moore, M. N., Lowe, D. M., and Widdows, J.,** Further studies on the effects of stress in the adult on the eggs of *Mytilus edulis, J. Mar. Biol. Assoc. U.K.,* 58, 825, 1978.

13. **Moore, M. N.,** Cellular responses to pollutants, *Mar. Pollut. Bull.,* 16, 134, 1985.

14. **Widdows, J.,** Physiological responses to pollution, *Mar. Pollut. Bull.,* 16, 129, 1985.

15. **Lee, R. F., Davies, J. M., Freeman, H. C., Ivanovici, A., Moore, M. N., Stegeman, J., and Uthe, J. F.,** Biochemical techniques for monitoring biological effects of pollution in the sea, in *Biological Effects of Marine Pollution and the Problems of Monitoring,* McIntyre, A. D. and Pearce, J. B., Eds., *Rapp. P.-V. Reun. Cons. Int. Explor. Mer.,* 179, 29, 1980.

16. **Livingstone, D. R.,** Biochemical measurements, in *The Effects of Stress and Pollution on Marine Animals,* Bayne, B. L. et al., Eds., Praeger Scientific, New York, 1985, 81.

17. **Livingstone, D. E.,** Biochemical differences in field populations of the common mussel *Mytilus edulis* L. exposed to hydrocarbons: some considerations of biochemical monitoring, in *Toxins, Drugs and Pollutants in Marine Animals,* Bollis, L., Zadunaisky, J., and Gilles, R., Eds., Springer-Verlag, Berlin, 1984, 161.

18. **Livingstone, D. R.,** General biochemical indices of sublethal stress, *Mar. Pollut. Bull.,* 13, 261, 1982.

19. **Stegeman, J. J.,** Benzo[a]pyrene oxidation and microsomal enzyme activity in the mussel *(Mytilus edulis)* and other bivalve mollusk species from the western North Atlantic, *Mar. Biol.,* 89, 21, 1985.

20. **Livingstone, D. R. and Farrar, S. V.,** Tissue and subcellular distribution of enzyme activities of mixed-function oxygenase and benzo[a]pyrene metabolism in the common mussel *Mytilus edulis* L., *Sci. Total Environ.,* 39, 209, 1984.

21. **Moore, M. N., Livingstone, D. R., Donkin, P., Bayne, B. L., Widdows, J., and Lowe, D. M.,** Mixed function oxygenases and xenobiotic detoxication/toxication systems in bivalve mollusks, *Helgol. Wiss. Meeresunters.,* 33, 278, 1980.

22. **Ade, P., Banchelli Soldanini, M. G., Castelli, M. G., Chiersara, E., Clementi, F., Fanelli, R., Funari, E., Ignesti, G., Marabini, A., Orunesu, M., Balmero, S., Pirisino, R., Ramundo Orlando, A., Silano, V., Viarengo, A., and Vittozzi, L.,** Comparative biochemical and morphological characterization of microsomal preparations from rat, quail, trout, mussel and *Daphnia magna,* in *Cytochrome P-450 Biochemistry, Biophysics and Environmental Implications,* Hietanen, E., Laitinen, M., and Hanninen, E., Eds., Elsevier, Amsterdam, 1982, 387.

23. **Gilelwicz, M., Guillame, J. R., Carles, D., Leveau, M., and Bertrand, J. C.,** Effects of petroleum hydrocarbons on the cytochrome P-450 content of the mollusk bivalve *Mytilus galloprovincialis, Mar. Biol.,* 80, 155, 1984.

24. **Krieger, R. I., Gee, S. J., Lim, L. O., Ross, J. H., and Wilson, A.,** Disposition of toxic substances in mussels *(Mytilus californianus):* preliminary metabolic and histologic studies, in *Pesticide and Xenobiotic Metabolism in Aquatic Organisms,* Khan, M. A. Q., Lech, J. J., and Menn, J. J., Eds., *Am. Chem. Soc. Symp. Ser.,* 99, 259, 1979.

25. **Livingstone, D. R., Moore, M. N., Lowe, D. M., Nasci, C., and Farrar, S. V.,** Responses of the cytochrome P-450 monooxygenase system to diesel oil in the common mussel, *Mytilus edulis* L., and the periwinkle, *Littorina littorea* L., *Aquat. Toxicol.,* 7, 79, 1985.

26. **Livingstone, D. R. and Farrar, S. V.,** Responses of the mixed function oxidase system of some bivalves and gastropod mollusks to exposure to polynuclear aromatic and other hydrocarbons, *Mar. Environ. Res.,* 17, 101, 1985.

27. **Livingstone, D. R.,** Seasonal responses to diesel oil and subsequent recovery of the cytochrome P-450 monooxygenase system to diesel oil in the common mussel, *Mytilus edulis* L., and the periwinkle, *Littorina littorea* L., *Sci. Total Environ.,* 65, 3, 1987.

28. **Suteau, P., Migaud, M. L., and Narbonne, J. F.,** Sex and seasonal variations of PAH detoxication/ toxication enzyme activities in bivalve mollusks, *Mar. Environ. Res.,* 17, 152, 1985.

29. **Kurelec, B.,** Exclusive activation of aromatic amines in the marine mussel *Mytilus edulis* by FAD-containing monooxygenase, *Biochem. Biophys. Res. Commun.,* 127, 773, 1985.

30. **Kurelec, B., Britvic, S., and Zahn, R. K.,** The activation of aromatic amines in some marine invertebrates, *Mar. Environ. Res.,* 17, 141, 1985.

31. **Bend, J. R., James, M. O., and Dansette, P. M.,** *In vitro* metabolism of xenobiotics in some marine animals, *Ann. N.Y. Acad. Sci.,* 298, 505, 1977.

32. **Leakey, J. E. A. and Dutton, G. J.,** Effects of phenobarbital on UDP-glucosyltransferase activity and phenolic glucosidation in the mollusk *Arion ater, Comp. Biochem. Physiol.,* 51C, 215, 1975.

33. **Van Cantfort, J., de Graeve, J., and Gielen, J. E.,** Radioactive assay for aryl hydrocarbon hydroxylase. Improved method and biological importance, *Biochem. Biophys. Res. Commun.,* 79, 505, 1977.

34. **Dehnen, W., Tomingas, R., and Roose, J.,** A modified method for the assay of benzo[a]pyrene hydroxylase, *Anal. Biochem.,* 53, 373, 1973.

35. **Holder, G., Yagi, H., Levin, W., Lu, A. Y. H., and Jerina, D. M.,** Effects of inducers and epoxide hydrase on the metabolism of benzo[a]pyrene by liver microsomes and reconstituted system: analysis by high pressure liquid chromatography, *Proc. Natl. Acad. Sci. U.S.A.,* 71, 4356, 1974.

36. **Livingstone, D. R. and Clarke, K. R.,** Seasonal changes in hexokinase from the mantle tissue of the common mussel *Mytilus edulis* L., *Comp. Biochem. Physiol.,* 74B, 691, 1983.

37. **O'Hara, S. C. M., Corner, E. D. S., Forsberg, T. E. V., and Moore, M. N.,** Studies on benzo[a]pyrene mono-oxygenase in the shore crab *Carcinus maenas, J. Mar. Biol. Assoc. U.K.,* 62, 339, 1982.

38. **Walters, J. M., Cain, R. B., Higgins, I. J., and Corner, E. D. S.,** Cell-free benzo[a]pyrene hydroxylase activity in marine zooplankton, *J. Mar. Biol. Assoc. U.K.,* 59, 553, 1979.

39. **Stegeman, J. J. and Kaplan, H. B.,** Mixed-function oxygenase activity and benz[a]pyrene metabolism in the barnacle *Balanus eburneau* (Crustacea: Cirripedia), *Comp. Biochem. Physiol.,* 68C, 55, 1981.

40. **Balk, L., Meijer, J., Seidegard, J., Morgenstein, R., and DePierre, J. W.,** Initial characterization of drug-metabolizing systems in the liver of the northern pike, *Esox lucius, Drug Metab. Dispos.,* 8, 98, 1980.

41. **Chang, K.-M., Wilkinson, C. F., Hetnarski, K., and Murray, M.,** Aryl hydrocarbon hydroxylase in larvae of the southern armyworm *(Spodoptera eridania), Insect Biochem.,* 13, 87, 1983.

42. **Lee, R. F., Furling, E., and Singer, S.,** Metabolism of hydrocarbons in marine invertebrates: aryl hydrocarbon hydroxylase from the tissues of the blue crab, *Callinectes sapidus* and the polychaete worm, *Nereis* sp., in *Pollutant Effects on Marine Organisms,* Giam, C. S., Ed., Lexington Books, Lexington, Kentucky 1977, 111.

43. **Anderson, R. S.,** Developing an invertebrate model for chemical carcinogenesis: metabolic activation of carcinogens, *Comp. Pathobiol.,* 4, 11, 1978.

44. **Stegeman, J. J.,** Polynuclear aromatic hydrocarbons and their metabolism in the marine environment, in *Polycyclic Hydrocarbons and Cancer,* Vol. 3, Gelboin, H. V. and Ts'O, P. O. P., Eds., Academic Press, New York, 1981, 1.

45. **Anderson, R. S.,** Benzo[a]pyrene metabolism in the American oyster *Crassostrea virginica, EPA Ecol. Res. Ser. Monogr.,* EPA-600/3-78-009,

46. **Anderson, R. S.,** Metabolism of a model environmental carcinogen by bivalve mollusks, *Mar. Environ. Res.,* 17, 137, 1985.

47. **Morgenstern, R., DePierre, J. W., Lind, C., Guthenberg, C., Mannervik, B., and Ernster, L.,** Benzo(α)pyrene quinones can be generated by lipid peroxidation and are conjugated with glutathione by glutathione S-transferase B from rat liver, *Biochem. Biophys. Res. Commun.,* 99, 682, 1981.

48. **Moore, M. N.,** Cytochemical determination of cellular responses to environmental stressors in marine organisms, in *Biological Effects of Marine Pollution and the Problems of Monitoring,* McIntyre, A. D. and Pearce, J. B., Eds., *Rapp. P.-V. Reun. Cons. Int. Explor. Mer,* 179, 7, 1980.

49. **Payne, J. F., Rahimtula, A., and Lim, S.,** Metabolism of petroleum hydrocarbons by marine organisms: the potential of bivalve mollusks, in *26th Annual Meeting of the Canadian Federation of Biological Societies,* Vol. 26 (No. PA-235), 1983.

50. **Dixon, D. R., Jones, I. M., and Harrison, F. L.,** Cytogenetic evidence of inducible processes linked with metabolism of a xenobiotic chemical in adult and larval *Mytilus edulis, Sci. Total Environ.,* 46, 1, 1985.

51. **Masters, B. S. S. and Okita, R. T.,** The history, properties and function of NADPH-cytochrome P-450 reductase, in *Hepatic Cytochrome P-450 Monooxygenase System,* Schenkman, J. B. and Kupfer, D., Eds., Pergamon Press, Oxford, 1982, 343.

52. **Moore, M. N., Livingstone, D. R., Widdows, J., Lowe, D. M., and Pipe, R. K.,** Molecular, cellular and physiological aspects of oil-derived hydrocarbons on mollusks and their use in impact assessment, *Philos. Trans. R. Soc. London Ser. B,* 316, 603, 1987.

53. **Livingstone, D. R., Stickle, W. B., Kapper, M., and Wang, S.,** Microsomal detoxication enzyme responses of the marine snail, *Thais haemastoma,* to laboratory oil exposure, *Bull. Environ. Contam. Toxicol.,* 36, 843, 1986.

54. **Sato, R. and Omura, T.,** *Cytochrome P-450,* Academic Press, New York, 1978.

55. **Lee, R. F., Singer, S. C., and Page, D. S.,** Responses of cytochrome P-450 systems in marine crab and polychaetes to organic pollutants, *Aquat. Toxicol.,* 1, 355, 1981.

56. **Quattrochi, L. and Lee, R. F.,** Purification and characterization of microsomal cytochrome P-450 from Spider Crabs, *Libinia* sp., *Mar. Environ. Res.,* 14, 399, 1984.

57. **Synder, R. and Remmer, H.,** Classes of hepatic microsomal mixed function oxidase inducers, in *Hepatic Cytochrome P-450 Monooxygenase System,* Schenkman, J. B. and Kupfer, D., Eds., Pergamon Press, Oxford, 1982, 227.

58. **James, M. O. and Little, P. J.,** 3-Methylcholanthrene does not induce *in vitro* xenobiotic metabolism in spiny lobster hepatopancreas, or affect *in vivo* disposition of benzo[a]pyrene, *Comp. Biochem. Physiol.,* 78C, 241, 1984.

59. **Schwen, R. J. and Mannering, G. J.,** Hepatic cytochrome P-450-dependent monoxygenase systems of the trout, frog and snake. III. Induction, *Comp. Biochem. Physiol.,* 71B, 445, 1982.

60. **Maemura, S. and Omura, T.,** Drug-oxidising monooxygenase system in liver of goldfish *(Carassius auratus), Comp. Biochem. Physiol.,* 76C, 45, 1983.

61. **Bend, J. R., James, M. O., Little, P. J., and Foureman, G. L.,** *In vitro* metabolism of benzo[a]pyrene by selected marine crustacean species, in *Phyletic Approaches to Cancer,* Harshbarger, J. C., Kondo, S., Sugimura, T., and Takayama, S., Eds., Japan Scientific Societies Press, Tokyo, 1981, 179.

62. **Moore, M. N., Pipe, R. K., Farrar, S. V., Thomson, S., and Donkin, P.,** Lysosomal and microsomal responses to oil-derived hydrocarbons in *Littorina littorea,* in *Oceanic Processes in Marine Pollution — Biological Processes and Waste in the Ocean,* Vol. 1, Capuzzo, J. M. and Kester, D. R., Eds., Krieger Publishing, Melbourne, Florida, 1986, 89.

63. **Srivastava, K. C. and Mustafa, T.,** Arachidonic acid metabolism and prostaglandins in lower animals, *Mol. Physiol.,* 5, 53, 1984.

64. **Srivastava, K. C. and Mustafa, T.,** Formation of prostaglandins and other comparable products during aerobic and anaerobic metabolism of [1-^{14}C] arachidonic acid in the tissues of sea mussels, *Mytilus edulis* L., *Mol. Physiol.,* 8, 101, 1985.

65. **Jarzebski, A.,** Major sterols of bivalve mollusks from the inner Puck Bay, southern Baltic, *Comp. Biochem. Physiol.,* 81B, 989, 1985.

66. **Palmork, K. H. and Solbakken, J. E.,** Distribution and elimination of [9-^{14}C] phenanthrene in the horse mussel *(Modiolus modiolus), Bull. Environ. Contam. Toxicol.,* 26, 196, 1981.

67. **Varanasi, U., Reichert, W. L., Stein, K. E., Brown, D. W., and Sanborn, H. R.,** Bioavailability and biotransformation of aromatic hydrocarbons in benthic organisms exposed to sediment from an urban estuary, *Environ. Sci. Technol.,* 19, 836, 1985.

68. **Riley, R. T., Mix, M. C., Schaffer, R. L., and Bunting, D. L.,** Uptake and accumulation of naphthalene by the oyster *Ostrea edulis* in a flow-through system, *Mar. Biol.,* 61, 267, 1981.

69. **Burton, D. T., Cooper, K. R., Goodfellow, W. L., Jr., and Rosenblatt, D. H.,** Uptake, elimination, and metabolism of ^{14}C-picric acid and ^{14}C-picramic acid in the American oyster *(Crassostrea virginica), Arch. Environ. Contam. Toxicol.,* 13, 653, 1984.

70. **Meyer, T. and Bakke, T.,** The metabolism of biphenyl. V. Phenolic metabolites in some marine organism, *Acta Pharmacol. Toxicol.,* 40, 201, 1977.

71. **Lee, R. F.,** Metabolism of tributyltin oxide by crabs, oysters and fish, *Mar. Environ. Res.,* 17, 145, 1985.

72. **Burns, K. A. and Smith, J. L.,** Biological monitoring of ambient water quality: the case for using bivalves as sentinel organisms for monitoring petroleum pollution in coastal water, *Estuarine Coastal Shelf Sci.,* 13, 433, 1981.

73. **Stegeman, J. J. and Teal, J. M.,** Accumulation, release and retention of petroleum hydrocarbons by the oyster *Crassostrea virginica, Mar. Biol.,* 22, 37, 1973.

74. **McLeese, D. W. and Burridge, L. E.,** Comparative accumulation of polynuclear aromatic hydrocarbons from water and from sediment by four marine invertebrates, in *Ocean Processes in Marine Pollution— Biological Processes and Waste in the Ocean,* Vol. 1, Capuzzo, J. M. and Kester, D. R., Eds., Krieger Publishing, Melbourne, Florida, 1986, 109.

75. **Newsholme, E. A. and Start, C.,** *Regulation in Metabolism,* John Wiley & Sons, London, 1973, 1.

76. **Lee, R. F., Ryan, C., and Neuhauser, M. L.,** Fate of petroleum hydrocarbons taken up from food and water by the blue crab *Callinectes spaidus, Mar. Biol.,* 37, 363, 1976.

77. **Livingstone, D. R., de Zwaan, A., Leopold, M., and Marteijn, E.,** Studies on the phylogenetic distribution of pyruvate oxidoreductases, *Biochem. Syst. Ecol.,* 11, 415, 1983.

78. **Zammit, V. A. and Newsholme, E. A.,** The maximum activities of hexokinase, phosphorylase, phosphofructokinase, glycerol phosphate dehydrogenases, lactate dehydrogenase, octopine dehydrogenase, phosphoenolpyruvate carboxykinase, nucleoside diphosphatekinase, glutamate-oxaloacetate transaminase and arginine kinase in relation to carbohydrate utilization in muscles from marine invertebrates, *Biochem. J.,* 160, 447, 1976.

79. **Dixon, D., Kadim, M., and Parry, J. M.,** The detection of mutagens in the marine environment using the mussel *Mytilus edulis,* in *Mutagens in Our Environment,* Sorsa, M. and Vainio, H., Eds., Alan R. Liss, New York, 1982, 297.

80. **Brown, R. S., Wolke, R. E., Brown, C. W., and Saila, S. B.,** Hydrocarbon pollution and the prevalence of neoplasia in New England soft-shell clams *(Mya arenaria),* in *Animals as Monitors of Environmental Pollutants,* National Academy of Sciences, Washington, D.C., 1979, 41.

81. **Mix, M. C., Trenholm, S. P., and King, K. I.,** Benzo[a]pyrene body burdens and the prevalence of proliferative disorders in mussels *(Mytilus edulis)* in Oregon, in *Animals as Monitors of Environmental Pollutants,* National Academy of Sciences, Washington, D.C., 1979, 55.

82. **Krieg, K.,** *Ampullarius australis* d'Orbigny (Mollusca, Gastropoda) as experimental animals in oncological research. A contribution to the study of cancerogenesis in invertebrates, *Neoplasma,* 19, 41, 1972.

83. **Dixon, D. R. and Pollard, D.**, Embryo abnormalities in the periwinkle, *Littorina 'saxatilis'*, as indicators of stress in polluted marine environments, *Mar. Pollut. Bull.*, 16, 29, 1985.

84. **Dixon, D. R., Moore, M. N., and Pipe, R. K.**, Environmentally induced embryonic abnormalities in the brood pouches of *Littorina saxatilis* from the region of Sullom Voe, Shetland: frequency, DNA levels and adult detoxication/toxication system, *Mar. Environ. Res.*, 17, 285, 105.

85. **Dixon, D. R.**, Aneuploidy in mussel embryos (*Mytilus edulis* L.) originating from a polluted dock, *Mar. Biol. Lett.*, 3, 155, 1982.

86. **Parry, J. M., Kadhim, M., Barnes, W., and Danford, N.**, Assays of marine organisms for the presence of mutagenic and/or carcinogenic chemicals, in *Phyletic Approaches to Cancer*, Dawe, J. J., Harshbarger, J. C., Kondo, S., Sugimura, T., and Takayama, S., Eds., Japan Scientific Societies Press, Tokyo, 1981, 141.

87. **Anderson, R. S. and Doos, J. E.**, Activation of mammalian carcinogens to bacterial mutagens by microsomal enzymes from a pelecypod mollusk, *Mercenaria mercenaria, Mutat. Res.*, 116, 247, 1983.

88. **Tynes, R. E. and Hodgson, E.**, The measurement of FAD-containing monooxygenase activity in microsomes containing cytochrome P-450, *Xenobiotica*, 14, 515, 1984.

89. **Masters, B. S. S. and Ziegler, D. M.**, The distinct nature and function of NADPH-cytochrome *c* reductase and the NADPH-dependent mixed-function amine oxidase of porcine liver microsomes, *Arch. Biochem. Biophys.*, 145, 348, 1971.

90. **Prough, R. A.**, The *N*-oxidation of alkylhydrazines catalyzed by the microsomal mixed-function amine oxidase, *Arch. Biochem. Biophys.*, 158, 442, 1973.

91. **Poulsen, L. L., Hyslop, R. M., and Ziegler, D. M.**, S-oxidation of thioureylenes catalyzed by a microsomal flavoprotein mixed-function oxidase, *Biochem. Pharmacol.*, 23, 3431, 1974.

92. **Ziegler, D. M. and Mitchell, C. H.**, Microsomal oxidase. IV. Properties of a mixed-function amine oxidase isolated from pig liver microsomes, *Arch. Biochem. Biophys.*, 150, 116, 1972.

93. **Bayne, B. L., Brown, D. A., Harrison, F., and Yevich, P. D.**, Mussel health, in *The International Mussel Watch*, National Academy of Sciences, Washington, D.C., 1980, 163.

94. **Kurelec, B., Britvic, S., Rijavec, M., Muller, W. E. G., and Zahn, R. K.**, Benzo[a]pyrene monooxygenase induction in marine fish — molecular response to oil pollution, *Mar. Biol.*, 44, 211, 1977.

95. **Davies, J. M., Bell, J. S., and Houghton, C.**, A comparison of the levels of hepatic aryl hydrocarbon hydroxylase in fish caught close to and distant from North Sea oil fields, *Mar. Environ. Res.*, 14, 23, 1984.

96. **Krieger, R. I., Gee, S. I., and Lim, L. O.**, Marine bivalves, particularly mussels *Mytilus* sp., for assessment of environmental quality, *Ecotoxicol. Environ. Saf.*, 5, 72, 1981.

97. **Trump, B. F. and Arstila, A. V.**, Cell membranes and disease processes, in *Pathobiology of Cell Membranes*, Vol. 1, Trump, B. F. and Arstila, A. V., Eds., Academic Press, New York, 1975, 1.

98. **Moore, M. N.**, Cellular responses to polycyclic aromatic hydrocarbons and phenobarbital in *Mytilus edulis, Mar. Environ, Res.*, 2, 255, 1979.

99. **Moore, M. N., Lowe, D. M., and Fieth, P. E. M.**, Lysosomal responses to experimentally injected anthracene in the digestive cells of *Nytilus edulis, Mar. Biol.*, 48, 297, 1978.

100. **Slater, T. F.**, Biochemical studies on liver injury, in *Biochemical Mechanisms of Liver Injury*, Slater, T. F., Ed., Academic Press, New York, 1978, 44.

101. **Allison, A. C.**, Lysosomes and cancer, in *Lysosomes in Biology and Pathology*, Vol. 2, Dingle, J. T. and Fell, H. B., Eds., Elsevier, Amsterdam, 1969, 178.

102. **Tzartsidze, M. A., Lomadze, B. A., and Shengelia, M. G.**, Characteristics of benzo(a)pyrene binding with lysosomal membrane, *Vopr. Med. Khim.*, 30, 17, 1984.

103. **Baccino, F. M.**, Selected patterns of lysosomal response in hepatocytic injury, in *Biochemical Mechanism of Liver Injury*, Slater, T. F., Ed., Academic Press, New York, 1978, 581.

104. **Hawkins, H. K.**, Reactions of lysosomes to cell injury, in *Pathobiology of Cell Membranes*, Vol. 2, Trump, B. F. and Arstila, A. V., Eds., Academic Press, New York, 1980, 252.

105. **Ericsson, J. L. E. and Brunk, U. T.**, Alterations in lysosomal membranes as related to disease processes, in *Pathobiology of Cell Membranes*, Vol. 1, Trump, B. F. and Arstila, A. V., Eds., Academic Press, New York, 1975, 217.

106. **Sumner, A. T.**, The distribution of some hydrolytic enzymes in the cells of the digestive gland of certain lamellibranchs and gastropods, *J. Zool. (London)*, 158, 277, 1969.

107. **Owen, G.**, Lysosomes, peroxisomes and bivalves, *Sci. Prog. Oxf.*, 60, 299, 1972.

108. **Moore, M. N.**, Lysosomes and environmental stress, *Mar. Pollut. Bull.*, 13, 42, 1982.

109. **Moore, M. N.**, Cytochemical demonstration of latency of lysosomal hydrolases in digestive cells of the common mussel, *Mytilus edulis*, and changes induced by thermal stress, *Cell Tissue Res.*, 175, 279, 1976.

110. **Moore, M. N., Bubel, A., and Lowe, D. M.**, Cytology and cytochemistry of the pericardial gland cells of *Mytilus edulis* and their lysosomal responses to injected horseradish peroxidase and anthracene, *J. Mar. Biol. Assoc. U.K.*, 60, 135, 1980.

111. **Lowe, D. M. and Moore, M. N.**, The cytochemical distributions of zinc (Zn II) and iron (Fe III) in the common mussel *Mytilus edulis*, and their relationship with lysosomes, *J. Mar. Biol. Assoc. U.K.*, 59, 851, 1979.

112. **Lowe, D. M., Moore, M. N., and Bayne, B. L.,** Aspects of gametogenesis in the marine mussel *Mytilus edulis, J. Mar. Biol. Assoc. U.K.,* 62, 133, 1982.

113. **Bayne, B. L., Bubel, A., Gabbott, P. A., Livingstone, D. R., Lowe, D. M., and Moore, M. N.,** Glycogen utilisation and gametogenesis in *Mytilus edulis* L., *Mar. Biol. Lett.,* 3, 89, 1982.

114. **Pipe, R. K. and Moore, M. N.,** The ultrastructural localization of acid hydrolases in developing oocytes of *Mytilus edulis, Histochem. J.,* 17, 939, 1985.

115. **Moore, M. N. and Stebbing, A. R. D.,** The quantitative cytochemical effects of three metal ions on a lysosomal hydrolase of a hydroid, *J. Mar. Biol. Assoc. U.K.,* 56, 995, 1976.

116. **Tiffon, Y., Rasmont, R., de Vos, L., and Bouillon, J.,** Digestion in lower metazoa, in *Lysosomes in Biology and Pathology,* Vol. 3, Dingle, J. T., Ed., Elsevier, Amsterdam, 1973, 49.

117. **Szego, C. M. and Pietras, R. J.,** Lysosomal functions in cellular activation: propagation of the actions of hormones and other effectors, in *International Review of Cytology,* Vol. 88, Bourne, G. M., Danielli, J. F., and Jeon, K. W., Eds., Academic Press, New York, 1984, 1.

118. **Viarengo, A., Zanicchi, G., Moore, M. N., and Orunesu, M.,** Accumulation and detoxication of copper by the mussel *Mytilus galloprovincialis* Lam: a study of the subcellular distribution in the digestive gland cells, *Aquat. Toxicol.,* 1, 147, 1981.

119. **Viarengo, A., Pertica, M., Mancinelli, G., Orunesu, M., Zanicchi, G., Moore, M. N., and Pipe, R. K.,** Possible role of lysosomes in the detoxication of copper in the digestive gland cells of metal-exposed mussels, *Mar. Environ. Res.,* 14, 469, 1984.

120. **George, S. G.,** Heavy metal detoxication in the mussel *Mytilus edulis* — composition of Cd-containing kidney granules (tertiary lysosomes), *Comp. Biochem. Physiol.,* 76C, 53, 1983.

121. **George, S. G.,** Heavy metal detoxication in *Mytilus* kidney — an *in vivo* study of Cd- and Zn-binding to isolated tertiary lysosomes, *Comp. Biochem. Physiol.,* 76C, 59, 1983.

122. **Harrison, F. L. and Berger, R.,** Effects of copper on the latency of lysosomal hexosaminidase in the digestive cells of *Mytilus edulis, Mar. Biol.,* 68, 109, 1982.

123. **Sternlieb, I. and Goldfischer, S.,** Heavy metals and lysosomes, in *Lysosomes in Biology and Pathology,* Vol. 5, Dingle, J. T. and Dean, R. T., Eds., Elsevier, Amsterdam, 1976, 185.

124. **Moore, M. N., Pipe, R. K., and Farrar, S. V.,** Lysosomal and microsomal responses to environmental factors in *Littorina littorea* from Sullom Voe, *Mar. Pollut. Bull.,* 13, 340, 1982.

125. **Moore, M. N., Koehn, R. K., and Bayne, B. L.,** Leucine aminopeptidase (aminopeptidase-1), N-acetyl-β-hexosaminidase and lysosomes in the mussel, *Mytilus edulis* L., in response to salinity changes, *J. Exp. Zool.,* 214, 239, 1980.

126. **Moore, M. N., Mayernik, J. A., and Giam, C. S.,** Lysosomal responses to a polynuclear aromatic hydrocarbon in a marine snail: effects of exposure to phenanthrene and recovery, *Mar. Environ. Res.,* 17, 230, 1985.

127. **Pipe, R. K. and Moore, M. N.,** An ultrastructural study on the effects of phenanthrene on lysosomal membranes and distribution of the lysosomal enzyme β-glucuronidase in digestive cells of the periwinkle *Littorina littorea, Aquat. Toxicol.,* 8, 65, 1986.

128. **Wolfe, D. A., Clark, R. C., Foster, C. A., Hawkes, J. W., and Macleod, W. D.,** Hydrocarbon accumulation and histopathology in bivalve mollusks transplanted to the Baie de Morlaix and the Rade de Brest, in *Amoco Cadiz: Fates and Effects of the Oil Spill,* CNEXO, Paris, 1981, 599.

129. **Brunk, U. T. and Collins, V. P.,** Lysosomes and age pigment in cultured cells, in *Age Pigments,* Sohal, R. S., Ed., Elsevier/North-Holland Biomedical Press, Amsterdam, 1981, 243.

130. **Lowe, D. M., Moore, M. N., and Clarke, K. R.,** Effects of oil on digestive cells in mussels: quantitative alterations in cellular and lysosomal structure, *Aquat. Toxicol.,* 1, 213, 1981.

131. **Moore, M. N. and Clarke, K. R.,** Use of microstereology and quantitative cytochemistry to determine the effects of crude oil-derived aromatic hydrocarbons on lysosomal structure and function in a marine bivalve mollusk *Mytilus edulis, Histochem. J.,* 14, 713, 1982.

132. **Couch, J. A.,** Atrophy of diverticular epithelium as an indicator of environmental irritants in the oyster *Crassostrea virginica, Mar. Environ. Res.,* 14, 525, 1984.

133. **Pipe, R. K. and Moore, M. N.,** Ultrastructural changes in the lysosomal-vacuolar system in digestive cells of *Mytilus edulis* as a response to increased salinity, *Mar. Biol.,* 87, 157, 1985.

134. **Bayne, B. L., Moore, M. N., and Koehn, R. K.,** Lysosomes and the response by *Mytilus edulis* to an increase in salinity, *Mar. Biol. Lett.,* 2, 193, 1981.

135. **Bitensky, L., Butcher, R. S., and Chayen, J.,** Quantitative cytochemistry in the study of lysosomal function, in *Lysosomes in Biology and Pathology,* Vol. 3, Dingle, J. T., Ed., Elsevier, Amsterdam, 1973, 465.

136. **Moore, M. N. and Lowe, D. M.,** Cytological and cytochemical measurements, in *The Effects of Stress and Pollution on Marine Animals,* Bayne, B. L. et al., Eds., Praeger Scientific, New York, 1985, 46.

137. **Lowe, D. M. and Moore, M. N.,** Cytological and cytochemical procedures, in *The Effects of Stress and Pollution on Marine Animals,* Bayne, B. L. et al., Eds., Praeger Scientific, New York, 1985, 179.

138. **Widdows, J., Bakke, T., Bayne, B. L., Donkin, P., Livingstone, D. R., Lowe, D. M., Moore, M. N., Evans, S. V., and Moore, S. L.,** Responses of *Mytilus edulis* L. on exposure to the water accommodated fraction of North Sea oil, *Mar. Biol.,* 67, 15, 1982.

139. **Moore, M. N. and Farrar, S. V.,** Effects of polynuclear aromatic hydrocarbons on lysosomal membranes in mollusks, *Mar. Environ. Res.,* 17, 222, 1985.

140. **Roubal, W. T. and Collier, T. K.,** Spin-labelling techniques for studying mode of action of petroleum hydrocarbons on marine organisms, *Fish. Bull.,* 73, 299, 1975.

141. **Nelson, A.,** Membrane-mimetic electro-chemistry in the marine sciences, *Mar. Environ. Res.,* 17, 306, 1985.

142. **Grossman, J. C. and Khan, M. A. Q.,** Metabolism of naphthalene by pigeon liver microsomes, *Comp. Biochem. Physiol.,* 63C, 251, 1979.

143. **Allison, A. C. and Young, M. R.,** Vital staining and fluorescence microscopy of lysosomes, in *Lysosomes in Biology and Pathology,* Vol. 2, Dingle, J. T. and Fell, H. B., Eds., Elsevier, Amsterdam, 1969, 600.

144. **Nott, J. A., Moore, M. N., Mavin, L. J., and Ryan, K. P.,** The fine structure of lysosomal membranes and endoplasmic reticulum in the digestive gland of *Mytilus edulis* exposed to anthracene and phenanthrene, *Mar. Environ. Res.,* 17, 226, 1985.

145. **Szego, C. M.,** Lysosomal function in nucleocytoplasmic communication, in *Lysosomes in Biology and Pathology,* Vol. 4, Dingle, J. T. and Dean, R., Eds., Elsevier, Amsterdam, 1975, 385.

146. **Tripp, M. R., Fries, C. R., Craven, M. A., and Grier, C. E.,** Histopathology of *Mercenaria mercenaria* as an indicator of pollutant stress, *Mar. Environ. Res.,* 14, 521, 1984.

147. **Bayne, B. L., Moore, M. N., Widdows, J., Livingstone, D. R., and Salkeld, P.,** Measurement of the responses of individuals to environmental stress and pollution, *Philos. Trans. R. Soc. London Ser. B,* 286, 563, 1979.

148. **Lowe, D. M. and Pipe, R. K.,** Cellular responses in the mussel *Mytilus edulis* following exposure to diesel oil emulsions: reproductive and nutrient storage cells, *Mar. Environ. Res.,* 17, 234, 1985.

149. **Stickle, W. B., Rice, S. D., and Moles, A.,** Bioenergetics and survival of the marine snail *Thais lima* during long-term oil exposure, *Mar. Biol.,* 80, 281, 1984.

150. **Stegeman, J. J. and Teal, J. M.,** Accumulation, release and retention of petroleum hydrocarbons by the oyster *Crassostrea virginica, Mar. Biol.,* 22, 37, 1973.

151. **Gilfillan, E. S.,** Decrease of net carbon flux in two species of mussels caused by extracts of crude oil, *Mar. Biol.,* 29, 53, 1975.

152. **Widdows, J., Donkin, P., and Evans, S. V.,** Recovery of *Mytilus edulis* L. from chronic oil exposure, *Mar. Environ. Res.,* 17, 250, 1985.

153. **Keck, R. T., Heess, R. C., Wehmiller, J., and Maurer, D.,** Sublethal effects of the water soluble fraction of Nigerian crude oil on the juvenile hard clams, *Mercenaria mercenaria* (Linn), *Environ. Pollut.,* 15, 109, 1978.

154. **Stekoll, M. S., Clement, L. E., and Shaw, D. G.,** Sublethal effects of chronic oil exposure on the intertidal clam, *Macoma balthica, Mar. Biol.,* 57, 51, 1980.

155. **Atema, J.,** Sublethal effects of petroleum fractions on the behaviour of the lobster, *Homarus americanus,* and the mudsnail, *Nassarius obsoletus,* in *Estuarine Processes,* Vol. 1, Wiley, M., Ed., Academic Press, New York, 1976, 302.

156. **Johnson, F. G.,** Sublethal biological effects of petroleum hydrocarbon exposure; bacteria, algae and invertebrates, in *Effects of Petroleum on Arctic and Subarctic Marine Environments,* Vol. 3, Malins, D. C., Ed., Academic Press, New York, 1977, 271.

157. **Hendry, B. M., Elliott, J. R., and Haydon, D. A.,** The actions of some narcotic aromatic hydrocarbons on the ionic currents of the squid giant axon, *Proc. R. Soc. London Ser. B,* 224, 389, 1985.

158. **Haydon, D. A., Requena, J., and Urban, B. W.,** Some effects of aliphatic hydrocarbons on the electrical capacity and ionic currents of the squid giant axon membrane, *J. Physiol. (London),* 309, 229, 1980.

159. **Widdows, J., Donkin, P., and Evans, S. V.,** Physiological responses of *Mytilus edulis* during chronic oil exposure and recovery, in preparation.

160. **Linden, O.,** Sublethal effects of oil on mollusk species from the Baltic Sea, *Water Air Soil Pollut.,* 8, 305, 1977.

161. **Gilfillan, E. S., Mayo, D. W., Page, D. S., Donovan, D., and Hanson, S.,** Effects of varying concentrations of petroleum hydrocarbons in sediments on carbon flux in *Mya arenaria,* in *Physiological Responses of Marine Biota to Pollutants,* Vernberg, F. J., Calabrese, A., Thurberg, F. P., and Vernberg, W. B., Eds., Academic Press, New York, 1977, 299.

162. **Stainken, D. M.,** A descriptive evaluation of the effects of no. 2 fuel oil on the tissues of the soft shell clam, *Mya arenaria* L., *Bull. Environ. Contam. Toxicol.,* 16, 730, 1976.

163. **Fong, W. C.,** Uptake and retention of Kuwait crude oil and its effects on oxygen uptake by the soft-shell clam, *Mya arenaria, J. Fish. Res. Board Can.,* 33, 2774, 1976.

164. **Gilfillan, E. S., Jiang, L. C., Donovan, D., Hanson, S., and Mayo, D. W.,** Reduction in carbon flux in *Mya arenaria* caused by a spill of no. 2 fuel oil, *Mar. Biol.,* 37, 115, 1976.

165. **Hargrave, B. T. and Newcombe, C. P.,** Crawling and respiration as indices of sub-lethal effects of oil and a dispersant on an intertidal snail, *Littorina littorea, J. Fish. Res. Board Can.,* 30, 1789, 1973.

166. **Dunning, A. and Major, C. W.,** The effects of cold seawater extracts of oil fractions upon the blue mussel, *Mytilus edulis,* in *Pollution and Physiology of Marine Organisms,* Vernberg, F. J. and Vernberg, W. B., Eds., Academic Press, New York, 1974, 349.

167. **Sabourin, T. D. and Tullis, R. E.,** Effects of three aromatic hydrocarbons on respiration and heart rate of the mussel, *Mytilus californius, Bull. Environ. Contam. Toxicol.,* 26, 729, 1981.

168. **Stainken, D. M.,** Effects of uptake and discharge of petroleum hydrocarbons on the respiration of the soft-shell clam, *Mya arenaria, J. Fish. Res. Board Can.,* 35, 637, 1978.

169. **Widdows, J., Moore, S. L., Clarke, K. R., and Donkin, P.,** Uptake, tissue distribution and elimination of [1-^{14}C] naphthalene in the mussel *Mytilus edulis, Mar. Biol.,* 76, 109, 1983.

170. **Palmork, K. H. and Solbakken, J. E.,** Distribution and elimination of 9-^{14}C phenanthrene in the horse mussel *(Modiolus modiolus), Bull. Environ. Contam. Toxicol.,* 26, 196, 1981.

171. **Riley, R. T., Mix, M. C., Schaffer, R. L., and Bunting, D. L.,** Uptake and accumulation of naphthalene by the oyster *Ostrea edulis* in a flow-through system, *Mar. Biol.,* 61, 267, 1981.

172. **Donkin, P. and Widdows, J.,** Hydrocarbon analysis of mussels *(Mytilus edulis)* to complement biological impact assessment, *Water Sci. Technol.,* in press.

173. **Widdows, J.,** Physiological responses to pollution, *Mar. Pollut. Bull.,* 16, 129, 1985.

174. **Read, A. D. and Blackman, R. A. F.,** Oily water discharges from offshore North Sea installations: a perspective, *Mar. Pollut. Bull.,* 11, 44, 1980.

175. **Gilfillan, E. S. and Vandermeulen, J. H.,** Alterations in growth and physiology in chronically oiled soft-shell clams, *Mya arenaria,* chronically oiled with Bunker C from Chedabucto Bay, Nova Scotia, 1970—1976, *J. Fish. Res. Board Can.,* 35, 630, 1978.

176. **McDonald, B. A. and Thomas, M. L. H.,** Growth reduction in the soft-shell clam *Mya arenaria* from a heavily oiled lagoon in Chedabucto Bay, Nova Scotia, *Mar. Environ. Res.,* 6, 145, 1982.

177. **Lowe, D. M. and Pipe, R. K.,** Mortality and quantitative aspects of storage cell utilization in mussels, *Mytilus edulis,* following exposure to diesel oil hydrocarbons, in preparation.

178. **Widdows, J.,** Sublethal biological effects monitoring in the region of Sullom Voe, Shetland, in 1983 Report, SOTEAG, 1984.

179. **Stickle, W. B., Rice, S. D., Villars, C., and Metcalf, W.,** Bioenergetics and survival of the marine mussel, *Mytilus edulis* L., during long-term exposure to the water-soluble fraction of Cook Inlet crude oil, in *Marine Pollution and Physiology,* Vernberg, F. J., Thurberg, F. P., Calabrese, A., and Vernberg, W. B., Eds., Belle Baruch Library in Marine Science No. 13, University of South Carolina Press, Columbia, 1985, 427.

180. **Bayne, B. L., Widdows, J., Moore, M. N., Salkeld, P. N., Worrall, C. M., and Donkin, P.,** Some ecological consequences of the physiological and biochemical effects of petroleum compounds on marine mollusks, *Philos. Trans. R. Soc. London Ser. B,* 297, 219, 1982.

181. **Grassle, J. F., Elmgren, R., and Grassle, J. P.,** Response of benthic communities in MERL experimental ecosystems to low level, chronic additions of no. 2 fuel oil, *Mar. Environ. Res.,* 4, 279, 1981.

182. **Bakke, T. and Sorensen, K.,** Oil in rocky shore mesocosms: a six years experiment, in *Proc. 1985 Oil Spill Conference American Petroleum Institute,* in press.

183. **O'Clair, C. E. and Rice, S. D.,** Depression of feeding and growth rates of the seastar *Evasterias troschelii* during long-term exposure to the water-soluble fraction of crude oil, *Mar. Biol.,* 84, 331, 1985.

184. **Craddock, D. R.,** Acute toxic effects of petroleum on arctic and subarctic marine organisms, in *Effects of Petroleum on Arctic and Subarctic Marine Environments and Organisms,* Vol. 2, Malins, D. C., Ed., Academic Press, New York, 1977, 1.

185. **Sanders, H. L., Grassle, J. F., Hampson, G. R., Morse, L. S., Garner-Price, S., and Jones, C. C.,** Anatomy of an oil spill: long term effects from the grounding of the barge Florida off W. Falmouth, Massachusetts, *J. Mar. Res.,* 38, 265, 1980.

186. **McLusky, D. S.,** The impact of petrochemical effluent on the fauna of an intertidal estuarine mudflat, *Estuarine Coastal Shelf Sci.,* 14, 489, 1982.

187. **Davies, J. M., Addy, J. M., Blackman, R. A., Blanchard, J. R., Ferbrache, J. E., Moore, D. C., Somerville, H. J., Whitehead, A., and Wilkinson, T.,** Environmental effects of the use of oil-based drilling muds in the North Sea, *Mar. Pollut. Bull.,* 15, 363, 1984.

188. **Pearson, T. H. and Rosenberg, R.,** Macrobenthic succession in relation to organic enrichment and pollution of the marine environment, *Oceanogr. Mar. Biol. Annu. Rev.,* 16, 229, 1978.

189. **Davies, J. M., Hardy, R., and McIntyre, A. D.,** Environmental effects of North Sea oil operations, *Mar. Pollut. Bull.,* 12, 412, 1981.

190. **Mann, K. H. and Clark, R. B.,** Long-term effects of oil spills on marine intertidal communities, *J. Fish. Res. Board Can.,* 35, 791, 1978.

191. **Vandermeulen, J. H.,** Some conclusions regarding long-term biological effects of some major oil spills, *Philos. Trans. R. Soc. London Ser. B,* 297, 335, 1982.

192. **Pruell, R. J., Hoffman, E. J., and Quinn, J. G.,** Total hydrocarbons, polycyclic aromatic hydrocarbons and synthetic organic compounds in the hard shell clam, *Mercenaria mercenaria,* purchased at commercial seafood stores, *Mar. Environ. Res.,* 11, 163, 1984.
193. **Wells, W. W. and Collins, C. A.,** Phosphorylation of lysosomal membrane components as a possible regulatory mechanism, in *Lysosomes in Biology and Pathology,* Vol. 7 Dingle, J. T., Dean, R. T., and Sly, W., Eds., Elsevier, Amsterdam, 1984, 119.
194. **Moore, M. N., Widdows, J., Cleary, J. J., Pipe, R. K., Salkeld, P. N., Donkin, R., Farrar, S. V., Evans, S. V., and Thomson, P. E.,** Responses of the mussel *Mytilus edulis* to copper and phenanthrene: interactive effects, *Mar. Environ. Res.,* 14, 167, 1984.
195. **Lowe, D. M. and Moore, M. N.,** Cytology and quantitative cytochemistry of a proliferative atypical hemocytic condition in *Mytilus edulis* (Bivalvia, Mollusca), *J. Natl. Cancer Inst.,* 60(6), 1455, 1978.
196. **Oprandy, J. J., Chang, P. W., Pronovost, A. D., Cooper, K. R., Brown, R. S., and Yates, V. J.,** Isolation of a viral agent causing hematopoietic neoplasia in the soft-shell clam, *Mya arenaria, J. Invertebr. Pathol.,* 38, 45, 1981.
197. **Mix, M. C.,** Haemic neoplasms of bay mussels, *Mytilus edulis* L., from Oregon: occurrence, prevalence, seasonality and histopathological progression, *J. Fish. Dis.,* 6, 239, 1983.
198. **Livingstone, D. R. and Farrar, S. V.,** unpublished data.
199. **Livingstone, D. R.,** unpublished data.
200. **Moore, M. N.,** unpublished data.
201. **Livingstone, D. R. and Widdows, J.,** unpublished data.